Insect Predators in Pest Management

Pests cause damage to the economic value of crops and stored products, while vectors are responsible for the transmission of disease-causing agents in human beings and livestock. Although application of synthetic pesticides in agriculture gives immediate relief, it also causes well-known side effects, leading to a consensus among entomologists and agriculturists to shift towards ecofriendly pest management methods. Natural enemies of insects including their predators, parasitoids and pathogens have attracted the attention of scientists across the globe. These natural enemies exist in agroecosystems and suppress the populations of pests. Predators are farmers' friends and the most successful group of natural enemies. Highly specialized/generalized in their prey choice, active stages of predators search for a suitable prey, attack or kill the prey and consume prey within a short handling time. Predatory ability is known to increase with increase in prey density. A single predator may devour several prey individuals. Exploiting this potency of predators may yield successful results in controlling notorious pests in an ecofriendly way.

This book provides information on the important biocontrol agents that are effective in pest suppression. It starts with insect predatory groups followed by specific groups of predators. It is hoped that the book presents a comprehensive account of beneficial predators and will be useful to undergraduate and postgraduate students of Entomology, Biological Control, Plant Protection, Agricultural Zoology and Zoology, besides those involved in competitive examinations and policy planning.

Features

- Each chapter has been authored by specialized senior professionals.
- Every chapter contains Learning Objectives and Points to Remember.
- This book offers comprehensive knowledge of predators and their application in pest management in a lucid way.

Insect Predators in Pest Management

Edited by
Omkar

CRC Press
Taylor & Francis Group
Boca Raton London New York

CRC Press is an imprint of the
Taylor & Francis Group, an **informa** business

First edition published 2023
by CRC Press
6000 Broken Sound Parkway NW, Suite 300, Boca Raton, FL 33487-2742

and by CRC Press
4 Park Square, Milton Park, Abingdon, Oxon, OX14 4RN

CRC Press is an imprint of Taylor & Francis Group, LLC

Library of Congress Cataloging-in-Publication Data
Names: Omkar, Omkar, editor.
Title: Insect predators in pest management / edited by Omkar.
Description: First edition | Boca Raton, FL : CRC Press, 2023. | Includes bibliographical references and index.
Identifiers: LCCN 2022055703 (print) | LCCN 2022055704 (ebook) |
Subjects: LCSH: Insect pests—Control. Classification: LCC SB931 .I44 2023 (print) | LCC SB931 (ebook) |
DDC 632/.7—dc23/eng/20230215
LC record available at https://lccn.loc.gov/2022055703
LC ebook record available at https://lccn.loc.gov/2022055704

ISBN: 9781032441764 (hbk)
ISBN: 9781032441771 (pbk)
ISBN: 9781003370864 (ebk)

DOI: 10.1201/9781003370864

Typeset in Times
by codeMantra

Contents

Preface

The class Insecta includes more than a million species, which are found in massive numbers across all habitats except oceans. Insects come in multiple shapes and sizes with interesting behaviour and biology. Despite being the most abundant and interesting neighbours, social conventions prime us with an anti-insect response, with most of us fearing them and wanting to exterminate them. This is not just true of social norms and conventions, but it is even true for academia, where most books on applied entomology concentrate largely on the control and management of insect pests.

This is despite the massive contribution of insects and the fact that only 1% of them are harmful to human beings, as pests and vectors. Also, insects form an integral part of our food webs and majority of them are beneficial to human beings by providing ecosystem services and serving as scavengers; yielding silks, lac, honey, propolis, wax and dyes; facilitating crop pollination; and providing pharmaceuticals and delicious food for the majority of human populations across the world. They are used in disease therapy and wound healing. They are also helpful in criminal investigations, as the best models for research and are quite useful in technology innovations. They also serve as ecofriendly tools as parasitoids and predators in the biological control of insect pests as an essential component in Integrated Pest Management. These natural enemies reduce the use of agrochemicals in modern agriculture and lead towards ecofriendly and sustainable agriculture by protecting the environment from synthetic chemicals. The predators, parasitoids and certain pathogens are recognized as farmers' friends as they play a major role in pest control.

Pests cause damage to the economic value of crops and stored products, and vectors are responsible for the transmission of disease-causing agents in human beings as well as livestock. To manage their populations, initially various ashes, inorganic chemicals and plant products were used. Gradually, after the discovery of organochlorinated, organophosphate and carbamate pesticides were used to suppress the pest and vector populations, and they gave immediate relief by minimizing pest/vector populations and reducing the losses caused by them. But, later various side effects of pesticides, viz., their persistence and non-degradable nature, accumulation in the environment, ecodegradation, effect on non-target organisms, especially the beneficial ones, resurgence of new pest species, resistance in target pest species, threat to biodiversity and human health hazards, have come into the light, which compelled the entomologists and agriculturists to go for a shift in the pest management strategies.

In the search for an ecofriendly, safe and alternative strategy, natural enemies of insects including predators and parasitoids attracted the attention of scientists; as they exist in natural agroecosystems, they can take care of populations of pests/vectors, and by conserving populations of insects' natural enemies and multiplying them in insectaries, they can be employed safely in pest management programmes. Thus, they form one of the essential components of Integrated Pest Management.

Insect predators are used as ecofriendly substitutes of synthetic pesticides in order to protect environment and human health and to increase crop production by decreasing losses caused by the insect pests. The predators can be immatures or adults, males or females or both. They can be generalists or specialists. They are generally larger than their prey. They attack, subdue and kill or consume the prey. They may attack the immature or adult prey. They are one of the important tools of biological control of pest and vector species. In fact, the biocontrol success story starts with the application of predators, e.g., vedalia beetle, in pest management. The present book, *Insect Predators in Pest Management*, highlights and provides an insight on important insect predators. It starts with a survey of different predatory insect groups, followed by the specific and effective insect predatory groups.

It is so designed that each chapter has been authored by senior and specialist scientists and university professors for better and valuable information. It is hoped that this book will give a comprehensive account of insect predators and present the positive aspects of insects for the readership.

It will be useful to undergraduate and postgraduate students of Entomology, Biological Control, Plant Protection, Agricultural Zoology and Zoology, those appearing in specific competitive examinations, and to those involved in the policy planning.

At the very outset, I am extremely thankful to the contributors of this book for accepting my invitation, sparing their invaluable time to complete the work in time to the best their ability and making revisions of their respective chapters as per suggestions. I am also thankful to Dr. Geetanjali Mishra for valuable inputs; Dr. Priya Singh for kind assistance in the cross-checking of references; and the members of our research team, Dr. Swati Saxena, Ms. Apoorva Shandilya, Ms. Chandni Verma, Ms. Tripti Yadav, Ms. Shriza Rai, Ms. Lata Verma, Ms. Deepali Gupta, Ms Gauranvita Singh, Ms. Priya Yadav, Mr. Md. Sharik and Mr. Sanjeev Dalal, for helping me in various ways during this work. I am thankful to my wife, Mrs. Kusum Upadhyay, for managing the home front and sparing me to complete the task in time.

Prof. Omkar
FNASc
Editor

Editor

Professor Omkar is a former Professor and Head, Department of Zoology, University of Lucknow, India. For his unique contribution to science, Professor Omkar has been conferred with Fellowship by The National Academy of Sciences, India (FNASc) along with other prestigious fellowships. With 25 PhDs supervised, Professor Omkar has completed 11 research projects funded by state and central agencies; published more than 300 papers/chapters (including one each in Nature Biotechnology & Conservation Biology), besides having two-thirds of research papers in international journals; written and edited more than 20 books published by well-recognized national and international publishers. He has also served as a Subject Expert for Swiss National Science Foundation, and many indigenous grant applications. In view of his academic contributions, Professor Omkar has been felicitated for his excellence in academics and awarded the prestigious Saraswati Samman by the Department of Higher Education, Government of Uttar Pradesh (2017); Professor T N Ananthakrishnan Foundation Award, Chennai (2012); Rescholar Award of Excellence in Agricultural Entomology, Patiala (2014); Prof. G. S. Shukla Gold Medal (2014); and Dr. S. Pradhan Memorial Lecture Award (2019). Known globally for identifying and harnessing the potential of beneficial insects, he has worked in collaboration with multiple international scientists and visited the School of Biological Sciences, University of East Anglia (UK) in 2006 for 3 months and the Department of Zoology, University of South Bohemia, Ceske Budejovice and Institute of Entomology, Czech Academy of Sciences, Czech Republic in 2016 for 4weeks. Professor Omkar is among Top 2% Scientists of World in the database of Stanford University published by Elsevier, Inc., USA (2021). He has RGScore 3565, H index30 and I_{10}97.

Contributors

Uzma Afaq
Department of Biosciences
Integral University
Lucknow, India

Archana Anokhe
Division of Entomology
ICAR-Indian Agricultural Research Institute
New Delhi, India

N. Bakthavatsalam
Division of Germplasm Conservation and Utilization
ICAR-National Bureau of Agricultural Insect Resources
Bengaluru, India

Chandish R. Ballal
ICAR-National Bureau of Agricultural Insect Resources
Bengaluru, India

S. N. Bhagyasree
Division of Entomology
ICAR-Indian Agricultural Research Institute
New Delhi, India

Mani Chellappan
AINPVPM, Department of Agricultural Entomology, College of Agriculture
Vellanikkara Kerala Agricultural University
Thrissur, Kerala

K. J. David
Division of Germplasm Collection & Characterisation
ICAR-National Bureau of Agricultural Insect Resources
Bengaluru, India

Yaghoub Fathipour
Department of Entomology
Faculty of Agriculture
Tarbiat Modares University
Tehran, Iran

Tripti Gupta
ICAR-National Bureau of Agricultural Insect Resources
Bengaluru, India

Sunil S. Joshi
Division of Germplasm Collection & Characterisation
ICAR-National Bureau of Agricultural Insect Resources
Bengaluru, India

Bhupendra Kumar
Department of Zoology
Banaras Hindu University
Varanasi, India

G. Mahendiran
Division of Germplasm Collection and Characterization
ICAR-National Bureau of Agricultural Insect Resources
Bengaluru, India

Omkar
Ladybird Research Laboratory
Department of Zoology
University of Lucknow
Lucknow, India

C. H. Patel
Lovely Professional University
Jalandhar, India

M. T. Ranjith
AINPVPM, Department of Agricultural Entomology, College of Agriculture
Vellanikkara Kerala Agricultural University
Thrissur, India

K. Sachin
Division of Germplasm Collection & Characterisation
ICAR-National Bureau of Agricultural Insect Resources
Bengaluru, India

S. Shalini
National Bureau of Agricultural Insect Resources
Bangalore, India

P. R. Shashank
Division of Entomology
ICAR-Indian Agricultural Research Institute
New Delhi, India

P. Sree Chandana
Division of Corp Protection
ICAR-Indian Institute of Horticultural Research
Bengaluru, India

Kolla Sreedevi
Division of Germplasm Collection and Characterization
ICAR-National Bureau of Agricultural Insect Resources
Bengaluru, India

K. A. Subramanian
Zoological Survey of India
Southern Regional Centre
Chennai, India

Richa Varshney
Division of Germplasm Conservation and Utilization
ICAR-National Bureau of Agricultural Insect Resources
Bengaluru, India

Shima Yazdanpanah
Department of Entomology
Faculty of Agriculture
Tarbiat Modares University
Tehran, Iran

1 Insect Predators

Mani Chellappan and M. T. Ranjith
Vellanikkara, Kerala Agricultural University

CONTENTS

DOI: 10.1201/9781003370864-1

LEARNING OBJECTIVES

1. To know insect predators as a group of entomophagous insects with predatory stages and their prey.
2. To have an idea of predator classification.
3. To learn predaceous groups in Insecta.
4. To understand prey specificity and potential of suppressing pests' populations in biocontrol programmes.

1.1 INTRODUCTION

Insect predators and parasitoids that attack and feed on other insects, particularly on insect pests of crop plants, are considered natural enemies of pests. Through this type of feeding, natural enemies contribute to a type of pest population regulation referred to as natural biological control and they are responsible for 33% of the natural pest control in agroecosystems (Metcalf and Flint, 1962). Typically, insect predators are characterised by a set of attributes that distinguish them from parasitoids, the other major group of entomophagous insects. Insect predators are large relative to their prey and require more than one prey individual to complete their development, they have free-living predatory immature stages, and many species of insect predators are predaceous in both immature and adult stages (Doutt, 1964, Hagen et al., 1976). Though predators occur in approximately 20 insect orders, the lists vary slightly, depending on the definition of scavenging and fortuitous predation. The orders of insects not known to contain predatory species are: Isoptera, Phasmatodea, Phthiraptera, Strepsiptera and Siphonaptera (Hagen, 1987).

Predaceous insects exhibit great variation in their dietary range. The vedalia beetle, *Rodolia cardinalis*, and the green lacewing, *Chrysopa slossonae*, are highly specific and feed on only one species of prey, whereas the aphid-feeding coccinellids, *Hippodamia convergens* and *Adalia bipunctata*, are stenophagous or oligophagous and restrict feeding to a range of related taxa. Finally, other predators, e.g., the bug *Podisus maculiventris*, and the ladybird beetle, *Coleomegilla maculata*, are polyphagous (general feeders) and consume a wide variety of prey and non-prey items, e.g., plant fluid and pollen. It is noteworthy that many species with wide prey ranges contain biotypes and populations that differ in their responses to prey; examples include web-building and jumping spiders and the green lacewing *Chrysopa quadripunctata* (Hedrick and Riechert, 1989; Tauber et al., 1995). In certain cases, symbiotic association with microorganisms may supply nutrients for development or reproduction. For example, the common green lacewing *Chrysoperla carnea* harbours yeasts (*Torulopsis*) in an enlarged oesophageal diverticulum; these yeasts provide essential amino acids that are lacking in the lacewing's diet (Hagen, 1987).

Predators can be classified according to the life stage of prey they attack (e.g. egg, larval and adult predators) and their foraging strategy (e.g. active searchers, ambush or filter feeders, web or bolas builders). Although these types of categories may be generally informative about the type of prey taken, they have no predictive value for the prey specificity of individual species. In this regard, it is of both practical and theoretical significance that predator phylogeny can be an important clue to prey breadth and prey preference. For example, among the Coccinellidae, the Chilocorinae prey on Homopteran scale insects, most species of Coccinellinae are aphid predators, and the Stethorinae has specialised on phytophagous mite species (Gordon, 1985). However, it is critical to note that sister-species may differ in the range of prey that they take within their preferred type of prey; e.g. sister-species may include a generalist and a specialist aphid-feeder (Albuquerque et al., 1996). Therefore, care must be exercised in using phylogenetic relatedness to generalise the prey breadth of specific taxa.

Adult predators may use a variety of cues, including volatile chemicals from both plants and insects, to find their prey. The role of chemical cues in locating prey is relatively well understood for the green lacewing *Chrysoperla carnea* (Hagen, 1987). A complex mixture of volatiles (an attractant and an arrestant) induces *Chrysoperla carnea* adults to fly upwind, towards a food source, and having reached the food, to cease movement and to oviposit. One of the volatile chemicals is a metabolite of oxidised tryptophan (most likely indole acetaldehyde), which occurs in aphid honeydew. Larval predators have evolved various means of locating and recognising their prey: phototactic and/or geotactic responses, vision, olfaction, sound or vibration detection and contact (Hodek, 1993). A predator's previous experience (learning) influences its searching behaviour as well as the type and proportion of prey taken (Begon et al., 1996). Plant structure may also affect the pattern of predator movement (Heinz and Zalom, 1996).

The prey specificity impinges directly on the use of predators in all types of biocontrol, *viz.*, importation as well as augmentation and manipulation (Hagen et al., 1976; Hagen, 1987). To become an integral component of pest management programmes, the selected natural enemies should have a high level of host specificity as a prime attribute. Indeed, the first modern success in biocontrol involved a highly specific scale predator, the vedalia beetle, *R. cardinalis* (Huffaker and Caltagirone 1986). However, prey specificity is not always necessary in pest management; many predatory species that are currently mass-produced and released in field crops and orchards have relatively broad prey ranges; examples include the anthocorids *Orius insidiosus* and *O. tristicolor*, and the green lacewings *Chrysoperla carnea* and *C. rufilabris*. The interaction between the predator and the target pest should be predictable; on the other hand, the nutritional requirements for an introduced predator may be complex and may include factors that are not readily ascertained, e.g., acceptance or preference for novel prey in the new habitat, obtaining nutrients from symbiotic microorganisms, utilisation of supplementary food, such as plant fluids, fungi, pollen or nectar (Hagen et al., 1976). A predator's use of these dietary resources has important implications for the outcome and stability of predator-prey dynamics (Begon et al., 1996). Predators belonging to various insect orders are successfully employed to manage the insect pests across the globe. Though most of the species are poorly understood, they do have potential in suppressing their insect prey. This chapter exclusively deals with species of insect predators belonging to the order Hemiptera, Neuroptera, Coleoptera Diptera and Lepidoptera, and their prospects in managing the insect pests in various crop ecosystems.

1.2 ORDER: HEMIPTERA

Hemipteran bugs are characterised by hemelytra (forewing half thin and half thick) and piercing-sucking type of mouthparts for sucking the body fluid of their prey or the sap of their host plant. Having an important role as predators, they do have various morphological adaptations, particularly the appendages for holding the prey as well as the number, shape and direction of style dentition for biting and killing the captured prey (Bal and Biswas, 2013). The suborder of Heteroptera contains predatory insect species that can be used as effective biocontrol agents for preventing the insect pests from damaging various crops. The characteristics of entomophagous families and their representatives (Table 1.1) under the suborder Heteroptera are discussed hereunder.

TABLE 1.1
Predator Species in Various Families under the Order Hemiptera and Host Insects

Sl. No.	Family	Predator Species	Host Insect
1	Pentatomidae	*Andrallus spinidens*	*Pamaram athias Pamara mathias* (Fabr.), *Sesamia inferens* (Walk.), *Spodoptera litura* (Fabr.) and *Melanitis leda ismene* in rice
		Amyotea malabaricus	*Leptocorisa acuta*
		Troilus luridus	Caterpillars
		Zicrona caerulea	*Spodoptera litura*, *Semiothisa pervolgata*, *Catopsilia pyranthe*, *Eurema hecabe*, *Anticarsia irrorata*, *Helicoverpa armigera* and *Pelopidas mathias*
		Eocanthecona concinna	*Biston suppressaria*
		Eocanthecona furcellata	*Pericallia ricini*, *Latoia lepida*, *Maruca vitrata*, *S. litura*, *Spodoptera frugiperda*
		Pentatoma rufipes	Egg, larva and pupa of insects
2	Nabidae	*Nabis capsiformis*	*Helicoverpa zea*, Pieris *rapae*, *mites*, *thrips*, *leafhoppers*
3	Anthocoridae	*Orius insidiosus*	Egg and larva of bollworm, aphids, corn earworm
		Orius niger	Thrips, aphids, whiteflies, mites and lepidoptera eggs
		Orius tantillus	*Haplothrips ganglbaueri* Schmutz, *Thrips palmi* Karny, *T. tabaci*, *Baliothrips biformis*, *Caliothrips indicus* (Bagn.)
		Tetraphleps spp.	Pineus pini
4	Lygaeidae	*Geocoris ochropterus*	*Myllocerus viridanus*, *Calocoris anguslatus*, *Pseudococcus* sp.,
		Geocoris jucundus	*Aphis gossypii*, *Acyrthosiphon pisum*
		Xylocoris spp.	*Megalurothrips* sp., *Apelaunothrips madrasensis* and *A. indicus*
5	Miridae	Cyrtorhinus lividipennis	White-backed plant hoppers and leafhoppers, brown plant hoppers
		Macrolophus spp.	Whitefly eggs, nymphs and pseudo-pupae, but especially nymphs, aphids, leafhoppers, leaf miners
		Nesidiocoris tenuis	Whiteflies, thrips
		Apolygus spinolae	Aphids
		Miris striatus	Eggs and larvae of various beetles and lepidopterans
		Blepharidopterus angulatus	Aphids, leafhoppers and small caterpillars
		Atractotomus spp.	Aphids. psyllids, small caterpillars and spider mites
6	Reduviidae	*Zelus renardii*	Caterpillars, leafhoppers, other bugs and aphids.
		Platymeris laevicollis	*Rhinoceros beetle*
		Rhynocoris marginatus	*H. armigera*, *Spodoptera litura*, *Dysdercus cingulatus*
		Rhynocoris fuscipes	Larvae of *H. armigera*, *Spodoptera litura*, *Semiothisa pervolgata*, *Eurema hecabe*, and *Catopsilia pyranthe* and beetles *Henosepilachana vigintioctopuncatata* and *Raphidopalpa foveicollis*
		Rhynocoris costalis	*Dysdercus koenigii*
		Blaptostethus pallescens	On primary and secondary stored product pests; *Rhopalosiphum maidis* (Fitch) on maize; onion thrips (*Thrips tabaci*); and mites
		Acanthaspis pedestris	*Pectinophora gossypiella* and *Odontotermes assumthi*
		Acanthas pissiva	*Honey bee*
		Cydnocoris gilvus	*S. litura*, *Okya nitidula*, *Odontotermes obesus*
		Endochus inornatus	*Stephanitis typica*

(*Continued*)

TABLE 1.1 (*Continued*)
Predator Species in Various Families under the Order Hemiptera and Host Insects

Sl. No.	Family	Predator Species	Host Insect
7	Gerridae	*Linnogonus (Linnogonus) nitidus*	Brown plant hopper of paddy
		Neogerris parvulus	Brown plant hopper of paddy

1.2.1 Family: Pentatomidae

Shield bugs (suborder Heteroptera, family Pentatomidae) are mainly medium-sized, stout, somewhat flattened, beetle-like insects, with a distinctly shield-shaped body outline, a distinct scutellum and wings (hemelytra) that cover most if not all of the abdomen. Although some species are predaceous, most of the shield bugs are phytophagous in nature (Eger et al., 2015).

Shield bugs lay barrel-shaped eggs in groups and the maternal female often guards over them until they hatch. First-instar nymphs are reliant on the imbibitions of plant sap or some other source of moisture in order to survive, and they do not become carnivorous until reaching their second instar. Shield bugs pass through five nymphal stages, with wingbuds becoming visible as small pads in the fifth instar stage. Such nymphs are often very colourful and may bear little or no resemblance to those of earlier or later instars or to the frequently more soberly patterned (often basically brown or green) adults (Powel, 2020).

Most species of shield bugs are phytophagous, but a few are predaceous in nature. Truly predaceous shield bugs (subfamily Asopinae) differ from other groups in possessing flexible mouthparts that are capable of being directed horizontally forward when impaling their prey (De Clercq, 2000).

Some predaceous shield bugs are known to target phytophagous insects, which accumulate plant-derived defence chemicals in their bodies, and may even be preferred as prey due to the sequestration of such chemicals. Known examples preyed upon by shield bugs include burnet moths (*Zygaena* spp.) and marsh fritillary butterflies (*Eurodrya saurinia*) (Hamilton and Heath, 1976; Tremewan, 1985; Konvicka et al., 2005). The various predatory species under subfamilies of Pentatomidae are given below.

1.2.1.1 Subfamily: Asopinae

1.2.1.1.1 Andrallus spinidens (Fabricius)

Andrallus spinidens is moderate in size with length of 13–16 mm and width of 8.5 mm between pronotal angles; blackish punctures in the form of a band can be seen on each side of the central lobe of head and a pale central smooth shiny line can be seen between the pronotal angles; pronotal lateral angles with two spines, posterior spine being very small (Plate 1.1). This species occurs on vegetation, especially on foliage (Mead and Eger, 1992).

This species preys upon *Pamara mathias* (Fabr.), *Sesamia inferens* (Walk.), *Spodoptera litura* (Fabr.) and *Melanitisleda ismene* (Zibaee et al., 2020).

1.2.1.1.2 Amyotea malabaricus (Fabricius)

The body length of *Amyotea malabaricus* varies from 12 to 14 mm; it has a basal spot on head, two to three spots on pronotum and two spots, one on each basal angle of scutellum. Wing membrane is black, and abdomen, rostrum and coxae femora are pale reddish in colour (Plate 1.2) (Santhamma, 2017). This species can be located in and around the rice fields. Although this species is basically phytophagous, it is also known to be a potential predator on rice bug *Leptocorisa acuta* (Thumb.) of the family Alydidae (Pati and Mathur, 1986).

Plate 1.1. *Andrallus spinidens*

Plate 1.2. *Amyotea malabaricus*

Plate 1.3 *Troilus luridus*

Plate 1.4 *Zicrona caerulea*

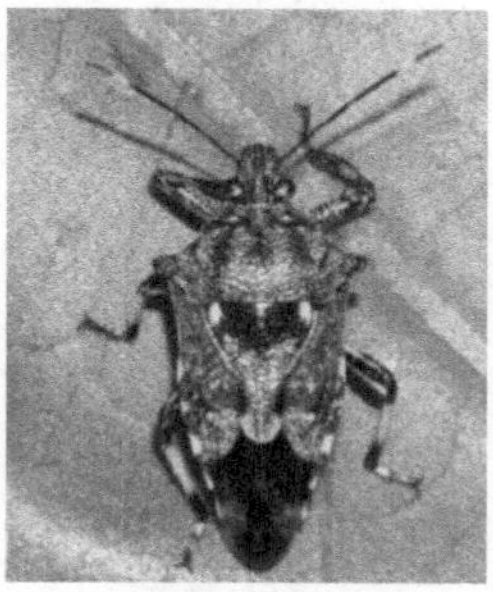

Plate 1.5. *Eocanthecona concinna*

Plate 1.6 *Eocanthecona furcellata*

PLATE 1.1 Predators in the order Hemiptera.

1.2.1.1.3 Troilus luridus (Bronze shield bug)

Troilus luridus (Fabricius) (=*Podisus luridus*) (Plate 1.3), also called the bronze shield bug, is a widely distributed Palearctic species. It is distributed in the northern part of India, especially in Himachal Pradesh and Uttarakhand (Salini et al., 2020). It is a brown-coloured bug with scattered ochraceous and black markings on the dorsum, and an orange band on the penultimate antennal segment. The legs are brown and the scutellum lacks an orange tip. There is only one generation per year, with nymphs feeding on plants as well as other insects, particularly caterpillars (Bantock and Botting, 2012).

1.2.1.1.4 Zicrona caerulea (Linnaeus) (Blue Shield Bug)

Zicrona caerulea is a medium-sized bug (Plate 1.4) with 9–10 mm in length; body is violaceous blue in colour; membrane and antennae are black; scutellum has a little elevation at the base; and the body surface is very fine and sparingly punctate. This species is known to occur on the foliage in the

Plate 1.7. *Pentatoma rufipes*

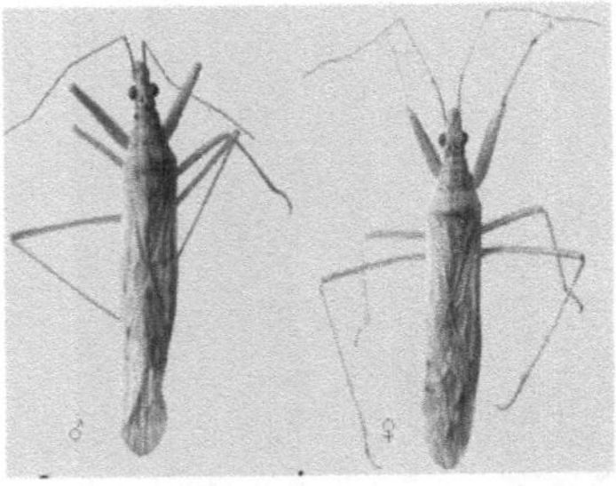

Plate 1.8. *Nabis capsiformis*

Plate 1.9 *Orius insidiosus*

Plate 1.10 *Orius niger*

Plate 1.11 *Orius tantillus*

Plate 1.12 *Tetraphleps galchanoides*

PLATE 1.1 (*Continued*) Predators in the order Hemiptera.

agroecosystem. Both adults and late inster nymphs are predaceous on larvae of *Spodoptera litura*, *Semiothisa pervolgata*, *Catopsilia pyranthe*, *Eurema hecabe*, *Anticarsia irrorata*, *Helicoverpa armigera* and *Pelopidas mathias* (Bal and Biswas, 2013).

1.2.1.1.5 Eocanthecona concinna

Eocanthecona concinna is a medium-sized bug, 11.16–11.57 mm in length, with chocolate brown–coloured body and pale ochraceous markings scattered over head, pronotum, scutellum and corium (Plate 1.5). Each basal angle of scutellum has a single, spherical, dark yellowish orange mark. It has metallic sheen over the head and pronotum, especially on humeral angles, with five-segmented antennae; the first segment is small, not reaching the apex of the head. Legs are ochraceous, thickly punctured, having numerous black spots. Tibiae are markedly sulcate down their whole length. Forelegs are different from midlegs and hindlegs. Tibiae of forelegs have blade-like protrusions and anterior

Plate 1.13. *Geocoris ochropterus*

Plate 1.14. *Geocoris jucundus*

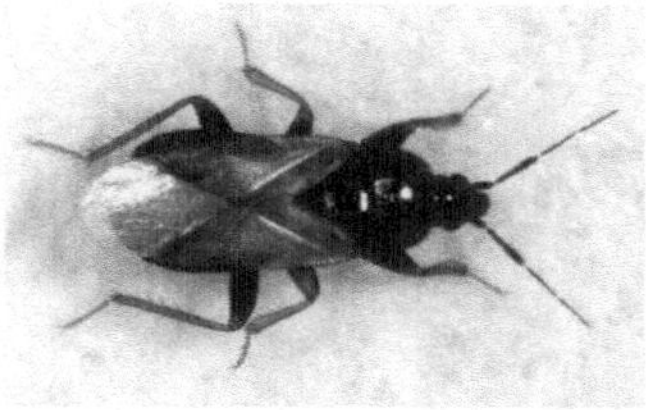

Plate 1.15 *Xylocoris flavipes*

Plate 1.16 *Cyrtorhynus lividipennis*

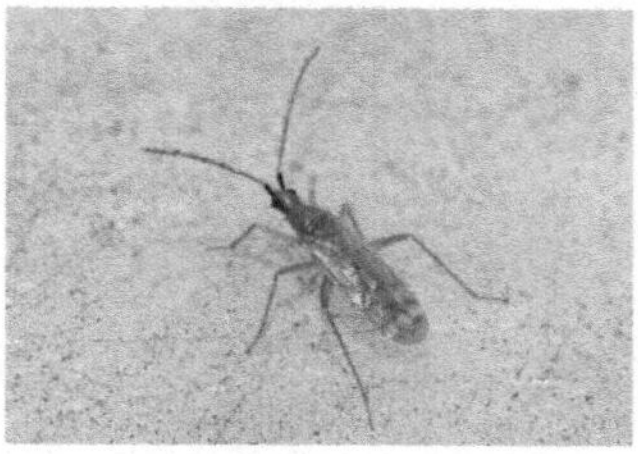

Plate 1.17 *Macrolophus melanotoma*

Plate 1.18 *Nesidiocoris tenuis*

PLATE 1.1 (*Continued*) Predators in the order Hemiptera.

femora with a prominent spine (Waghmare and Gaikwad, 2017; Srikumar et al., 2018). Members of this genus feed upon other insects, especially on lepidopterous larvae, and act as a biocontrol agent against several pests. Some asopines even feed on other pentatomid species, which are herbivorous (De Clercq, 2008). Srikumar et al. (2018) reported the predatory efficacy of *Eocanthecona concinna* (Walker, 1867) on tea looper, *Biston suppressaria*, for the first time in India.

1.2.1.1.6 Eocanthecona furcellata (Wolff) (=Canthecona furcellata, =Cantheconidea furcellata)

This bug may attain a body length of 15–16 mm, with greyish yellow colour and dark bronzy punctures; the pronotum with a broken transverse fascia on the anterior portion, some more or less apparent longitudinal fasciae on the front half, and bronzy black lateral angles (Plate 1.6). This species is known to occur on vegetation, especially on foliage (Bal and Biswas, 2013).

The early instar nymphs feed on host plants, and later they become predaceous on other insects. Both nymphs and adults approach insect caterpillars from behind, place the rostrum in between the two anal prolegs and thrust the stylets into the body of the larvae and suck the body fluid. The larvae get killed and hang their head downwards. In India, *Eocanthecona furcellata* has been identified as an important predator on larvae of *Pericallia ricini* (Lepidoptera: Arctiidae) (now known

Plate 1.19 *Apolygus spinolae*

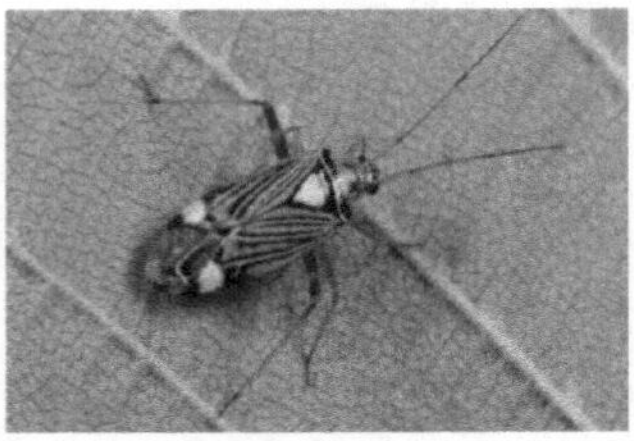

Plate 1.20 *Miris striatus*

Plate 1.21 *Blepharidopterus angulatus*

Plate 1.22. *Atractotomus mali*

Plate 1.23. *Zelus renardii*

Plate 1.24. *Platymeris laevicollis*

PLATE 1.1 (*Continued*) Predators in the order Hemiptera.

as *Olepa ricini* (Fabricius) on cotton) (Shophiya and Sahayaraj, 2014), *Latoia lepida* (Lepidoptera: Limacodidae) (now known as *Parasa lepida* (Cramer)) (Senrayan, 1988), *Maruca vitrata* (Fabricius) (Lepidoptera: Pyralidae), and *S. litura* Fabricius (Lepidoptera: Noctuidae) (Tiwari et al., 2017) and fall armyworm, *Spodoptera frugiperda* (J. E. Smith) (Keerthi et al., 2020).

1.2.1.2 Subfamily: Pentatominae

1.2.1.2.1 Pentatoma rufipes (L.)

Pentatoma rufipes is a large pentatomid, with adults typically 12–15 mm long (Plate 1.7). The dorsal surface of the adult is glossy dark brown in colour, but the legs, tip of the scutellum and sometimes the basal segments of the antennae are vibrant orange or red. The margins of the abdomen are marked with alternating black and cream/orange bands. The hooked front corners of the pronotum have a tiny pointed barb or tooth on each side of the body, which is particularly characteristic of this species. First-instar nymphs are of dark colour, but the nymphs become lighter with age. Later nymphal stages are mottled with areas of yellow or green and brown or black and can be well camouflaged against a tree bark (Powell, 2020). *P. rufipes* are adaptable and opportunistic in their predatory behaviour, and are able to feed on the eggs, larvae, pupae and adults of many insect

Plate 1. 25. *Rhynocoris marginatus*

Plate 1. 26. *Rhynocoris fuscipes*

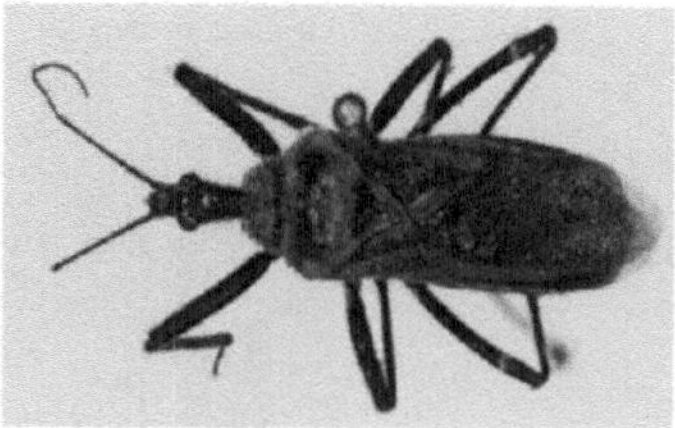

Plate 1. 27. *Rhynocoris costalis*

Plate 1. 28. *Blaptostethus pallescens*

Plate 1. 29. *Acanthaspis pedestris*

Plate 1. 30. *Acanthaspis siva*

PLATE 1.1 (*Continued*) Predators in the order Hemiptera.

species (König, 2015). Adults of *P. rufipes* have been found to be carnivorous (Bantock, 2018) and predation by nymphs may also occur.

1.2.2 Family: Nabidae

Damsel bugs (family Nabidae) form a group of predaceous, mainly brown-bodied heteropterans, with four-segmented antennae and a curved, four-segmented beak or rostrum, which is not held against the body when in repose. They include both fully (macropterous) and partly winged (brachypterous) forms and are exemplified by two well-known genera, *viz.*, *Himacerus* and *Nabis*. The predator under the genus *Nabis* is recorded from various locations in India.

1.2.2.1 Subfamily: Nabinae

1.2.2.1.1 Genus: Nabis

Damsel bugs of the genus *Nabis* are the most common. They are generalist predators, catching any insect smaller than themselves and cannibalising each other when no other food is available. Some species of *Nabis* work as an important biocontrol agent (predators) in various cropping systems

Plate 1.31. *Cydnocoris gilvus*

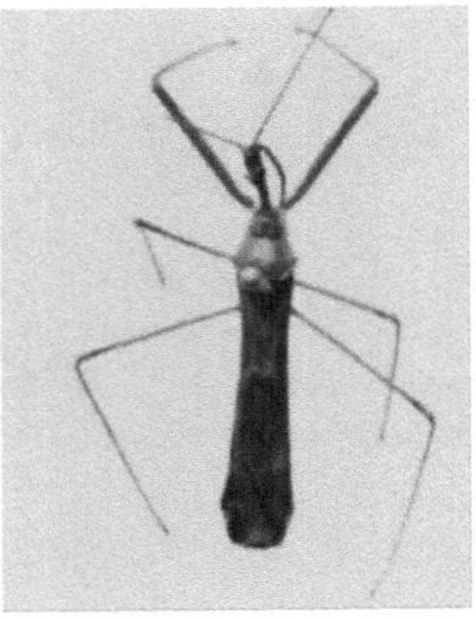

Plate 1.32. *Endochus inornatus*

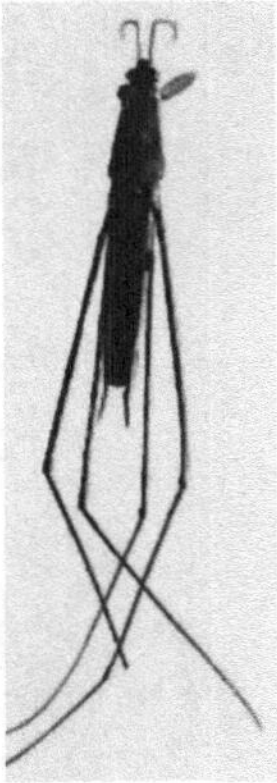

Plate 1.33. *Linnogonus (Linnogonus) nitidus*

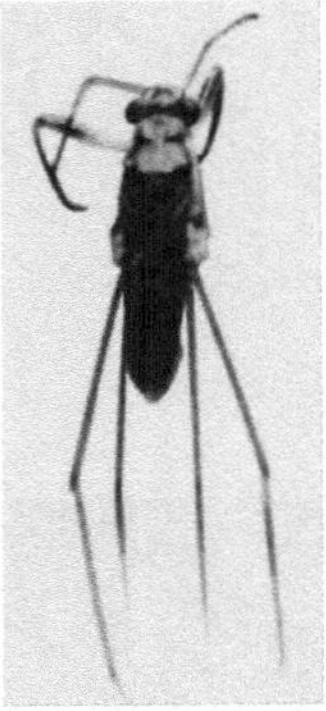

Plate 1.34. *Neogerris parvula*

PLATE 1.1 (*Continued*) Predators in the order Hemiptera.

(Braman, 2000). Among them, *Nabis capsiformis* is a cosmopolitan species and has been recovered from various locations of India (Kerzhner, 1983) (Plate 1.8). The adults of *N. capsiformis* are generally pale to light brown in colour. Their body is elongated and covered with fine white hairs. The head is narrow and covered sparsely with whitish hairs, which are more abundant ventrally. They have a brown area between the eyes and antennae that run laterally. Antennae are long and slender with small hairs. The mouthparts are long, needle-like and tucked under the head and body at rest but are flexible and can be positioned in front of the head when feeding. The pronotum is covered with long soft hairs with a brown stripe in the middle (in some specimens, this stripe is diffused) (Cornelis et al., 2012).

N. capsiformis (Plate 1.8) is a predator of aphids, moth eggs and small caterpillars, including corn earworm (*Helicoverpa zea*), imported cabbageworm (*Pieris rapae*) and some armyworm species (Ashby, 1974; Buschman et al., 1977; Godfrey et al., 1989). Other prey may include mites, thrips, leafhoppers, tarnished plant bug nymphs, asparagus beetle (*Crioceris asparagi*) and Colorado potato beetle (*Leptinotarsa decemlineata*) eggs and larvae (Jessep, 1964; Samson and Blood, 1980). Previously, Distant (1910) reported four species, *viz.*, *N. capsiformis*, *N. funebris*, *N. indicus* and *N. nigrescens* from India; however, Chandra et al. (2014) reported *Nabis tibialis* Distant from Chhattisgarh, which is a new addition to the fauna of India.

1.2.3 Family: Anthocoridae (Pirate Bug)

Anthocorid bugs (family Anthocoridae) are small (<5 mm long), mainly predaceous, flat, elongate oval to oval, primarily black, orange-brown to dark brown insects, with four-segmented antennae and a needle-like three-segmented beak or labium. Although similar in appearance to members of the family Miridae, adult anthocorids differ in possessing a pair of ocelli. Reliable determination of anthocorids to species is often difficult and may require microscopic examination of the genitalia. Anthocorid predators are recognised as potential biocontrol agents and are represented in all the zoogeographical regions of the world. They feed on small lepidopteran larvae, small grubs, psocids, mites, thrips, aphids and storage pests, and are commonly known as minute flower bugs or minute pirate bugs (Muraleedharan and Ananthakrishnan, 1978).

1.2.3.1 *Orius insidiosus* (Insidious Flower Bug)

The adults of *O. insidiosus* are small sized, approximately 3 mm long, and are oval in shape and black with white patches on wings (Plate 1.9) (Lattin, 1999). The wings extend beyond the abdomen. Adults have a long proboscis used for sucking plant juices and consuming prey (Wright, 1994). The insidious flower bug, *O. insidiosus* Say, is a common predator of a wide variety of small, soft-bodied arthropods, such as thrips, spider mites and small caterpillars. They also feed on the eggs and larvae of the bollworm, spotted tobacco aphids, corn earworm, European corn borers (*Ostrinia nubilalis*), corn leaf aphids (*Rhopalosiphum maidis*), potato aphids (*Macrosiphum euphorbiae*) and potato leafhopper (*Empoasca fabae*) nymphs (Lattin 1999). Although a generalist predator, *O. insidiosus* has a preference for flower thrips (Thysanoptera: Thripidae) (Baez et al., 2004). Ramachandran et al. (2002) revealed the capacity of the *O. insidiosus* to suppress populations of thrips, *Frankliniella* spp. in field pepper during the spring even when thrips are rapidly colonising and reproducing. However, Rishi et al. (2014) found that *O. insidiosus* supresses the population of *Thrips tabaci* in Bt cotton hybrids, which confer resistance against primary lepidopteran pest.

1.2.3.2 *Orius niger*

Orius niger is found in the Palearctic from Eastern Europe to Siberia and across central Asia to China and India. The adult bugs are 1.8–2.3 mm in length, completely black in colour, with only the antennae and the tibia of the forelegs light in colour. The membranes of the hemelytra are dark, and there is a single hair at the front and back corners of the pronotum (Greathead and Greathead, 1992). The females lay eggs in flowers of herbaceous plants. In favourable years, there are two generations. In laboratory conditions, the larval development of *O. niger* takes 14 days at a temperature of 25°C; females have a longevity of 60 days and can lay up to 150 eggs (Pehlivan and Atakan, 2017).

O. niger (Plate 1.10) is widespread in Eastern Europe and Northern Asia and is usually found on flowers and leaves of various cultivated or other plants. They are predaceous usually on thrips, aphids, whiteflies, mites and lepidoptera eggs. They are reported as major predators of Onion thrips, *T. tabaci*, and two spotted spider mites, *Tetranychus urticae* in potato (Fathi and Ganabalani, 2009) and western flower thrips, *Frankliniella occidentalis* in green houses (Bahsi and Tunc, 2008).

1.2.3.3 *Orius tantillus* (Motschulsky)

Orius tantillus (Plate 1.11) is widely distributed in India, Sri Lanka, Thailand, Malaysia, China, Australia, Micronesia, Solomon Islands, Kenya, Nigeria and Tanzania (Hernandez, 1999). The adult body length is 1.48 mm. Head, pronotum, last two segments of antennae, scutellum, base of clavus, sternum and abdomen are brown to dark brown in colour; pronotal disc containing shallow callosities is separated by a few punctures; anterior tibiae in males have a row of peg-like teeth; long setae on pronotal angles is absent; and lateral margins of pronotum are wide. The similarity in external appearance of *O. tantillus* (Mats.), *Orius indicus* (Reuter) and *Orius ianthe* (Dish) might be due to their overlapping distribution. The only isolation between them is apparently sexual. Hence, their only differentiating feature is in the shape of the aedeagus (Bal and Biswas, 2013).

It is known as predator of insects, like thrips, aphids and mites, which are infesting the cultivated crops. These predators are found attacking thrips species, like *Haplothrips ganglbaueri* Schmutz,

Thrips palmi Karny, *Anaphothrips sudanensis* Tryb., *Caliothrips graminicola* (Bagn. & Cam.), *Scirtothrips dorsalis*, *T. tabaci*, *Baliothrips biformis*, *Caliothrips indicus* (Bagn.), *Microcephalothrips abdominalis* (D. L. Crawford), *Megalurothrips usitatus* Bagnall and *Thrips hawaiiensis* (Morgan) (Ananthakrishnan and Thangavelu, 1976; Muraleedharan and Ananthakrishnan, 1978).

1.2.3.4 *Tetraphleps* spp.

Members of the genus *Tetraphleps* are small-sized bugs, 3–4 mm in length, and two predatory species, *Tetraphleps galchanoides* Ghauri (Plate 1.12) and *Tetraphleps raoi* Ghauri, have been recorded from different parts of India (Balla et al., 2018). *T. raoi* is reported to inhabit Benguet pine (*Pinus kesiya*), feeding on *Pineus* sp. (Adelgidae) (Ghauri, 1965; Chacko, 1973). This species was sent from India to Kenya for release against *Pineus pini* Macquart on several spp. and it was established there, resulting in the decline of *P. pini* population (Karanja and Aloo, 1990).

1.2.4 Family: Geocoridae

1.2.4.1 *Geocoris ochropterus* (Fieber)

Geocoris ochropterus (Fieber) (Plate 1.13) is the most widely occurring species from Myanmar to Pakistan across North Indian foothills and plains up to the Coimbatore district of Tamil Nadu, stretching through the Peninsular India.

This species grows up to 3.5–4 mm in length; head is brownish yellow in colour; pronotum, scutellum and body are black; wing membrane is transparent; abdomen, corium, coxae, legs, lateral margins of sternum, pronotum and rostrum are yellow with slight brownish tinge; and the first joint of rostrum is little longer than the second (Bal and Biswas, 2013).

G. ochropterus has been found attacking the sedentary pests, like jassids, flies, thrips and also small caterpillars. They also hunt by actively chasing, as well as using the wait and capture method. The dual feeding habit, i.e., partial phytophagic and narcophagic habits, gives this species a special characteristic to survive at the time of extreme shortage of live prey.

G.s ochropterus is predaceous on weevil, *Myllocerus viridanus*, the mired bugs, *Calocoris angustatus* and *Ragmus importunitas*, pentatomid bugs, *Coptosoma cribraria*, the apids, *Aphid gossypii* and *Aphis nerii*, and the coccids, *Pseudococcus* sp. and *Dactylopius nipae* (Bal and Biswas, 2013).

1.2.4.2 *Geocoris jucundus* (Fieber)

Geocoris jucundus (Plate 1.14) is abundant and widely distributed in various agroclimatic regions of India (Bihar and West Bengal) and Myanmar. The adult bugs are small in size with a length of 3 mm; the head is black in colour, with spots on both sides in front of the eyes; the pronotum is black, and the lateral angular area is yellow with slight brownish tinge; scutellum and corium are pale yellowish brown with a brown spot at the anal margin; the rostrum is yellowish brown with base of first, third joint and the apex is black.

This species is an effective predator in the natural enemy complexes of crop pest, *viz.*, tomato aphid, *Aphis gossypii*, and lucerne aphid, *Acyrthosiphon pisum* (Pal, 1974).

1.2.4.3 *Xylocoris* spp.

There are approximately 50 described species of *Xylocoris* across the globe, in which both phytophagous and zoophagous species exist. In India, the two species, *Xylocoris flavipes* and *X. clarus*, have been reported as major predators of stored grains and phytophagous insects. *X. flavipes* (Plate 1.15) is found mainly in stored grains, such as wheat and rice, and feeds on several storage pests, including *Tribolium castaneum* (Mukherjee et al., 1971; Ballal and Yamada, 2016) and bruchids. The species has been reported to prey on 13 species of insects belonging to three orders (Arbogast, 1976). In addition to being noted in storage containers and warehouses, it was also noted in India on *Butea monosperma*, *Ficus* sp., and maize, where it was found in association with *Megalurothrips* sp. and a few mealybugs (*Planococcus citri*) (Ballal et al., 2018). On the contrary, *X. clarus* is a leaf-litter inhabiting species feeding on insects and mites including the thrips species *Apelaunothrips madrasensis* and *A. indicus* (Muraleedharan and Ananthakrishnan, 1978).

1.2.5 Family: Miridae

Mirid bugs (family: Miridae) form the main group of heteropterans and include an eclectic range of small to medium-sized bugs with three-segmented tarsi, four-segmented antennae and a four-segmented rostrum, and the ocelli is absent. Despite the possibility of being brachypterous, adult mirids are often macropterous; their hemelytra each have a distinct census and a membrane with two enclosed cells. Unlike other heteroptera, mirids also possess sensory setae, known as trichobothria, located on the middle and the hind femora (Schuh, 1975). Some mirids are facultative if not obligate predators, and (as in apple orchards or in greenhouses where biological control or integrated pest management (IPM) strategies are adopted) these are often of economic importance as natural enemies of crop pests and may outnumber harmful ones (Ravn et al., 1995).

1.2.5.1 *Cyrtorhinus lividipennis* (Reuter)

Cyrtorhinus lividipennis is distributed in various parts of India, *viz*., Andaman and Nicobar Island, Andhra Pradesh, Bihar, Madhya Pradesh, Karnataka, and neighbouring countries, such as Myanmar and Sri Lanka. The adult bug measures 2.75–3.50mm in length; the body is black in colour; base of head, rostrum, apex of the first joint of the antennae, anterior callosities, a broad central region to posterior area of the pronotum, a big lateral spot on each side of the scutellum, body beneath, and legs flavescent; a lateral spot to sternum, which is black in colour (Bal and Biswas, 2013). *Cyrtorhinus lividipennis* (Plate 1.16) has been recorded from the rice agroecosystem. Prey of this species is known to be white-backed plant hoppers and leafhoppers and brown plant hoppers (Manjunath et al., 1977; Heong et al., 1990).

1.2.5.2 *Macrolophus* spp.

The genus *Macrolophus* contains several predaceous mirids of economic importance, including *M. melanotoma* (formerly known as *M. caliginosus*) (Plate 1.17) and *M. pygmaeus* (formerly known as *M. nubilis*), both of which (notably in the Mediterranean region) are highly effective predators of a range of harmful insects, especially whiteflies. The former mirid preys upon the two-spotted spider mite (*Tetranychus urticae*) (Foglar et al., 1990). A further species (*M. costalis*, formerly known as *M. rubi*) is of particular significance as a predator of aphids, including the peach/potato aphid on tobacco plants in Greece (Margaritopoulos et al., 2003). It is also an effective predator of the chrysanthemum aphid (*Macrosiphoniella sanborni*) (Brezinski, 1988). *Macrolophus costalis* and *M. melanotoma* are also useful predators of thrips and the former being particularly effective as a predator of onion thrips in Bulgarian tobacco crops (Dimitrov, 1975). The nymphs of *M. pygmaeus* are the most efficient predator of aphid, *Myzus persicae* foraging on the middle or lower leaves of tomato plants (Sarmah et al., 2019).

1.2.5.3 *Nesidiocoris tenuis*

Nesidiocoris tenuis (Plate 1.18) is a multivoltine species of Mediterranean origin, which continues to breed throughout the year and does not enter diapause. Development is rapid, and in particularly favourable areas, the bug can complete up to six summer and two winter generations annually. *N. tenuis* is a voracious natural enemy of whiteflies, including the tomato whitefly (*Bemisia tabaci*) (Sridhar et al., 2012). It also feeds on thrips and has proved to be an effective biocontrol agent for combating tomato borer moth infestations on both outdoor and protected crops (Shaltiel-Harpaz et al., 2016). It also preys on eggs and neonate larvae of fruit borer, leaf eating caterpillar, stem borer and tomato pin worm, and aphids and mites (Bhatt and Patel, 2018). Although considered an important biocontrol agent, as a zoophytophage *N. tenuis* can also cause direct damage to crops, primarily when suitable prey is in short supply or absent (Perdikis et al., 2009; Arno et al., 2010).

1.2.5.4 *Apolygus spinolae*

Apolygus spinolae can be found throughout Europe, except for Albania, Estonia, Liechtenstein, Malta, and Portugal and not in the extreme south; then it is found towards east across the Palearctic to Siberia, and through central Asia to China and Japan. Members of the genus *Apolygus* are essentially phytophagous and one species (*A. spinolae*) (Plate 1.19), although associated mainly with nettles, thistles and other low-growing plants, and are known to damage hopbines. The same species also feeds on aphids (Sunderland et al., 1987). *A. spinolae* is rather broad-bodied and mainly pale green, with a black-tipped cuneus, dark red eyes and a smoky-black hemelytra membrane. Adults are active from June to September, and there is just one generation annually, with eggs forming in the overwintering stage.

1.2.5.5 *Miris striatus* (British bugs)

Miris striatus (Plate 1.20) is a relatively large (9–11 mm long), mainly black, yellow-streaked species, associated mainly with hawthorn and oak, where it is predaceous on a range of insects, including aphids and the eggs and larvae of various beetles and lepidopterans. On oak, for example, it is known to attack larvae of the green oak tortrix moth, *Tortrix viridana* (Gulde, 1921). Eggs, 3.5 mm in length (Butler. 1923), are surprisingly large, and they form in the overwintering stage and hatch in the spring from late April onwards. The mainly black or reddish-black, red-legged cylindrical nymphs are distinctly ant-like and similar in appearance to those of *Pilophorus* spp. They feed for about 6 weeks before finally reaching the adult stage, usually in late May or early June. Eggs are laid in the tissue of 'host' plants, there being just one generation annually. Adults usually survive no later than mid- or late July, with numbers reaching their peak in June.

1.2.5.6 *Blepharidopterus angulatus* (black-kneed capsid)

Blepharidopterus angulatus is found in north Africa, Europe, towards the east across the Palearctic, to central Asia and North America. It (Plate 1.21) is a rather narrow-bodied, generally common Holarctic species, with a prominent black mark at the base of each tibia, a feature displayed in both adults and nymphs. Known widely as the black-kneed capsid, *B. angulatus* occurs on a range of deciduous trees (including alder, apple, ash, birch, lime, oak and willow) and is of particular significance as a predator of the fruit tree red spider mite in apple orchards (Collyer, 1952). For many years, it has been a major focus of attention in research relating to biocontrol of mite pests (Wheeler, 2000). In addition to spider mites, other soft-bodied arthropods are also preyed upon. These include tree-inhabiting aphids, leafhoppers and small caterpillars as well as the eggs of the codling moth (*Cydia pomonella*) (Glen, 1977). *B. angulatus* is also partly phytophagous, but any damage caused in apple orchards is of little or no consequence and is outweighed by the mirid's beneficial role as a predator.

1.2.5.7 *Atractotomus* spp.

Three species of *Atractotomus* occur in Europe, northern Asia and North America, one of which (*A. mali*) (Plate 1.22) is found mainly on apple and hawthorn and two of which (*A. magnicornis* and *A. parvulus*) are associated with conifers. All three species are small (adults with a body length often less than 3 mm), with *A. parvulus*, at 3.2–3.7 mm in length, being the smallest. Adults and nymphs feature much enlarged (swollen: cylindrical or spindle-shaped) second antennal segments. *A. mali*, the well-known black apple capsid, is sometimes regarded as a pest in apple and pear orchards and in common with certain other mirids, such as *Campylomma verbasci*, occasionally pierces developing fruitlets. Such feeding triggers the subsequent development of ill-shaped, callused fruits, which are then unmarketable (van Frankenhuizen et al., 2002; Alford, 2014). *A. mali* is also recorded in the webs of the common small ermine moth (*Y. padella*) (Giard, 1900) that commonly occurs in, for example, blackthorn and hawthorn hedges. It is univoltine, with nymphs developing mainly throughout June. Adults are present mainly from June to late August, and summer-laid eggs overwinter.

1.2.6 Family: Reduviidae

Reduviidae are members of the suborder Heteroptera of the order Hemiptera. Reduviids (Insecta, Hemiptera, Reduviidae) are general, polyphagous predatory insects and cosmopolitan biocontrol agents worldwide. The family is almost all predatory, except for a minority that are blood-sucking species of importance as disease vectors. About 7,000 species have been described, making it one of the largest families in the Hemiptera. They dwell in all plantation crops including forest and social forest. Reduviids are efficient predators of insect pests of crops, playing a significant role in keeping pest populations in check. Some important biocontrol agents are *Platymeris laevicollis* Distant, *Zelus renardii* Kolenati, *Rhynocoris marginatus* (Fab.), *Rhynocoris kumarii* Ambrose and Livingstone, *Rhynocoris fuscipes* (Fab.), *Pristhesancus plagipennis* Walker, *Blaptostethus pallescens* Poppius, *Acanthaspis pedestris* (Stål), *Catamiarus brevipennis* (Serville), *Ectomocoris tibialis* (Distant), etc. (Sahayaraj and Balasubramanian, 2016).

1.2.6.1 *Zelus renardii* (leafhopper assassin bug)

Assassin bug adults and nymphs are slender, colourful insects, often blackish, reddish or brown. They have long legs, a long narrow head, round beady eyes and an extended, three-segmented, needle-like beak (Plate 1.23). Nymphs are quite small, 5 mm (1/4 inch) in length, when they hatch and grow to an adult size measuring approximately 2 cm (3/4 inch). Insects in this order undergo incomplete metamorphosis. Eggs of *Zelus* spp. are barrel-shaped, dark brown with a white cap and are laid openly in groups on plant surfaces. Adults are poor fliers, and both adults and nymphs move rapidly when disturbed. All assassin bugs are predators; some species feed on insects while others feed on the blood of mammals. Insect-feeding species eat a wide variety of small to medium-sized insect prey including caterpillars, leafhoppers, other bugs and aphids. They also feed on beneficial species, such as lacewings. Nymphs and adults are often seen stalking or lying in wait for their prey, which they inject with venom once they have caught. Assassin bugs are common natural enemies on many plants including row and tree crops (Sahayaraj and Balasubramanian, 2016).

1.2.6.2 *Platymeris laevicollis* (red-eyed assassin bug)

Platymeris laevicollis is a venomous predatory true bug from central Africa that can be found in forests, scrublands, grasslands and croplands. It is an efficient predator and is used by farmers on coconut plantations to control herbivorous pests, such as the rhinoceros beetle. As a true bug of the order Hemiptera, it has needle-like mouthparts designed for sucking juices out of plants or other insects instead of chewing. *P. laevicollis* (Plate 1.24) has sharp stylets in its proboscis or rostrum used to pierce the exoskeleton of its prey. Saliva is then injected into the prey, which liquefies its tissues, and the rostrum is used to suck out the digested fluids. If disturbed, it is capable of a defensive bite considered to be more painful than a bee sting (Sahayaraj, 2014a).

1.2.6.3 *Rhynocoris marginatus*

Rhynocoris marginatus is a predator of other insects and is found in Asia. Crops in India on which it has been found feeding on pests include sugarcane, pigeon pea, cardamom, cotton, tea and peanuts. The insects are potentially useful in biocontrol because they are more resistant to pesticides than are the pests on which they feed (Sahayaraj, 2014b).

From the family Reduviidae, *R. marginatus*' (Plate 1.25) success as a predator in agricultural systems is due to unique morphological and physiological adaptations for the predation of other insects. Most notable among these adaptations is their extra-oral digestion feeding mechanism, as well as a three-segmented rostrum, or barb, armed with a number of hair-like mechanical and chemical sensors, used not only to sense and determine the suitability of their prey but also to deliver venomous saliva to paralyse their victims (Kumar and Sahayaraj, 2012). Unique to *R. marginatus* are their accessory salivary glands, which function as water recapturing organs that recirculate water from the body to keep a steady flow of watery saliva going as they feed, helping the predator to flush out predigested food from the body of its prey (Sahayaraj, 2014b).

Assassin bugs are hemimetabolous and do not undergo metamorphosis. Instead, the egg hatches into a miniature version of the adult form, and at each growth stage, the nymph becomes more and more like the adult form (Sahayaraj et al., 2010).

R. marginatus is a generalist predator, feeding on at least 20 species of crop pests, occurring naturally in cotton, groundnut and soybean crops. Different life stages have different food preferences, but in general, when offered various choices, the larva of the tobacco cutworm *S. litura* (a moth larva) was the first choice, followed by *H. armigera*, *Aproaerema modicella* and *Amsacta albistriga* in that order (Sahayaraj, 1999). *R. marginatus* is one of the most common and widespread predatory insects in Indian agricultural areas. This assassin bug is attracted by the kairomones emitted by its prey; even the shed scales of adult moths provide an odour that attracts the bugs (Sahayaraj et al., 2010).

Among pests of cotton crops, three species of moth larvae are more attractive to the assassin bug than are the sap-sucking hemipteran, *Dysdercus cingulatus*, and two species of beetles in the genus *Mylabris*. Feeding on the moth larvae results in the assassin bug growing faster than on other diets and research suggests that the bug may be effective in biocontrol of moth larvae in cotton (Sahayaraj, 2014a).

1.2.6.4 *Rhynocoris fuscipes* (Fabricius)

Rhynocoris fuscipes (Plate 1.26) is a generalist predator of cotton pests and is commonly found inhabiting cotton-growing regions in Assam, Maharashtra, southern India and Sri Lanka (Tomson et al., 2017). Its body is 14–16 mm long and coral red in colour; rostrum, antennae, anterior area of the posterior lobe of pronotum, disk of scutellum, an oblong spot between antennae, upper surface of post ocular area, legs, two spots to pronotum are black in colour. The pronotum has an anterior lobe that is distinctly sculptured and a posterior margin that is pale fuliginous..Harpactorine bugs occur on vegetation, especially on the foliage. They may occupy a variety of microhabitats, such as underneath the bark, in litter and under boulders.

It is predaceous on larvae of *H. armigera*, *Spodoptera litura*, *Semiothisa pervolgata*, *E. hecabae* and *Catopsilia pyranthe*, and beetles *Henosepilachana vigintioctopunctata* and *Raphidopalpa foveicollis*.

1.2.6.5 *Rhynocoris costalis* (Stal.)

The length of an adult *Rhynocoris costalis* is between 12 and 14 mm. Its body is black, lateral margin and basal area of posterior pronotal lobe, lateral areas of corium, coxae, trochanters, abdomen beneath, apex of scutellum is coral red in colour; the anterior lobe of pronotum is sculptured, whereas posterior lobe is not impressed; a spot between and a lateral spot behind eyes is coral red; ventral side of head and spots beneath the femora is brownish yellowin colour (Plate 1.27).

Although this species occurs in the vegetable gardens and other agricultural fields on vegetation, especially on foliage, they may be seen in the cotton fields preying upon cotton bugs, *Dysdercus koenigii* (David and Ananthakrishnan, 2004).

1.2.6.6 *Blaptostethus pallescens* Poppius

Blaptostethus pallescens (Plate 1.28) is found in Celebes, Egypt, Madagascar and the Indian states of Karnataka, Maharashtra and Tamil Nadu. It preys on primary and secondary stored product pests, namely, *R. maidis* (Fitch) on maize, onion thrips (*Thrips tabaci*) and mites. Its biology, feeding potential and rearing techniques were studied by Ballal et al. (2003). The common laboratory host, *Corcyra cephalonica* (Stainton), could be used for mass producing this predator.

1.2.6.7 *Acanthaspis pedestris* (Stal.)

Acanthaspis pedestris (Plate 1.29) is found in India, especially Tamil Nadu in South India and West Bengal, and also found predominantly in the semi-arid zones, scrub jungles and on the foothills of forests and their bordering agroecosystems.

A pitch-black species, it grows up to 13.5 mm in length, with a basal angular and an irregular shaped creamy white spot on corium, another creamy white spot at the apex of the membrane; anterior lobe of pronotum sculptured, posterior lobe finely wrinkled with a central and two narrow lateral spots, blackish red in colour.

This species kills higher number of prey with increasing prey density. The prey includes *Pectinophora gossypiella* and *Odontotermes assumthi.*

1.2.6.8 *Acanthaspis siva Distant*

Acanthaspis siva Distant is predominantly found in south India and the north-central provinces of India. A moderate to large size is about 18 mm in length. Generally, it is a black or dark species with an angular creamy white spot on the basal area and an irregular rounded spot on the subapical area of corium; apex of the membrane is also creamy white enclosing a smoky brown spot, anterior lobe of the pronotum is sculptured, lateral angle is dentate and recurved, the disc is unarmed and the scutellar spine is obliquely ascendant (Plate 1.30).

Both nymphs and adults freely lurk around the hives, in crevices or in cracks, in tree trunks near the locations of the nests of the Indian bees. The adult bugs are active during September to November. Generally, this species is found in the entire three major ecosystems, i.e., tropical rain forests, scrub jungles and semi-arid zones as well as agroecosystems

The brownish spherical eggs are laid singly in crevices on hive stand or decaying leaf moulds accumulated in the vicinity of the hive. Incubation period is 17–28 days. Nymph passes through five nymphal instars to become an adult. Total nymphal period is 107–186 days. The longevity of male and female is 202 and 115 days, respectively. Frequent cleaning of base boards and surrounding of the hives help in keeping away these bugs from attacking the honey bees.

1.2.6.9 *Cydnocoris gilvus* (Bunnester)

Cydnocoris gilvus is found in India, Java, Sumatra, Sri Lanka and Myanmar. It is of moderate to large size, i.e., 16–18 mm. Its body is blood red in colour; anterior spine to head is curved; antennae, eyes, legs, membrane, coxae and trochanters are black in colour; lateral area of abdomen and posterior area of pro-, meso- and metasternum are black in colour (Plate 1.31). Although the Harpactorines prefer the rainforests followed by semi-arid zones and scrub jungle ecosystems, this species is recorded from the agroecosystems. It is found on vegetation, especially on foliage. It also occupies a variety of microhabitats, such as underneath bark, in litter under boulders and in termitaria. This species is known to attack *S. litura*, *Oxya nitidula* and *Odontotermes obesus.*

1.2.6.10 *Endochus inornatus* Stal.

Endochus inornatus (Plate 1.32) is found almost all over India. This species attains a length of 20–21 mm. Body is dark cinnamon brown; legs are pale red except the base of the posterior tibiae and the apex of the posterior femora; head is about the same length of the pronotum with a short spine behind each antenna; anterior pronotal lobe is longitudinally impressed and the posterior lobe is finely wrinkled; lateral spines to the pronotum are horizontally produced, and the pronotum between the lateral spines are transversely ridged.

This species prefers the scrub jungles, semi-arid zones and adjacent agroecosystems, especially foliage. They may occur under the bark, in litter and under boulders. David and Ananthakrishnan (2004) have reported this species as a potential predator of banana lacewing bug, *Stephanitis typica* (Fletcher).

1.2.7 Family – Gerridae

1.2.7.1 *Limnogonus nitidus* (Mayr.)

Limnogonus nitidus is distributed in India, Indonesia, Malaysia, Maldives Islands, Myanmar, Nepal, Singapore, Sri Lanka, Thailand and Vietnam. Body length of adult insects may be 6.0–8.0 mm. Betweenthe eyes, head has the distinctive pair of yellowish longitudinal lines.. First and fourth antennal segments are longest and subequal to each other, while the second and third segments are shortest and also subequal to each other in length (Plate 1.33). Posterior margin of dorsal vesicle plate of pronotum is remarkably angular. The seventh abdominal segments are provided with fairly prominent connexival spines in both the genders.

This species has been widely found in temporary pools, rice fields flooded with water, ponds and lakes. It also has been recorded at an altitude of 1,000 m above the sea level from Nilgiri Hills, Southern India. This species is very much attracted by light. It preys upon small aquatic organisms that fall in its way. This species is also one of the predators of the Brown plant hopper of paddy *Nilaparvata lugens* (Stal) of the family Delphacidae, which is a serious pest of paddy (Bal and Biswas, 2013).

1.2.7.2 *Neogerris parvula* (Stal.)

Neogerris parvula is found in India, China, Indonesia, Iran, Japan, Malaysia, Myanmar, New Guinea, Oman, Philippines, Pakistan, Solomon Islands, Sri Lanka, Taiwan, Thailand and Vietnam. Body length of the adult insect may be 6.0–6.5 mm. Presence of a large rounded yellowish spot on the pronotum makes this species unique from all other known species. Antennal first segment is the longest and as long as or a little longer than the second and third segments together, and the fourth segment is almost half the length of the first segment. In males, the hind margin of pygopore is straight while the hind margin of the seventh ventral abdominal segment is bisinuate in females. The species is represented by both apterous and macropterus forms (Plate 1.34).

This species is very common in both lentic and lotic waterbodies, like reservoirs, ponds, rain-fed pools, flooded rice fields and slow running streams throughout the Indian subcontinent. They prey upon whatever small aquatic animalcules they find on their way, like *Limnogonus* (L.) *nitidus*. This species is also a notable predator of the Brown planthopper of paddy, *Nilaparvata lugens* (Stal)—a serious pest of paddy—and performs as a natural controlling agent (Sahayaraj, 2014a). Earlier this species was known as *Gerris tristan* as reported by Distant.

1.3 ORDER: NEUROPTERA (LACEWINGS AND ALLIED INSECTS)

Lacewings (order Neuroptera) are small to large, mainly predatory insects, with two pairs of transparent wings and long, thread-like (filiform) or bead-like (moniliform) antennae. Wing venation is often complex, with numerous cross-veins and as in the families Chrysopidae and Hemerobiidae, a ladder-like arrangement of short veins running along the leading edge of each wing; in addition, the veins may be forked (bifurcated) close to the wing margin. When at rest, the wings of lacewings are held over the body in a sloping, roof-like posture. Although considered somewhat primitive, lacewings are holometabolous insects, and they develop to adulthood through egg, larval and pupal stages. Pupae of lacewings are exarate, with the antennae, mandibles, legs and other appendages 'free': in addition, the mandibles of pupae are moveable (articulated) and are used to bite through the pupal cocoon prior to the emergence of the adults.

Members of three Neuroptera families (namely, Chrysopidae, Hemerobiidae and Coniopterygidae) are universally recognised as beneficial predators of a range of soft-bodied arthropods, including aphids, scale insects, thrips, whiteflies, lepidopteran eggs and phytophagous mites (Table 1.2). In Asia, however, as a notable exception, commercial producers of shellac have reason to regard lacewings (specifically *Chrysopa lacciperda* and *C. madestes*) as pests, since they prey upon and destroy cultured lac scales (*Kerria lacca*), the insects from which shellac is obtained (Singh et al., 2010).

TABLE 1.2
Predator Species in Various Families under the Order Neuroptera and Host Insects

Sl. No.	Family	Predator Species	Host Insect
1	Chrysopidae	*Chrysoperla carnea*	Egg and larva of lepidopteran pest, mites
2	Hemerobiidae	*Hemerobius sp.*	Aphids and spider mites
3	Coniopterygidae	*Coniopteryx* sp.	Red spider mites
4	Mantispidae	*Mantispa styriaca*	Spider eggs, soft-bodied insects
5	Raphidiidae	*Raphidia ophiopsis*	Eggs of lepidopteran moths

1.3.1 Family: Chrysopidae (Green Lacewings)

Green lacewings, with their often bright green bodies and golden compound eyes, are familiar insects, distinguished morphologically from other lacewings by their overall appearance and distinctive wing venation. They are exemplified by the widely distributed and generally abundant Holarctic insect, *Chrysoperla carnea* (formerly known as *Chrysopa carnea* and universally dubbed the common green lacewing). This insect was first described in 1835 by British entomologist James Francis Stephens, and for many years it was accepted as a distinct species.

Eggs of green lacewings, which hatch within a week or two, are laid on leaves and other such surfaces, usually singly (as in the case of *Chrysoperla carnea*) or in small groups, with a thread-like stalk (pedicel) at the tip of which an egg is laid. Stalk-borne eggs are probably less likely to fall victim to predators (Ruzicka, 1997a). In some species, the eggs are arranged in close-knit batches (e.g. *Dichochrysa flavifrons* and *Nineta flava*), and the stalks may then be tightly or loosely intertwined. Egg-laying green lacewing females typically mark surfaces with chemicals (known as allomones), which deter other females from laying eggs in the same vicinity (Ruzicka, 1997b, 1998).

The larvae of green lacewings are highly mobile and formidable predators well capable of tracking down and overcoming their prey, especially aphids. Once grasped, prey items are often lifted into the air. As in the case of *Chrysoperla carnea* (Plate 2.1), a single larva may consume 200–500 aphids during the course of its development; larvae have also been shown to kill around 30–50 fruit tree red spider mites (*Panonychus ulmi*) in just an hour (Principi and Canard, 1974). The piercing mouthparts of green lacewing larvae include a long, sickle-like pair of inwardly curved and sharply pointed jaws. Each jaw is formed from an interlocked mandible and maxilla, with a food channel sandwiched between them; the maxilla also includes a small venom canal. Such jaws are used to impale the prey, to inject venomous saliva and then to imbibe the victim's liquefied body contents. The larvae typically possess prominent thoracic and abdominal tubercles, from which arise clusters of long setae.

Adult green lacewings are active mainly at night. However, *C. pallida* commences flying during the afternoon (Szentkirlyi, 2001). In summer, following their emergence from the pupal cocoon, new adults of *C. carnea* have been observed to undertake obligatory flights downwind for two to three successive nights at heights above 3 m (described as 'adaptive dispersal flights'). Such activity takes them well away from emergence sites. Also, newly emerged adults of *C. carnea* (s. str.) (unlike those of *C. lucasina* and *C. pallida*) often undertake high-altitude, preovipositional flights that may extend for a few hundred kilometres (Chapman et al., 2006). Subsequently, the lacewings again fly downwind, but closer to the ground (described as 'appetitive downwind flight'); they are then prompted into landing by the odour of honeydew. Having once landed, the lacewings subsequently fly upwind towards the source of the honeydew (described as 'appetitive upwind flight') (Duelli, 1980).

Plate 2.1 *Chrysoperla carnea*

Plate 2.2 *Hemerobius* sp

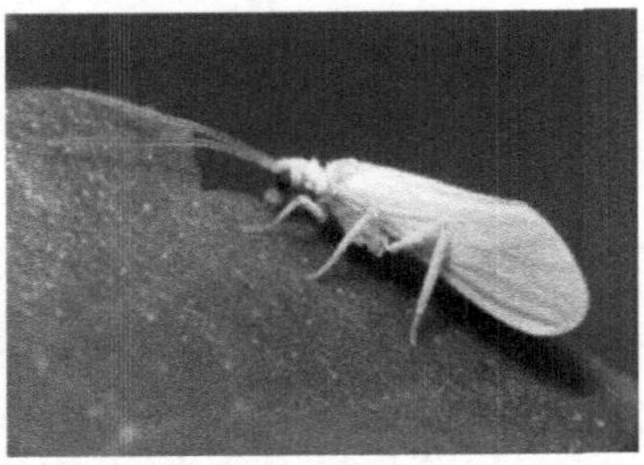

Plate 2.3 *Coniopterigid* sp

Plate 2.4 *Mantispa styriaca*

Plate 2.5 *Subilla confinis*

PLATE 1.2 Predators in the order Nueroptera.

1.3.2 Family: Hemerobiidae (Brown Lacewings)

Brown lacewings are mainly brown or greyish, black-marked insects, with numerous small hairs on the wing membrane. Also, compared with green lacewings, there are relatively few cross-veins. Larvae are usually slender-bodied, with a tapered, more or less pointed, abdomen, and the head is fully exposed (i.e. not retracted into the prothorax). Such forms are described as 'type l' larvae (Oswald and Tauber, 2001), and are typical of *Drepanepteryx phalaenoides* a rare species, and *Hemerobius*, *Micromus* and *Wesmaelius* spp., all of which are highly active predators. 'Type 2' larvae typified by relatively sedentary species, such as *Psectra diptera*, are more sluggish. They have a somewhat swollen, less noticeably tapered body, and the head is partly retracted into the prothorax. Brown lacewing larvae, whether 'type I or 'type 2', lack the lateral tubercles found in chrysopidae larvae, and possess a few body hairs; the jaws are also shorter, stouter and less strongly curved, and the head is noticeably smaller. Brown lacewings are predaceous as both adults and larvae, but adults also feed on honeydew and pollen.

The eggs of brown lacewings are oval, and they possess an often very prominent disc-shaped or knob-like micropyle. They are laid singly or in small groups and placed directly onto leaves and other surfaces. There is no pedicel. Following egg hatch, the larvae attack a wide range of soft-bodied, immobile or relatively immobile arthropods, including aphids and spider mites.

Adults are similarly voracious feeders and during the course of their 2-month life span, males and females have been estimated to consume more than 13,000 and 17,000 adelgids, respectively (Laidlaw, 1936). Pupal cocoons, in common with those of other brown lacewings, are more or less elliptical, double-walled, flimsy and transparent, and they are usually hidden away under loose bark or found amongst crinkled leaves, moss or other shelter. The occupying pupa within the cocoon is usually clearly visible.

1.3.3 Family: Coniopterygidae (Powdery Lacewings)

Powdery lacewings, also known as dusty-wings and wax-flies, are tiny insects (body length of 3 mm) with a wingspan of about 7–8 mm. Characteristically, the body and wings are coated in white or greyish mealy wax, which gives the insects a superficial whitefly-like appearance (Plate 2.3). Compared with other lacewings, wing venation is relatively simple, and cross-veins on both the forewings and hindwings are few in number.

Powdery lacewings are predaceous as both adults and larvae. Their prey includes small, sedentary or slow-moving, soft-bodied phytophagous arthropods, such as aphids, phylloxera, scale insects and mites, including their eggs. The adults also feed on honeydew. In spite of their small size, powdery lacewings are considered at least potentially of value as predators of various arboreal pests, especially spider mites and also well-known predators of bryobia mites (Monserrat and Marin, 2001).

They lay tiny oval eggs singly on the leaves of plants, sometimes close to a future food source. The eggs hatch within 1 week; the larvae are more or less fusiform to globular in appearance and no more than a few millimetres long. They are also relatively long-legged, and often distinctly coloured or patterned. Unlike those of brown lacewings and green lacewings, the larvae have straight, rather than sickle-shaped jaws. The head is also largely inflexible, so 'captured' prey cannot be lifted into the air. As in orchards, larvae of some species, including *Coniopteryx* sp., may still be active after leaf-fall. They are then able to complete their development by feeding on the winter eggs of fruit tree red spider mites (Collyer, 1951).

1.3.4 Family: Mantispidae (Mantid-flies)

Lesser-known neuropterans include mantid-flies (family Mantispidae), unusual insects found mainly in the tropics and subtropics. The predaceous adults are opportunist feeders and use their much modified, mantid-like forelegs to catch their prey. These include *Mantispa styriaca* (Plate 2.4) whose larvae (in common with those of other members of the subfamily Mantispinae) feed on eggs that develop in the egg sacs of spiders, especially wolf spiders and soft-bodied insects (Redborg, 1998). Although sometimes mistaken for true mantids, adult mantid-flies are readily distinguished by their net-like wing venation and other structural differences. Although adults might fortuitously include pest insects in their diets, these are not specifically targeted.

1.3.5 Family: Raphidiidae and Inocelliidae (Snake-flies)

Snake-flies are distinguished by their significantly extended prothorax and lengthy ovipositors in the females. The wing venation is somewhat lacewing like but includes a distinct pterostigma on the leading edge (costal margin) of each wing. Most species are included in the family Raphidiidae, but a few (known as square-headed or inocelliid snake-flies) are placed in the family Inocelliidae.

Snake-flies are mainly arboreal-based predators, but they tend to inhabit the tops of older trees and are often overlooked. The two species, *Atlantoraphidia maculicollis* and *Subilla confinis* (Plate 2.5) occur on conifers, whereas *Phaeostigma notata* is associated primarily with oak.

Both the larval and adult snake-flies feed mainly on insects, such as aphids, and other members of the hemipteran suborder Sternorrhyncha (Aspock, 2002).

Snake-flies are rarely found in cropping environments, other than in commercial timber forests where larvae of the species, such as *Raphidia ophiopsis*, are known to attack eggs of the gregarious, spruce-feeding saw-fly *Cephalcia abietis* (Pschorn-Walcher, 1982). Similarly, a species of *Raphidia* attacks the egg rafts of the black arches moth (*Lymantria monacha*) (Bogenschtz, 1978).

Snake-fly larvae also feed under loose bark, where they attack various coleopteran pests, including bark beetles. Exceptionally, *Parainocellia bicolor* (family Inocelliidae), a Mediterranean species, is an abundant and indiscriminate predator of arthropod pests in Italian vineyards (Pantaleoni and Alma, 2001).

1.4 ORDER: COLEOPTERA

Beetles are minute to large insects, with biting mouthparts adapted for chewing. The body is usually covered by a hard, often brightly coloured, metallic-looking exoskeleton, and the forewings are typically modified into horny or leathery elytra, which usually meet in a straight line along the back. By contrast, the hindwings (although often reduced or absent, as in flightless species) are membranous, and folded beneath the elytra when in repose. Beetles are holometabolous insects, in which metamorphosis is complete, and they develop to adulthood through egg, larval and pupal stages. Beetle larvae usually have a distinct head and three pairs of thoracic legs, but are sometimes apodous. Beetles constitute the largest order of insects, with more than a quarter of a million species worldwide, many of which are phytophagous, highly destructive pests of cultivated plants, e.g., bark beetles, flea beetles, jewel beetles, leaf beetles, longhorn beetles and various kinds of weevil. Some beetles also act as vectors (carriers and transmitters) of harmful plant diseases, especially viruses. Redressing the balance, many beetles are voracious, economically important predators, and these are often utilised as biocontrol agents in pest management programmes, to combat harmful insects, mites and other plant-infesting organisms. Beneficial species include members of some well-known predatory families, Carabidae (ground beetles), Coccinellidae (ladybirds) and to a lesser extent, Staphylinidae (rove beetles).

1.4.1 Family: Carabidae (Ground Beetles)

Ground beetles (also known as carabid beetles) form a very large group of small to large beetles, with over 40,000 known species worldwide (Lovei and Sunderland, 1996), many of which are epigenetic (ground-dwelling), highly active and voracious predators. However, overall, although playing an important predatory role, ground beetles might be more appropriately described as omnivores, since many (if not most) species act, at least in part, as scavengers or seed-feeders, roles that have not always been fully recognised. Nevertheless, in combination with other factors, mainly seed-feeding species, such as *Amara* and *Harpalus* spp., may be effective biocontrol agents of certain weeds (Kromp, 1999).

Ground beetle larvae are campodeiform, with well-developed mouthparts, three pairs of thoracic legs, ten abdominal segments, and a pair of cerci protruding from the final abdominal segment. Larvae have a more restricted diet than adults and tend to be more carnivorous (Lovei and Sunderland, 1996). The non-predatory, phytophagous larvae of the cereal ground beetle (*Zabrus tenebrioides*), however, are destructive pests of cereals and grasses, and a notable exception. Ground beetle larvae develop through three instars and pupate in the ground, each in a purposely formed earthen cavity. Irrespective of their lifestyle, species in temperate Europe typically complete just one generation annually.

1.4.1.1 *Bembidion* spp.

These species are small ground beetles (body length 3–6 mm), characterised by maxillary palps in which the apical (last) segment is very small. Adults vary in appearance, according to the species, ranging from dark bronze (e.g. *B. lampros* and *B. properans*) to those in which each elytron includes a pair of irregular yellowish patches (e.g. *B. tetracolum* and *B. quadrimaculatum*). Both brachypterous and macropterous adults are known. *Bembidion* spp. are diurnal, and are often common and frequently dominant or subdominant in cultivated farmland. Adults overwinter in hedgerows and under grasslands and become active in the early spring, ahead of many other predators. *B. lampros* (Plate 3.1) is a spring-breeder, although in more northerly regions, such as Sweden, it is considered an early summer breeder (Wallin, 1989). New adults appear from mid-summer onwards, before eventually overwintering. A period of summer aestivation may also occur in the lifecycle. Larvae feed in the soil, and also on the surface. *Bembidion* spp., of which *B. lampros* is a prime example, are useful predators of various crop pests, including cereal aphids, and dipteran eggs, such as those of cabbage root fly (*Delia radicum*) (Wright et al., 1960). The predatory species, such as *Bembidion*

Plate 3.1. *Bembidion lampros*

Plate 3.2. *Calosoma sycophanta*

Plate 3.3. *Carabus auratus*

Plate 3.4. *Harpalus rufipes*

Plate 3.5. *Loricera pilicornis*

Plate 3.6. *Nebria brevicollis*

PLATE 1.3 Predators in the order Coleoptera.

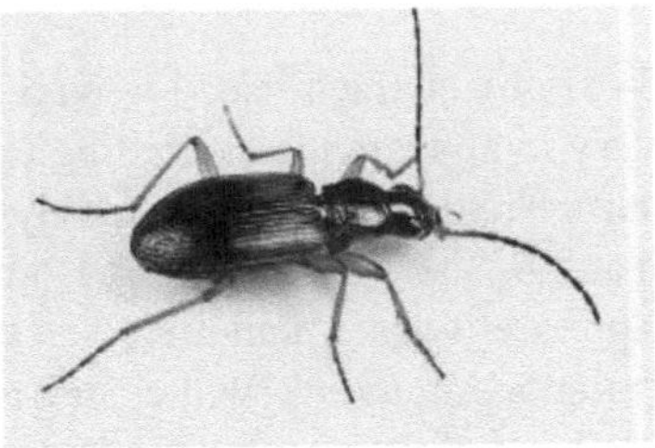

Plate 3.7. *Anchomenus dorsalis*

Plate 3.8. *Pterostichus madidus*

Plate 3.9. *Amara aenea*

Plate 3.10. *Dendroxena quadrimaculata*

Plate 3.11. *Aleochara bilineata*

Plate 3.12. *Dalotia coriaria*

PLATE 1.3 (*Continued*) Predators in the order Coleoptera.

conforme, *B. tetracolum* (Thakur and Parikh, 2016), *Bembidion sobrinum* Boheman and *Bembidion niloticum* (Kushwaha et al., 2017), have been reported from different parts of India.

1.4.1.2 *Calosoma* spp.

The adult beetles are very large (body length 24–30 mm), and metallic black to blue-black, with golden-green to purplish or golden-red elytra with a short, characteristically rounded pronotum, which has a raised gutter-like margin (Plate 3.2). *Calosoma sycophanta* (the forest caterpillar hunter) is a renowned predator species in the forest ecosystem. They complete just one generation annually. In the early 1900s, *C. sycophanta* was introduced into the eastern United States in order to combat outbreaks of two alien pests of European origin: the brown-tail moth and European

Plate 3.13. *Cantharis fusca*

Plate 3.14. *Rhagonycha fulva*

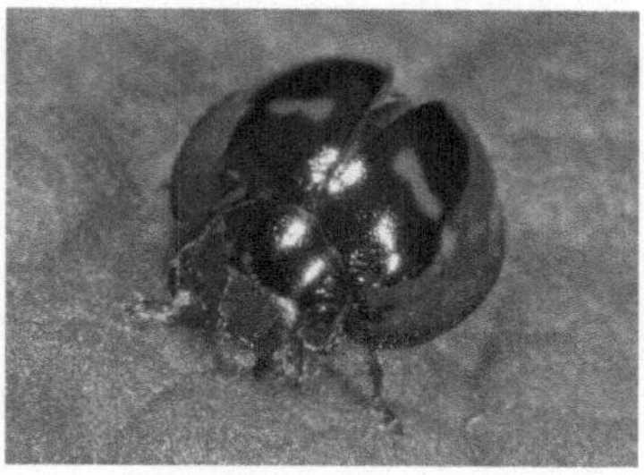

Plate 3.15. *Chilocorus bipustulatus*

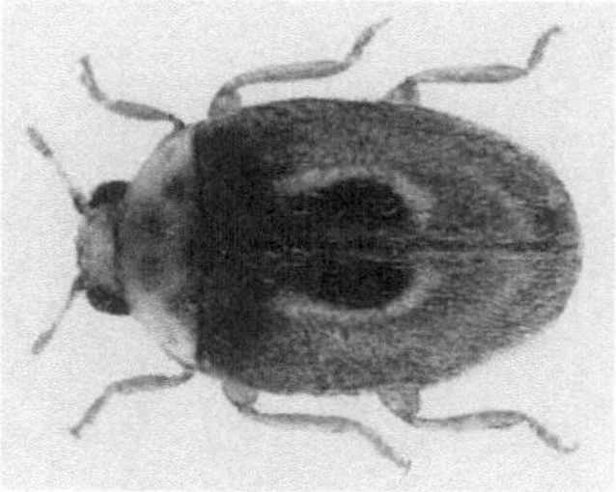

Plate 3.16. *Clitostethus arcuatus*

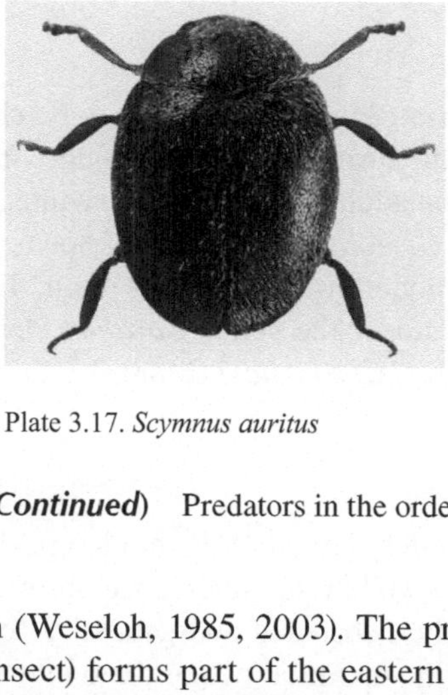

Plate 3.17. *Scymnus auritus*

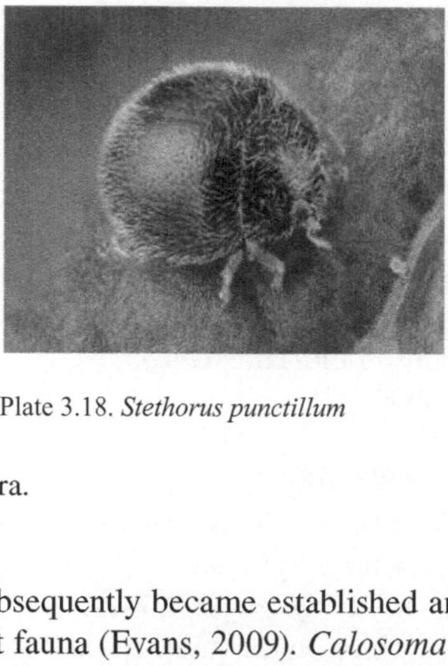

Plate 3.18. *Stethorus punctillum*

PLATE 1.3 (*Continued*) Predators in the order Coleoptera.

gypsy moth (Weseloh, 1985, 2003). The predator subsequently became established and now (as a beneficial insect) forms part of the eastern US insect fauna (Evans, 2009). *Calosoma maderae*, F. var. *indicum* Hope is widely distributed in northern India and both the grubs and adults of this species are usually found in cracks in soil and under clods, and feed on the caterpillars of *Cirphis* and *Plusia* (Noctuidae). Similarly, another species, *Calosoma olivieri* Dej., is widely distributed in the Oriental region and it feeds on young locusts, *Schistocerca gregaria* Forsk. (Acridida), and on the larvae of *Cirphis unipuncta* Haw. (Noctuidae) (Rahman, 1940).

1.4.1.3 *Carabus* spp.

Several species of *Carabus* are associated with cultivated farmland. These include some particularly large species that may reach, if not exceed, a body length of 30mm: *C. auratus*, (Plate 3.3) *C. cancellatus* and *C. violaceus* are examples. *Carabus* spp., are long-legged black-bodied ground beetles, often with a metallic-blue, copper, greenish or purplish elytral sheen and prefer to reside

Plate 3.19. *Coccinella septumpunctata*

Plate 3.20. *Cryptolaemus montrouzieri*

Plate 3.21. *Harmonia axyridis*

PLATE 1.3 (*Continued*) Predators in the order Coleoptera.

on moist conditions. In addition, as in the case of *C. granulatus*, the elytra may be characteristically sculptured. Adults of *Carabus* spp. are typically nocturnal and feed on slugs, snails, earthworms and various insects. They are also scavengers. The adults and larvae overwinter, the former often surviving for more than 2 years. *C. nemoralis* and *C. violaceus*, the latter known as the violet ground beetle, have been considered familiar garden-inhabiting species (Moreton, 1969; Helyer et al., 2014). The different species of predatory beetles under the genus *Carabus* identified from various locations in India are: *Carabus indicus*, *C. everesti*, *C. sanchari* (Mani, 1968) and *C. boysi* (Singh and Banyal, 2013).

1.4.1.4 *Harpalus* spp.

Harpalus rufipes (Plate 3.4) is a medium-sized to moderately large (10–17 mm long), short-legged, fully winged, nocturnal ground beetle, with an indistinct yellowish pubescence on the elytra. The smaller *H. affinis* (body length 8.5–12.0 mm) is a widely distributed and generally common species, but one in which elytral pubescence is restricted to the marginal two or three intervals. Unlike females, which have a matt-black appearance, males are shiny black, blue, copper or green. Adult beetles are diurnal, fully winged and most of them are seed feeders. However, the soil-inhabiting larvae are carnivorous. *H. affinis* is a common inhabitant in gardens and cultivated farmland including field vegetable crops, cereals and oilseed rape, where it may be dominant or subdominant (Warner et al., 2000). Mili et al. (2018) found a higher proportion of *Harpalus griseus* followed by *H. calceatus* and *H. rufipes* in a survey conducted at Gladiolus and Gerbera fields in Assam, India.

1.4.1.5 *Loricera pilicornis*

Loricera pilicornis (Plate 3.5) is a small (6.0–8.5 mm long), mainly bronzy-black ground beetle, with protruding eyes and three pit-like depressions (foveate depressions) on each elytron. Also, the basal antennal segment is long and robust, and segments two to six are distinctly setose.

Setae are also present on the underside of the head. Adults, which may be found throughout the year, are rather lively and frequently fly. They overwinter in hedgerows, woodlands and other such sites, and become active in the early spring. They often invade arable fields where they begin to feed and breed. The carnivorous larvae (body length up to ca 9 mm, plus a pair of very long anal cerci) develop from spring onwards. They eventually pupate, and in mid- to late summer, new adults appear. In the autumn, these finally take up their winter quarters. Adults and larvae of *L. pilicornis* are usually associated with disturbed soils, and are specialist predators of springtails and soil-inhabiting mites. In oilseed rape fields, where they may prove to be a dominant or subdominant species, the fully fed larvae of soil-pupating pests, such as brassica pod midge (*Dasineura brassicae*) and pollen beetle (*Meligethes aeneus*), are also attacked (Warner et al., 2000).

1.4.1.6 *Nebria brevicollis*

Nebria brevicollis (Plate 3.6) is one of the most abundant ground beetles, and a well-known predator of small invertebrates, including aphids, various dipterans, springtails and soil mites (Penney, 1966). They also kill small slugs, although they tend to feed on dead ones, as this avoids the problems associated with overcoming molluscan slime (mucus) (Mair and Port, 2001). The beetles favour damp, disturbed sites and are often dominant in arable crops. They are also common in grasslands and gardens. Adults and larvae are highly mobile surface-feeders and have a strong affinity with hedgerows (Joyce et al., 1999). However, although fully winged, the adults rarely fly (Nelemans, 1987). Adults are active from April or May onwards, but include a period of summer diapause prior to egg laying (Nelemans et al., 1989). Larvae occur from autumn onwards, and may continue to feed throughout the winter, before pupating in the spring.

1.4.1.7 *Anchomenus dorsalis*

Anchomenus dorsalis (Plate 3.7) is a common, widely distributed ground beetle, which often aggregates under stones during the daytime. The adults (body length 6–9 mm) are winged, characterised by their colourful appearance, the elytra being greenish-black with reddish or orange patches on the shoulder; the distinctly narrow pronotum is metallic green. *A. dorsalis* is a fast-moving, long-legged species, which is often common in arable crops in spring and summer, usually having overwintered nearby. As in winter oilseed rape, a positive spatio-temporal (time and space) relationship between adult beetles (when active on the soil surface) and dropped larvae of the brassica pod midge (i.e. fully fed larvae that have dropped to the ground in order to pupate in the soil) has been observed (Warner et al., 2000: Williams et al., 2010). *A. dorsalis* is also a useful predator of other fully fed midge larvae, including, in cereal crops, larvae of both the orange wheat blossom midge (*Sitodiplosis mosellana*) and yellow wheat blossom midge (*Contarinia tritici*). Aphids that have fallen to the ground from their host plants are also attacked.

1.4.1.8 *Pterostichus* spp.

The genus *Pterostichus* includes a range of large, mainly wingless, black-bodied, ground-based beetiles that are often common in fields and gardens. Species, such as *P. madidus* (Plate 3.8) (body length 15–20 mm) and *P. melanarius* (body length 13–17 mm), are well-known examples. Both species are phytophagous and carnivorous, and regularly include small earthworms and molluscs in their diets. However, at least in the case of *P. madidus*, the beetles apparently have difficulty coping with molluscan slime. Accordingly, they tend to favour dead slugs, although (as in the case of *Nebria brevicollis*: Section 5.1.8) they are able to kill small molluscan prey (Mair and Port, 2001). The beetles also attack and consume aphids, caterpillars and other insects encountered during their typical nocturnal activities. Other species associated with arable crops include *P. niger* (body length 15–21 mm), which is a black, narrow-bodied species, and *P. oblongopunctatus* (body length 10–12 mm), a smaller, bronzy-black species, with four small, pit-like depressions on the elytra (Williams et al., 2010).

1.4.1.9 *Amara* spp.

The genus *Amara* includes numerous small to medium-sized, short-legged species (body length 6–9 mm), with a broad pronotum, an oval body outline and a mainly black, metallic appearance. The beetles often occur in cultivated fields and gardens, and several species (namely *A. aenea* (Plate 3.9), *A. communist, A. eurynota, A. familiaris, A. ovata, A. plebeja* and *A. similata*) have, at one time or another, proved dominant or subdominant in European oil seed rape crops (Buchs and Alford, 2003). *Amara* spp. completes one or two generations annually, are typically omnivorous and adults often climb plants in search of seeds and other food. *A. ovata* and *A. similata*, for example, ascend oilseed rape plants to feed on developing seeds and also brassica pod midge larvae, gaining access to them by biting through the pod walls (Luka et al., 1998). In cereal fields, dipteran pests, such as the wheat bulb fly (*Delia coarctata*), are also attacked (Dobson, 1961).

1.4.2 Family: Silphidae (Carrion Beetles)

Carrion beetles or burying beetles comprise not only scavenging, carcass-disposing beetles but also various, essentially epigeic species in which adults are partially predaceous. Several species (including *Nicrophorus vespillo, Silpha obscura, S. tristis* and *Thanatophilus sinuatus*) have been recorded in arable fields where oilseed rape forms part of the crop rotation in Germany (Tischler, 1958). In such fields, predaceous carrion beetles might well encounter the soil-inhabiting stages of pests, such as pollen beetles and Ceutorhynchus weevils.

1.4.2.1 *Dendroxena quadrimaculata*

Dendroxena quadrimaculata (Plate 3.10), known as the four-spotted burying beetle, is predaceous and also a scavenger. The arboreal adults, which are frequently associated with oak trees, feed on various lepidopteran and sawfly larvae including those of important European forestry pests, such as the European gypsy moth, lunar marbled brown moth, oak procession art moth and winter moth (*Operophtera brumata*) (Jahn, 1978). The adults of *D. quadrimaculata* are 12–14 mm long and black, with broad brownish orange elytra, each marked with two black spots. They are active from May to July (Dekeirsschieter et al., 2011). The black, rather elongate, woodlouse-like larvae develop in the soil, where they feed on a range of mainly soft-bodied prey, including insect eggs and pupae. In the 1920s, specimens of *D. quadrimaculata* from Czechoslovakia, Hungary and Spain were sent to the United States as a potential biocontrol agent for both the brown tail moth and European gypsy moth (Burgess and Crossman, 1929).

1.4.3 Family: Staphylinidae (Rove Beetles)

Rove beetles form one of the largest coleopteran families, the Staphylinidae, and are characterised by their short elytra that leave most of the abdomen fully exposed. The hindwings are large, but when not in use they are folded away beneath the elytra. This task is achieved by movements of the highly flexible abdomen. Most species are good fliers. Antennae are usually filiform and rarely clubbed. Rove beetles (which range in size from 1 to over 30 mm in length) are mostly predaceous, as both adults and larvae, but they may also be partly or temporarily (if not entirely) coprophagous, fungivorous, saprophagous, etc. (Good and Giller, 1991). The larvae of rove beetles are narrow-bodied and elongated, and distinguishable from those of most other coleopterans, including ground beetles, by a pair of articulated appendages (known as urogomphi) that arise from the tip of the ninth abdominal segment. Taxonomic difficulties continue to hamper the study of rove beetles and relatively little is known about the bionomics and pre-adult stages of the vast majority of the species.

1.4.3.1 *Aleochara* spp.

Members of the genus *Aleochara* are minute to small, mainly black beetles, some little more than a couple of millimetres or so in length. One species, *A. bilineata* (adults 2.5–5.5 mm long) (Plate 3.11), is an important combatant of the cabbage root fly, not only as a predator but also as a parasitoid, and both pest and natural enemy occur in most parts of the world where brassicaceous field vegetable crops are grown. In view of its importance, *A. bilineata* has also been mass-reared as a biocontrol agent for the cabbage root fly, in both Europe and North America (Hertveldt et al., 1984). The parasitoid/predator also attacks other pests, including bean seed flies (*D. florilege* and *D. platura*), onionfly (*D. antiqua*) and turnip rootfly (*D. floralis*).

1.4.3.2 *Dalotia coriaria*

Dalotia coriaria (Plate 3.12) (formerly known as *Atheta coriaria*) is a tiny rove beetle, which has proven to be a highly effective biocontrol agent for use against the soil-dwelling stages of various greenhouse pests (Carney et al., 2002). Target pests include fully fed second-instar nymphs, prepupa and pupa of western flower thrips (*Frankliniella occidentalis*), eggs, larvae and pupae of fungus gnats (i.e. sciarid flies), and eggs, larvae and pupae of glasshouse wing-spot flies (i.e. shore flies) (specifically, *Scatella stagnalis* and *S. tenuicosta*). The beetle also has potential as a useful predator of other soil pests including the cabbage root fly (Bennison et al., 2011). Adults of *D. coriaria* (body length 3–4 mm) are dark brown to black and extremely active. Their larvae, which are whitish to yellowish-brown and up to 5 mm long, are also highly active predators. *D. coriaria* deposits eggs in the soil or other growing medium, and they hatch a few days later. Larvae then feed for about a week before forming cocoons of silk and soil particles and pupating. Adults emerge shortly afterwards; the cycle from egg to adult takes about 2–3 weeks, depending on temperatures. Breeding is continuous, so long as conditions remain favourable, with no indication of a diapause (Helyer et al., 2014).

1.4.4 Family: Cantharidae (Soldier Beetles)

Soldier beetles form a group of small to medium-sized, somewhat dorsoventrally flattened beetles, with soft, narrow, mainly yellowish-orange or black, straight-sided elytra and long, filiform antennae. Adults are mainly diurnal and although several species are at least partly phytophagous, the beetles are primarily predaceous on other soft-bodied insects. Their soil-inhabiting larvae are also predaceous.

1.4.4.1 *Cantharis* spp.

Members of the genus *Cantharis*, in common with other soldier beetles, overwinter in the larval stage, and eventually pupate in the spring. Adults emerge from late spring or early summer onwards, there being just one generation annually. Several species of *Cantharis* are useful predators of pests in agricultural crops, and more than 20 larvae per square metre have been found in oilseed rape fields in Germany, where they are of particular significance in the late autumn and winter, as predators of peach/potato aphids (*M. persicae*) (Buchs, 2003). Cantharid species reported in European oilseed rape fields include *C. fusca* (Plate 3.13), *C. lateralis* and *C. livida* (Schwenke, 1974).

1.4.4.2 *Rhagonycha* spp.

Rhagonycha fulva (Plate 3.14), the common red soldier beetle, is a small (7–10 mm long), widely distributed cantharid beetle, characterised by the noticeably black-tipped, orange-red elytra. *R. fulva* is primarily a predator of aphids and other small, soft-bodied insects, which include chrysomelid beetle larvae, including those of the red poplar leaf beetle, *Chrysomela populi* (Maisner, 1974). In addition, *R. fulva* has been highlighted as a predator of the invasive psyllid, *Acizzia jamatonica*, an insect that is becoming an important alien pest of ornamental silky acacia trees (Harizanova et al., 2012). Another species, *R. lignosa*, sometimes occurs as a minor predator in apple orchards (Steiner, 1974). This is an arboreal species and is smaller than *R. fulva*. Also, the thorax and head are black, and the elytra and legs are testaceous.

1.4.5 Family: Coccinellidae (Ladybird beetles)

Ladybird beetles are well-known, small to medium-sized, more or less hemispherical beetles, often adorned with black, white or creamy white spots/strips on often highly colourful (orange, red or yellow) elytra. The number spots in the nominate form is frequently reflected in both the vernacular and the formal names of a species; the two-spot ladybird (*A. bipunctata*, i.e., *A. 2-punctata*), seven-spot ladybird (*Coccinella septempunctata*, i.e., *Coccinella 7-punctata*) and 24-spot ladybird (*Subcoccinella vigintiquattuoripunctata*, *Subcoccinella 24-punctata*) are examples. However, there is often considerable variation within a species, in both colour and pattern (including the number of spots) (Majerus, 1994), and this is spectacularly portrayed by the harlequin ladybird (*Harmonia axyridis*), a species described in North America as the multicoloured Asian lady beetle (Koch, 2003). Although there are exceptions, most ladybirds are carnivorous, feeding as both adults and larvae on a wide range of phytophagous insects and mites, including aphids, psyllids, scale insects, thrips, mealybugs, whiteflies and spider mites. Selectivity in the choice of prey also features.

Ladybird beetle larvae are active, voracious predators, with well-developed thoracic (true) legs, and a sucker-like pseudopod at the tip of the highly flexible abdomen. Each larva commences active life by devouring its egg shell (chorion), having first burst out of the egg using an egg-buster (a tiny spatulate structure on the head and/or on the prothorax of the first-instar larva). The number of generations within a species may also vary according to local climate and weather conditions. Ladybirds are highly promiscuous, and multiple mating within populations is common occurrence. Also, intergenerational mating frequently occurs. This is highlighted where, as in certain bivoltine (or partially bivoltine) species, overwintered (i.e. 'old') adult females mate with 'young' males reared during the summer. In some such species, mating between 'old' males and 'young' females may also take place (Majerus, 1994).

1.4.5.1 *Chilocorus* spp.

Two species of Chilocorus, *C. bipustulatus* (Plate 3.15) (the heather ladybird) and *C. renipustulatus* (the kidney-spot ladybird), are useful arboreal predators in apple and pear orchards, where both species attack pests, such as the pear scale (*Aspidiotus leperii*), mussel scale (*Lepidos aphes ulmi*), pear oyster-shell scale (*Diaspidiotus piri*) and San Jose scale (*Comstockaspis perniciosus*) (Iperti, 1974). *C. bipustulatus* is also known to attack the Mediterranean black scale (*Saissetia oleae*) in olive groves. *Chilocorus* spp. similarly prey upon a wide range of forestry pests including the Brown scale (*Parthenoleca-nium corni*), mulberry scale (*Pseudaulacaspis pentagona*), nut scale (*Eulecanium tiliae*), oleander scale, pear oyster-shell scale, poplar scale (*D. gigas*), San José scale and willow scale (*Chionaspis salicis*) (Kahradník, 1972; Schmutterer, 1972).

C. bipustulatus (adults 3.5–5.5 mm long) and the somewhat larger *C. renipustulatus* (adults 4.5–5.5 mm long) are both shiny-black, dome-shaped ladybirds, with a distinctly flat, fringe-like elytral margin, and a redspot on each elytron (*Exochomus quadripustulatus*, Section 5.5.3). Larvae of both species are adorned with long black spines, each arising from a large, black tubercle. The larvae of *C. bipustulatus* are further characterised by the presence of a pale, belt-like cross band. When pupating, *Chilocorus* spp., does not discard their final-instar larval skin. Instead, the sloughed skin remains as a partial protective coat, enclosing all but the upper-most part of the underlying pupa. Aggregations of pupae are often formed on the trunks of inhabited trees (Majerus, 1994). Eleven species of *Chilocorus* have been reported from various parts of India (Omkar and Pervez, 2004). Among them, *Chilocorus nigritus*, which originated from the Indian subregion, feeds on a variety of insect pests, such as aphids (*Lipaphis erysimi* (Kaltenbach), *Brevicoryne brassicae* Linnaeus, *Myzus persicae* (Sulzer), *A. nerii* Boyer and *Aphis craccivora* Koch) and scale insects (*Coccus viridis, Aleurodicus dispersus, Aspidiotus destructor* (Signoret), *Ceroplastes actiniformis* Green, *Vinsonia stellifera* (Westwood), *Coccus acutissimus* Green, *Coccus hesperidum* Linnaeus, *Pseudococcus citriculus* Green, *Palmicultor* sp. and *Pseudaulacaspis* sp.) (Omkar and Pervez, 2003).

1.4.5.2 *Clitostethus arcuatus*

Clitostethus arcuatus (Plate 3.16), often described as the inconspicuous ladybird, is an important predator of whiteflies, such as ash whitefly (*Siphoninus phillyreae*) (notably on pomegranate crops), cabbage whitefly (*Aleyrodes proletella*), citrus whitefly (*Dialeurodes citri*) and honeysuckle whitefly (*A. lonicera*). All whitefly stages (eggs, nymphs, pseudo pupae and adults) are attacked.

The white-bodied larvae of *C. arcuatus* are no more than a few millimetres in length and are easily overlooked owing to the presence of whitefly secreted wax. Adults too are tiny (up to 1.5 mm long) and are similarly often overlooked. *C. arcuatus* is multibrooded and active throughout much of the year (Brown and Whitehead, 2012).

1.4.5.3 *Scymnus* spp.

Adults of *Scymnus* spp. are small (up to 3 mm long), mainly black ladybirds, with (if present) variable, often indistinct, bronze-coloured elytral markings. Some species, however, are entirely reddish-brown. In common with other members of the Coccidulinae, the adults are pubescent, and the larvae typically bear numerous tubercles, each coated with a white waxen secretion. *Scymnus* spp. are often found in association with aphids and spider mites, especially in apple and pear orchards.

Scymnus spp. are also natural enemies of forestry pests. European examples include *S. abietis* (which preys upon the coccid *Physokermes hemicryphus*), *S. auritus* (Plate 3.17) (which attacks the phylloxeran, *Phylloxera coccinea*), *S. impexus* (which feeds mainly on adelgids) and *S. suturalis* (which is associated with both adelgids and coccids). *S. suturalis* also attacks the phytophagous aradid bug, *Aradus cinnamomeus* (Hoberlandt, 1972).

Of the abovementioned ladybirds, the mainly conifer-inhabiting predator *S. impexus* has sometimes been mass-reared as a potential biocontrol agent. Commencing in the early 1950s and again about a decade later, for example, thousands of the beetles were released in Canada and North Carolina, to combat the balsam woolly adelgid (*Adelges piceae*). The latter is an adventive pest, accidentally introduced from Europe on nursery stock in about 1900. However, the predator, although initially becoming established in Canada, did not survive (Majka and McCorquodale, 2006). *Scymnus coccivora* has been recorded earlier from South India, which is widely distributed, being found in other parts of the country. This species is an important predator of scale insects. However, the larvae of the species have also been recorded to feed upon the eggs of mealybugs (Kapur, 1972).

1.4.5.4 *Stethorus* spp.

Stethorus punctillum (Plate 3.18) (the minute black ladybird) is a widely distributed predator of spider mites, including fruit tree red spider mites and two-spotted spider mite (*T. urticae*) (Jary, 1935). The prey range of the species of *Stethorus* reported in India include *Oligonychus indicus, Raoiella indica, Tetranychus neocaledonicus, Eutetranychus orientalis, T. ludeni* (*Stethorus pauperculus* Wiese), *T. aurantii* and red spider mites (*Stethorus gilvifrons* Mulsant) and *T. cinnabarinus* (*Stethorus rani* Kapur) (Omkar and Pervez, 2004).

1.4.5.5 *Coccinella* spp.

These include *C. septempunctata* (Plate 3.19) (the seven-spotted ladybird), one of the best known and most abundant ladybirds. They occur widely in the Palaearctic region, cosmopolitan in distribution, and they prey on aphids, mealybugs, scale insects, whiteflies, etc. (Omkar and Pervez, 2000, 2004), with aphids being the most accepted prey. *C. septempunctata* is a moderately sized (ca 5.5–8.0 mm long), mainly red, black-spotted ladybird that overwinters in various situations, often clustered in small groups amongst dead or crinkled leaves (as on rose bushes) or sheltered by leaf litter at the base of the plants upon which they were reared. Conversely, large aggregations of adults may overwinter in fully exposed situations in the west. Overwintered adults become active from March or April onwards. Eventually, after feeding and mating, the first eggs are laid. In common with many other species, *C. septempunctata* completes just one generation annually, but a partial second generation may be initiated by precocious young males and females. Particularly during the summer, *C. septempunctata* adults and larvae may be found on many aphid-infested herbaceous plants (Blackman, 1965, 1967).

Coccinella undecimpunctata, the 11-spotted ladybird, is another generalist predator that feeds mainly on aphids, especially those infesting herbaceous plants (including cereal crops). The adults of *C. undecimpunctata* (with an overall length of 3.5–5.0 mm) are noticeably smaller than those of *C. septempunctata* and in the typical form possess 11 black spots on their otherwise bright red elytra. Originally restricted to the western Palaearctic region, *C. undecimpunctata* has become established as an adventive species in North America. It also occurs as an introduced coccinellid in Western Australia and Tasmania (Hangay and Zborowski, 2010), where it is considered a useful biocontrol agent. Although now widely distributed in New Zealand (Martin, 2016), the much-cited belief that it was originally introduced to that country back in the 1800s is unfounded (Galbreath and Cameron, 2015).

1.4.5.6 *Cryptolaemus montrouzieri*

Cryptolaemus montrouzieri (Plate 3.20), known as the Australian ladybird, is a voracious predator of mealybugs. It was first introduced into Europe to control infestations of glasshouse mealybug (*P. citri*). The Australian ladybird is also an effective predator of other Pseudococcidae and it is now used against various mealybugs on a range of cultivated plants including ornamentals. *C. montrouzieri* also attacks aphids and soft-scale insects. Albeit infrequently, in the Côte d'Azur, France, the predator has been known, for example, to attack the woolly apple aphid in apple orchards (Iperti, 1974).

C. montrouzieri is a small ladybird (4 mm long), with an orange head, pronotum and abdomen, and pubescent, dark brown to black elytra. The surprisingly large larvae (up to 13 mm long when fully grown) are mealybug-like, and they are covered in white, flocculent wax. They also bear numerous, filamentous, segmentally arranged waxen projections. When feeding within colonies of mealybugs, such larvae are easily overlooked, particularly when young. *C. montrouzieri* is a multivoltine species, capable of completing several generations annually. However, feeding, breeding and development are slowed or curtailed at low temperatures. In suitably warmer conditions (ca 25°C–30°C), development from egg to adult takes about a month (Kairo et al., 2013).

1.4.5.7 *Harmonia axyridis*

H. axyridis (Plate 3.21), the harlequin ladybird, is a native of Asia. *H. axyridis* is an effective predator of a wide range of aphids and similar pests. However, in a search for moisture during dry weather, the adults bite into and thereby cause damage to ripening or ripened grapes, raspberries and other soft-skinned fruits (Alford, 2011).

Larvae arc mainly black, with (in their final instar) five pairs of orange tubercles towards the front of the abdomen (each bearing a prominent branched spine) and a separate group of four pale spines dorsally. Pupae are mainly orange, with black patches. As a primarily aphidophagous predator, *H. axyridis* breeds on a wide range of aphid-infested trees and usually completes two generations annually (Koch, 2003). Being a generalist predator, *H. axyridis* has a broad prey range and it includes Aphids (*M. persicae*, *Aphis fabae*, *A. gossypii*, *Rhopalosiphum prunifoliae*, *Chaitophorus populialbae*), scale insects (*Matsucoccus massoniana*, *M. matsumurae* (Kuwana), *Pseudaulacaspis pentagona* (Targioni) and *M. thunbergianae*) and mealybugs (*Nesticoccus sinensis* Tang and, *Rhizococcus transversus*) (Pervez and Omkar, 2006).

1.4.5.8 *Adalia bipunctata* (Two-Spotted Lady Bird)

A. bipunctata (Linnaeus) (Plate 3.22) is an aphidophagous ladybird endemic to Europe, Central Asia and North America (Majerus, 1994). *A. bipunctata* is a tree-dwelling species. It commonly feeds on more mobile prey with long legs than coccidophagous ladybirds. It is a polyphagous species with a wide range of aphid prey. It uses coccids and diaspids as its alternative prey and also indulges in pollenvory, which supports survival in the absence of prey. The host range includes aphids, *viz.*, *A. fabae* Scopoli, *A. craccivora* Koch, *A. nerii* Boy., *Aphis sambuci* (Linnaeus), *Megoura viciae* Buckton and *Megoura albifrons* (Omkar and Pervez, 2005). A series of attempts to establish *A. bipunctata* in new pest-prevalent habitats have failed in the past, which suggest that *A. bipunctata* is not a successful biocontrol agent, unlike some coccidophagous ladybirds and parasitoids (Omkar and Pervez, 2003). Some early releases in the form of eggs failed, as eggs did not hatch, probably because of poor weather conditions. It was out-competed by indigenous *H. axyridis* in Japan and *C. septempunctata* in India and other South Asian countries (Omkar and Pervez, 2005). However, some field releases of larvae and eggs of *A. bipunctata* against the rosy apple aphid, *Dysaphis plantaginea* in Switzerland, were successful in predator establishment and aphid biocontrol (Wyss et al., 1999).

1.4.5.9 *Propylea* spp.

Propylea dissecta (Plate 3.23) is aphidophagous and abundant in the Oriental region, and found particularly in India, Bangladesh and Nepal (Poorani, 2004). *Propylea japonica*, also referred to as the Japanese ladybird, is largely aphidophagous (Agarwala et al., 2003) but also preys on whiteflies. It is found in abundance in the arid and semi-arid areas of China (Liu et al., 2004). Another species, *Propylea quatuordecimpunctata* (or *P. 14-punctata* or P-14) is native to and widespread in the Old World. The most suitable prey of *P. dissecta* is *A. craccivora* followed by *A. gossypii*, *Uroleucon compositae*, *Rhopalosiphum maidis*, *L. erysimi*, *M. persicae* and *A. nerii* (Omkar and Mishra, 2005).

1.5 ORDER: DIPTERA

True flies (order Diptera) are minute to large insects, with a single pair of membranous wings. Characteristically, the hindwings are reduced to knobbed, drumstick-like stalks known as halteres that function as balancing organs during flight. Diptera mouthparts are usually suctorial, but are sometimes adapted for piercing. Metamorphosis is complete (i.e. development is holometabolous) and usually includes egg, larval, pupal and adult stages. Some flies, however, are ovoviviparous, and their eggs hatch within the maternal body prior to being deposited as larvae. With few exceptions, fly larvae are usually apodous and maggot-like, with a reduced, inconspicuous head.

True flies include many predaceous or parasitic species, and these often fulfil an important role as natural enemies of harmful or potentially harmful insects and other invertebrates. This is particularly so in the case of plant pests, where dipterans may act as important biocontrol agents in pest management or pest control programmes. Some predatory dipterans are generalist feeders as adults or larvae (if not both), whereas others may be more specialised, as in the case of aphidophagous midge larvae (family Cecidomyiidae) and hoverfly larvae (family Syrphidae). Parasitic dipterans, primarily tachinid flies (family Tachinidae), are specialists, the overwhelming majority being larval parasitoids of lepidopteran caterpillars or pupae (Chandler, 2014).

1.5.1 Family: Cecidomyiidae (Gall midges)

Gall midges (family Cecidomyiidae) are small or tiny, delicate insects, with broad, often hairy wings and a much-reduced wing venation. The antennae are typically bead-like (moniliform), and each so-called segment (strictly speaking, each flagellomere) bears a basal whorl of hairs. Legs are long and thin and the male genitalia are clearly visible. Midge larvae have a small, inconspicuous head, and the body is narrowed both anteriorly and posteriorly. Furthermore, they usually possess a characteristic sternal spatula, often known as an anchor-process or breastbone. Midge larvae develop through three instars and usually pupate inside small silken cocoons (Chandler, 2014).

1.5.1.1 *Aphidoletes* spp.

Four species of *Aphidoletes* are known in Europe, two of which are predaceous on aphids and two are predaceous on adelgids. Of these, *A. aphidimyza* (Plate 4.1) is the most common and best-known species, a name under which a large number of supposedly distinct 'species' have been synonymised (Gagné, 2010). The adults of *A. aphidimyza* are small, gnat-like creatures, although relatively large for the genus (females have a wing length of ca 2.5–3.0 mm) (Harris, 1966).

Plate 4.1. *Aphidoletes aphidimyza*

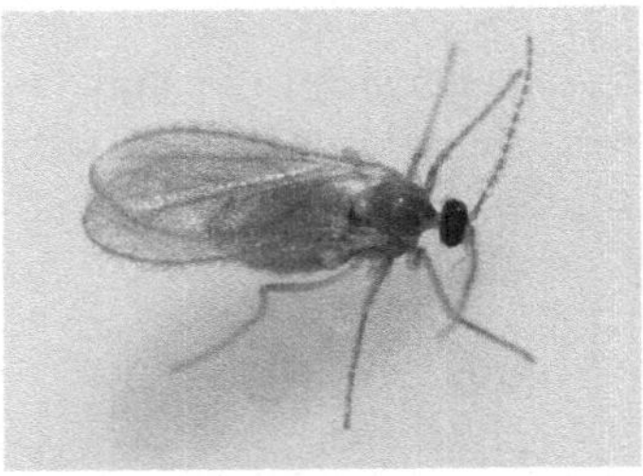

Plate 4.2. *Lestodiplosis trifolii*

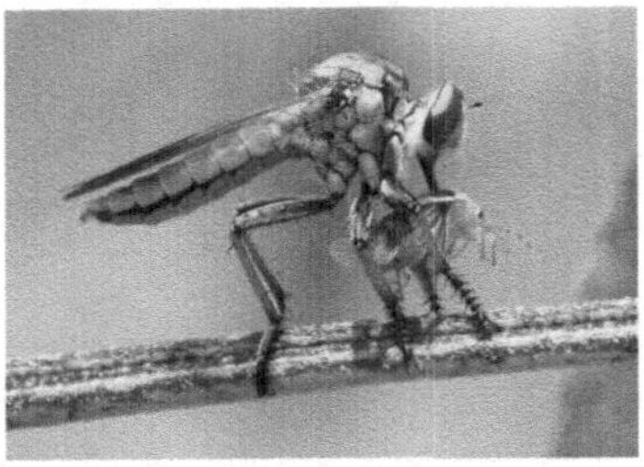

Plate 4.3. *Rhadiurgus variabilis*

Plate 4.4. *Dolichopus acuticornis*

Plate 4.5. *Borophaga incrassata*

Plate 4.6. *Epistrophe eligans*

PLATE 1.4 Predators in the order Diptera.

A. aphidimyza is polyphagous on aphids, including the full range of commonly encountered pest species associated with greenhouse crops. The midge is also associated with aphids on a wide range of outdoor plants, both cultivated and non-cultivated. Victims encompass many aphids' economic importance, including significant wax-secreting species, such as the cabbage aphid (*B. brassicae*) on brassicaceous crops and the mealy plum aphid (*Hyalopterus pruni*) on plum trees. It is also exploited as a predator of aphids on ornamental (or amenity) trees, shrubs and flowers, and in orchards, *A. aphidimyza* is suitable for use in augmentative and inundative biocontrol programmes. Its many biological attributes make it easy to culture, distribute and introduce (Harris, 1982).

Aphidoletes urticaria is another polyphagous European predator of aphids, but it appears to have a more northerly distribution than the slightly smaller *A. aphidimyza*. It is not found, for example, in the Mediterranean region. Recorded prey includes the black bean aphid (*A. fabae*), cherry blackfly (*Myzus cerasi*), fly-honeysuckle aphid (*Hyadaphis foeniculi*), green apple aphid

Plate 4.7. *Eupeodes corollae*

Plate 4.8. *Syrphus ribesii*

Plate 4.9. *Chlorops pumilionis*

Plate 4.10. *Agria mamillata*

Plate 4.11. *Phaonia trimaculata*

PLATE 1.4 (*Continued*) Predators in the order Diptera.

(*Aphis pomi*), pea aphid (*Acyrthosiphon pisum*), peach/potato aphid (*M. persicae*), red currant blister aphid (*Cryptomyzus ribis*), sycamore periphyllus aphid (*Periphyllusa cericola*) and thistle aphid (*Brachycaudus cardui*) (Harris, 1973; Harris and Bennett, 2003). One of the far less frequently encountered species is *Aphidoletes abietis*, whose larvae feed as predators in the pineapple-like galls of *Adelges abietis* on young spruce trees (including those grown commercially for the Christmas market). Larvae of the midge force their way into the galls after these have closed. They then seek out the host adelgids upon which to feed. Later in the year, the final-instar midge larvae cause the galls to open prematurely, thereby enabling the fully fed predators to escape and eventually pupate (Mitchell and Maksymov, 1977). Adults of *Aphidoletes abietis* have been reared in the spring from autumn-collected larvae (Nijveldt, 1969). There is just one generation annually. Another species of *Aphidoletes* (*A. thompsoni*) is also associated with adelgids (specifically *A. piceae* on silver-fir trees). *A. thompsoni* has also been found attacking Scots pine adelgid (*Pineuspini*) on Scots pine trees in the Isle of Man (Harris and Bennett, 2003). Unlike *Aphidoletes abietis*, *A. thompsoni* is multivoltine and completes three to five generations annually (Harris, 1973).

1.5.1.2 *Lestodiplosis* spp.

The genus *Lestodiplosis* embraces a group of predaceous midges, whose larvae most frequently attack the larvae of other gall midges. *Lestodiplosis* larvae are also known to prey upon small beetles and mites (Harris, 1966). By way of example, as on clover seed crops, several species of gall midge are associated with the flower heads, including phytophagous species, such as the clover leaf midge (*Dasineura trifolii*), clover seed midge (*D. leguminicola*) and white clover seed midge (*D. gentneri*), that are generally regarded as pests. In addition, midge-infested clover flower heads may also be occupied by larvae of predaceous midges (Milne, 1960). These natural enemies include *Lestodiplosis trifolii* (Plate 4.2) a predator of both the clover seed midge and white clover seed midge. Larvae of *Lestodiplosis* have also been found on oak trees, where they were not only associated with phytophagous gall midge larvae and phytophagous gall mites but were also considered predators of lepidopterous larvae, in particular, those of the green oak tortrix moth (*T. viridana*) and winter moth (*O. brumata*) (Cole and Harris, 2002).

1.5.2 Family: Asilidae (Robber flies)

Robber flies (family Asilidae), of which the northern robber fly (*Rhadiurgus variabilis*) (Plate 4.3), is a typical example form a large family of predaceous dipterans. As adults, they are superficially similar in appearance to stiletto flies, but are readily distinguished by the distinctly grooved vertex (the top part of the head between the compound eyes) and more robust legs. Robber flies are opportunist predators and typically attack flying insects, which they chase and then pounce upon whilst in flight. Captured prey is immediately stabbed, using the much-modified piercing mouthparts. The victim is then sucked dry, having been injected with not only neurotoxins that quickly cause paralysis but also proteolytic enzymes that breakdown and liquefy the body contents, which are particularly active in hot, sunny conditions. They are purposeful hunters and often tackle insects even larger than themselves including potentially powerful adversaries, such as bees and wasps (Wood, 1981). Although known to occur in field crops, grasslands and open forests, robber flies are minor and non-selective predators of crop pests. Their carnivorous, soil- or dung-inhabiting larvae feed on various insect larvae and take from 1 to 3 years to complete their development (Dennis et al., 2013).

1.5.3 Family: Dolichopodidae (Long-Legged Flies)

Long-legged flies (family Dolichopodidae) are mainly small, bluish-metallic or greenish-metallic predatory flies, with long, stilt-like legs and usually prominently exposed male genitalia. As both adults and larvae, they prey mainly on tiny insects, including psocids, springtails, thrips and also other Diptera. In northern Germany, several adults per square metre have been recorded in emergence traps sited in fields of oilseed rape (Prescher and Buchs, 1998) and adults of many species of long-legged fly were found in water traps placed in flowering oilseed rape in France (Brunel et al., 1989). In oilseed rape, long-legged flies are considered as at least potentially useful predators of the brassica pod midge. *Dolichopus acuticornis* (adults 4 mm long) (Plate 4.4) and the much smaller *Medetera micacea* (adults ca 2 mm long) are examples. Long-legged flies (specifically, *M. truncorum*) are also predators of adult fruit flies (Jones, 1969), although *Medetera* spp. are associated mainly with thrips (Laurence, 1951).

1.5.4 Family: Phoridae (Scuttle flies)

Scuttle flies (family Phoridae), also commonly known as phorid flies, are small or very small diptrans, with a distinctly humped thorax and a characteristically reduced wing venation. The adults are renowned for their habit, when disturbed, of running (scuttling) over surfaces rather than taking flight. The larvae of scuttle flies have various feeding habits (Disney, 1983, 1989). Those of several species, for example, are saprophagous or fungivorous. In addition, some act as parasites, parasitoids and predators, a few of which are considered of potential value as natural enemies of plant pests. The following are examples: *Borophagain crassate* (Plate 4.5) is parasitic on larvae of the St Mark's fly (*Bibiomarci*) (Morris, 1922) and predaceous on the sugar beet root aphid (*Pemphigus fuscicornis*) (Toth et al., 2004).

1.5.5 Family: Syrphidae (Hoverflies)

Hoverflies (family Syrphidae) are readily distinguished from other dipterans by the presence of a so-called false vein, the vena spuria, formed from the thickened remnants of a former longitudinal fold in the wing membrane. Although not a unique feature of hoverflies, the wing also has a false margin, formed by two submarginal cross-veins that run parallel to the true wing margin. Hoverflies often mimic bees or wasps and frequently hover in the air whilst perhaps emitting a shrill, bee-like hum. Adult hoverflies, also known as flower flies, feed avidly on nectar and pollen, and also on honeydew, and are widely recognised as playing a beneficial role as pollinators of wild flowers and also important flowering crops.

Although the predaceous larvae of hoverflies usually feed on aphids (and less frequently on other soft-bodied invertebrates, such as coccids and psyllids), a few attack other kinds of prey. Larvae of several other hoverflies, including *Heringia heringi*, *Meliscaeva auricollis* and *Scaeva mecogramma*, are associated with psyllids. This is particularly so in southern Europe, where the hoverfly larvae are known to prey upon well-known pests, such as the bay sucker (*Inoza alacris*) and olive sucker (*Euphyllura olivina*). In the case of *M. auricollis*, which overwinters as an adult, larval development in southern Europe is completed in 2 weeks and adults emerge from pupae just over a week later (Rojo et al., 1999). *S. mecogramma* has a similarly rapid rate of development. *H. heringi*, which overwinters in the larval stage and is more polyphagous, not only attacks the olive sucker but (at least in Italy) is also an important predator of the cherry blackfly and peach leaf-roll aphid (*Myzus varians*) on cherry, leaf-curling plum aphid (*Brachycaudus helichrysi*) on plum, pear/bedstraw aphid (*Dysaphis pyri*) on pear and rosy apple aphid (*D. plantaginea*) on apple (Niccoli and Fagnani, 1994).

1.5.5.1 *Epistrophe* spp.

Species of *Epistrophe* are medium-sized, mainly forest or scrubland-inhabiting hoverflies, with typically green, dorsoventrally flattened larvae. Several species are present in Europe including the widely distributed *E. eligans* (Plate 4.6). Its larvae attack aphids on a range of deciduous plants, including fruit trees and ornamental or specimen trees and shrubs. Larvae have also been found in association with black bean aphids (*A. fabae*) on *Vicia beans* (Wojciechowicz-Zytko and Wnuk, 2012) and with cereal aphids on winter wheat (Chambers et al., 1986). Adults of *E. eligans* are active from April to June, earlier or later, according to local conditions and are often found in the vicinity of black thorn and haw thorn bushes. Their eggs and larvae develop rapidly, the former hatching in 2–3 days, and the latter usually becoming fully fed in about 8 days. The larvae then enter an extended period of obligatory diapause, before eventually overwintering and pupating in the spring (Schneider, 1948). Several other species of *Epistrophe* in European apple orchards, including *E. melanostoma*, *E. nitidicollis* and possibly *E. ochrostoma*, are known to feed on aphids. Their prey includes the apple/grass aphid (*Rhopalosiphum insertum*), woolly apple aphid (*Eriosoma lanigerum*), green apple aphid (*A. pomi*) and rosy apple aphid (Lyon and Goldlin de Tiefenau, 1974).

1.5.5.2 *Eupeodes* spp.

Species of *Eupeodes* were formerly included in the genus *Metasyrphus* (Speight, 2011). Several species are present in Europe, of which two (*E. corollae* and *E. luniger*) are of particular significance as natural enemies of aphids on cultivated garden plants and farm crops. Sometimes, *E. corollae* (Plate 4.7) is also found in greenhouses (Chambers et al., 1986). Adults of both of these species are similar in appearance, with three pairs of creamy-white, comma-shaped markings (lunules) on the abdomen, which either reach (as in *E. corollae*) or fall just short of (as in *E. luniger*) the tergite lateral margin. Although most species of *Eupeodes* are associated with conifer-infesting arachnid aphids, *E. corollae* and *E. zuniga* have bucked the trend and both attack aphids on a wide range of herbaceous and non-coniferous plants. They are also associated with aphids on broad-leaved bushes, shrubs and, sometimes, fruit trees. As on cereal crops, *Eupeodes* larvae are voracious carnivores and they help to destroy or limit the build-up of many important pests including the bird-cherry aphid (*Rhopalosiphum padi*), cereal leaf aphid (*R. maidis*) and grain aphid (*Sitobion avenae*) (Marcos-Garcia, 1981; Chambers et al., 1986). They also play a beneficial role as natural enemies of aphids on many other field crops.

1.5.5.3 *Syrphus* spp.

This genus, within which nine European species are currently recognised, includes *Syrphus ribesii* (Plate 4.8), one of the most common, widely distributed and economically important aphidophagous hoverfly. Often known as the currant hoverfly, *S. ribesii* is found in most anthropomorphic habitats (including parks, gardens, orchards and soft-fruit plantations) as well as in broad-leaved woodlands and coniferous forests. Third-instar larvae undergo diapause during the winter in leaf litters before they pupate in the spring. There they follow up to a maximum of four, but usually two to three annual generations, which classify *S. ribesii* as an oligovoltine species. At a temperature of 21°C, eggs hatch in 2–3 days, larvae feed for 9 days and the pupal stage lasts for a further 10 days (Schneider, 1948). However, in nature, a period of facultative summer diapause by larvae may extend this timeframe and thereby reduce the number of annual generations. In northern Europe, adults are active from April to November, with peaks of activity in May and August (Speight, 2011). However, adults appear about a month earlier in southern Europe. *S. ribesii*, along with many other species, is a well-known migrant. In addition to *S. ribesii*, two other species of *Syrphus* are cited as predators of aphids in European apple orchards: *S. torvus* and *S. vitripennis* (Lyon and Goldlin de Tiefenau, 1974; Rossi et al., 2006).

1.5.6 Family: Chloropidae (Grass Flies)

Grass flies (family Chloropidae) are small or very small, shiny, often yellow, black-marked insects, associated primarily with Poaceae. As larvae, many species are phytophagous, including economically important grassland and cereal pests, such as the fritfly, gout fly (*Chlorops pumilionis*) (Plate 4.9) and yellow cereal fly (*Opomyza florum*). Larvae of a few species, however, are predaceous. These include *Thaumatomyia glabra* and *T. notata*, both of which as larvae feed on root aphids. Their prey includes injurious species, such as the lettuce root aphid (*Pemphigus bursarius*) and sugar beet root aphid (Toth et al., 2006). Adults of *Thaumato myia* overwinter as adults. Before hibernating, they frequently invade buildings in vast swarms in the autumn (Kotrba, 2010). This phenomenon was first formally documented in September 1831, when tens of thousands of the flies invaded an upper room of the Provost's Lodge in King's College, Cambridge, entering through an open window. (Jenyns, 1832). Adults emerge from hibernation in late March or early April, and eventually locate suitable hosts upon which to breed. There are usually two or more generations annually, before young adults seek suitable overwintering sites.

1.5.7 Family: Sarcophagidae (Flesh Flies)

Flesh flies (family Sarcophagidae) are blowfly-like insects, formerly regarded as a subfamily within the Calliphoridae. Several species are parasitic or predaceous including *Agria mamillata* (Plate 4.10) a natural enemy of small ermine moths (*Yponomeuta* spp.). Recorded hosts of *A. mamillata* range from the apple small ermine moth (*Y. malinellus*), bird-cherry small ermine moth (*Y. evonymella*), common small ermine moth (*Y. padella*) (on hawthorn, etc.) and spindle small ermine moth (*Y. cagnatella*), to the willow small ermine moth (*Y. rorella*), all of which inhabit communal silken webs (tents) formed on their respective arboreal hosts. *A. mamillata* is univoltine, with adults emerging from overwintered pupae in late spring or early summer. Following an extended pre-oviposition period, the ovoviviparous females commence egg-laying on final-instar larvae and pupae. The eggs hatch immediately and the larvae then attack their victims. Each maggot-like predator kills about five host larvae or pupae during the course of its development, with the period of feeding extending from ca 9 to 15 days (Kuhlmann, 1995). Fully fed *Agria* larvae are about 10 mm long and usually vacate the host's web in order to pupate in the ground, their escape often being triggered by rainfall.

A second, widely distributed Holarctic species (*Agria affinis*) is also a predator of lepidopterans including forestry pests, such as the black arches moth (*Lymantria monacha*), European gypsy moth (*L. dispar*) and lackey moth (*Malacosoma neustria*) (Schwenke, 1974). In Canada, the fly is a well-known predator of the non-European eastern spruce budworm (*Choristoneura fumiferana*) (Coppel et al., 1959).

1.5.8 Family: Muscidae (Muscid Flies)

As a group, muscid flies (family Muscidae) include various economically important pests, some of which are of worldwide significance. House flies (*Musca domestica*) and false stable flies (*Muscina stabulans*) (Plate 4.11) are well-known examples. Larvae of muscid flies display an extraordinarily diverse range of feeding habits, with not only coprophagous and saprophagous species but also (for example) carnivorous, endoparasitic and phytophagous species included in the mix (Kutty et al., 2014). The larvae of some species, exemplified by the western Palaearctic species *Phaonia trimaculata* (Plate 4.11), may be predaceous on plant pests (Fritzsche, 1955; Krivesheina, 2013). They are justifiably regarded as beneficial insects for their role as biocontrol agents. The larvae of most carnivorous muscid species are monomorphic, in that they are released from the egg in their final (i.e. third) instar. However, carnivorous larvae of some species (then described as dimorphic) hatch and commence feeding as second instars. Victims of *P. trimaculata* include the following crop pests: Cabbage root fly (*Delia radicum*), a major pest of brassicaceous crops, Cabbage stem weevil (*Ceutorhynchus pallidactylus*), a pest of oilseed rape and other bras solanaceous crops.

1.6 MISCELLANEOUS PREDATORS

1.6.1 Dragonflies

Although well-known predatory insects, particularly as adults but also as aquatic nymphs, dragonflies (order Odonata) (Plate 5.1), including damselflies (suborder Zygoptera), are essentially opportunist feeders. Foraging adults, for example, pounce on any potential airborne prey that happen to enter their field of vision, and even bulky, highly active insects, such as bumblebees may be captured by larger species. Adult and larval dragonflies are usually considered no more than fortuitous natural enemies of plant-infesting insects and mosquito larvae. However, the damselfly *Ischnur aposita* (family Coenagrionidae) represents an infrequent and minor exception. This tiny Zygoptera (with a body length of about 25 mm) has been known to visit asparagus beds, where it swoops down and wrenches young asparagus beetle (*Crioceris asparagi*) larvae from the fronds (Chittenden, 1898).

1.6.2 Mantises

Mantises or praying mantises (order Mantodea: family Mantidae) are found in many warmer parts of the world, especially in the tropics, where (as both nymphs and adults) they act as indiscriminate predators of insects and other invertebrates. Larger species are even known to catch vertebrate prey,

Plate 5.1. Dragonfly

Plate 5.2. *Mantis religiosa*

Plate 5.3. *Forficula auricularia*

Plate 5.4. *Aelothrips melaleuca*

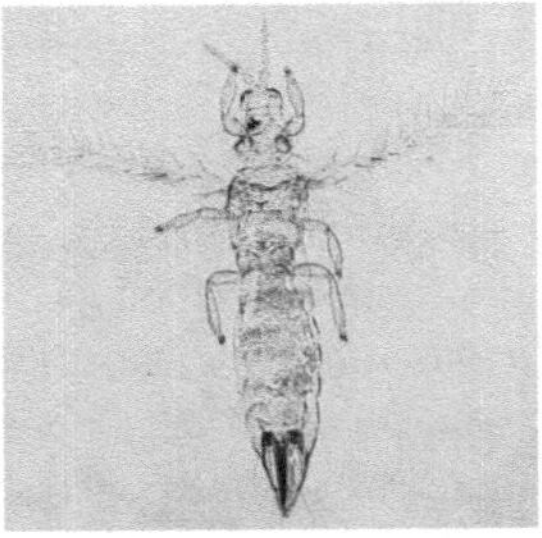

Plate 5.5. *Scolothrips sexmaculatus*

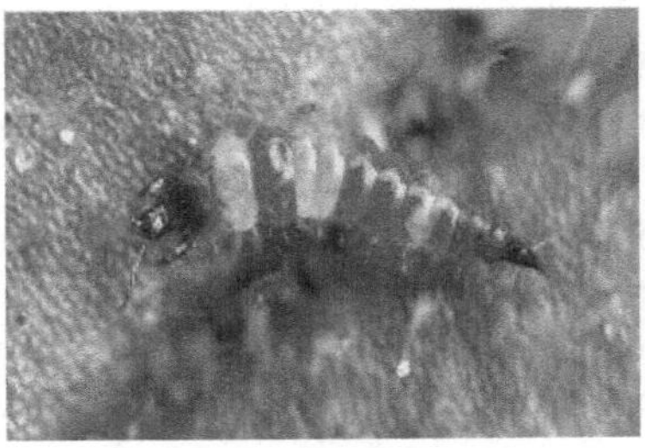

Plate 5.6. *Haplothrips subtilissimus*

Plate 5.7. Lycaenid butterfly

PLATE 1.5 Micselleaneous predators.

including hummingbirds. Mantises, with their characteristically elongated prothorax, arise from eggs deposited in distinctive, purse-like oothecae. They develop through several nymphal stages before finally emerging into adults. Mantises possess raptorial, usually distinctly spiny, forelegs. These are used to capture prey and when in repose are typically held together as if in prayer. Hence, they are termed as 'praying mantises'.

European praying mantis (*Mantis religiosa*) (Plate 5.2) is a well-known Old-World species that includes southern Europe in its natural range. It has also become naturalised in North America, joining the ranks of several indigenous US species. The Chinese praying mantis (*Tenodera sinensis*) (formerly known as *Tenodera aridifolia sinensis*) is yet another imported predator to have become naturalised in North America. It is now reared commercially and marketed as a biocontrol agent by several US-based biocontrol companies. Although available to farmers as a pest management tool on a range of crops, Chinese praying mantises are used mainly by amateur gardeners (Clausen, 1940).

1.6.3 Earwigs

European earwigs (*Forficula auricularia*) (order Dermaptera: family Forficulidae) are omnivorous feeders and as such, they are at least potentially injurious to a wide range of wild and cultivated plants. European earwigs not only cause direct damage to plants by feeding on roots, stems, buds, flowers and fruits but also contaminate and disfigure their feeding sites with pellets of black frass. Setting their pest status aside, earwigs are useful as predators of aphids and other harmful invertebrates. However, overall, their beneficial predatory role (as in orchard IPM systems, for example) tends to be limited since they invade plants only after having developed into third-instar nymphs. In Europe, this is usually not until late May at the earliest.

By way of example, the absence of earwigs on orchard trees in the spring means that they can play no part in combating the spring build-up of important pests, such as the green apple aphid (Clausen, 1940).

1.6.4 Thrips

Thrips (order Thysanoptera) are tiny, narrow-bodied insects, often no more than a millimetre or so in length, with two pairs of strap-like wings bearing long hair fringes. In some instances, wings maybe greatly reduced or absent. Thrips develop to adulthood through an egg and two active nymphal stages, followed by a resting prepupal stage and pupal stages (suborder Terebrantia: e.g. families Acolothripidae and Thripidae) or two (suborder Tubulifera: e.g. family Phacothripidae) sedentary pupal stages. Nymphs are similar in appearance to adults but wingless and often noticeably paler in colour; they also have fewer antennal segments. As metamorphosis is slight, in spite of the presence of a pupal stage, development is considered intermediate between that of hemimetabolous and holometabolous insects (Clausen, 1940).

1.6.4.1 Family: Aeolothripidae

Within the family Aeolothripidae, species of *Aeolothrips* (commonly known as banded thrips) are predaceous on phytophagous thrips, mites and other small arthropods. They are, therefore, regarded as beneficial insects. With their broad, apically rounded black- and white-banded forewings and relatively large size (adult females nearly 2 mm in length), they are readily distinguishable from most other thrips. Many species are associated with herbaceous plants. These are frequently found on yellow-flowering Asteraceae, Brassicaceae, Fabaceae and Linaceae. However, at least in eastern Europe, *A. intermedius* (perhaps the most frequently encountered species) occurs on plants from 30 plant families (Trdan et al., 2005), indicative of a wide floral host range. Such thrips are not only predaceous but also feed on pollen. They are, therefore, facultative, as opposed to obligate predators. In addition, the thrips pierce flower petals in order to obtain moisture (Bournier et al., 1979). Depending on temperature, *A. intermedius* completes up to three or more generations annually (Conti, 2009). Its impact as a biocontrol agent, however, is limited since the thrips seem to restrict their activities to the floral parts of plants, even when the vegetative parts of their hosts are well-infested with suitable prey. Other common species of *Aeolothrips* include *A. melaleuca* (Plate 5.4) and *A. versicolor*, both of which inhabit rosaceous fruit trees and other broad-leaved trees. Another predaceous species,

A. vittatus (recognisable by having just one black cross-band on each forewing), occurs on pine trees; its prey includes *Oxythrips bicolor*, a thrips associated with the male pine cones. In mainland Europe, *A. vittatus* is also known to prey upon larch thrips (*Taeniothrips laricivorus*) (Nolte, 1951).

1.6.4.2 Family: Thripidae

Apart from members of the essentially subtropical and entirely predaceous genus *Scolothrips*, *S. sexmaculatus* (Plate 5.5), which is reared commercially as a biocontrol agent to combat mites, most thrips in the family Thripidae are phytophagous. However, a few primarily phytophagous species are also opportunistic predators of spider mite eggs, especially as second-instar nymphs, which have the greatest impact on prey populations. Onion thrips (*Thrips tabaci*) and western flower thrips (*F. occidentalis*), albeit both being notorious plant pests, are frequently cited as examples. Victims of onion thrips include fruit tree red spider mite (*P. ulmi*) in apple orchards (Collyer, 1953) and more generally, two-spotted spider mite (*T. urticae*) on a range of annual and perennial crops (Wilson et al., 1996). Onion thrips seem to show a preference for plant hosts infested by spider mites, and fecundity may be increased when mite eggs are included in their diet (Milne and Walter, 1998).

1.6.4.3 Family: Phlaeothripidae

A few examples of predatory thrips are found within the family Phlaeothripidae, including the widely distributed species *Haplothrips subtilissimus*. This insect inhabits the bark of oak and other deciduous trees and is predaceous on the eggs of mites and certain other invertebrates. However, records of the thrips as a predator of the oriental fruit moth (*Cydia molesta*) and fruit tree red spider mite in Canadian peach orchards are the result of confusion with the closely related species *H. kurdjumovi*, formerly known (at least in North America) as *H. faurei* (Putnam, 1942; Putnam, 1965).

1.6.5 Butterflies and Moths

The larvae of certain lycaenid butterflies (Plate 5.7) are known to be carnivorous on the eggs, larvae or pupae of ants. They are described as myrmecophagous. In some parts of the world, carnivorous lepidopteran larvae may initially act as ectoparasites before adopting a predatory lifestyle. This is particularly highlighted in two families (the Cyclotornidae and the Epipyropidae, both in the super family Zygaenidae). Hosts of such parasites/predators are mainly hemipterans, especially plant hoppers. Uniquely, in the Hawaiian Islands, larvae of various species of *Eupithecia* (family Geometridae: subfamily Larentiinae) have evolved into entirely predaceous carnivores, having long-since abandoned their original plant-based diets. Such larvae have become structurally and behaviourally adapted to ambushing, capturing and dispatching their prey, which includes flies and various other small invertebrates (Montgomery, 1982; Montgomery, 1983). Elsewhere in the world, species of *Eupithecia* (commonly known as 'pug moths', an epithet also applied to members of other closely related genera, e.g., *Chloroclystis* and *Gymnoscelis*) have remained phytophagous as larvae and show no carnivorous tendencies. Certain case-bearing larvae in the family Coniopterigidae have become specialist predators of tiny, leaf-inhabiting snails, which they firstly ensnare using silk. The recently described species *Hyposmocoma molluscivora* (Rubinoff and Haines, 2005) is an example.

1.7 CONCLUSIONS

- The insect predators are characterised by a set of attributes that distinguish them from parasitoids and they form a major group of entomophagous insects with large size relative to their prey and require more than one prey individual to complete development. They have free-living predatory immature stages, and many species of insect predators are predaceous as both immatures and adults.

- Predators can be classified according to the life stage of the prey they attack (e.g. egg, larval and adult predators) and their foraging strategy (e.g. active searchers, ambush or filter feeders, web or bolas builders). Although these categories may be generally informative about the type of prey taken, they have no predictive value for the prey specificity of individual species.
- Predators occur in approximately 20 insect orders; lists vary slightly, depending on the definition of scavenging and fortuitous predation. The orders of insects that are not known to contain predatory species are Isoptera, Phasmatodea, Phthiraptera, Strepsiptera and Siphonaptera.
- This chapter exclusively deals with insect predators under various orders under five major heads. Characteristics of predatory insect species in the seven families in Hemiptera (Pentatomidae, Nabidae, Anthocoridae, Lygaeidae, Miridae, Reduviidae and Gerridae), five families in Neuroptera (Chrysopidae, Hemerobiidae, Coniopterygidae, Mantispidae and Raphidiidae), five families in Coleoptera (Carabidae, Silphidae, Staphylinidae, Cantharidae and Coccinellidae), seven families in Diptera (cecidomyiidae, Asilidae, Dolichopodidae, Phoridae, Syrphidae, Chlopropidae, Sarcophaegidae and Muscidae) and miscellaneous predators (Dragonflies, Mantises, Earwigs, Thrips and Butterflies and moths) are briefly discussed above.
- Though most of the species belong to the order Hemiptera, Diptera, Coleoptera, Lepidoptera and Neuroptera are highly specific and capable of supressing their host insect populations. More focused research work needs be carried out to exploit them as successful biocontrol agents under integrated pest management programmes.

POINTS TO REMEMBER

i. Insect predators are one of the effective natural enemies; they are capable of searching and killing their prey insect directly from the surrounding and become an inevitable component in integrated pest management programmes.
ii. Predators are classified according to the life stage of prey they attack (e.g. egg, larval and adult predators) and their foraging strategy (e.g. active searchers, ambush or filter feeders, web or bolas builders).
iii. Since predators are poorly understood groups, an overview of different under predatory species in the order Hemiptera (seven families), Neuroptera (five families) Coleoptera (five families), Diptera (seven families) and miscellaneous predators (dragonflies, mantises, earwigs, thrips and butterflies and moths), and their role in suppression of insect pest across the globe has been deduced from available scientific literature.
iv. These predators with active host searching ability and preference towards their prey insects are capable of suppressing those populations of insect pests under various crop ecosystems and have the potential to become an inevitable component in biocontrol programmes.

REFERENCES

Agarwala, B.K., Yasuda, H. & Kajita, Y. (2003). Effect of conspecific and heterospecific feceson foraging and oviposition of two predatory ladybirds: role of fecal cues in predatoravoidance. *Journal of Chemical Ecology*, 29: 357–376.

Albuquerque, G. S., Tauber, C. A. & Tauber. M. J. (1996). Postmating reproductive isolation between *Chrysopa quadripunctata* and *Chrysopa slossonae*: mechanisms and geographic variation. *Evolution*, 50: 1598–1606.

Alford, D. V. (2011). *Plant Pests*. Collins: London, UK, 400p.

Alford, D. V. (2014). *Pests of Fruit Crops: A Colour Handbook* (2nd ed). CRC Press: Boca Raton, FL, 462p.

Ananthakrishnan, T. N. & Thangavelu, K. (1976). The cereal thrips, *Haplothrips ganglbaueri* Schmutz with particular reference to the trends of infestation on *Oryza sativa* and the weed *Echinochloacrusgalli*. *Proceedings of the Indian Academy of Sciences. Section B*, 83(5): 196–201.

Arbogast, R. T. (1976). Suppression of *Oryzaephilus surinamensis* (L.) (Coleoptera: Cucujidae) on shelled corn by the predator *Xylocoris flavipes* (Reuter) (Hemiptera: Anthocoridae). *Journal of Georgia Entomological Society*, 11: 67–71.

Arno, J., Casta, C., Riudavets, J. & Gabarra, R. (2010). Risk of damage to tomato crops by the generalist zoophytophagous predator *Nesidiocoris tenuis* (Reuter) (Hemiptera: Miridae). *Bulletin of Entomological Research*, 100: 105–115.

Ashby, J. W. (1974). A study of arthropod predation of *Pieris rapae* L. using serological and exclusion techniques. *Journal of Applied Ecology*, 11: 419–425.

Aspock, H. (2002). The biology of Raphidioptera: a review of present knowledge. *Acta Zoologica Academiae Scientiarum Hungaricae*, 48 (Suppl. 2): 35–50.

Baez, I., Reitz, S. R. & Funderburk, J. E. (2004). Predation by *Orius insidiosus* (Heteroptera: Anthocoridae) on life stages and species of *Frankliniella* flower thrips (Thysanoptera: Thripidae) in pepper flowers. *Environmental Entomology*, 33: 662–670.

Bahsi, S. U. & Tunc, I. (2008). Development, survival and reproduction of *Orius niger* (Hemiptera: Anthocoridae) under different photoperiod and temperature regimes, *Biocontrol Science and Technology*, 18(8): 767–778.

Bal, A. & Biswas, B. (2013). *Handbook on Major Hemipteran Predators of India*. Zoological Survey of India, Kolkata, 44p.

Ballal, C., Akbar, S., Yamada, K., Wachkoo, A. & Varshney, R. (2018). Annotated catalogue of the flower bugs from India (Heteroptera: Anthocoridae; Lasiochilidae). *Acta Entomologica Musei Nationalis Pragae*, 58(1): 207–226.

Ballal, C. R. & Yamada, K. (2016). Anthocorid predators. In: *Ecofriendly Pest Management for Food Security* (Ed. Omkar) Elsevier: London, pp. 329–366.

Ballal, C. R., Singh, S. P., Poorani, J. & Gupta, T. (2003). Biology and rearing requirements of an anthocorid predator, Blaptostethus pallescens Poppius (Heteroptera: Anthocoridae). *Journal of Biological Control*, 17(1): 29–33.

Bantock, T. (2018). Provisional atlas of shield bugs allies. Available from: https://www.britishbugs.org.uk/Provisional_atlas_of_shieldbugs_and_allies_2018.pdf [accessed 21 April 2021].

Bantock, T. & Botting, J. (2012). *Troilus luridus Bronze Shieldbug*. British Bugs - An online identification guide to UK (Retrieved on 4th May 2021).

Begon, M., Haper, J. L. & Townsend. C. R. (1996). *Ecology* (3rd ed.). Blackwell Science Ltd.: Oxford, UK. *Bioscience*, 46(6): 401–453.

Bennison, J., Lole, M. & Pope, T. (2011). Potential control of cabbage root fly (*Delia radicum*) with the predatory staphylinid beetle *Atheta coriaria*. *OEPP/EPPO Bulletin*, 65: 1721.

Bhatt, N. A. & Patel, M. V. (2018). Tomato bug, Nesidiocoris tenuis (Reuter): A zoo phytophagous insect. *Journal of Entomology and Zoology Studies*, 6(4): 1550–1556.

Blackman, R. L. (1965). Studies on specificity in Coccinellidae. *Annals of Applied Biology*, 56: 336–338.

Blackman, R. L. (1967). The effects of different aphid foods on *Adalia bipunctata* L. and *Coccinella 7punctata* L. *Annals of Applied Biology*, 59: 207–219.

Bogenschtz, H. (1978). Tortricinae. In: *Die Forstschdlinge Europas*. (Ed. Dritter, B. & Schwenke, W). Paul Parey: Hamburg and Berlin, Germany, pp. 55–89.

Bournier, A., Lacasa, A. & Pivot, Y. (1979). Rgimealimentairedun thrips prdateur *Aeolothrips intermedius* Bagnall [Thys.: Aeolothripidae]. *Entomophaga*, 24: 353–361.

Braman, S. K. (2000). Damsel plant bugs (Nabidae). In: *Heteroptera of Economic Importance* (Ed. Schaefer, C. W. & Panizzi, A. R.).CRC Press: Boco Raton, FL, pp. 639–656.

Brown, A. & Whitehead, P. F. (2012). *Clitostethus arcuatus* (Rossi, 1794) (Coleoptera: Coccinellidae) breeding in the Kidderminster area of Worcestershire: overwintering strategies and breeding biology. *Worcestershire Record*, 33: 20–22.

Brunel, E., Grootaert, P. & Medquida, J. (1989). Entomofaune associate. la floraison du colza (*Brassica napus* L.): note preliminaire sur les Dolichopodidae et les Empididae (Insectes: Diptera). *Mededelingen van de Faculteit Landbouwwetenschappen Rijksuniversiteit* Gent, 54: 727–737.

Brezinski, K. (1988). Prospects for using *Macrolophus costalis* Fieb. (Heteroptera: Miridae) in integrated programs for controlling aphids on glasshouse crops. In: *Ecology and Effectiveness of Aphidophaga* (Ed. Riemsdyk, E. & Dixon A. F. G.). SPB, The Hague, pp. 285–288.

Buchs, W. (2003). Predators as biocontrol agents of oilseed rape pests. In: *Biocontrol of Oilseed Rape Pests* (Ed. Alford, D.). Blackwell Publishing, Oxford, pp. 279–298.

Buchs, W. Alford, D. V. (2003). Predators of oilseed rape pests. In: *Biocontrol of Oilseed Rape Pests* (Ed. Alford, D.). Blackwell Publishing, Oxford, pp. 181–199.

Burgess, A. F. & Crossman, S. S. (1929). Imported insect enemies of the gypsy moth and the brown-tail moth. *Technical Bulletin No. 86*, USDA, Washington, DC, 148p.

Buschman, L. L., Whitcomb, W. H., Hemenway, R. C., Mays, D. L., Ru, N., Leppla, N. C. & Smittle, B. J. (1977). Predators of velvet bean caterpillar eggs in Florida soybeans. *Environmental Entomology*, 6: 403–407.

Butler, E. A. (1923). *A Biology of the British Hemiptera-Heteroptera*. H.F.& G. Witherby, London, 682p.

Carney, V. A., Diamond, J. C., Murphy, G. D. & Marshall, D. (2002). The potential of *Atheta coriana* Kraatz (Coleoptera: Staphylinidae) as a biological control agent for use in greenhouse crops. *IOBC/WPRS Bulletin*, 25: 37–40.

Chacko, M. J. (1973). Observations on some natural enemies of *Pineus* sp. (Hem.: Adelgidae) at Shillong (Meghalaya), India, with special reference to *Tetraphleps raoi* Ghauri (Hemiptera: Anthocoridae). *Technical Bulletin of the Commonwealth Institute of Biological Control*, 16: 41–46.

Chambers, R. J., Sunderland, K. D., Stacey, D. L. & Wyatt, I. J. (1986). Control of cereal aphids in winter wheat by natural enemies: aphid-specific predators, parasitoids and pathogenic fungi. *Annals of Applied Biology*, 108: 219–231.

Chandler, P. J. (2014). An update of the 1998 checklist of Diptera of the British Isles, p. 340, *Dipterists Forum*, Oxford, UK. Available from: www.dipteristsforum.org.uk/documents/BRITISH_ISLES_CHECKLIST.pdf [accessed 23 December 2018].

Chandra, K., Kushwaha, S., Biswas, B. & Mukherjee, P. (2014). *Nabis tibialis* distant (Hemiptera: Nabidae), a first record from India. *Munis Entomology & Zoology*, 9(1): 178–181.

Chapman, J. W., Reynolds, D. R., Brooks, S. J., Smith, A. D. & Woiwod, I. P. (2006). Seasonal variation in the migration strategies of the green lacewing *Chrysoperla carnea* species complex. *Ecological Entomology*, 31: 378–388.

Chittenden, F. H. (1898). Insects that affect *Asparagus*. *Bulletin of the U.S. Department of Agriculture. Division of Entomology (N.S.)*, 10: 54–59.

Clausen, C. P. (1940). *Entomophagous Insects*. McGraw Hill, New York and London, UK, 688p.

Cole, L. R. & Harris, K. M. (2002). *Lestodiplosis* spp. (Dipt., Cecidomyiidae) predaceous on larvae of Lepidoptera and other insects on oaks (*Quercus robur* L.) in southern England. *Entomologists Monthly Magazine*, 138: 1–10.

Collyer, E. (1951). The separation of *Conwentzia pineticola* end. from *Conwentzia psociformis* (Curt.) and notes on their biology. *Bulletin of Entomological Research*, 42: 555–564.

Collyer, E. (1952). Biology of some predatory insects and mites associated with the fruit tree red spider mite (*Metatetranychus ulmi* (Koch)) in south-eastern England I The biology of *Blepharidopterus angulatus* (Fall) (Hemiptera-Heteroptera, Miridae). *Journal of Horticultural Science*, 27: 117–129.

Collyer, E. (1953). Biology of some predatory insects and mites associated with the fruit tree red spider mite (*Metatetranychus ulmi* (Koch)) in south-eastern England II Some important predators of the mite. *Journal of Horticultural Science*, 28: 85–97.

Conti, B. (2009). Notes on the presence of *Aeolothrips intermedius* in northwestern Tuscany and on its development under laboratory conditions. *Bulletin of Insectology*, 62: 107–112.

Coppel, H. C., House, H. L. & Maw, M. G. (1959). Studies on dipterous parasites of the spruce budworm, *Choristoneura fumiferana* (Clem.) (Lepidoptera: Tortricidae): VII. *Agria affinis* (Fall.) (Diptera: Sarcophagidae). *Canadian Journal of Zoology*, 37: 817–830.

Cornelis, M., Quirán, E. & Coscarón, M. C. (2012). Description of some immature stages of *Nabis (Tropiconabis) capsiformis* (Hemiptera: Nabidae). *Revista Mexicana de Biodiversidad*, 83: 1009–1012.

David, B. V. & Ananthakrishnan, T. N. (2004). *General and Applied Entomology*. Tata McGraw-Hill Publishing Company Limited, 877p.

De Clercq, P. (2000). Predaceous stink bugs (Pentatomidae: Asopinae). In: *Heteroptera of Economic Importance* (Ed. Schaefer, C. W. & Panizzi, A. R.). CRC Press, Boca Raton, FL, pp. 737–789.

De Clercq, P. (2008). Predatory stink bugs (Hemiptera: Pentatomidae, Asopinae). In: *Encyclopedia of Insects* (2nd ed.) (Ed. Capinera, J. L.). Kluwer Academic Publishers, Dortrecht, pp. 3042–3045.

Dekeirsschieter, J., Verheggen, F., Lognay, G. & Haubruge, E. (2011). Large carrion beetles (Coleoptera, Silphidae) in Western Europe: a review. *Biotechnology, Agronomy, Society and Environment*, 15: 435–447.

Dennis, D. S., Barnes, J. K. & Knutson, L. (2013). Review and analysis of information on the biology and morphology of immature stages of robber flies (Diptera: Asilidae). *Zootaxa*, 3671: 1–64.

Dimitrov, A. (1975). Control of tobacco thrips with *Macrolophus costalis*. *Rastitelna Zashchita*, 23(6): 34–37.

Disney, R. H. L. (1983). Scuttle flies. Diptera, Phoridae (except Megaselia). *Handbooks for the Identification of British Insects*, 10(6): 1–81.

Disney, R. H. L. (1989). Scuttle flies. Diptera Phoridae genus Megaselia. *Handbooks for the Identification of British Insects*, 10(8): 1–155.

Distant, W. L. (1910). The fauna of British India including Ceylon and Burma. *Rhynchota*, II: 185–188.

Dobson, R. M. (1961). Observations on natural mortality, parasites and predators of wheat bulb fly, *Leptohylemyiacoarctata* (Fall.). *Bulletin of Entomological Research*, 52: 281–291.

Doutt, R. L. (1964). Biological characteristics of entomophagous adults. In: *Biological Control of Insect Pests and Weeds* (Ed. DeBach, P.). Reinhold Publ. Co., New York, pp. 145–167.

Duelli, P. (1980). Adaptive dispersal and appetitive flight in the green lacewing, *Chrysopa cornea*. *Ecological Entomology*, 5: 213–220.

Eger, J., Barcellos, A. & Weiler, L. (2015). Shield bugs (Scutelleridae). In: *True Bugs of the Neotropics* (Ed. Panizzi, A. R. & Grazia, J.). Springer, New York, pp. 757–788.

Evans, A. V. (2009). The forest caterpillar hunter, *Calosoma sycophanta*, an old-world species confirmed as part of the Virginia beetle fauna (Coleoptera: Carabidae). *Banisteria*, 34: 33–37.

Fathi, S. A. A. & Ganbalani, G. N. (2009). Prey preference of *Oriusniger* (Wolf.) and *O. minutus* (L.) from *Thrips tabaci* (Lind.) and *Tetranychus urticae* (Koch.). *Journal of Entomology*, 6: 42–48.

Foglar, H., Malusa, J. C. & Wajnberg, E. (1990). The functional response and preference of *Macrolophus caliginosus* [Heteroptera: Miridae] for two of its prey: *Myzus persicae* and *Tetranychus urticae*. *Entomophaga*, 35: 465–474.

Fritzsche, R. (1955). *Phaoniatri maculata* Bouchals Parasit des Groen Kohltriebrlers *Ceuthorhynchus napi* Gyll. und des Gefleckten Kohltriebrlers *Ceuthorhynchus quadridens* Panz. Nachrichtenblattfr den DeutschenPflanzenschutzdienst, 9: 35–36.

Gagné, R. J. (2010). Update for a catalog of the Cecidomyiidae (Diptera) of the world. http://www.ars.usda.gov/SP2UserFiles/Place/12754100/Gagne_2010_World_Catalog_Cecidomyiidae.pdf [accessed 04 May 2021].

Galbreath, R. A. & Cameron, P. J. (2015). The introduction of the eleven-spotted ladybird *Coccinella undecimpunctata* L (Coleoptera: Coccinellidae) to New Zealand in 1874: Aspurious record created by cumulative misreporting. *New Zealand Entomologist*, 38: 7–9.

Ghauri, M. S. K. (1965). Notes on the Hemiptera from Pakistan and adjoining areas. *Annals and Magazine of Natural History-Series*, 13(7): 673–688.

Giard, A. (1900). Sur un Hmiptre (*Atractotomus mali* Mey.) parasite des chenilles d *Hyponomeutamalinellus* Zeller et *H. padellus* L. *Bulletin de la Socit. Entomologique de France*, 5(18): 359–360.

Glen, D. M. (1977). Predation of codling moth eggs, *Cydia pomonella*, the predators responsible and their alternative prey. *Journal of Applied Ecology*, 14: 445–456.

Godfrey, K. E., Whitcomb, W. H. & Stimac, J. L. (1989). Arthropod predators of velvetbean caterpillar, *Anticarsia gemmatalis* (Lepidoptera: Noctuidae), eggs and larvae. *Environmental Entomology*, 18: 118–123.

Good, J. A. & Giller, P. S. (1991). The diet of predatory staphylinid beetles - a review of records. *Entomologists Monthly Magazine*, 127: 77–89.

Gordon, R. D. (1985). The Coccinellidae (Coleoptera) of America north of Mexico. *Journal of the New York Entomological Society*, 93: 1–912.

Greathead, D. J. & Greathead, A. H. (1992). Biological control of insect pests by insect parasitoids and predators: the BIOCAT database. *Biocontrol News and Information*, 13(4): 61–68.

Gulde, J. (1921). Die Wanzen (Hemiptera-Heteroptera) der Umgebung von Frankfurt a. M. und des Beckens. *Abhandlungen der SeneckenbergischeNaturforschende Gesellschaft*, 37: 327–503.

Hagen, K. S. (1987). Nutritional ecology of terrestrial insect predators. In: *Nutritional Ecology of Insects, Mites, Spiders, and Related Invertebrates* (Ed. Slansky, F. & Rodriquez, J. G.), John Wiley & Sons, New York, pp. 533–577.

Hagen, K. S., Bombosch, S. & McMurtry, J. A. (1976). The biology and impact of predators. In: *Theory and Practice of Biological Control* (Ed. Huffaker, C. B & Messenger, P. S.), Academic Press, New York, pp. 93–142.

Hamilton, I. & Heath, J. (1976). Predation of *Pentatoma rufipes* (L.) (Heteroptera: Pentatomidae) upon *Zygaena filipendulae* (L.) (Lepidoptera: Zygaenidae). *Irish Naturalists - Journal*, 18: 337–350.

Hangay, G. & Zborowski, P. (2010). *A Guide to the Beetles of Australia*. CSIRO Publishing, Collingwood, Australia, 238p.

Harizanova, V., Stoeva, A. & Mohamedova, M. (2012). Preliminary study on the invasive *Acizzia jamatonica* (Hemiptera: Psyllidae) and its predators in Bulgaria. *Agricultural Science and Technology*, 4(1): 56–61.

Harris, K. M. (1966). Gall midge genera of economic importance (Diptera: Cecidomyiidae). Part 1: Introduction and subfamily Cecidomyiinae; supertribe Cecidomyiidi. *Transactions of the Royal Entomological Society of London*, 118: 313–358.

Harris, K. M. (1973). Aphidophagous Cecidomyiidae (Diptera): taxonomy, biology and assessments of field populations. *Bulletin of Entomological Research*, 63: 305–325.

Harris, K. M. (1982). The apple midge: a brief history. *Antenna*, 6: 286–289.

Harris, K. M. & Bennett, F. D. (2003). Records of gall midges (Diptera.: Cecidomyiidae) from the Isle of Man. *Entomologists Record and Journal of Variation*, 115: 109–115.

Hedrick, A. V. & Riechert. S. E. (1989). Genetically-based variation between two spider populations in foraging behavior. *Oecologica*, 80: 532–539.

Heinz, K. M. & Zalom, F. G. (1996). Performance of the predator *Delphastus pusillus* on *Bemesia* resistant and susceptible tomato lines. *Entomologia Experimentalis et Applicata*, 81: 345–352.

Helyer, N., Cattlin, N. D., & Brown, K. C. (2014). *Biological Control in Plant Protection: A Colour Handbook* (2nd ed.). CRC Press, Boca Raton, FL, 267p.

Heong, K. L., Bleih, S. & Lazaro, A. A. (1990). Predation of *Cyrtorhinus lividipennis* Reuter on eggs of the green leafhopper and brown planthopper in rice. *Researches on Population Ecology*, 32: 255–262.

Hernandez, L. M. (1999). A review of the economically important species of the genus *Orius* (Heteroptera: Anthocoridae) in East Africa. *Journal of Natural History*, 33: 543–568.

Hertveldt, L., Keymeulen, M. van. & Pelerents, C. (1984). Large scale rearing of the entomophagous rove beetle *Aleocha rabilineata* (Coleoptera: Staphylinidae). *Mitteilungenaus der BiologischenBundesanstaltfr Land- und Forstwirtschaft Berlin-Dahlem*, 218: 70–75.

Hoberlandt, L. (1972). Ordnung Heteroptera, Wanzen. In: *Die Forstschdlinge Europas. Erster Band Paul Pared* (Ed. Schwenke, W.). Paul Parey: Hamburg and Berlin, Germany, pp. 114–125.

Hodek, I. (1993). Habitat and food specificity in aphidophagous predators. *Biocontrol Science and Technology*, 3: 91–100.

Huffaker, C. B. & Caltagirone, L. E. (1986). The impact of biological control on the development of the Pacific. *Agriculture, Ecosystems & Environment*, 15: 95–107.

Iperti, G. (1974). Les coccinelles. In: *Les organisms auxiliairesen verger de pommiers* (Ed. Brader, L.). OILB/SROP, Brochure No. 3, pp. 111–121.

Jahn, E. (1978). Notodontidae, Zahnspinner. In: *Die Forstschdlinge Europas. Dritter Band* (Ed. Schwenke, W). Paul Parey, Hamburg and Berlin, Germany, pp. 404–420.

Jary, S. G. (1935). Some observations upon the œred spider, *Tetranychus telarius* L., on hops and its control, with notes on some predatory insects. *Annals of Applied Biology*, 22: 538–548.

Jenyns, L. (1832). Account of an extraordinary swarm of flies. *Annals and Magazine of Natural History*, 5(series 1): 302.

Jessep, C. T. (1964). A note on the feeding habits of *Nabis capsiformis* Germar. *New Zealand Entomologist*, 3: 23–34.

Jones, M. G. (1969). The effect of weather on fruit fly and its predators. *Journal of Applied Ecology*, 6: 425–441.

Joyce, K. A., Holland, J. M. & Doncaster, C. P. (1999). Influence of hedgerow intersections and gaps on the movement of carabid beetles. *Bulletin of Entomological Research*, 89: 523–531.

Kahradník, J. (1972). Diaspididae, Deckelschildluse. Armoured scales. In: *Die Forstschdlinge Europas* (Ed. Schwenke, W). Paul Parey, Hamburg and Berlin, Germany, pp. 422–446.

Kairo, M. T. K., Paraiso, O., Gautam, R. D. & Peterkin, D. D. (2013). *Cryptolaemus montrouzieri* (Mulsant) (Coccinellidae: Scymninae): a review of biology, ecology, and used in biological control with particular reference to potential impact on non-target organisms. *CAB Reviews*, 8: 1–20.

Kapur, A. P. (1972). The Coccinellidae (Coleoptera) of Goa. *Records of Zoological Survey of India*, 66(1–4): 310–320.

Karanja, M. K. & Aloo, T. C. (1990). The introduction and establishment of *Tetraphleps raoi* Ghauri as a control agent of woolly aphid in Kenya. *Technical Note - Kenya Forestry Research Institute*, 12: 1–12.

Keerthi, M. C., Sravika, A., Mahesha, H. S., Gupta, A., Bhargavi, H. A. & Ahmed, S. (2020). Performance of the native predatory bug, *Eocanthecona furcellata* (Wolff) (Hemiptera: Pentatomidae), on the fall armyworm, *Spodoptera frugiperda* (J. E. Smith) (Lepidoptera: Noctuidae), and its limitation under field condition. *Egyptian Journal of Biological Pest Control*, 30: 1–4.

Kerzhner, I. M. (1983). Airborne *Nabis capsiformis* (Heteroptera: Nabidae) from the Atlantic, Indian and Pacific oceans. *International Journal of Entomology*, 25: 273–275.

Koch, R. L. (2003). The multicolored Asian lady beetle, *Harmonia axyridis*: a review of its biology, uses in biological control, and non-target impacts. *Journal of Insect Science*, 3: 1–16.

König, V. (2015) Monitoring der Rotbeinigen Baumwanze. *Pflanzengesundheit*, 4: 19– 22.

Konvicka, M., Hula, V. & Fric, Z. (2005). *Picromerus bidens* (Heteroptera: Pentatomidae) as predator of the checkerspot *Euphydrya saurinia* (Lepidoptera: Nymphalidae). *Entomologica Fennica*, 16: 233–236.

Kotrba, M. (2010). Massenauftreten der Gemeinen RasenhalmfliegeThaumatomyianotata (Diptera, Chloropidae) an Gebuden. *Denisia*, 30: 451–456.

Krivesheina, N. P. (2013). On the ecology of *Phaonia* larvae (Diptera, Muscidae). *Ecological Review*, 93: 324–333.

Kromp, B. (1999). Carabid beetles in sustainable agriculture: a review on pest control efficacy, cultivation impacts and enhancement. *Agriculture, Ecosystems & Environment*, 74: 187–228.

Kuhlmann, U. (1995). Biology and predation rate of the sarcophagid fly, *Agria mamillata*, a predator of European small ermine moths. *International Journal of Pest Management*, 41: 67–73.

Kumar, S. M. & Sahayaraj, K. (2012). Gross Morphology and Histology of Head and Salivary Apparatus of the Predatory Bug. *Journal of Insect Science*, 12(19): 1–12.

Kushwaha, R., Chakraborti, U., Roy, S., Biswas, O. & Mitra, B. (2017). Diversity of Ground beetles (Insecta: Coleoptera: Carabidae) from Indian Sundarban with seven new records. *Bioscience Discovery*, 8: 335–339.

Kutty, S. N., Pont, A. C., Meier, R. & Pape, T. (2014). Complete tribal sampling reveals basal split in Muscidae (Diptera), confirms saprophagy as ancestral feeding mode, and reveals an evolutionary correlation between instar numbers and carnivory. *Molecular Phylogenetics and Evolution*, 78: 349–364.

Laidlaw, W. B. R. (1936). The brown lacewing flies (Hemerobiidae): their importance as controls of *Adelges cooleyi* Gillette. *Entomologist's Monthly Magazine*, 72: 164–174.

Lattin, J. D. (1999). Bionomics of the Anthocoridae. *Annual Review of Entomology*, 44: 207–231.

Laurence, B. R. (1951). The prey of some tree trunk frequenting Empididae and Dolichopodidae (Dipt.). *Entomologist's Monthly Magazine*, 87: 166–169.

Liu, J., Wu, K. M., Hopper, K. R. & Zhao, K.J. (2004). Population dynamics of *Aphis glycines* (Homoptera: Aphididae) and its natural enemies in soybean in Northern China. *Annals of the Entomological Society of America*, 97: 235–239

Lovei, G. L. & Sunderland, K. D. (1996). Ecology and behavior of ground beetles (Coleoptera: Carabidae). *Annual Review of Entomology*, 41: 231–236.

Luka, H., Pfiffner, W. & Wyss, E. (1998). *Amara ovata* und *A. similata* (Coleoptera, Carabidae), zweiphytophage Lauf k.␣ferarten in Rapsfeldern. *Mitteilungen der SchweizerischenEntomologischen Gesellschaft*, 71: 125–131.

Lyon, J. P. & Gldlin de Tiefenau, P. (1974). Les syrphespr. dateurs des pucerons. In: *Les organisms auxiliairesen verger de pommiers* (Ed. Brader, L.). OILB/SROP Brochure No. 3, pp. 163–170.

Mair, J. & Port, G. R. (2001). Predation by the carabid beetles *Pterostichus madidus* and *Nebria brevicollis* is affected by size and condition of the prey slug *Deroceras reticulatum*. *Agricultural and Forest Entomology*, 3: 99–106.

Maisner, N. (1974). Chrysomelidae. Blattkfer. In: *Die ForstschdlingeEuropas. Zweiter Band* (Ed. Schwenke, W.). Paul Parey, Hamburg and Berlin, Germany, pp. 202–236.

Majerus, M. E. N. (1994). *Ladybirds*. HarperCollins: London, UK, 354p.

Majka, C. G. & McCorquodale, D. B. (2006). The Coccinellidae (Coleoptera) of the maritime provinces of Canada: new records, biographical notes, and conservation concerns. *Zootaxa*, 1154: 49–68.

Mani, M. S. (1968). *Ecology and Biogeography of High-Altitude Insects*. Springer Science, Berlin, 526p.

Manjunath, T. M., Naidu, B. S. & Prasad, N. K. (1977). The rice brown planthopper and its predator, *Cyrtorhinus lividipennis*, on some rice cultivars in India. *International Rice Research Newsletter*, 2(3): 5.

Marcos-Garcia, M. (1981). Contribucin al conocimiento de los Syrphidae (Dipt) de la zona noreste de la provincia de Salamanca. *Boletin de la Asociacin Espaola Entomologa*, 4: 157–171.

Margaritopoulos, J. T., Tsitsipis, J. A. & Perdikis, D. C. (2003). Biological characteristics of the mirids *Macrolophus costalis* and *Macrolophus pygmaeus* preying on the tobacco form of *Myzus persicae* (Hemiptera: Aphididade). *Bulletin of Entomological Research*, 93: 39–45.

Martin, N. A. (2016). Eleven-spotted ladybird – *Coccinella undecimpunctata*. Interesting insects and other invertebrates. New Zealand Arthropod Factsheet Series Number 21. ISSN 1179-643X. http://nzacfactsheets.landcareresearch.co.nz/Index.html [accessed 05 April 2021].

Mead, F. W. & Eger, J. E. (1992). The predatory stink bug, *Andrallus spinidens* (Fabricius) in Florida. Attached to *Tri-ology Technical Report*, 31(3): 1.

Metcalf, C. L. & Flint, W. P. (1962). *Destructive and Useful Insects*. McGraw-Hill Co., New York, 1087p.

Mili, P., Devee, A. & Saikia, D. K. (2018). Diversity of carabids in Gerbera and Gladiolus. *Journal of Entomology and Zoology Studies*, 6(2): 116–124.

Milne, D. L. (1960). The gall midges (Diptera: Cecidomyidae) of clover flower-heads. *Transactions of the Royal Entomological Society of London*, 112: 73–108.

Milne, M. & Walter, G. H. (1998). Significance of mite prey in the diet of the onion thrips *Thripstabaci* Lindeman (Thysanoptera: Thripidae). *Australian Journal of Entomology*, 37: 120–124.

Mitchell, R. G. & Maksymov, J. K. (1977). Observations of predation on spruce gall aphids within galls. *Entomophaga*, 22: 179–186.

Monserrat, V. J. & Marin, F. (2001). Comparative plant substrate specificity of Iberian Hemeerobiidae, Coniopterygidae and Chrysopidae. In: *Lacewings in the Crop Environment* (Ed. McEwen, P. K., New, T. R. &. Whittington A. E.). Cambridge University Press, Cambridge, UK, pp. 424–434.

Montgomery, S. L. (1982). Biogeography of the moth genus *Eupithecia* in Oceania and the evolution of ambush predation in Hawaiian caterpillars (Lepidoptera: Geometridae). *Entomologia Generalis*, 8: 27–34.

Montgomery, S. L. (1983). Carnivorous caterpillars: the behavior, biogeography and conservation of *Eupithecia* (Lepidoptera: Geometridae) in the Hawaiian Islands. *GeoJournal*, 7: 549–556.

Moreton, B. D. (1969). *Beneficial Insects and Mites*. Ministry of Agriculture, Fisheries and Food, Bulletin 20. HMSO, London, UK, 45p.

Morris, H. M. (1922). On the larva and pupa of a parasitic phorid fly-*Hypocera incrassata*. *Parasitology*, 14: 70–74.

Mukherjee, A. B., Chaudhury, A. K. S., Bronniman, H. & Chu Y. (1971). Artificial diets for rearing of *Xylocoris flavipes* (Reuter), a predator of some pests of stored cereals. *Indian Journal of Entomology*, 33: 356–358.

Muraleedharan, N. & Ananthakrishnan, T. N. (1978). Bioecology of four species of Anthocoridae (Hemiptera: Insecta) predaceous on thrips, with key to genera of anthocorids from India. *Occasional Paper, Records of the Zoological Survey of India*, 11, 1–32.

Nelemans, M. N. E. (1987). Possibilities for flight in the carabid beetle *Nebria brevicollis* (F.): the importance of food during larval growth. *Oecologia*, 72: 502–509.

Nelemans, M. N. E., den Boer, P. J. & Spee, A. (1989). Recruitment and summer diapause in the dynamics of a population of *Nebria brevicollis* (Coleoptera: Carabidae). *Oikos*, 56: 157–169.

Niccoli, A. & Fagnani, F. (1994). Indaginisull-andamentodelleinfestazioni di *Myzus varians* (Davidson) in peschetidella Toscana. *Redia*, 77: 175–187.

Nijveldt, W. (1969). *Gall Midges of Economic Importance. Vol. VIII: Miscellaneous*. Crosby Lockwood & Son, London, UK, 221p.

Nolte, H. W. (1951). BeitrgezurMorphologie und Biologie des Lrchenblasenfu.Ÿes (*Taeniothrips laricivorus* Krat.). *Beitrgezur Entomologie*, 1: 110–139.

Omkar & Mishra, G. (2005). Evolutionary significance of promiscuity in an aphidophagous ladybird, *Propylea dissecta* (Coleoptera: Coccinellidae). *Bulletin of Entomological Research*, 95, 527–533.

Omkar & Pervez, A. (2000). Biodiversity in predaceous coccinellids (Coleoptera: Coccinellidae) in India: a review. *Journal of Aphidology*, 14: 41–66.

Omkar & Pervez, A. (2003). Ecology and biocontrol potential of a scale-predator, *Chilocorus nigritus*. *Biocontrol Science and Technology*, 13(4): 379–390.

Omkar & Pervez, A. (2004). Predaceous Coccinellids in India: Predator-prey catalogue (Coleoptera: Coccinellidae).*Oriental Insects*, 38(1): 27–61.

Omkar & Pervez, A. (2005). Ecology of two-spotted ladybird, *Adalia bipunctata:* a review. *Journal of Applied Entomology*, 129(9/10): 465–474.

Oswald, J. D. & Tauber, C. A. (2001). Preimaginal stages of the family Hemerobidae. In: *Lacewings in the Crop Environment* (Ed. McEwen, P. K., New T. R. & Whittington, A. E.).Cambridge University Press: Cambridge, UK, pp. 50–60.

Pal, S. (1974). *Geocoris jucundus* Fieb. (Lygaeidae: Hemiptera) as predator of Lucerne and Tomato Aphids in the Indian Desert. *Current Science*, 43(17): 564–564.

Pantaleoni, R. A. & Alma, A. (2001). Lacewings and snake-flies in Piedmont vineyards (northwestern Italy). In: *Lacewings in the Crop Environment* (Ed. McEwen, P. K., New T. R. & Whittington, A. E.). Cambridge University Press, Cambridge, UK, pp. 471–480.

Pati, P. & Mathur, K. C. (1986). *Amyotea (Asopus) malabarica* (Fabricius), a predatory bug on leaf feeding pests of rice. *Oryza*, 23(3): 200–201.

Pehlivan, S. & Atakan, E. (2017). Some biological parameters of *Orius niger* (Wolff, 1811) (Hemiptera: Anthocoridae) under outdoor conditions in Turkey. *Turkish Journal of Entomology*, 2: 231–241.

Penney, M. M. (1966). Studies on certain aspects of the ecology of *Nebria brevicollis* (F.) (Coleoptera, Carabidae). *Journal of Animal Ecology*, 35: 505–512.

Perdikis, D., Fantinou, A., Garantonakis, N., Kitsis, P., Maselov, D. & Panagakis, S. (2009). Studies on the damage potential of the predator *Nesidiocoris tenuis* on tomato plants. *Bulletin of Insectology*, 62: 41–46.

Pervez, A. & Omkar. (2006). Ecology and biological control application of multicoloured Asian ladybird, *Harmonia axyridis*: a review. *Biocontrol Science and Technology*, 16(2): 111–128.

Poorani, J. (2004). An annotated checklist of the Coccinellidae (Coleoptera) of the Indian Subregion. *Oriental Insects*, 36(1): 307–383.

Powel, G. (2020). The biology and control of an emerging shield bug pest, *Pentatoma rufipes* (L.) (Hemiptera: Pentatomidae). *Agricultural and Forest Entomology*, 22(4): 298–308.

Powell, G. (2020). A review of current knowledge of the forest bug, *Pentatoma rufipes*. https://projectblue.blob.core.windows.net/media/Default/Horticulture/AHDB%20Forest%20Bug%20Review-1.pdf [accessed 20 April 2020].

Prescher, S. & Buchs, W. (1998). Der Einfuabgestufter Extensivierungsmanahmenim Raps- und Erbsenanbau auf ntzlicheFliegen (Diptera, Brachycera). *Gesunde Pflanzen*, 50: 213–218.

Principi, M. M. & Canard, M. (1974). Les Nvroptres. In: *Les organismes auxilliairesen verger de pommiers* (Ed. Brader, L.). OILB/SROP Brochure No. 3, pp. 151–162.

Pschorn-Walcher, H. (1982). Familienreihe Megalodontoidea. In: *Die Forstschdlinge Europas* (Ed. Schwenke, W.). Paul Parey, Hamburg, Germany, pp. 22–57.

Putnam, W. L. (1942). Notes on the predaceous thrips *Haplothrips subtilissimus* Hal. and *Aeolothrips melaleucus* Hal. *The Canadian Entomologist*, 74: 37–43.

Putnam, W. L. (1965). The predaceous thrips *Haplothrips faurei* Hood (Thysanoptera: Phloeothripidae) in Ontario peach orchards. *The Canadian Entomologist*, 97: 1208–1221.

Rahman, K. A. (1940). Important insect predators of India. *Proceedings of the Indian Academy of Sciences - Section B*, 12: 67–74.

Ramachandran, S. Funderburk, J., Stavisky, J. & Olson, S. (2002). Population abundance and movement of *Frankliniella* species and *Orius insidiosus* in field pepper. *Agricultural and Forest Entomology*, 3(2): 129–137.

Ravn, P. H., Rasmussen, A. N. & Alford, D. V. (1995). Tgeproblemerikrnefrugt. 12. Danske Plantevrnskonference, pp. 209–222.

Redborg, K. E. (1998). Biology of the Mantispidae. *Annual Review of Entomology*, 43: 175–194.

Rishi, K., Jun-Ce, T., Steven, E. N. & Anthony, M. S. (2014). Effects of Bt Cotton on *Thrips tabaci* (Thysanoptera: Thripidae) and Its Predator, *Orius insidiosus* (Hemiptera: Anthocoridae). *Journal of Economic Entomology*, 107(3): 927–932.

Rojo, S., Prez-Ban, C. & Marcos-Garca, M. (1999). First observations on the biology of *Scaeva mecogramma* (Bigot, 1860) (Diptera, Syrphidae) and notes on some other syrphids preying on psyllids (Hemiptera, Aphalaridae and Triozidae). *Volucella*, 4: 105–111.

Rossi, J., Gamba, U., Pinna, M., Spagnolo, S., Visentin, C. & Alma, A. (2006). Hoverflies in organic apple orchards in north-western Italy. *Bulletin of Insectology*, 59: 111–114.

Rubinoff, D. & Haines, W. P. (2005). Web-spinning caterpillar stalks snails. *Science*, 309: 575.

Ruzicka, Z. (1997a). Protective role of the egg stalk in Chrysopidae (Neuroptera: Chrysopidae). *European Journal of Entomology*, 94: 111–114.

Ruzicka, Z. (1997b). Recognition of oviposition-deterring allomones by aphidophagous predators (Neuroptera: Chrysopidae, Coleoptera: Coccinellidae). *European Journal of Entomology*, 94: 431–434.

Ruzicka, Z. (1998). Further evidence of oviposition-deterring allomone in chrysopids (Neuroptera: Chrysopidae). *European Journal of Entomology*, 95: 35–39.

Sahayaraj, K. & Balasubramanian, M. (2016). Response of the reduviid bug, *Rhynocoris marginatus* (Heteroptera: Reduviidae) to six different species of cotton pests. *European Journal of Entomology*, 113: 29–36.

Sahayaraj, K. (1999). Effect of prey and their ages on the feeding preference of *Rhynocoris marginatus* (Fab.). *International Arachis Newsletter*, 19: 39–41.

Sahayaraj, K. (2014a). *Basic and Applied Aspects of Biopesticides*. Springer, New Delhi, 384p.

Sahayaraj, K. (2014b). Reduviids and their merits in biological control. In: *Basic and Applied Aspects of Biopesticides* (Ed. Sahayaraj, K.). Springer India, New Delhi, India, pp.195–214.

Sahayaraj, K., Kanna, A., Vinoth, K. & Subramanian, M. (2010). Gross morphology of feeding canal, salivary apparatus and digestive enzymes of salivary gland of Catamirus brevipennis (Servile) (Hemiptera: Reduviidae). *Journal of the Entomological Research Society*, 12 (2): 37–50.

Salini, S., David, K. J. & Rabbani, M. K. (2020). Beneficial fauna of Stink bugs of India. ICAR- National Bureau of Agricultural Insect Resources. https://www.nbair.res.in/sites/default/files/202101/Folder%2009_0.pdf.

Samson, P.R. & Blood, P.R.B. (1980). Voracity and searching ability of *Chrysopa signata* (Neuroptera: Chrysopidae), *Micromus tasmaniae* (Neuroptera: Hemerobiidae), and *Tropicona biscapsiformis* (Hemiptera: Nabidae). *Australian Journal of Zoology*, 28: 575–580.

Santhamma, S. (2017). Redescription of a predatory stink bug, *Amyoteama labarica* (Fabricius, 1775) (Hemiptera: Pentatomidae: Asopinae). *Journal of Biological Control*, 30: 240–247.

Sarmah, N., Devee, A. & Perdikis, D. (2019) *Macrolophus pygmaeus* (Hemiptera: Miridae) foraging on tomato leaves from different plant strata. *Phytoparasitica*, 47: 663–670.

Schmutterer, H. (1972). Coccidae (Lecaniidae), Napfschildluse. Soft scales. In: *Die Forstschdlinge Europas, Erster Band* (Ed. Schwenke, W). Paul Parey, Hamburg and Berlin, Germany, pp. 405–418.

Schneider, F. (1948). BeitragzurKenntnis der Generationsverhltnisse und Diapause rauberischer Schwebfliegen (Syrphidae, Dipt.). *Mitteilungen der Schweizenischen Entomologischen Gesellschaft*, 21: 249–285.

Schuh, R. T. (1975). The structure, distribution, and taxonomic importance of *Trichobothria* in the Miridiae (Hemiptera). *American Museum Novitates Number 2585*, pp. 1–26.

Schwenke, W. (1974). *Die Forstschädlinge Europas: Band II Käfer*. Parey, Hamburg, Berlin, 468p.

Senrayan, R. (1988) Functional response of *Eocanthecona furcellata* (Wolff) (Heteroptera: Pentatomidae) in relation to prey *Latoia lepida* (Cramer) (Lepidoptera: Limacodidae). *Proceedings of the Indian Academy of Sciences (Animal Science)*, 97: 339–345.

Shaltiel-Harpaz, L., Gerling, D. & Graph, S. (2016). Control of the tomato leafminer, *Tuta absoluta* (Lepidoptera: Gelechiidae), in open-field tomatoes by indigenous natural enemies occurring in Israel. *Journal of Economic Entomology*, 109: 120–131.

Shophiya, N. J. & Sahayaraj, K. (2014) Biocontrol potential of entomophagous predator *Eocanthecona furcellata* (Wolff) against *Pericallia ricini* (Fab.) larvae. *International Journal of Current Research*, 6(10): 9052–9056.

Singh, J. P., Jaiswal, A. K., Monobrullah, M. & Bhattacharya, A. (2010). Effect of selected pesticides on larval mortality of the neuropteran predator, *Chrysopa lacciperda* Kimmins, of the lac insect, *Kerria lacca* (Kerr). *Journal of Asia-Pacific Entomology*, 13: 69–72.

Singh, V. & Banyal, H. S. (2013). Insect fauna of Khajjiar Lake of Chamba District, Himachal Pradesh, India. *Pakistan Journal of Zoology*, 45(4): 1053–1061.

Speight, M. C. D. (2011). Species accounts of European Syrphidae (Diptera). In: *The Database of European Syrphidae* (Ed. Glasgow). The Syrph Net Publications, Dublin, Ireland, Volume 65, 285p.

Sridhar, V., Jayashankar, M. & Vinesh, L. S. (2012). Population dynamics of Zoo phytophagous mirid bug, *Nesidiocoris tenuis* (Reuter) (Heteroptera: Miridae) and its prey, *Bemisia tabaci* Genn. (Homoptera: Aleyrodidae) on tomato (*Solanum lycopersicum* Mill). *Pest Management in Horticultural Ecosystems*, 18(1): 35–38.

Srikumar, K. K., Santhamma, S., Smitha, S., Kumar, B. B. & Radhakrishnan, B. (2018). Taxonomy, bionomics and predatory potential of *Eocanthecona concinna* (Walker) (Hemiptera: Pentatomidae: Asopinae). *Journal of Biological Control*, 32: 81–86.

Steiner, H. (1974). Carabidae, Staphylinidae, Cantharidae. In: *Les organismes auxiliaries en verger de pommiers* (Ed. Brader, L.). Wageningen IOBC, WPRS, Brochure No. 3, pp. 123–124.

Sunderland, K. D., Crook, N. E., Stacey, D. L. & Fuller, B. J. (1987). A study of feeding by polyphagous predators on cereal aphids using ELISA and gut dissection. *Journal of Applied Ecology*, 24: 907–933.

Szentkirlyi, F. (2001). Lacewings in fruit and nut crops. In: *Lacewings in the Crop Environment* (Ed. McEwen, P. K., New, T. R. & Whittington, A. E.). Cambridge University Press, Cambridge, UK, pp. 172–238.

Tauber, C. A., Tauber, M. J. & Milbrath, L. R. (1995). Individual repeatability and geographical variation in the larval behaviour of the generalist predator, *Chrysopa quadripunctata*. *Animal Behaviour*, 50: 1391–1403.

Thakur, B. & Parikh, P. H. (2016). A study on diversity and abundance of coleopterans in Gujarat, India. *Journal of Entomology and Zoology Studies*, 4(5):1082–1089.

Tischler, W. (1958). Synkologische Untersuchungen an der Fauna der Felder und Feldgehlze. *Zeitschrift fr Morphologie und kologieTiere*, 47: 54–114.

Tiwari, S., Maurya, R. P. & Pandey, A. K. (2017) Effect of different insect hosts on biology and predation efficiency of *Eocanthecona furcellata* (Wolff) (Hemiptera: Pentatomidae). *Bioscan*, 12(1):193–197.

Tomson, M., Sahayaraj, K., Kumar, V., Avery, P. B., McKenzie, C. L. & Osborne, L. S. (2017). Mass rearing and augmentative biological control evaluation of *Rhynocoris fuscipes* (Hemiptera: Reduviidae) against multiple pests of cotton. *Pest Management Science*, 73(8):1743–1752.

Toth, P., Tancik. J. I., Tthov, M. & Pauta, V. (2006). Distribution, host plants and natural enemies of sugar beet root aphid (*Pemphigus fuscicornis*) in Slovakia. *Proceedings for Natural Sciences*, 110: 221–226.

Toth, P., Tthov, M. & Tancik. J. (2004). First records of *Pemphigus fuscicornis* (Homoptera, Pemphigidae) from Slovakia. *Biologia, Bratislava,* 59: 271–272.

Trdan, J., Andjus, L., Raspudi, E. and Ka, M. (2005). Distributon of *Aeolothrips intermedius* Bagnall (Thysanoptera: Aeolothripidae) and its prey Thysanoptera species on different cultivated host plants. *Pesticide Science,* 78: 217–226.

Tremewan, W. G. (1985). Zygaenidae. In: *The Moths and Butterflies of Great Britain and Ireland* (Ed. Heath, J.). Harley Books, Colchester, UK, pp. 74–123.

van Frankenhuizen, A., Stigter, H. & Dickler, E. (2002). *Schdliche und ntzlicheInsekten und Milben an Kern und Steinobst in Mitteleuropa.* Ulmer, Stuttgart, Germany, 288p.

Waghmare, S. & Gaikwad, S. (2017). First record of the predatory stinkbug *Eocanthecona concinna* (Walker, 1867) (Pentatomidae: Asopinae) from India. *Journal of Threatened Taxa*, 9(2): 9870–9873.

Wallin, H. (1989). Habitat selection, reproduction and survival of two small carabid species on arable land: a comparison between *Trechussecalis* and *Bembidion lampros*. *Holartctic Ecology*, 12: 193–200.

Warner, D. J., Allen-Williams, I. J., Ferguson, A. W. & Williams, I. H. (2000). Pest predator spatial relationships in winter rape: implications for integrated crop management. *Pest Management Science*, 56: 977–982.

Weseloh, R. M. (1985). Predation by *Calosoma sycophanta* L. (Coleoptera: Carabidae): evidence for a large impact on gypsy moth, *Lymantria dispar* L. (Lepidoptera: Lymantriidae), pupae. *The Canadian Entomologist*, 117: 1117–1126.

Weseloh, R. M. (2003). People and the gypsy moth: a story of human interactions with an invasive species. *American Entomologist*, 49: 180–190.

Wheeler Jr, A. G. (2000). Predacious plant bugs (Miridae). In: *Heteroptera of Economic Importance* (Ed. Schaefer, C. W. & Panizzi, A. R.), CRC Press, Boca Raton, FL, pp. 657–693.

Williams, I. H., Ferguson, A. W., Kruss, M., et al. (2010). Ground beetles as predators of oilseed rape pests: incidence, spatio-temporal distributions and feeding. In: *Biocontrol-Based Integrated Management of Oilseed Rape Pests* (Ed. Williams, I. H.). Springer, Dordrecht, The Netherlands, pp. 115–149.

Wilson, L. J., Bauer, L. J. & Walter, G. H. (1996). -Phytophagous thrips are facultative predators of two spotted spider mites (Acari: Tetranychidae) on cotton in Australia. *Bulletin of Entomological Research*, 86: 297–305.

Wojciechowicz-Zytko, E. & Wnuk, A. (2012). The occurrence of Syrphidae in Aphis fabae Scop (Hemiptera) colonies on broad bean intercropped with Phacelia (part II). *Journal of Plant Protection Research*, 52: 196–201.

Wood, G. C. (1981). Asilidae. In: *Manual of Nearctic Diptera*. Volume 1. Monograph No. 27 (Ed. McAlpine, J. F., Peterson, G. E. & Shewell, B. V.). Research Branch Agriculture Canada, Ottawa, Canada, pp. 549–573.

Wright, B. (1994). Know your friends: minute pirate bugs. *Midwest Biological Control News Online*, University of Nebraska-Lincoln. https://edis.ifas.ufl.edu/publication/IN1148 [accessed 06 May 2021].

Wright, D. W., Hughes, R. D. & Worrall, J. (1960). The effect of certain predators on the number of cabbage root fly (*Erioischia brassicae* (Bouch) and on the subsequent damage caused by the pest. *Annals of Applied Biology*, 48: 756–763.

Wyss, E., Villiger, M., Hemptinne, J. L. & Muller-Scharer, H. (1999). Effects of augmentative releases of eggs and larvae of the ladybird beetle, *Adalia bipunctata*, on the abundance of the rosy apple aphid, *Dysaphis plantaginea*, in organic apple orchards. *Entomologia Experimentalis et Applicata*, 90: 167–173.

Zibaee, A., Ramzi, S. & Hoda, H. (2020). The stink bug, *Andrallus spinidens* (Hemiptera: Pentatomidae), a potential predator for effective biological control. *Journal of Plant Protection Research*, 60(4): 336–344.

2 Reduviid Predators

S. N. Bhagyasree, Archana Anokhe, and P. R. Shashank
ICAR-Indian Agricultural Research Institute

C. H. Patel
Lovely Professional University

CONTENTS

LEARNING OBJECTIVES

1. To provide an insight on generalist heteropteran predators and their unique advantages that a specialist might not provide.
2. To promote exploration, taxonomy and conservation efforts of heteropteran predators, with special reference to predatory clade of Reduviidae.
3. To promote utilisation of assassin bugs in augmentative biocontrol in the field and protected Agri-Horti Ecosystem similar to Coccinellidae and Chrysopidae.
4. To bring mass rearing of assassin bugs in practice, making them commercially available.
5. To develop the habitat manipulating technique to enhance/conserve the assassin bugs to incorporate in the Integrated Pest Management (IPM) module.

DOI: 10.1201/9781003370864-2

2.1 INTRODUCTION

Traditionally, biocontrol of insect pests by predators has been predominantly referred to only in Coccinellidae and Chrysopidae for augmentation, whereas other predatory groups of insects, like heteropterans, are highly ignored due to poor understanding on their life history attributes and ecology. In addition to that, problems associated with measuring the biological control provided by the generalist heteropteran predators are also partially responsible for limited research in this area (Stanley and Gregg, 1994). In the complex agroecosystem, where the main/key pest is obsolescent after suppression or is characterised by migratory behaviour, heteropterans, being generalist predators, have a survival advantage over host-specific parasitoids or predators by being able to switch to secondary prey during the absence of main pest species.

A few members of the suborder Heteroptera are entirely predaceous, which include Reduviidae, Nabidae, Geocoridae (presently Geocorinae) and subfamily Asopinae of the family Pentatomidae. Among them, Reduviidae, commonly known as assassin bugs, are found to be promising ones with great abundance and diversity both in natural and agroecosystems. This is the second most megadiverse family in Heteroptera after Miridae and one of the largest predatory clades in the animal kingdom with 7,000 described species under 26 subfamilies (Weirauch et al., 2014) in the world. In India, it consists of 459 species under 146 genera and 14 subfamilies (Centrocnemidinae, Ectrichodinae, Peiratinae, Reduvinae, Salyavatinae, Triatominae, Harpactorinae, Phymatinae, Physoderinae, Emesinae, Saicinae, Holoptilinae, Stenopodinae and Tribelocephalinae); all of them are predaceous except the members of the subfamily Triatominae, which are blood-sucking species. Reduviids are identified by elongated head, large compound eyes, distinct narrowed neck behind postocular region and three curved, short and stout segmented labia, which fit into a striated prosternal groove (Figure 2.1). Assassin bugs are either specialised (stenophagy) on certain groups of prey

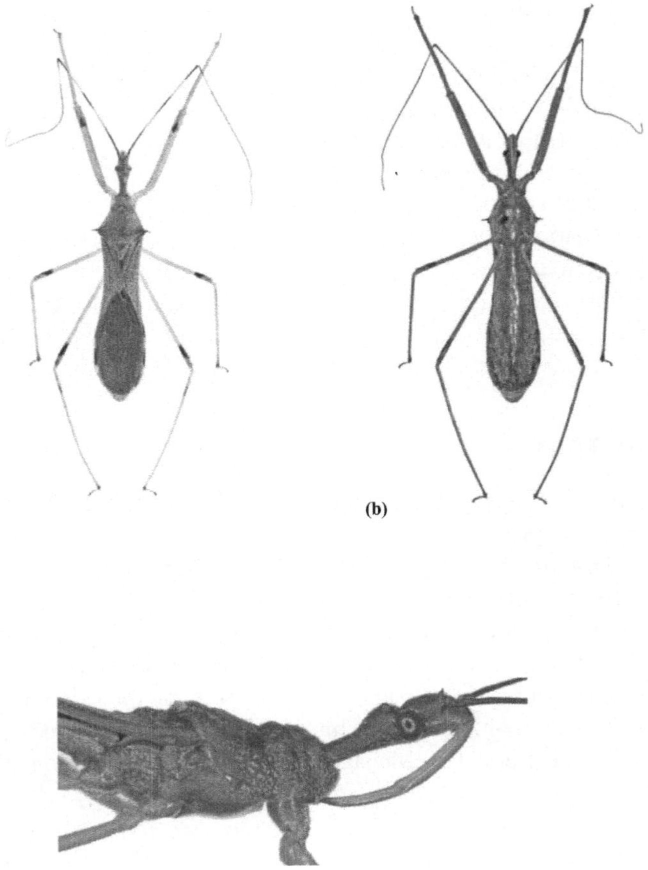

(a) (b) (c)

organisms or relatively unspecialised (euryphagy). In India, considerable amount of information has been generated by Dr. Dunston P. Ambrose and his co-workers on prey records, biology, feeding behaviour, host and stage preference, insecticidal impact and augmentation of potential ones in small-scale trials. Although the basic work has been addressed, reduviids are not yet commercially recommended due to unavailability of sound and low-cost rearing technique, mass multiplication protocol and taxonomic confusion.

2.2 TAXONOMY

Reduviids provide a major ecosystem service as generalist predators against a wide array of insect pests. To utilise assassin bugs in biocontrol, proper taxonomic identification is very important. Despite serious taxonomic and phylogenetic efforts made during the past two-and-half centuries, the classification of Reduviidae is far from being settled and numerous taxa at the level of genus, tribe or subfamily are in need of modern systematic revision.

Globally, intense taxonomic work in the family Reduviidae was done in 19th and 20th centuries by the contribution of Stål, Distant and Breddin. Fabricius (1775) described *Reduvius personatus* as the first species in Reduviidae; later, Latreille (1807) described the family Reduviidae to indicate a group above the genus level. Amyot and Serville (1843) first classified the 'Reduvides' as a tribe in the Cimicidae. They considered the Ectrichodides, Emesides, Harpactorides, Peiratides, Phymatides, Salyavatides, Sphaeridopides and Stenopides to be groups of equal rank. Stål (1874) placed these groups as subfamilies in the family Reduviidae. Four taxonomists, who together contributed more than 40% of all valid species in Reduviidae, were Miller, Schouteden, Villiers and Wygodzinsky. Higher-level classification of Assassin bugs was settled by Amyot and Serville, who identified many subfamilies. Published phylogenies using morphological (Davis, 1957, 1961; Weirauch, 2008), molecular (Weirauch and Munro, 2009; Hwang and Weirauch, 2012) and combined datasets (Schuh et al., 2009) show high support for the monophyly of the family Reduviidae, with Pachynomidae as their sister group.

In India, earliest work on the fauna of Reduviidae was by Distant (1904, 1910) in the monumental 'Fauna of British India' series. He described 298 species under 106 genera in 13 subfamilies from the Indian faunal limits. Maldonado catalogued the world Reduviidae in which he recognised 23 subfamilies, 912 genera and 6,224 species (Maldonado, 1990). A few decades ago, Livingstone, Ravichandran, Murugan and Ambrose described and documented many species from India. Most of their studies were mainly based on the external colouration and morphology. Ambrose (2006) listed 464 species under 144 genera and 14 subfamilies (excluding Phymatinae). Though claimed as from India, he also included species from Indochina and Cochinchina, which has increased the species number. Recently, Bhagyasree (2017) revised the checklist and documented 459 species under 146 genera and 14 subfamilies from India excluding species from Indochina and Cochinchina and also included the excluded subfamily Phymatinae. Apart from this, taxonomic problems or poor taxonomy is creating a gap between the fundamental aspects and applied biology, especially augmentative biocontrol in India. Since two decades, by realising the importance of assassin bugs in applied aspects, Dr. Christiane Weirauch, University of California; Dr. M. B. Malipatil, Department of Environment and Primary Industries Victoria, Australia; Dr. Dimitri Forero, Pontificia Universidad Javeriana, Colombia; Wanzhi Cai, China Agricultural University; and David Redei, Hungarian Natural History Museum were working intensively on settling these bugs in basic aspects, like taxonomy, phylogeny, predatory behaviour, morphology with functional dynamics and distribution to bridge the gap between applied and basic aspects of reduviids (Bhagyasree, 2017).

2.3 DIVERSITY OF REDUVIIDS

Analysis of the diversity by Ambrose (1999) reveals that the distribution and diversity of reduviids in relation to their microhabitat are well pronounced. For instance, out of 370 commonly known

Indian species, microhabitat is known for 238 species only. Among them, 82 (34%) species of reduviids live exclusively under boulders, 30 (13%) species on shrubs, 14 species (6%) under bark and 8 species (3%) in litter. Diversity in adaptive ecotypism and polymorphism of *Rhynocoris marginatus* (Fabricius) has been studied in relation to biocontrol. The niger (black) morphs were more diverse and efficient in predation with insecticidal resistance and good biological attributes than the sanguineous (red) and nigrosanguineous (black and red combined) morphs (Ambrose, 1999). Bhagyasree (2017) analysed the diversity of various species of the subfamily Harpactorinae across various microhabitats in south India, namely, foliage, bark, boulders and litter using Shannon–Weiner Index and revealed that, among the various microhabitats, foliage dwelling bugs were the most diverse (H-2.66), followed by bark (H-1.27), leaf litter (H-1.15) and the least in boulders (H-0.89). Foliage dwelling as a lifestyle, in contrast to diversity, occurs in more derived clades (e.g. *Brassavola, Euagorus, Irantha, Isyndus, Macracanthopsis, Occamus, Panthous, Polididus, Rhynocoris, Rhirbus, Sphedanolestes, Serendib, Sycanus* and *Vesbius*). It shows the scope of using Harpactorinae in conservative and augmentative biological control.

2.4 BIOLOGY AND BEHAVIOUR

In India, compared to any other aspect of the family Reduviidae, biology is very well studied, as it is highly evitable for utilising bugs in conservation and augmentative biological control. Nearly 100 research articles addressed the biology, *viz*., fecundity, sex ratio, per cent hatching, eclosion, per cent survival, number of instars, incubation period of eggs, nymphal period, camouflage, longevity of adult male and female, preoviposition period, oviposition period and feeding potential. In general, assassin bugs have three stages, i.e., egg, five nymphal instars and adult stage.

Reduviids lay 150–400 elongated eggs in clusters of 8–15 on twigs and barks of plants. Egg laying site, fecundity, survival and patterns greatly vary with the dwelling location of the bugs. Assassin bugs have five nymphal instars. Newly hatched first nymphal instars are fragile, their cuticle gradually becomes hard after few hours of emergence and they start feeding on soft-bodied insects and neonate larvae. Mortality of first instar nymphs is very high due to their fragile nature, inability to feed and lack of intestinal content of dead and parental bugs. Nymphal instars can be easily identified by long spider-like legs and slender body. Early instar reduviids exhibit crowding behaviour and disperse as they grow. Fifth instar has the longest developmental period with high feeding potential. The total developmental period from eggs to adult varies across the species. It ranges from 73.58±2.12 days for *Sycanus collaris* (Fabricius) (Rajan et al., 2017); 36.8±0.6 days for *Euagoras plagiatus* Burmeister; 57.5±1.8 days for *Endochus albomaculatus* Sta°l; 43±0.2 days for *Epidaus bicolor* Distant; 52.6±0.4 days for *Irantha armipes* Sta°l; 53.3±3.4 days for *Panthous bimaculatus* Distant and 44.6±0.7 days for *Sphedanolestes signatus* Distant (Srikumar et al., 2014).

Sex ratio: Bugs have both male- and female-biased sex ratio; in few species, like *I. armipes, S. signatus* and *P. bimaculatus*, female-biased sex ratio was observed, whereas in *E. Plagiatus, E. bicolor and E albomaculatus* male-biased sex ratio was observed. As females are long lived, they indulge in multiple matings with males of different ages, and male-biased sex ratio is prevalent in assassin bugs (Ambrose, 1999, 2007; Srikumar et al., 2014).

Feeding behaviour: Reduviids exhibited a 'pin and jab' kind of predation in a sequence of acts, *viz*., arousal, approaching, prey capture with their raptorial legs, thrust/piercing of labium, paralysing/tranquilising the prey and haemolymph sucking. During arousal, prey motion is an important stimulus. Sensory hairs on the forelegs, antennae and tibial pads help in the perception of kairamones and allomones while capturing the prey. Predators hold the prey with their forelegs, pin and jab on the dorsal side of the host with their elongated labium and paralyse by injecting the venom. After paralysing, the bugs drag the prey to a suitable and safe place and suck the sap (Ambrose 1999; Putchkova, 1979; Rani and Wakamura, 1993).

Mating behaviour: Assassin bugs are polygamous and polyandrous. The act of mating in reduviids involves arousal and approach, riding over and copulation. After sighting the female bugs of their species, virgin males get aroused instantly and exhibit excitation patterns and approach females with their antennae, extended labium and place their legs over the female, which is termed as arousal and approach. After successful approach, males clasp the female bugs and press the pterothorax region with labial palps and ride dorsoventrally over females for longer time compared to other behaviours. After culminating the riding, males stretch their phallus out and establish connection with female genitals and become motionless (Srikumar et al., 2014; Ambrose 1999; Das and Ambrose, 2008).

2.5 UNIQUE FEATURES OF ASSASSIN BUGS IN FAVOUR OF BIOCONTROL

Members of reduviids exhibit all desirable characteristic features of predators, like positive functional and numerical response, good host searching efficiency, multiply faster with high fecundity, high pest suppression efficiency, amenable for mass rearing in laboratory, resistant to insecticides, high adaptability to fluctuating and new climatic conditions and they are completely free from natural enemies. In addition to that, reduviids being evolved as predators from their phytophagous ancestors during the course of evolution are also evolved with additional features, which facilitate the biocontrol, which includes:

a. **Morphological adaptations, i.e., advanced true raptorial forelegs**: Forelegs of reduviids show apparently a unique modification in relation to prey type and strategies involved in prey capture. Two types of legs are noticed in Reduviidae: (1) tibiarolate legs and (2) non-tibiarolate legs. Primitive tibiarolate legs with plesiomorphic fossula spongiosa (tibial pad) are present in the subfamily Reduviinae, Triatominae, Centrocnemidinae, Peiratinae, Ectrichodinae and Salyavationae for effective grabbing and grasping of the wide array of prey. Advanced non-tibiaroliate legs, like chelated forelegs in Phymatinae, are used for motionless catching; sticky trap in Harpactorinae (i.e. apex of the forelegs are coated with resin) is used for entrapping soft-bodied insects; and legs with spine as in Stenopodainae are used for gripping prey. Their examples are given in Table 2.1 and Figure 2.2.

TABLE 2.1
Leg Pattern Observed in Different South Indian Genera of Reduviidae

	Leg Type	Subfamily	Genera
		Centrocnemidinae	*Centrocnemis* sp. and Paracentrocnemis sp.
		Ectrichodiinae	*Eriximachus* sp., *Haematorrhophus* sp., *Scadra* sp. and *Vilius* sp.
		Peiratinae	*Androclus* sp., *Catamiarus* sp., *Ectomocoris* sp., *Pirates* sp. and *Sirthenea* sp.
A	**Tibiaroliate leg**	Reduvinae	*Acanthaspis* sp., *Edocla* sp., *Reduvius* sp. and *Pasira* sp.
		Salyavatinae	*Lisarda* sp.
		Triatominae	*Linshcosteus* sp. and *Triatoma* sp.
B	**Non-tibiaroliate leg**		
1	Sticky trap type	Harpactorinae	*Brassavola* sp., *Sycanus* sp. and *Vesbius* sp.
2	Chelated/subchelated	Phymatinae	*Carcinocoris* sp., *Amblythyreus* sp. and *Cnizocoris* sp.
3	Raptorial type	Emesinae	*Stenolemus* sp., *Bagauda* sp., *Schidium* sp. and *Empicoris* sp.
		Saicinae	*Gallobelgicus* sp. and *Polytoxus* sp.
4	Feathery leg	Holoptilinae	*Holoptilus* sp. and *Ptilocerus* sp.
5	Legs with spines	Stenopodinae	*Aulacogenia* sp., *Canthesaneus* sp., *Oncocephalus* sp., *Pygolampis* sp., *Staccia* sp. and *Thodelmus* sp.

Source: Bhagyasree and Khan (2016).

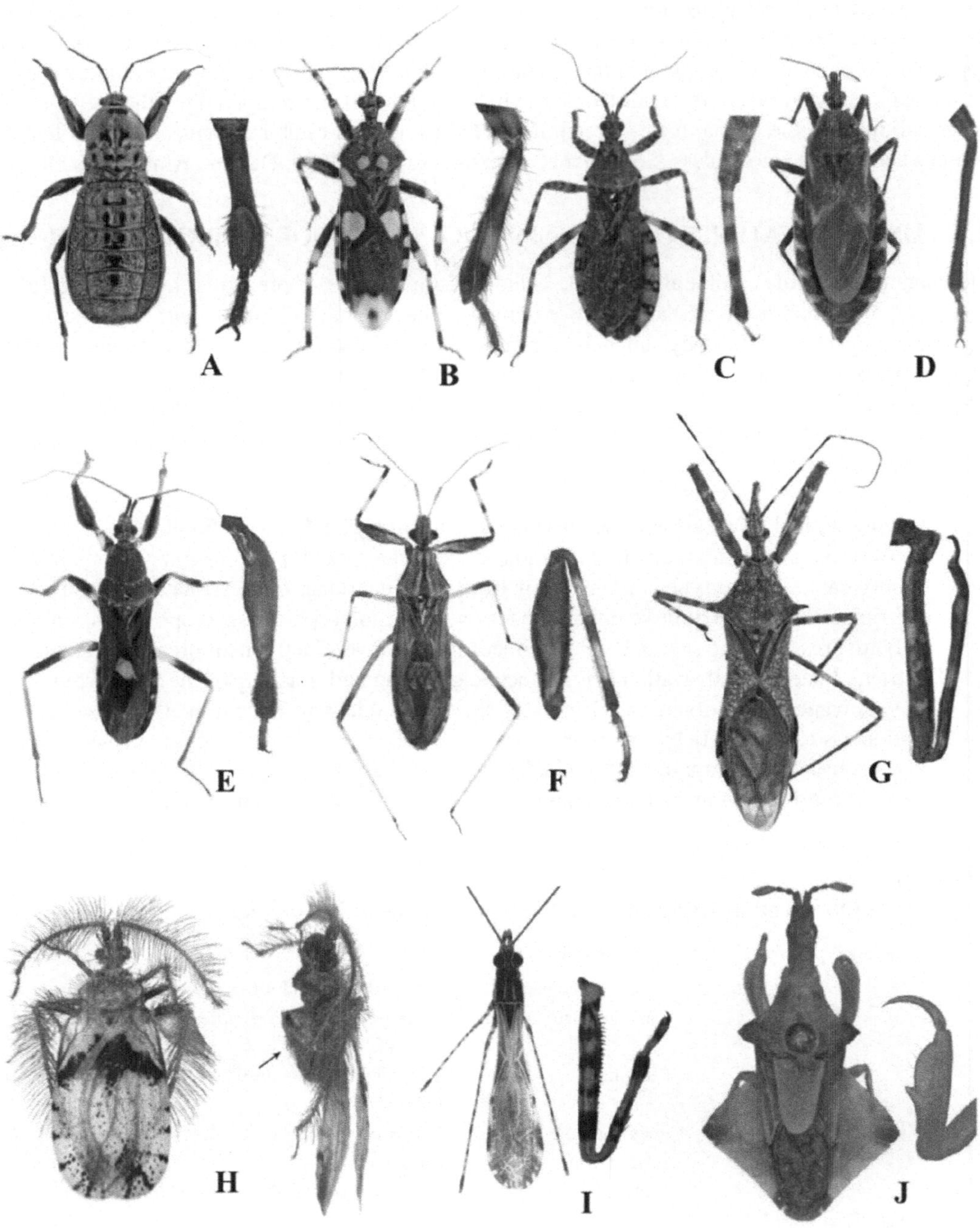

b. **Innovative prey capturing strategies**: Reduviidae are diversified in prey choice and have developed a wide repertoire of innovative prey capture strategies. For example, some Emesinae (thread-legged bugs) cut through webs to reach their psocids and spiders through aggressive mimicry (Soley et al., 2011). Members of the Harpactorinae coat their forelegs with plant resins for prey capture (Forero et al., 2011) and the Holoptilinae, feather-legged bugs, attract ants to imbibe their paralysing secretions of their trichomes before killing their prey (Jacobson, 1911).
c. **Venomous saliva:** Presence of disulphide-rich peptides, bioactive phospholipids, small molecules, such as N, N-dimethylaniline and 1,2,5-trithiepane, and toxic enzymes, such

as phospholipase A2, is identified in the reduviid venom (Walker et al., 2016). Saliva of reduviids is relatively homologous to the calcium channel blockers ω –conotoxins, which are analogous to venom of snakes, spiders and marine cone snails. So, immediately after piercing, prey gets paralysed and dies after sometime even though they escape. In this way, they are known to kill more prey than they actually consume (Corzo et al., 2001).

d. **Behaviour:** As the bugs are armed with strong forelegs and poisonous saliva, they are habituated to attack the prey larger than their body size (Ambrose, 1999).

2.6 POTENTIAL ASSASSIN BUGS FOR AUGMENTATIVE BIOCONTROL

In India, during 1979 itself, *Platymeris laevicollis* Distant set an example for utilising assassin bugs as classical biocontrol agents. Antony et al. (1979) released the laboratory reared *P. laevicollis* in the coconut ecosystem in Kerala, Lakshadweep and Karnataka for the management of Rhinoceros Beetle, *Oryctes rhinoceros* Linnaeus, and they found the colonisation of bugs and the suppression of Rhinoceros beetles.

Functional response studies of reduviids, *viz.*, *Sycanus indagator* Stål (Bass and Shepard, 1974); *Zelus renardii* Kolenati (Ables, 1978); *Rhynocoris fuscipes* (Fabricius), *R. marginatus* and *Rhynocoris kumara* Ambrose and Livingstone (Ambrose and Kumaraswami, 1990); and *Coranus obscurus* (Kumar, 1993) revealed their prey suppression efficiency, and functional response confirmed the type II model of Holling. High consumption, increased searching capacity, decreased handling time, high fecundity and hatchability were observed with increasing prey density. Prey record of the reduviids is large and diverse; nearly 9–10 subfamilies that have been observed to feed on various types of arthropods abundantly in the Indian ecosystem are discussed below.

2.6.1 Subfamily Harpactorinae (Figure 2.3a)

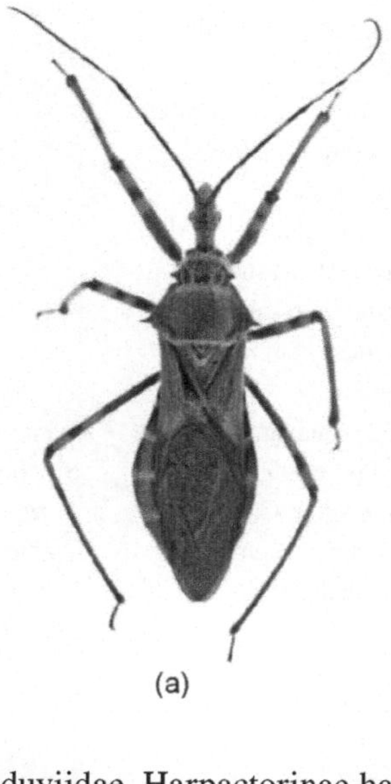

(a)

Among all the 14 subfamilies of Reduviidae, Harpactorinae has been highly studied for biological control. Harpactorinae, the largest subfamily, consists of more than 89 species with cosmopolitan distribution; they can be easily identified by cylindrical head, strikingly elongate scape, hemelytron with quadrate cell formed by anterior and posterior cross vein between Cu and Pcu, fore tarsi two segmented and pedicel of articulatory apparatus short or reduced. These bugs consist predominantly of generalist predators, which are amenable for mass production. They are found in many terrestrial ecosystems and microhabitats, ranging from boulders to foliage. They are being successfully explored in an integrated pest management system, which includes *Pristhesancus plagipennis* (Walker) against cotton bollworm (Grundy and Maelzer, 2000); *Creontiades* spp. (Miridae) (Grundy, 2007), *Zelus* and *Sinea* as generalist predators against bugs, caterpillars and boll weevils (Cogni et al., 2002); *Endochus albomaculatus* Stål, *Epidaus bicolor* Distant, *Euagoras plagiatus* Burmeister, *Irantha armipes* Stål, *Panthous bimaculatus* Distant and *Sphedanolestes signatus* Distant against tea mosquito bugs; and *Helopeltis antonii* Signoret in Cashew (Srikumar et al., 2014). Apart from these, a few more promising candidates for biocontrol are given in Table 2.2.

TABLE 2.2
Most Promising Indian Assassin Bugs for Biological Control

Sl No.	Scientific Name	Distribution	Prey Record	References
1.	*Brassoivola hystrix* Distant 1904	Karnataka: Chikkaballapura (Distant 1919).	*Achaea janata* L., *Dysdercus koenigii* (F.), *Lipaphis erysimi* Kalt., *Macrotermes estherae* (Desneux), *Sylepta derogata* (F.)	Lakkundi (1989)
2.	*Coranus siva* Kirkaldy, 1891	Jharkhand: Ranchi. Kolkatta. Meghalaya: North Khasi hills, Maharastra: Bor ghat. Nagaland: Naga hills. Sikkim (Distant, 1904), Tamil Nadu: Chandrapuram, Malumichampatti, Manimuthar, Pilavaikal (Ambrose, 2006), Srivilliputhur (Ambrose and Vennison, 1989), Sunkankadai, Palayankottai, Pilavaikal, Tirunelveli. Kerala: Walayar, Wayanad (Ambrose, 2006).	*Helicoverpa armigera* (Hübner), *Leptocorisa varicornis* Fabricius, *Spodoptera litura* (F.) and *Utetheisa pulchella* L.	Ambrose and Claver (2001); Ambrose (2003)
3.	*Cydnocoris gilvus* (Burmeister 1837)	Kerala: Walayar, Vandiperiya, Thenmalai. Tamil Nadu: Courtallam, Keeriparai, Pechiparai (Ambrose, 2006).	*Helopeltis antonii* Signoret. and *Pyrausta machaeralis* Walker.	Patil and Thontadarya (1983); Srikumar et al. (2014)

(*Continued*)

TABLE 2.2 (*Continued*)
Most Promising Indian Assassin Bugs for Biological Control

Sl No.	Scientific Name	Distribution	Prey Record	References
4.	*Irantha armipes* (Stål, 1855)	Sikkim (Distant, 1910). Kerala: Pandippara, Panthanthodu, Thuthampara-Pothumala, Thuthampara-Pothumala, Wayanad hills, Walayar, Tamil Nadu: Alamarathupatti, Azhagarkoli, Courtallam, Godayar, Kalikesam, Kolipatti, Pechiparai, Sunkankadai, Thanniparai, Vandavasi (Ambrose, 2006).	*H. antonii* Signoret.	Sundararaju (1984); Ambrose (2003)
5.	*Isyndus heros heros* (Fabricius, 1803)	Meghalaya: North Khasi Hills, Maharastra: Bor Ghat, Karnataka: Bengaluru (Distant, 1904).	Dysdercus *koenigii* (Fabricius), E. *insulana*, *H. armigera* and *S. litura*.	Ambrose (2007).
6.	*Rhynocoris fuscipes* (Fabricius, 1787)	Karnataka: Chikkamagaluru (Usman and Puttarudraiah, 1955). Maharashtra: Bor Ghat. Mumbai (Distant, 1904). Kerala: Mannar (Miller,1959). Tamil Nadu: Aliyar, Aralvaymozhi, Arockia natha puram, Azhagar Koli, Chandrapuram, Coimbatore, Donavoor, Killikulam, Marunthuvazhmalai, Melapattam, Muthur, Muppanthal, Nanguneri, Pabanasam, Pechiparai; Reddiarpatti, Sawyerpuram, Sivakasi, Sivanthipatti, Thiruthangal, Vagaikulam, Vandaloor, Veerakeralamputhur, Yalagiri hills (Ambrose, 2006).	*Calocoris angustatus* Leth., *Dicladispa armigera* (Oliver), *H. armigera*, *H. puera*, *Myllocerus curvicoronis* F., *Myzus persicae* (Sulz.), *N. viridula* (L.) var. smaragdula (F.), *S. litura*.	Rao (1974); Mohanadas (1996); Hiremath and Thontadarya (1983); Ponnamma et al. (1979); Nagarkatti (1982); Singh (1985); Singh and Singh (1987); Sathiamma et al. (1999).

(*Continued*)

TABLE 2.2 (*Continued*)
Most Promising Indian Assassin Bugs for Biological Control

Sl No.	Scientific Name	Distribution	Prey Record	References
7.	*Rhynocoris marginatus* (Fabricius, 1794)	Andra pradesh: Visakhapatnam. North India (Distant, 1904). Tamil Nadu: Aliyar, Chandrapuram, Courtallam, Kangeyampalayam, Kovaipudur, Manimuthar, Kovilpatti, Manimuthar, Mathukarai, Melapattam, Muthur, Pabanasam, Palayankottai, Reddiarpatti, Sivanthipatti, Thiruvannamalai (Ambrose, 2006).	*Dysdercus cingulatus* (Fabricius), *Exelastis atomosa* (Walsingham), Earias insulana (Boisd.), *Earias vittella* (Fab), *H. armigera*, Mylabris *indica* (Thunberg), *Mylabris pustulata* (Thunberg), *Oxycarenus hyalinipennis* (Costa), *Pectinophora gossypiella* (Saunders), *S. litura*.	Nayar et al. (1976); Ambrose (1980, 1999); Bhatnagar et al. (1983); Pawar et al. (1986); Kumaraswami (1991); Ambrose (2003, 2007)
8.	*Sycanus affinis affinis* Reuter, 1881	Assam: Margherita, Sadiya. Kerala: Malabar, Thiruvananthapuram. West Bengal: Darjeeling (Distant, 1904).	*Acherontia styx* Westwood, *Amsacta albistriga* Walker, *Opisina arenosella* Walker, *Papilio demoleus* Linnaeus, S. litura and *Sesamia inferens* Walker.	Satpathy et al. (1975); Haines (1991)
9.	*Sycanus collaris* (Fabricius 1785)	Bengal (Distant 1910). Meghalaya: North Khasi Hills (Distant, 1904).	*Spodoptera litura* (Fabricius).	Rajan et al. (2017).

2.6.2 Subfamily Reduviinae (Figure 2.3b)

(b)

Most of the species are generalist predators with cosmopolitan distribution. They are nocturnal and found to be associated with bark of trees and some species of Genus *Acanthaspis* are known to feed on bees. Bugs are primarily recognised by the absence of characteristics associated with other subfamilies of Reduviidae, such as oblong body, very large and prominent eyes and ocelli, pronotum broadly impressed and sulcate and fossula spongiosa very well developed in fore- and midlegs (Schuh and Slater, 1995).

2.6.3 Subfamily Ectrichodiinae (Figure 2.3c)

(c)

Ectrichodiinae are nocturnal and have a cosmopolitan distribution. They are specialised to feed on millipedes. Sexual dimorphism is very common in this group. Females often are with reduced wings and males are fully winged. These bugs can be easily identified by their robust built body with red and black colouration, labrum subdivided by transverse membrane, ventrally notched labial segment II, scutellum with two short posteriorly projecting prongs and corium restricted to proximal part of wings; female eight tergum is reduced to narrow strap and vertically oriented.

2.6.4 Subfamily Holoptilinae (Figure 2.3d)

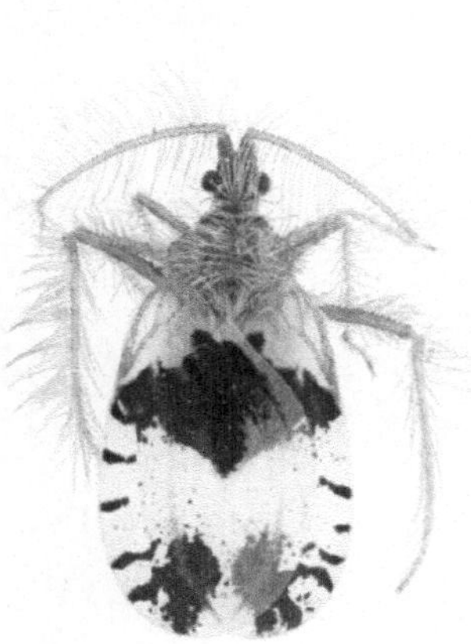

(d)

Holoptilinae are commonly called as feather-legged bugs with southern Palearctic distribution, specialised to feed on ants. They are known to tranquilise their prey by the secretion of trichomes present on the undersurface of their body. Feather-legged bugs are typically identified by their depressed body. They are densely covered with thick, rigid, serrated and long setae, particularly on antennae and hind tibiae. On the ventral abdominal base, they have specialised trichome. In hemelytra, corium is restricted to one-fourth of the extreme base of hemelytra.

2.6.5 Subfamily Peiratinae (Figure 2.3e)

(e)

Peiratinae have a cosmopolitan distribution, appear to be primarily ground dwelling and feed on a variety of ground-dwelling arthropods. They can be identified by antennae with the scape shorter than the pedicel. On the pronotum, the anterior pronotal lobe is longer than the posterior lobe, and transverse sulcus on the pronotum is situated before the middle. Forecoxa is greatly swollen, and the scutellum is acuminate or spinously produced.

2.6.6 Subfamily Emesinae (Figure 2.3f)

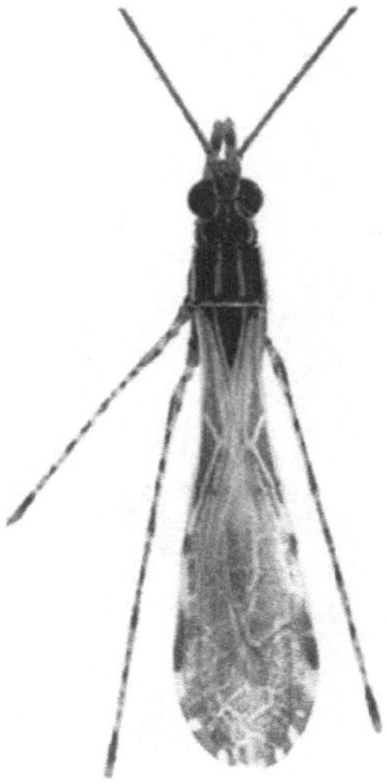

(f)

Emesinae are commonly called as thread-legged bugs, with a cosmopolitan distribution, but their diversity is high in the tropical region. These bugs exhibit aggressive mimicry and are specialised to feed on spiders. They are typically diagnosed by their elongated and slender body, long and thread-like appendages, hence they are called as thread-legged bugs. Ocelli are absent in all Indian species, and labium segment II is ventrally notched; fore acetabulum opens anteriorly, legs are raptorial, claws are greatly modified. Sometimes toothed and triangularly projected, the ejaculatory duct is unilobed.

2.6.7 Subfamily Phymatinae (Figure 2.3g)

(g)

These bugs have a cosmopolitan distribution, commonly called as ambush bugs because of their ambush (hide and attack) nature of predation. Phymatinae have distinct morphology compared to other groups of reduviids. Bugs can be identified by the absence of foretibia and tarsus and greatly enlarged forefemur; connexivum greatly expanded beyond hemelytra and long pedicel.

2.6.8 Subfamily Salyavatinae (Figure 2.3h)

(h)

Salyavatinae are distributed in the oriental region, and they exhibit cryptic colouration (usually brown) and camouflage with the dwelling surface. These bugs are specialised to feed on termites. They can be identified by their relatively small head with short anteocular and postocular region, somewhat globose behind the eyes; antennae inserted on anteriorly directed prominent antenniferous tubercle, basi and disiti flagellomere very short; pronotum, scutellum and abdomen with strongly acuminate spine; fore tarsi two segmented.

2.6.9 Subfamily Stenopodainae (Figure 2.3i)

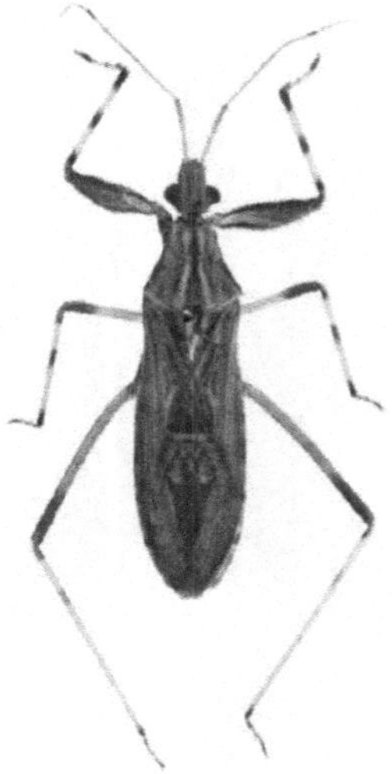

(i)

These bugs are distributed worldwide, dwell in leaf litter or on soil and are known to feed on soil-dwelling insects. These bugs can be identified by their elongated and dull coloured body; legs long and slender; scapo-pedicillar articulation bent, so flagellum and pedicel held underneath head; corium with pentagonal cell formed by cubital and postcubital veins.

2.7 MASS MULTIPLICATION AND FIELD EVALUATION

2.7.1 Mass Multiplication

Lakkundi and Prashad (1987) made the first preliminary attempt for mass rearing of reduviids on refrigerated *Corcyra cephalonica* larvae by pasting the heads on cello tape with zigzag twists in a plastic tube or jars to provide more space for the movement of bugs and to reduce the labour cost for handling the bugs. After this, Sahayaraj (1991) and Kumaraswami (1991) mass reared the bugs with quite advanced techniques; they crushed the head of *C. cephalonica* larvae to prevent their metamorphosis and altered the microhabitat of a rearing plastic trough to provide the required environment for bugs by adding small stones and sand. This method reduced the entangling of reduviids in the *C. cephalonica* webs and reduced the cost of multiplication in comparison to the technique of Lakkundi and Prashad (1987).

Sahayaraj et al. (2006) proposed an innovative idea to mass rear *R. marginatus* on insects and meat-based liquid oligidic diet on different carrier materials, *viz*, cotton, meat, paraffin capsule, foam and cavity slide. Results showed that *R. marginatus* preferred pig liver–based (meat-based) artificial diet for consumption with cotton as the carrying material, but these bugs failed to lay eggs after mating. Employment of insect-based diet prolongs the developmental period and reduces the survival rate.

2.7.2 Field Evaluation

Field cage evaluation studies of *R. fuscipes* on rice hispa (Singh, 1985) and Bihar hairy caterpillar (Singh and Gangrade, 1975); *Acanthaspis pedestris*, *Ectomocoris tibialis* and *Catamiarus brevipennis* on *Helicoverpa armigera* (Sahayaraj, 1991); and *R. kumari* and *R. marginatus* on *H. armigera*, *Dysdercus cingulatus* and *Spodoptera litura* further supported the pest suppression efficiency even at lower density of predators. Grundy (2007) also found the successful suppression of *Helicoverpa* spp. and *Creontiades* spp. (Miridae) by *P. plagipennis* in the cotton ecosystem with low release rate in Australia. After these, Ambrose (1999), Lakkundi (1989) and Schaefer (1988) felt the potential of these bugs and emphasised on understanding and evaluating the reduviids bugs for their biocontrol efficiency. Till date, research work was restricted to laboratory and small-scale release in the experimental field cages. Even though results obtained so far were promising, these predators could be effectively used only after conducting intensive and consistent studies on mass multiplication and after their evaluation in the field conditions.

2.8 CONCLUSIONS

Reduviids provide major ecosystem services as generalist predators against a wide array of insect pests. Considerable amount of information has been generated by Dr. Dunston P. Ambrose and his co-workers on prey record, biology, feeding behaviour, host and stage preference, insecticidal impact and augmentation in small-scale trials. They suggested that reduviids are one of most feasible and potential heteropteran bugs for insect pest supression. And, being a generalist predator, reduviids have survival advantage over host-specific parasitoids or predators by being able to switch their dynamics to secondary prey during the absence of the main pest. Assassin bugs possess all the feasible characteristic features for biocontrol similar to Coccinellidae and Chrysopidae. During the course of evolution from phytophagy to predation, even these bugs were evolved with additional features, like innovative prey capturing strategies, poisonus venom, specialised forelegs with respect to prey type and behaviour of attacking large-sized prey and more numbers of prey. Despite their positivity towards biocontrol, the techonology required for utilisation of reduviids is straightforward and easy. Currently, the reduviids are not commercially available for augmentative release. The basic prerequisite for an effective biocontrol agent is the availability of sound and low-cost rearing technique, and standard procedure for mass multiplication, but it is lacking. Against this backdrop, problems associated with measuring the biocontrol provided by the generalist heteropteran predators are also partially responsible for limited research in this area. In order to fill these identified lacunae for better utilisation of assassin bugs, comprehensive studies in relation to biocontrol is required.

Future Line of Work

i. Survey and screening of potential reduviid bugs for biocontrol of pests.
ii. Studies on suitability of the different prey for mass multiplication and field release.
iii. Evaluation of reduviids for their utilisation in augmentative biocontrol of insect pests in vegetable ecosystem.
iv. Identified reduviids should be made available in IPM modules for various crops, which harbour pest complexes.

2.9 POINTS TO REMEMBER

i. Reduviids are identified by their distinct narrow neck behind the postocular region and three curved, short and stout segmented labia which fit into striated prosternal groove.
ii. Reduviidae is the largest predatory clade in the animal kingdom with 7,000 described species.

iii. Assassin bugs have advanced true raptorial legs, which are highly modified according to the prey they capture. For example, (1) chelate forelegs in Phymatinae, (2) sticky trap in Harpactorinae and (3) feathery in Holoptilinae.
iv. Venom of reduviids is very poisonous. They are calcium channel blockers ω–conotoxins, which are analogous to the venom of snakes, spiders and marine cone snails.
v. Harpactorinae is the biggest subfamily of Reduviidae, and they possess all desirable characteristics of a good biocontrol agent.

ACKNOWLEDGEMENTS

Authors are thankful to Dr. Yeshwanth H M, Research Assistant, for helping in taking and arranging photographs.

REFERENCES

Ables, J. R., 1978. Feeding behaviour of an Assassin bugs *Zelus renardis* Kolenati. *Annual Entomological Society of America*, 71: 476–478.

Ambrose, D. P. and Kumaraswami, N. S., 1990. Functional respone of reduviid predator *Rhinocoris marginatus* Fabr. on the Cotton Stainer *Dysdrecus cingulatus* Fabr. *Journal of Biological Control*, 4: 22–25.

Ambrose, D. P. and Vennison, S. J., 1989. A new species of *Coranus* from south India (Heteroptera, Reduviidae, Harpactorinae). *Journal of Bombay Natural History Society*, 86: 67–71.

Ambrose, D. P., 1980. Bioecology, Ecophysiology and Ethology of Reduviids (Heteroptera) of the Scrub Jungles of Tamil Nadu, India (Ph.D. thesis). University of Madras, Madras, India.

Ambrose, D. P., 1999. *Assassin Bugs*. Oxford & IBH Publ. Co. Pvt. Ltd., New Delhi, India and Science Publishers, Inc., New Hampshire, USA, 337pp.

Ambrose, D. P., 2003. Biocontrol potential of assassin bugs (Hemiptera: Reduviidae). *Indian Journal of Experimental Zoology*, 6(1): 1–44.

Ambrose, D. P., 2006. A checklist of Indian assassin bugs (Insecta: Hemiptera: Reduviidae) with taxonomic status, distribution and diagnostic morphological characteristics. *Zoos' Print Journal*, 21(9): 2388–2406.

Ambrose, D. P., 2007. *The Insects: Beneficial and Harmful Aspects*. Kalyani Publishers, New Delhi, 801pp.

Ambrose, D. P. and Claver, M. A., 2001. Survey of reduviid predators in seven pigeonpea agroecosystems in Tirunelveli, Tamil Nadu, India. *International Chickpea and Pigeonpea Newsletter*, 8: 44–45.

Amyot, C. J. B. and Serville, A., 1843. *Histoire Naturelle des Insectes. Hemipteres*. Libraire Encyclopedique de Roret, Paris: Fain et Thunot, 675pp.

Antony, J., Danial, M., Kurian, C. and Pillai, G. B., 1979. Attempts on introduction and colonization of the exotic reduviid predator *Platymeris laevicollis* Distant for the biological suppression of the coconut Rhinocerous beetle, *Oryctes rhinocerous*. *Plantation Crop Symposium*, 11: 445–454.

Bass, J. A. and Shepard, M., 1974. Predation by *Sycanus indagator* on larvae on *Galleria mellonella* and *Spodoptera frudgiperda*. *Entomologia experimentalis et applicata*, 17: 143–148.

Bhagyasree, S. N., 2017. Taxonomic Studies of Assassin Bugs (Reduviidae: Hemiptera) of South India. Ph. D. Thesis, University of Agricultural Sciences, GKVK, Bengaluru.

Bhagyasree, S. N. and Khan, H. K., 2016. Morphological adaptations of forelegs associated with prey capture in Assassin bugs (Reduviidae: Heteroptera). *Mysore Journal of Agricultural Sciences*, 50(2): 223–228.

Bhatnagar, V. S., Sithanantham, S. S., Pawar, C. S., Jadhav, D., Rao, V. R., Reed, W., 1983. Conservation and augmentation of natural enemies with reference to integrated pest managements in chickpea (*Cicer arietinum* L.) and pigeonpea (*Cajanus cajan* (L.). Millsp.). In: *Proceedings of the International Workshop on Integrated Pest Control for Grain Legumes*, Goiania, Brazil, pp. 157–180.

Cogni, R., Freitas, A. V. L. and Amaral, F. B. F., 2002. Influence of prey size on predation success by *Zelus longipes L.* (Heteroptera: Reduviidae). *Journal of Applied Entomology*, 126: 74–78.

Corzo, G., Adachi-Akahane, S., Nagao, T., Kusui, Y. and Nakajima, T., 2001. Novel peptides from assassin bugs (Hemiptera: Reduviidae): Isolation, chemical and biological characterization. *FEBS Letters*, 499: 256–261.

Das, S. S. M. and D. P. Ambrose. 2008. Rediscription biology and behaviour of the harpatocorine assassin bug *Irantha armipes* (Stal) (Hemiptera: Reduviidae). *Acta Entomologica Slovenia*, 16: 37–56.

Davis, N. T., 1957, Contributions to the morphology and phylogeny of the Reduvioidea (Hemiptera: Heteroptera). Part I. The morphology of the abdomen and genitalia of Phymatinae. *Annual Entomological Society of America*, 50(4): 432–443.

Davis, N. T., 1961, Morphology and phylogeny of the Reduvioidea (Hemiptera: Heteroptera). Part II. Wing venation. *Annual Entomological Society of America*, 54(3): 340–354.

Distant, W. L., 1904, *The Fauna of British India Including Ceylon and Burma. Rhynchota*. Volume 2, Taylor and Francis, London, 503pp.

Distant, W. L., 1910. *The Fauna of British India Including Ceylon and Burma. Rhynchota*. Volume 5. Taylor and Francis, London, pp. 112–362.

Distant, W. L., 1919. Descriptions of new species and genera of the heteropterus family Reduviidae from British India. *Annual Magazine of Natural History*, 9(4): 71–79.

Fabricius, J. C., 1775. Systema entomologiae. *Flensburgi and Lipsiae*, 5: 1–832.

Forero, D., Choe, D. H. and Weirauch, C., 2011. Resin gathering in neotropical resin bugs (Insecta: Hemiptera: Reduviidae): Functional and comparative morphology. *Journal of Morphology*, 272: 204–229.

Grundy, P. and Maelzer, D., 2000. Assessment of *Pristhesancus plagipennis* (Walker) (Hemiptera: Reduviidae) as an augmented biological control in cotton and soybean crops. *Australian Journal of Entomology*, 39: 305–309.

Grundy, P. R., 2007. Utilizing the assassin bug, *Pristhesancus plagipennis* (Hemiptera: Reduviidae), as a biological control agent within an integrated pest management programme for *Helicoverpa* spp. (Lepidoptera: Noctuidae) and *Creontiades* spp. (Hemiptera: Miridae) in cotton. *Bulletin of Entomological Research*, 97: 281–290.

Haines, C.P., 1991. *Insects and Arachnids of Tropical Stored Products: Their Biology and Identification (A Training Manual)*, 2nd ed. Natural Resource Institute, Chatham, UK, 246pp.

Hiremath, I. G. and Thontadarya, T. S., 1983. Natural enemies of sorghum earhead bug *Calocoris angustatus* Lethierry (Hemiptera: Miridae). *Current Research*, 2: 10–11.

Hwang, W. S. and Weirauch, C., 2012. Evolutionary history of assassin bugs (Insecta: Hemiptera: Reduviidae): Insights from divergence dating and ancestral state reconstruction. *PLoS One*, 7(9): e45523. doi:10.1371/journal.pone.0045523

Jacobson, E., 1911. Biological notes on the hemipteron *Ptilocerus ochraceus*. *Tijdschrift voor entomologie*, 54: 175–179.

Kumar, S. P., 1993. Biology and Behaviour of Chosen Assassin Bugs (Insecta: Hemiptera: Reduviidae). Ph.D. thesis, Madurai Kamaraj University, Madurai, India.

Kumaraswami, N. S., 1991. Bioecology and Ethology of Chosen Predatory Bugs and Their Potential in Biological Control. Ph.D. thesis, Madurai Kamaraj University, Madurai, India.

Lakkundi, N. H. and Prashad, B., 1987. A technique for mass multiplication of Predator with sucking type of mouth parts with special reference to reduviids. *Journal of Soil Biology and Ecology*, 7: 65–69.

Lakkundi, N. S., 1989. Assessment of Reduviids for Their Predation and Possibilities of Their Utilization in Biological Control. Ph. D. Thesis, Indian Agricultural Research Institute, New Delhi.

Latreille, P. A., 1807. Genera crustaciorum et insectorum secundum ordnium naturalem in families disposita, inconibus exemplisque plurimis explicate. *Armand Koenin, Paris*, 3: 1–258.

Maldonado, J. C., 1990. *Systematic Catalogue of the Reduviidae of the World (Insecta: Heteroptera). A Special Edition of Caribbean Journal of Science*, University of Puerto Rico, Puerto Rico, 694pp.

Miller, N. C. E., 1959. New genera and species of Old World Reduviidae (Hemiptera - Heteroptera). *Records of the Indian Museum*, 54: 9–22 (1956).

Mohanadas, 1996. New records of some natural enemies of the teak defoliator *Hyblaea puera* Crammer (Lepidoptera: Hybiaeidae) from Kerala, India. *Entomon*, 21: 251–254.

Nagarkatti, S., 1982. The utilization of biological control of *Heliothis* management in India. In: Reed, W., Kumble, V. (Eds.), *Proceedings of International Workshop on Heliothis Management*. ICRISAT, Patencheru, India, pp. 159–167.

Nayar, K. K., Ananthakrishnan, T. N., David, B. V., 1976. *General and Applied Entomology*. Tata McGraw-Hill Publ. Co. Ltd., New Delhi, India, 589pp.

Patil, B. V., Thontadarya, T. S., 1983. Natural enemy complex of the teak skeletonizer *Pyrausta mechaeralis* Walker (Lepidoptera: Pyralididae) in Karnataka. *Entomon*, 8: 249–255.

Pawar, C. S., Bhatnagar, V. S. and Jadhav, D. R., 1986. *Heliothis* species and their natural enemies, with their potential in biological control. *Proceedings of Indian Academy of Science (Animal Sciences)*, 95: 695–703.

Ponnamma, K. N., Kurian, C. and Koya, K. M. A., 1979. Record of *Rhinocoris fuscipes* (Fabr.) (Heteroptera: Reduviidae) as a predator of *Myllocerus curicornis* (F.) (Coleoptera: Curculionidae) the ash weevil, pest of the coconut palm. *Agriculture Research Journal of Kerala*, 17: 91–92.

Putchkova, L. V., 1979. Adaptive features of leg structures in the Heteroptera. *Entomological Review*, 58: 189–196.

Rajan, S. J., Suneetha, N. and Sathish, R., 2017. Biology and predatory behaviour of an assassin bug, *Sycanus collaris* (Fabricius) on rice meal moth, *Corcyra cephalonica* (Stainton) and leaf armyworm, *Spodoptera litura* (Fabricius). *Agricultural Update*, 12: 1181–1186.

Rani, P. U. and Wakamura, S., 1993. Host acceptance behaviour of a predatory pentatomid, *Eocanthecona furcellata* (Wolff.) (Heteroptera: Pentatomidae) towards larvae of *Spodoptera litura* (Lepidoptera: Noctuidae). *Insect Science and Its Application*, 14: 141–147.

Rao, V. P., 1974. Biology and Breeding Techniques for Parasites and Predators of *Ostrina* sp. and Heliothis sp. CIBC Final Technical Report PL 480 project, Bangalore, India, p. 86.

Sahayaraj, K., 1991. Bioecology, Ecophysiology and Ethology of Chosen Predatory Hemipterans and the Potential in Biological Control (Insecta: Heteroptera: Reduviidae). Ph.D. Thesis, Madurai Kamaraj University, Madurai, India.

Sahayaraj, K., Martin, P., Selvaraj, C. and Raju, M., 2006. Feeding behaviour of reduviid predator on meat and insect-based artificial diet. *Belgian Journal of Entomology*, 8: 55–65.

Sathiamma, B., Nair, K. R. C. and Soniya, V. P., 1999. Additional record of insect predators of coconut lace bug *Stephanitis typical* (Distant) and studies on biology and feeding potential of *Euagorus plagiatus* Burmeister (Hemiptera: Reduviidae). *Entomon*, 24: 165–171.

Satpathy, J. M., Patnaik, N. C. and Samalo, A.P., 1975. Observations on the biology and habits of *Sycanus affinis* Reuter (Hemiptera: Reduviidae) and its status as a predator. *Journal of Bombay Natural History Society*, 72: 589–595.

Schaefer, C. W., 1988. Reduviidae (Hemiptera: Heteroptera) as agent of biological control. In: Ananthasubramanian, K. S., Venkatesan, P., Sivaraman, S. (Eds.). *Bicovas.1.* Loyola College, Madras, India, pp. 27–33.

Schuh, R. T. and Slater, J. A., 1995. *True Bugs of the World (Hemiptera: Heteroptera) Classification and Natural History*. Cornell University Press, Ithaca, 336pp.

Schuh, R. T., Weirauch, C. and Wheeler, W. C., 2009. Phylogenetic relationships within the Cimicomorpha (Hemiptera: Heteroptera): A total-evidence analysis. *Systematic Entomology*, 34: 15–48.

Singh, O. P., 1985. New record of Rhynocoris fuscipes Fabr. as a predator of *Dicaldispa armigera* (Oliver). *Agricultural Science Digest*, 5: 179–180.

Singh, O. P. and Gangrade, G. A., 1975. Parasites, predator and disease of larvae of *Diacrisia obligna* Walker (Lepidoptera: Arctidae) on Soybean. *Current Science*, 44: 481–482.

Singh, O. P. and Singh, K. J., 1987. Record of *Rhynocoris fuscipes* Fabricius as a predator of Green stink bug, *Nezara viridula* Linn., infesting soybean in India. *Journal of Biological Control*, 1: 143–146.

Soley, F. G., Jackson, R. R. and Taylor, P. W., 2011. Biology of *Stenolemus giraffe* (Hemiptera: Reduviidae), a web invading, araneophagic assassin bug from Australia. *New Zealand Journal of Zoology*, 38: 297–316.

Srikumar, K. K., Bhat, P. S., Raviprasad, T. N., Vanitha, K., Saroj, P. L. and Ambrose, D. P., 2014. Biology and behavior of six species of Reduviids (Hemiptera: Reduviidae: Harpactorinae) in a cashew ecosystem. *Journal of Agriculture and Urban Entomology*, 30(1): 65–81.

Stål, C., 1874. Enumeratio Reduviidarum Europae, Africae, Asiae, et Australiae. *Enumeratio Hemipterorum. Kongl. Svenska Vetenskaps-Akademiens Handlingar*, 4: 3–97.

Stanley, J. and Gregg, P., 1994. *Helicoverpa* Predator: Do We Know Anything about Them? *Proceedings of Australian Cotton Conference*, Broadbeach, Queensland, Australian Cotton Growers Association, pp. 49–51.

Sundararaju, D., 1984. Cashew pests and their natural enemies in Goa. *Journal of Plant Crops*, 12: 38–46.

Usman, S. and Puttarudraiah, M., 1955. *A List of Insects of Mysore Including the Mites*. Entomology Series Bulletin No. 16, Department of Agriculture, Mysore State, 104pp.

Walker, A. A., Weirauch, C., Bryan, G. F. and King, G. F., 2016. Venoms of heteropteran insects: A treasure trove of diverse pharmacological toolkits. *Toxins*, 1: 32.

Weirauch, C., 2008. Cladistic analysis of Reduviidae (Heteroptera: Cimicomorpha) based on morphological characters. *Systematic Entomology*, 33(2): 229–274.

Weirauch, C., Bérenger, J. M., Berniker, L., Forero, D., Forthman, M., Frankenberg, S., Freedman, A., Gordon, E., Hoey-Chamberlain, R., Hwang, W.S., Marshall, S. A., Michael, A., Paiero, S. M., Udah, O., Watson, C., Yeo, M., Zhang, G. and Zhang, J., 2014. An illustrated identification key to assassin bug subfamilies and tribes (Hemiptera: Reduviidae). *Canadian Journal Arthropod Identification*, 26: 1–115.

Weirauch, C. and Munro, J. B., 2009. Molecular phylogeny of the assassin bugs (Hemiptera: Reduviidae), based on mitochondrial and nuclear ribosomal genes. *Molecular Phylogeny and Evolution*, 53: 287–299.

3 The Flower Bugs (Anthocoridae)

Chandish R. Ballal, Tripti Gupta, and Richa Varshney
ICAR-National Bureau of Agricultural Insect Resources

CONTENTS

LEARNING OBJECTIVES

1. Anthocorid bugs are considered as important biocontrol agents as the adult and nymphal stages of majority of the species of anthocoridae can predate on eggs of lepidopteran pests and all stages of sucking pests, viz., thrips, aphids, mites and mealybugs.
2. Diversity of Anthocoridae is reported to be the highest in the tropics and the Holarctic region while Anthocorid predators of the Northern hemisphere are better studied than the southern Hemisphere.

DOI: 10.1201/9781003370864-3

3. If this group of predators has to be documented, conserved and utilized effectively as biocontrol agents globally, extensive exploratory surveys have to be carried out in unexplored geographical regions and in-depth research should go into the taxonomy and ecological aspects of anthocoridae.
4. Biocontrol researchers have developed protocols for the mass rearing of some of the most promising anthocorid predators, which can be utilized for the management of several insect pests, especially polyhouse pests.

3.1 INTRODUCTION

Anthocorid bugs are also referred to as minute pirate bugs or flower bugs. They occupy different natural and disturbed habitats. Except a few, most of the anthocorids are predaceous and they feed on small, soft-bodied arthropods belonging to different taxonomic groups. Hence, they are recorded to be major contributors in biological control of pest arthropods infesting varied habitats, including orchards, row crops, forests, ornamental plants, greenhouses and granaries. In several countries, including France, the United Kingdom, the Netherlands, Germany and some European countries, anthocorid bugs are commercially available for pest management, especially sucking pests infesting polyhouses. In India, earlier studies had focused on identifying the indigenous anthocorid predators and recording their predatory status in nature. However, recent studies have focused on identifying those anthocorid predators, which can be reared and released against pests of agricultural and horticultural crops. Globally, most of the research till date has focused on the temperate species, indicating overwintering in the adult stage or embryonic diapause in some of the species and the preponderance of multivoltine species. There are no such studies on the tropical or subtropical species (Lattin, 1999). The systematic position of the Anthocoridae and its close relatives remains unsettled, and there is a distinct need for additional taxonomic and phylogenetic work. A checklist of the flower bug families Anthocoridae and Lasiochilidae in India was published by Ballal et al. (2018a), which provides baseline data for future studies on Anthocoridae and Lasiochilidae. However, a large part of the country remains to be explored for anthocorid predators. Though considerable amount of work has also been done on taxonomy, biology, prey preference, predatory efficiency, development of rearing protocols and evaluation of anthocorid predators under laboratory, semi-field and field situations, future emphasis has to be on the commercial production of anthocorid predators so that they can be widely used in field conditions, especially in polyhouses.

3.2 DIVERSITY, SYSTEMATIC POSITION AND TAXONOMY

Anthocoridae (*sensu stricto*) is a large and diverse family within Cimicoidea. The family Anthocoridae includes between 500 and 600 species in 80–100 genera worldwide. Diversity of Anthocoridae is reported to be the highest in the tropics and the Holarctic Region. Anthocorid predators of the northern hemisphere are better studied than those of the southern Hemisphere. In comparison to the Nearctic region, lesser number of new species has been described in the Palaearctic and Oriental regions (Carpintero, 2002; Horton, 2008). Reuter (1884) published a monograph of the world Anthocoridae and Schuh and Slater (1995) published an exhaustive overview of Anthocoridae and its related families. Several researchers brought out regional monographs and catalogs for Hemiptera in general or Anthocoridae in particular: for Pakistan by Ghauri (1964, 1965); the West-Palaearctic Region by Péricart (1972, 1996); Turkey by Önder (1982); the United States and Canada by Henry (1988) and Horton et al. (2016); Australia by Cassis and Gross (1995); Palaearctic Region by Péricart (1996) and Aukema et al. (2013); China by Bu and Zheng (2001); Neotropical region by Carpintero (2002, 2015); Iran by Falamarzhi et al. (2000), Ghahari et al. (2009) and Ostovan et al. (2017); Korea by Jung et al. (2013); and Japan by Yamada et al. (2016). Distant (1904, 1906, 1910) and Poppius (1909, 1913) conducted the initial faunal studies on the anthocorids of the Oriental Region, which covered the taxa known from India, Ceylon (Sri Lanka), and Burma (Myanmar).

This was followed by publications by Ghauri (1964, 1972), Rajasekhara (1973), Muraleedharan (1975, 1977a, b, c), and Muraleedharan and Ananthakrishnan (1974a, b, c, 1978a, b). Muraleedharan and Ananthakrishnan (1978b) reported approximately 50 species recorded from the Oriental region, 20 species from Japan and Formosa and 15 species from Polynesian islands. After the 1970s, taxonomic studies on Indian Anthocoridae were done after 2000, which include studies by Yamada et al. (2008, 2010a, b, c, 2011) and Ballal and Yamada (2014). Some of the new species/records in India include *Buchananiella pseudococci pseudococci* (Wagner, 1951) (earlier identified as *Cardiastethus pseudococci pseudococci* by Yamada et al., 2008); *Anthocoris muraleedharani* described from Karnataka by Yamada et al. (2010a) and *Montandoniola indica* Yamada, 2011, a predator of *Liothrips karnyi* infesting black-pepper leaves in Kerala and of *Gynaikothrips uzeli* Zimmerman infesting *Ficus* in Karnataka (Yamada et al., 2011; Ballal et al., 2012b). It is also felt that the Indian *Montandoniola moraguesi* (Puton), which was reported to occur in agroecosystems in India and the bionomics of which was studied by Muraleedharan and Ananthakrishnan (1971), probably refer to *M. indica*.

A recent checklist by Ballal et al. (2018a) records that the Indian fauna of flower bugs are represented by 73 species belonging to 26 genera under eight tribes of two families. The 11 species recorded from India for the first time include *Amphiareus ruficollaris* Yamada & Hirowatari, 2003; *Anthocoris dimorphus* Zheng, 1984; *Bilia burma* Yasunaga & Yamada, 2016; *Cardiastethus kathmandu* Yamada, 2016; *Lippomanus brevicornis* Yamada & Hirowatari, 2004; *Montandoniola bellatula* Yamada, 2007; *Physopleurella armata* Poppius, 1909; *P. flava* Carayon, 1958; *P. pessoni* Carayon, 1956; *Rajburicoris stysi* Carpintero & Dellapé, 2008; and *Xylocoris* (*Proxylocoris*) *cerealis* Yamada & Yasunaga, 2006. The most diverse genera in terms of species are *Orius*, *Anthocoris*, *Physopleurella*, *Xylocoris* and *Buchananiella*. Six out of the 26 genera (*Galchana*, *Indocoris*, *Carayonocoris*, *Crytosternum*, *Odontobrachys* and *Pachytarsus*) and 28 out of the 73 species have so far been reported only from India. Thorough faunistic studies for some geographic regions are lacking, and estimates of diversity will require revision (Horton, 2008).

Anthocorids occur in a variety of habitats. They could be partially plant feeders (chiefly feeding on pollen), and could occur below ground in ant nests, in litter layers, and on forbs, shrubs and trees. It has also been reported that plant architecture has a great role to play in species richness and diversity of insects in general. Studies have indicated that the natural diversity and species richness of anthocoridae in specific countries could have been altered through introductions and colonization of non-native species. Lattin (1999) reported that out of the 16 species of bugs, which were introduced into America, nine were introduced accidentally. Specimens of Anthocoridae, Lyctocoridae and Lasiochilidae were intercepted on a variety of commodities, including ornamental plants, cut flowers, bouquets, agricultural produce, ceramic tiles and wood products at various ports-of-entry and housed at the Animal and Plant Health Inspection Services (APHIS) Miami Plant Inspection Station (Horton et al., 2013). Detailed studies on the intercepted specimens indicated that 14 of the identified species were not occurring in the continental United States at that point of time. It was interesting to observe that ten species were intercepted on shipments arriving from countries not previously listed as being part of their known geographic ranges. For example, an old world species, *Cardiastethus affinis*, known previously only from East Africa and India, was intercepted in two separate shipments arriving from Central America and the West Indies. Thus, besides deliberate introductions for the purpose of biological control, commercial activities and widespread habitat alterations were thought to be responsible for the accidental introductions leading to alterations in the diversity of anthocoridae in different countries.

Classification of the Anthocoridae has been extensively studied and revised since the family was erected in the 1800s, not only with reference to placement of related families but also with respect to higher-level groupings of taxa within the Anthocoridae. Classification of Anthocoridae has been through varied treatments by different taxonomists. The Anthocoridae is placed in the infraorder Cimicomorpha within the suborder Heteroptera. Horton (2008) has summarized the classification of Anthocoridae by different taxonomists and the allocation of the genera of pirate bugs to seven or eight tribes. Carayon (1972) and Cassis and Gross (1995) included three subfamilies (Lasiochilinae, Lyctocorinae and Anthocorinae) within the family Anthocoridae, while Schuh and Štys (1991) and Schuh and Slater (1995) stated that the family Anthocoridae *sensu lato* (referred to as minute pirate bugs or flower bugs) include three families, viz., Anthocoridae *sensu stricto,* Lasiochilidae and Lyctocoridae. The four tribes Almeidini, Cardiastethini, Scolopini and Xylocorini included within the Lyctocorinae by Carayon (1972) were placed in the Anthocorinae by Cassis and Gross (1995), or in the Anthocoridae by Schuh and Slater (1995).

Male genitalia characters are generally utilized to identify anthocorids. Southwood (1956) and Sands (1957) have attempted to use egg structure (especially egg operculum) and characters of immature stages for taxonomic studies of anthocorids. Based on structural differences in egg opercula, Schuldiner-Harpaz and Coll (2012) identified four species of *Orius.* Muraleedharan and Ananthakrishnan (1978b) stated that the following characters – the produced head, elongate clypeus, four-segmented antennae, three-segmented rostrum and tarsi, ocelli in alate forms, metathoracic scent glands, forewings with an embolium and an incomplete cuneus, membrane without basal cell, and asymmetrical male genitalia – as the outstanding features serving to distinguish the anthocorids. Ke and Bu (2007) reported that female copulatory tubes can be used as a reliable character to identify females and analyze phylogenetic relationships in the genus *Anthocoris.*

Besides traditional taxonomy, molecular techniques have also been used for identifying anthocorids. Gómez-Polo et al. (2013) could discriminate between seven *Orius* species commonly present in Mediterranean vegetable crops, using molecular tools. A combination of different tools has also been utilized effectively by some researchers to identify anthocorids. Shapiro et al. (2010) utilized male and female genitalia characters, DNA sequences, and the effects of within-species and between-species pairings on egg production and egg development to understand the relationship between the two sympatric species *Orius insidiosus* (Say) and *Orius pumilio* (Champion) in north Florida. Hassanzadeh-Avval et al. (2020) could identify three species of *Anthocoris* including *A. minki* (two subspecies: *A. minki minki* and *A. minki pistaciae*), *A. guentheri* and *A. nemoralis* (predators of psyllid species in northeastern Iran) morphologically and by mitochondrial cytochrome oxidase I (COI) sequence analyses. Jung and Lee (2011) conducted a phylogenetic and systematic study of *Orius* species from Korea using both morphological and molecular characters.

Accurate identification and geographical and biological data are essential to understand the role played by anthocorid predators in nature and also to incorporate them in biological control programmes. The phytophagous *P. laeviusculus* (perhaps misidentified to be *Paratriphleps pallida* (Reuter)) was introduced from Peru into Texas for the control of bollworms. If it had established in Texas, the mis-identification would have proved to be costly (Lattin, 1999). Based on the studies conducted by Horton et al. (2016), it emerged that the *Orius* fauna of the western United States is composed of a mixture of two described species and possibly five undescribed cryptic species. Certain of the phenotypes appeared to be relatively specialized on a few non-crop plant taxa, indicating that even large densities of *Orius* in habitats adjacent to crops may not provide practical benefit to agriculture. Thus, there is a great need to conduct detailed investigations on anthocorid fauna, especially in tropical and subtropical regions, to be followed by additional in-depth taxonomic and phylogenetic studies.

3.3 MORPHOLOGY

Anthocorid bugs are relatively small (1.5–5 mm), having an oval or elongated oval shape and are brownish or blackish in general coloration. They have a flattened body, and may be either glabrous or pubescent. The head extends horizontally, with large compound eyes on either side. Paired ocelli are present near eyes in the adult and not present in nymphs. Antennae are four-segmented and inserted anterior of the eyes. Mouth parts are of the piercing-sucking type, in the form of a slender beak or labium, which may be short or long, have three visible segments and house the four piercing stylets. The pronotum is trapezoidal, bearing paired callosities, which are prominent in some species and inconspicuous in others. As characteristic of true bugs, anthocorids have two pairs of wings: the membranous hindwings and the thickened forewings (hemelytra). Most of the taxa are fully winged, while few species in some genera (e.g. *Xylocoris*, *Temnostethus*, *Anthocoris* and *Elatophilus*) exhibit brachyptery. The forewing has a triangular cuneus, which is a unique character in most Heteroptera. Genitalia in the male are asymmetrical. The ovipositor is well-developed in most species but is greatly reduced in some taxa (e.g. *Cardiastethus*).

3.4 BASIC BIOLOGY AND LIFE HISTORY

A better understanding of anthocorid biology has contributed greatly to understanding their role as natural enemies for the management of pest (thrips, mites, aphids, mealybugs, lepidopterans, stored insect pests, etc.) populations. The pirate bugs have hemimetabolous development, in which the immatures (nymphs) undergo incomplete metamorphosis, and resemble the adults morphologically except in lacking wings, ocelli and reproductive structures. The number of nymphal instars is generally reported as five; however, some species, such as *Temnostethus gracilis* Horváth, were reported to have four instars (Péricart, 1972; Štys and Vilimová, 1989). Thontadarya and Rao (1987) reported four instars in *Orius maxidentex* Ghauri, while Devi et al. (2007) and Kemasa et al. (2008) observed five nymphal instars in *O. maxidentex* when fed on *Bemisia tabaci* (Gennadius). Hence, the number of instars in the above-mentioned species needs to be rechecked.

The anthocorid species *Orius*, *Blaptostethus* and *Anthocoris* insert their eggs into plant tissues, while species, such as *Xylocoris*, *Scoloposcelis*, *Cardiastethus* and *Amphiareus*, lay loose eggs, which are placed under bark, plant debris and soil (Péricart, 1972; Lattin, 1999; Ballal et al., 2003a, b, 2012b, d; 2019; Musolin and Ito, 2008; Gupta and Ballal, 2015; Ballal and Yamada, 2016). Egg viability depends on the oviposition sites. An egg-laying substrate with hard plant tissues in stem and fruit with adequate humidity is generally the preferred site for oviposition (van den Meiracker and Sabelis, 1999; Shapiro and Ferkovich, 2006).

The biological parameters of anthocorid predators are reported to be affected by the diet supplied during rearing, and several diets have been evaluated for raising different species of anthocorids. Kiman and Yeargan (1985) reported that *Orius insidiosus* (Say) can survive on diets of pollen, vegetable juice and water. However, studies have clearly indicated that anthocorids require animal prey to ensure effective reproduction and fertility (Shapiro and Ferkovich, 2006; Zambrano, 2009). Several studies have evaluated the effect of diets based on eggs, larvae and adults of different insect species, and it has emerged that lepidopteran eggs are the ideal diet, which optimizes the biological parameters of anthocorid species (Mendes et al., 2002; Ballal et al., 2003a, b, 2012a; Saini et al., 2003; Zambrano, 2009; Sobhy et al., 2010; Gupta and Ballal, 2006, 2015; Gupta et al., 2018).

The duration of nymphal development depends primarily on the temperature, the quantity and quality of food (Hodgson and Aveling, 1988; Ballal et al., 2003a, b) and also on the photoperiodic conditions (Musolin and Ito, 2008). Devi et al. (2007) recorded a total developmental period of 27 days and longevity of 38 days for *O. maxidentex* adults, while Kemasa et al. (2008) reported the total developmental period of *O. maxidentex* to be 17.06 days and male and female longevity was 10.70, and 11.78 days, respectively, with *B. tabaci* as a host. Variation in terms of duration of developmental period and longevity may be due to different rearing conditions. Temperature and photoperiod regimes have a significant impact on the developmental times, adult longevity and other biological parameters of *Orius* species, viz., *Orius sauteri* (Poppius) (Sun et al., 2017), *Anthocoris minki* Dohrn (Pourali et al., 2012) and *Orius similis* Zheng (Zhang et al., 2012;).

The effect of different temperature regimes on the biological parameters of *Orius tantillus* (Motschulsky) and *Blaptostethus pallescens* Poppius was studied (Ballal et al., 2018b; Gupta, 2009; Gupta et al., 2018). Maximum longevity of *O. tantillus* was recorded at 24°C and progeny production at 24°C and 28°C. Reproduction rate, intrinsic rate of increase and the weekly multiplication rate were also higher at 28°C. In the case of *B. pallescens*, progeny production and longevity were highest at 28°C.

Laboratory studies on comparative biology of *Xylocoris flavipes* (Reuter) was studied using Ultraviolet (UV)-irradiated eggs of *Sitotroga cerealella* (Ol.), *Ephestia cautella* (Walker) and *Corcyra cephalonica* Stainton as host material. No significant differences were observed in the fecundity, nymphal period and longevity of *X. flavipes* when fed on the eggs of the three laboratory hosts (Gupta and Ballal, 2015). Sarker et al. (2016) reported a reduction in the total development time, and adult longevity of *X. flavipes* when fed on pupae of *Cryptolestes pusillus* (Schon.) and an increase when fed on second instar larvae of *C. pusillus*. These studies help in identifying the ideal host for rearing specific anthocorid predators.

Ballal et al. (2019) studied the biology of *Amphiareus constrictus* (Stål) at different temperature regimes. Rearing temperatures of 25°C and 30°C were found to be suitable for hatching and nymphal survival to the adult stage. Life table parameters, like net reproductive rate (R_0), the precise intrinsic rate of increase (r_m), finite rate of increase (λ) and hypothetical F_2 females were higher at 25°C.

Photic conditions along with temperature affect the biological parameters, like development and reproduction of minute pirate bugs. Detailed information on overwintering of various anthocorid species from temperate region and embryonic diapause has been discussed by Saulich and Musolin (2009). In several *Orius* species, only fertilized females can properly accumulate fat body and enter into diapause to ensure successful overwintering, whereas males usually mate before winter and do not survive until spring (Ito and Nakata, 1998a, b; Shimizu and Kawasaki, 2001; Kobayashi and Osakabe, 2009; Saulich and Musolin, 2009). Bonte et al. (2016) investigated the cold hardiness and overwintering potential of the southern African pirate bugs, *Orius thripoborus* (Hesse) and *Orius naivashae* (Poppius). A 12-hour photoperiod and 18°C induced reproductive diapause in 84% of *O. naivashae* females and 42% of *O. thripoborus* females.

Meng et al. (2017) studied the impact of temperature and photoperiod regimes on the life table of *O. sauteri*. The low-temperature and short-photoperiod regime resulted in significant increases in the developmental durations of the egg and nymphal stages, and the total development duration. Female adult longevity was significantly increased, whereas male adult longevity was significantly decreased, at 21°C, L8:D16 hours. The adult preoviposition period, the total oviposition period and the sex ratio were all increased under this regime. Fecundity decreased, whereas the mean generation time (T) increased, as temperature and photoperiod decreased. The intrinsic rate of increase (r_m), finite rate of increase (λ) and net reproductive rate (Ro) increased as the temperature decreased under diapause induction.

3.5 ECOLOGY AND BEHAVIOUR

3.5.1 Seasonal Cycle and Overwintering

Anthocorid bugs from the temperate zone overwinter at the adult stage. However, embryonic diapause has been reported in *Tetraphleps abdulghanii*, *Temnostethus pusillus* and *T. gracilis* and a homodynamic seasonal cycle in *Lyctocoris campestris* and in few species of *Xylocoris*. In a number of species in the genera *Anthocoris* and *Orius*, it has been reported that copulation occurs before overwintering and only females survive winter. Multivoltinism with up to eight generations has been reported in some species. Daylength and temperature are the factors, which control the development of multivoltine species. Owing to the obligatory onset of diapause, the northern population always develops in a single generation. The females of the multivoltine anthocorids of the temperate zone reproduce under the long-day conditions and enter diapause under the short-day conditions (Saulich and Musolin, 2009). However, in the southern region, weak response to photoperiod has been observed and some populations do not enter into diapause even under short-day conditions. Though information is not available on the factors that can trigger termination of diapause in anthocoridae, research data indicate that for a number of species low-temperature treatment for a few weeks can induce oviposition. Summer diapause is unusual for the family, except for the summer–winter embryonic diapause reported in *Tetraphleps abdulghanii* (Krishnaswamy and Yaseen, 1972). Some populations of *Anthocoris nemorum* and *A. nemoralis* exhibit a seasonal change of host plants, which is an unusual trait in Heteroptera. Colour polymorphism is not commonly reported; however, Yasunaga (1997) reported colour variants of *Orius sauteri* adults, which was assumed to be related to the temperature conditions. Though seasonality of tropical and subtropical species has not been well studied, majority of the species develop without winter diapause, hence are considered as potential agents for biological control attempts. If anthocorid bugs have to be utilized successfully for Integrated Pest Management (IPM) programmes, in-depth studies have to be done on their ecophysiology, seasonal development and identifying the factors regulating the seasonal development.

3.5.2 Movement

Lattin (1999) reported that wing polymorphism is evident in anthocoridae and the type of wing development probably affects flight. Most of the species are fully winged, some species are dimorphic with both long- and short-winged forms, and a few have only reduced wings. Generally, long-winged forms are known to be found in exposed habitats and short-winged forms in concealed habitats; however, there are clear exceptions to this rule. Saulich and Musolin (2009) report that seasonal alary polymorphism or polymorphism governed by external ecological factors is rare in anthocorid species. Lattin (1999) has given a clear account of the movement and flight in anthocorids, which could be trivial or migratory and the former is reported to be more common. Trivial movement aided by air movements is expected to have led to long distance movement (Southwood, 1960). Seasonal migrations and aggregation behaviour have not been reported in Anthocoridae. The endemic species of anthocoridae in the Hawaiian Islands and a similar speciation in the Seychelle Islands were thought to be due to long-term movement over water and further colonization in these islands. Considering the above, the adaptive significance of wing polymorphism in anthocoridae is not known.

3.5.3 Predator Prey Interactions

In complex ecosystems, species interact with each other in many ways. Among all interactions, predator–prey interaction is of great importance, as it is the factor determining significant reduction in insect pest populations (Sarmento et al., 2007; El-Basha et al., 2012). In general, the efficiency of a predator to control a pest depends upon its functional response (i.e. changes in attack rate in response to variations of prey populations × number of prey consumed per unit time in relation to prey density (Butt and Xaaceph, 2015; Rehman et al., 2020)). Functional response describes the ability of a predator to consume prey at different prey densities, and it is used to study the potential of a predator to regulate prey populations (Murdoch and Oaten, 1975).

Several studies have reported type II functional response for different species of anthocorid bugs (Gitonga et al., 2002; Montserrat et al., 2000; Coll and Ridgway, 1995; Donnelly and Phillips, 2001; Isenhour and Yeargan, 1981; Kheradmand et al., 2017; Shahpouri et al., 2019; Varshney et al., 2020), while there are reports on type III functional response for a few anthocorid species (*B. pallescens*, *Amphiareus constrictus* (Stal), *O. niger* (Wolf.), *Montandoniola indica* Yamada, *Orius albidipennis* Reuter and *O. insidiosus*) at certain temperature regimes (McCaffrey and Horsburgh, 1986; Jalalizand et al., 2011; Queiroz et al., 2015; Shahpouri et al., 2019; Varshney et al., 2020). Wang et al. (2019) reported type II response in *Orius sauteri* (Poppius) and its ability to control fungivorous sciarid fly *Lycoriella pleuroti* (Yang and Zhang), a pest of the oyster mushroom *Pleurotus ostreatus* (Jacq.) P. Kumm. cultivation.

3.5.4 Intraguild Predation

As general predators, anthocorids feed on various kinds of insects and might be engaged in different ecological interactions when they share the same resources with the same or other species. These interactions could be competition, intraguild predation, cannibalism, etc. Some studies inferred that these interactions may disrupt biological control (Rosenheim et al., 1993, 1999; Snyder and Ives, 2001; Finke and Denno, 2003) and increase pest population while others have reported that intraguild predation could have a positive effect on biocontrol of pests (Schausberger and Walzer, 2001; Snyder et al., 2004; Gardiner and Landis, 2007).

A number of basic studies have been conducted to investigate these interactions in anthocorid bugs. These interactions either enhance or decrease the potential of biocontrol. Both *O. laevigatus* and *Macrolophus pygmaeus* Rambur when released together were able to persist in one crop for a sufficiently long period and they complemented each other in the control of thrips and aphids in greenhouse-grown sweet pepper (Messelink and Janssen, 2014). Fathi et al. (2008) reported that intensity of intraguild predation is usually more when prey density is low. They further indicated that the combined use of *Orius niger* (Wolf) and *Aeolothrips intermedius* (Bagnall) against a high population density of onion thrips was a successful strategy in integrated pest management. *Orius majusculus* (Reuter) acted as intraguild predator (IGP) as it preyed on the predatory mite *Phytoseiulus persimilis* Athias-Henriot, thus delaying the control of the spider mite *Tetranychus urticae* Koch (Brødsgaard and Enkegaard, 1995). *Orius insidiosus* showed no apparent preference for *A. swirskii* and adult or immature stages of *F. occidentalis* within rose flowers or canopies but tended to switch over to the most abundant prey (Chow et al., 2010). There are several studies that have investigated the combinations of *Orius* species with predatory mites to manage *F. occidentalis* and the results indicated that these combinations could be complimentary, viz., *Orius tristicolor* (White) and *Amblyseius* (= *Neoseiulus*) *cucumeris* (Oudemans) (Gillespie and Quiring, 1992); *Orius laevigatus* (Fieber) and *Amblyseius degenerans* Berlese (Wittmann and Leather, 1997); and *Orius majusculus* (Reuter) and *A. degenerans* (Brødsgaard and Enkegaard, 2005). On the contrary, some other studies conclude that combinations of *Orius* and phytoseid mites cannot be recommended for field releases, viz., *O. laevigatus* and *A. cucumeris* (Wittmann and Leather, 1997; Magalhães et al., 2004); *O. insidiosus* and *A. cucumeris* (Sanderson et al., 2005); and *O. insidiosus* and *A. degenerans* (Chow et al., 2008) because of their incompatibility to manage *F. occidentalis.* Urbaneja et al. (2003) reported that when *O. laevigatus* and *N. cucumeris* were released together to target *F. occidentalis* on sweet peppers, the predatory mite served as an alternate food for the anthocorid bug and aided in their establishment.

Besides intraguild predation between two predators, there are some studies where anthocorid bugs act as IGP on immature stages of parasitoids (Naranjo, 2007). *Orius majusculus* (Reuter) is reported as IGP on *Encarsia formosa* (Gahan) parasitized *Bemisia tabaci* (Gennadius) (Sohrabi et al., 2013). *Orius albidipennis* did not show any recognizable preference for parasitized over unparasitized whitefly nymphs (Pirzadfard et al., 2020). Studies conducted by Gupta and Ballal (2007) indicated the complementarity between some anthocorid predators and *Trichogramma chilonis* Ishii. *Blaptostethus pallescens* and *O. tantillus* preferred to feed on un-parasitized eggs when given a choice of unparasitized eggs of *Helicoverpa armigera* and *Corcyra cephalonica* and those parasitized by *T.chilonis.*

In a few studies, *Orius* and other anthocorid species act as IGPs, but there are some reports where they act as intraguild prey for other predators (Rosenheim, 2005; van den Bosch et al., 1956; Whitcomb and Bell, 1964). Intraguild predation can also influence oviposition preferences. For instance, *Orius insidiosus* and *Nabis americoferus* are able to detect the presence of one another, and adjust their reproductive decisions, to avoid potential competitive interactions (Pumariño et al., 2011).

3.5.5 Effect of Host Plant Traits

Host plant traits can affect interactions between anthocorids and their prey (Coll and Ridgway, 1995; Coll et al., 1997; Coll and Izraylevich, 1997). Madadi et al. (2008) studied the effect of host plants (cucumber, eggplant, and sweet pepper) with different leaf trichome densities on predation of *N. cucumeris* and *Thrips tabaci* Lindeman by *O. albidipennis. Orius albidipennis* significantly reduced thrips populations on sweet pepper and eggplant but not on cucumber (Madadi et al., 2009). Some studies have focused on the role of pollen on intraguild predation among anthocorid bugs and phytoseiid mites. Shakya et al. (2009) reported that the availability of pollen increased intraguild predation of *N. cucumeris* by *O. laevigatus* and reduced suppression of their shared extraguild prey, *F. occidentalis*. Similarly, Skirvin et al. (2006) reported that the availability of pollen on chrysanthemum resulted in a 40–50% reduction in the predation of *F. occidentalis* by *O. laevigatus* and *N. cucumeris*. Fransen and Tolsma (1992) reported that *O. insidiosus* at the rate of single adult per chrysanthemum plant was able to reduce damage by *F. occidentalis* by 5.0–20.0%, while Parrella and Murphy (1996) observed that *O. insidiosus* was not effective in reducing thrips population on some ornamental plants, which was attributed to the host plant traits.

3.5.6 Responses to Plant Volatiles

Omnivores are likely to respond to both constitutive and herbivore-induced plant volatiles (Stenberg et al., 2010). *Anthocoris nemorum* (L.) detects and avoids the pathogen *B. bassiana* when it forages on host plants with which it is adapted but not on soil surfaces (Meyling and Pell, 2006). Traczyk et al. (2020) found that western flower thrips deposit non-volatile semiochemicals that can be utilized for behaviour manipulation in *O. insidiosus*, which may help in early control of thrips populations in fields and greenhouses. Drukker et al. (2000) investigated the acquired preference by *A. nemorum* to methyl salicylate with prey. However, this preference did not depend upon volatile experience during the juvenile or adult stage. Lehrman et al. (2013) examined the attraction of *A. nemorum* to plant volatiles released from three contrasting Salix clones and concluded that *A. nemorum*'s preferences for different clones changed when there was a genotype-specific systemic change in the volatile profile induced through plant feeding by leaf beetle *Phratora vulgatissima* (Linnaeus). The attraction of *Orius* species to the volatiles emitted by prey-infested plants has also been reported (Janssen et al., 1998; Venzon et al., 1999; van Laerhoven et al., 2000; Tillman, 2006; Aldrich et al., 2007; Arab et al., 2007; Mochizuki and Yano, 2007; Lundgren et al., 2009; Carvalho et al., 2011). Zeng et al. (2021) reported that *O. similis* is a promising candidate for the biological control of Fall Armyworm (FAW), *Spodoptera frugiperda* (J. E. Smith), specifically based on the observation that this predator was attracted to FAW-damaged leaves, which would enable it to locate the target prey rapidly. The phytophagous nature of some of the *Orius* spp. is known to trigger plant defense mechanism, which can repel the pests or attract natural enemies. When *Orius laevigatus* (Fieber) punctured sweet pepper plants, it was observed to repel *Bemisia tabaci* and *F. occidentalis* and attract the whitefly parasitoid *Encarsia formosa* (Bouagga et al., 2018). *Orius majusculus* had an innate preference for *Zyginidia scutellaris* (Herrich-Schäffer) and *Spodoptera littoralis* (Boisduval)–induced volatiles, and this preference could be modified through experience (Ardanuy et al., 2016).

3.5.7 Prey Preference

Prey preferences and switching behaviour of predators are important strategies for coping with prey scarcity. This behaviour has often been proposed to stabilize predator–prey dynamics (Gentleman et al., 2003). *Anthocoris minki* has a strong preference for psyllids over aphids (Hassanzadeh-Avval et al., 2018). However, it did not show significant preference between psyllid species (*Psyllopsis repens* vs *Agonoscena pistaciae*). Similarly, Sigsgaard (2010) reported preference for pear psyllid to green apple aphid by *Anthocoris nemoralis* (Linnaeus), while *A. nemorum* preferred green apple aphid. Both species preferred psyllids to spider mites. This preference could be due to olfactory cues, as reported by Drukker et al. (1995) and Scutareanu et al. (1997, 1999). *Orius albidipennis* exhibited preference for *Tetranychus turkestani* Ugarov & Nikolski compared to *B. tabaci* on both sweet pepper and cucumber, but its preference for *T. turkestani* on sweet pepper was significantly greater than that on cucumber (Banihashemi et al., 2017), indicating that the effectiveness as a predator can depend not only on the prey but also on the host plant. *Orius laevigatus* and *O. majusculus* were able to feed on whitefly eggs, nymphs and adults, and complete their development. However, when both whitefly and *F. occidentalis* were available as prey, the preference was for thrips (Arnó et al., 2008).

3.6 ANTHOCORIDS AS EFFECTIVE PREDATORS

Species of anthocorid predators have proved to cause significant mortality of some notorious pests in natural field situations. Several studies have indicated that the anthocorid predators on pests of low-growing forbs and shrub crops belong to the genus *Orius*. *Orius* species are reported as the most promising amongst anthocorids. Around 70 species of *Orius* are known all over the world. *Orius minutus* was recorded in China in apple orchards (Qin, 1985) and *Orius sauteri* (Poppius) in Japan, China, Korea and Russian Far East; both are important predators of sucking pests and lepidopteran eggs in field crops and orchards (Yano and van Lenteren, 1996).

The pirate bugs are recorded as important biological control agents in annual and perennial crops, in forests, in greenhouse crops, on ornamental plants and in stored products. Several species of *Anthocoris* are important bioagents of psyllids and aphids in temperate fruit orchards; *Orius* spp. of thrips, mites, aphids and eggs, and neonates of Lepidoptera on field crops or greenhouses; *Xylocoris* and *Lyctocoris* spp. of beetle and moth pests in stored products and granaries; *Elatophilus*, *Tetraphleps*, *Melanocoris* and *Acompocoris* spp. of aphid and scale insects of coniferous trees; *Xylocoris*, *Scoloposcelis* and *Lyctocoris* spp. of bark beetle pests; *Montandoniola*, *Macrotrachelia* and *Macrotracheliella* on ornamental fig infested by Cuban laurel thrips (*Gynaikothrips ficorum* (Marchal)) and other *Gynaikothrips* spp. Table 3.1 lists the important Indian anthocorid predators and their associated pest insects and host plants.

The Anthocoridae have also been successfully used in a number of successful classical biological control programmes (Horton, 2008). *Anthocoris nemoralis*, indigenous to Europe, was released into western North America in the early 1960s to target an invasive psyllid pest of pears (*Cacopsylla pyricola* (Förster)). The predator established in British Columbia, Washington, Oregon and California. The North American *Orius insidiosus* was released in Europe and Hawaii for controlling thrips and lepidopteran pests, and established in Hawaii. *Montandoniola moraguesi* (Puton), native to southeast Asia, was released in southern California, Texas, Bermuda and Hawaii to control Cuban laurel thrips and established in Bermuda and Hawaii.

In situations where the natural populations of the predators were not able to manage the pests, augmentative releases of mass-reared anthocorid bugs can be made.

3.6.1 Predators of Thrips

In Switzerland, *O. majusculus* could suppress the populations of thrips on sweet pepper and cucumber (Fischer et al., 1992). Kawai (1995) observed that by releasing *Orius* spp., fruit damage by *Thrips palmi* Karny in egg plant could be significantly reduced. *Orius pumilio* (Champion) and *O. insidiosus* are major predators of *Frankliniella bispinosa* (Morgan) infesting Queen Anne's lace (*Daucus carota* L.) and False Queen Anne's lace (*Ammi majus* L.) in Florida (Shapiro et al., 2009). Two releases of *O. albidipennis* and *O. majusculus* at the rate of ten adult females per square meter

TABLE 3.1
Some Potential Anthocorid Predators Recorded in India

Anthocorid	Place	Host Plant	Host Insect	Source
		On Thrips		
Orius sp.	New Delhi	Groundnut	Thrips	Singh et al. (1991)
Orius albidipennis (Reut.) (could be *O. maxidentex*)	India	Onion	*Thrips tabaci* Lind.	Saxena, (1975;1981b)
Orius bifilarus Ghauri	Uttar Pradesh Himachal Pradesh	Polyphagous	*Thrips flavus* Schrank and *Thrips hawaiiensis* (Morgan)	Veer (1984) Devi and Gupta (2010)
		Jasmine	Thrips	Devi and Gupta (2010)
Orius niger (Wolff)	Himachal Pradesh	Jasmine	Thrips	
Orius indicus	India		*Taeniothrips nigricornis*	Rajasekhara and Chatterji (1970)
Orius maxidentex Ghauri	Tamil Nadu	Sesame	*Thrips palmi* Karny *Frankliniella schultzei* (Tryb.)	Kumar and Ananthakrishnan (1984)
		Croton sparsiflorus Morong	*Thrips palmi*	Ananthakrishnan and Thangavelu (1976)
		Panicum maximum Jacq.	*Anaphothrips sudanensis* Tryb. *Caliothrips graminicola* (Bagn. and Cam.)	
		Prosopis spicigera L.	*Scirtothrips dorsalis* Hood	
		rice	*Haplothrips ganglbaueri* Schmutz.	
Orius maxidentex Ghauri and *O. tantillus* Motsch.	India	Onion and garlic	*Thrips tabaci* Lind.	Muraleedharan and Ananthakrishnan (1978b)
		Chilli, tea and other plants	*Scirtothrips dorsalis* Hood	
		Rice	*Baliothrips biformis* (Bagn.)	
		Groundnut, sesbania and bajra	*Caliothrips indicus* (Bagn.)	
		Chrysanthemum, dahlia and marygold	*Microcephalothrips abdominalis* (D. L. Crawford)	
		Various plants	*Haplothrips ganglbaueri*	
		On Thrips		
Orius minutus (L.)	Tamil Nadu	*Gliricidia sepium* (*maculata*) (Jacq.)	*Megalurothrips distalis* (Karny) *Frankliniella schultzei* (Tryb.) *Haplothrips ganglbaueri*	Viswanathan and Ananthakrishnan (1974)

(*Continued*)

TABLE 3.1 (*Continued*)
Some Potential Anthocorid Predators Recorded in India

Anthocorid	Place	Host Plant	Host Insect	Source
Carayonocoris indicus Muraleedharan	India	*Cassia marginata* Roxb. and mango	*Haplothrips ganglbaueri* *Frankliniella schultzei* (Tryb.)	Muraleedharan and Ananthakrishnan (1978b) Kumar and Ananthakrishnan (1984)
Montadoniola moraguesi (Put.) (could be *Montandoniola indica*)	India	Different plants	gall makers, *Gynaikothrips flaviantennatus* Moulton *Schedothrips orientalis* (Ananthakrishnan)	Ananthakrishnan and Varadarasan (1977)
		Black pepper	*Gynaikothrips bengalensis* Ananthakrishnan	Muraleedharan and Ananthakrishnan (1978b)
			Liothrips karnyi	Devasahayam and Koya (1994) Devasahayam (2000)
Montandoniola indica (Yamada)	Karnataka Kerala	Ficus Pepper	*Gynaikothrips uzeli* *Liothrips karnyi*	Ballal et al. (2012b) Yamada et al. (2011)
Rajburicoris keralanus Yamada	Kerala	Black pepper	*Liothrips karnyi*	Yamada et al. (2010b)
Scoloposcelis parallelus (Motsch.)	Tamil Nadu	Decaying bark of *Erythrina indica* Lam.	Fungus thrips, *Ecacanthothrips sanguinens* Bagn.	Viswanathan and Ananthakrishnan (1973) and Muraleedharan and Ananthakrishnan (1978b)
Xylocoris clarus (Dist.)	India	Leaf litter	Litter insects and mites including *Dexiothrips madrasensis* (Ananthakrishnan) and *Apelaunothrips indicus* (Ananthakrishnan)	Muraleedharan and Ananthakrishnan (1978b)
		On Midges		
Orius maxidentex Ghauri	Karnataka	Sorghum	*Contarinia sorghicola* Coquillett	Thontadarya and Rao (1987)
		On Midges		
Orius albidipennis (Reut.) (Could be *O. maxidentex*)	Delhi	Sorghum	*Contarinia sorghicola*	Rao (1976)
Orius sp.	Gujarat	Sorghum	*Contarinia sorghicola*	Patel et al. (1975)

(*Continued*)

TABLE 3.1 (*Continued*)
Some Potential Anthocorid Predators Recorded in India

Anthocorid	Place	Host Plant	Host Insect	Source
		On Stored Grain Pests		
Orius sp.	Punjab	wheat in grain storage structures	*Sitophilus oryzae* (L.) *Rhyzopertha dominica* (F.) *Tribolium castaneum* (Hbst.) *Trogoderma granarium* Everts *Sitotroga cerealella* (Ol.) *Ephestia cautella* (Wlk.) (*Cadra cautella*) *Cryptolestes pusillus* (Schonh.) (*Laemophloeus minutus* (Ol.)) *Liposcelis* sp.	Dhaliwal (1976; 1977); Battu et al. (1975)
Xylocoris flavipes (Reut.)	India	Stored cereals	*Tribolium castaneum* (Hbst.)	Mukherjee et al. (1971)
		On Lepidopterans		
Cardiastethus sp.	Kerala	Coconut	*Opisina arenosella* Walker	Abdurahiman et al. (1982)
Cardiastethus exiguus Poppius, *Cardiastethus affinis* Poppius, *Bucchananiella sodalis* Buchanan-White				Nasser and Abdurahman (1990)
Cardiastethus exiguus pauliana *C. pygmaeus pauliniae Lansbury*				Pillai and Nair (1993)
Buchananiella pseudococci pseudococci Wagner (as *Cardiastethus pseudococci pseudococci*)				Yamada et al. (2008)
Cardiastethus affinis Poppius	Kerala	Mango	*Orthaga exvinacea* (Hampson)	Yamada et al. (2008)
		On Aphids		
Tetraphleps raoi Ghauri	Shillong	Benguet pine	*Pineus* sp.	Chacko (1973)

(*Continued*)

TABLE 3.1 (*Continued*)
Some Potential Anthocorid Predators Recorded in India

Anthocorid	Place	Host Plant	Host Insect	Source
Bilia castanea (Carvalho) *Orius albidipennis* (Reut.) (could be *O. maxidentex*)	West Bengal	*Brassica nigra* (L.)	*Lipaphis erysimi* Kaltenbach	Ghosh et al. (1981) Saxena (1981a)
Blaptostethus pallescens Poppius	Karnataka	Maize	*Rhopalosiphum maidis*	Ballal et al. (2009)
Anthocoris flavipes Reuter	Jammu and Kashmir	Several agricultural, horticultural and forest plants	Several aphid genera, viz., *Anuraphis, Amphicercidus, Aphis, Brachycaudus, Macrosiphoniella* and *Pemphigus*	Bhagat (2015)
Temnostethus (*Ectemnus*) *paradoxus* (Hutchinson)	Jammu and Kashmir	Several economically important crops	Several aphid genera, viz., *Anuraphis, Amphicercidus, Aphis, Brachycaudus, Macrosiphoniella* and *Pemphigus*	Bhagat (2015)
		On Mealybugs		
Cardiastethus exiguus Poppius	Karnataka	Papaya	Papaya mealybug	Chandish R Ballal (unpublished)
Anthocoris muraleedharani Yamada	Karnataka	*Bauhinia purpurea*	*Ferrisia virgata*	Yamada et al. (2010a); Ballal and Yamada (2016)
		On Spider Mites		
Blaptostethus pallescens Poppius	Karnataka	Castor	*Tetranychus urticae*	Ballal et al. (2009)
Bilia castanea Carvalho	Karnataka	Grape *Nyctanthes arbor-tristis*	*Tetranychus urticae*	Chandish R Ballal (unpublished)
		On Adelgids		
Tetraphleps galchanoides Ghauri	Sikkim	Hemlock	*Adelges* sp.	Ghauri (1972)
		On Several Pests		
Amphiareus constrictus (Stål.)	Kerala	Hay stacks, sheaths of palms, nests of birds, leaf litter,	*Apelaunothrips consimilis*	Muraleedharan and Ananthakrishnan (1978b)
	Karnataka	sugarcane, mango	Thrips and mites	Ballal et al. (2018a)
Buchananiella pseudococci pseudococci (Wagner 1951)	Kerala	Coconut	*Opisina arenosella* *Sacharicoccus sacchari*	Yamada et al. (2008)
	Tamil Nadu	Mulberry	Thrips	Ballal et al. (2018a)

Source: Excerpted from Ballal and Yamada (2016) and updated.

were found to be effective against *F. occidentalis* on Saintpaulia flowers in glasshouse (Blaeser et al., 2004). *Orius albidipennis* was also found to be effective as a biological control agent of *F. occidentalis* by Sanchez et al. (2000) in vegetable culture in the Mediterranean. Sanchez and Lacasa (2002) recommended three releases of 0.75–0.25 *Orius*/plant from March till May for the complete control of thrips. Hulshof and Linnamaki (2002) found *O. insidiosus* to be superior to *O. laevigatus* for controlling thrips. In the Mediterranean basin, *O. albidipennis* performed better than O. *laevigatus* in covered sweet pepper (Sanchez and Lacasa, 2002; Urbaneja et al., 2003). *Orius strigicollis* were field released on eggplant and adzuki bean grown in open fields in central and southern Taiwan. In the treatment plots, the densities of thrips gradually declined in 4–6 weeks. However, in the chemical control plots, though there was an immediate drop in the thrips population just after the application of insecticides, there was a rapid increase in the pest population later (Wang et al., 2001). Laboratory studies indicated that *Cardiastethus affinis* (originally recorded as a predator of *O. arenosella*) and *Montandoniola indica* (originally recorded to prey on *Ficus* thrips) could also prey on *Frankliniella schultzei* (Varshney et al., 2020). Hence, it would also be worth evaluating some of the potential anthocorid bugs against the major sucking pests in polyhouses and open fields, especially against thrips. Earlier studies have indicated that releases of predators from the temperate region were not always effective against *Ficus* thrips; hence, the Indian *M. indica*, which is a tropical species, would be an ideal candidate for evaluation trials against Cuban Laurel thrips, which is a serious pest on ornamental *Ficus* in several countries.

3.6.2 Predators of Lepidopterans

Orius insidiosus was observed to prey heavily on *Helicoverpa virescens* (Fabricius), *Ostrinia nubilalis* (Hübner) and *Helicoverpa zea* (Boddie) in the United States (Lingren, 1977; Lingren et al., 1978; McDaniel et al., 1981; Reid, 1991). Al-Jboory et al. (2012) recorded *O. albidipennis* and *Orius* sp. as predators of *Tuta absoluta* (Meyrick) in Jordan. *Cardiastethus exiguus* was recorded as an effective bioagent against eggs and neonate larvae of coconut black-headed caterpillar in the states of Karnataka and Kerala in India (Lyla et al., 2006; Venkatesan et al., 2008). In India, *Xylocoris afer* was collected from dry *Ficus* fruits, and found to be highly amenable to rearing on alternate laboratory host eggs. Although it is considered a predator of stored grains, Vieira et al. (2018) demonstrated its predatory efficiency against *Plutella xylostella* and inferred that the use of this predator may be important not only for managing warehouse pests but also pests infesting agricultural systems.

3.6.3 Predators of Whiteflies

Adults of *O. tristicolor* (Say), an important predator of *B. tabaci* in Arizona cotton crops (Hagler and Naranjo, 1994), exhibited a clear preference towards whitefly adults rather than on nymphs or eggs of *B. tabaci* (Hagler et al., 2004).

3.6.4 Predators of Aphids

The predatory bugs, *Anthocoris nemorum* (L.) and *Anthocoris nemoralis* (Fabricius), exhibited a higher predation rate on *Myzus persicae* (Sulzer) than on other aphid species (Meyling et al., 2003). *Orius similis* Zheng, which originated in China, was recorded as an excellent predator of *Aphis gossypii* Glover (Sengonca et al., 2008). Gómez-Polo et al. (2012) reported that *O. majusculus* could effectively prey upon the aphid *Nasonovia ribisnigri* (Mosely) besides the western flower thrips on lettuce crops. *Orius sauteri* was evaluated against two species of walnut aphids. Though *O. sauteri* was observed to be an efficient predator of both the species, the efficiency of preying declined when the two species were present together on the same leaf (Wang et al., 2021).

3.6.5 Predators of Scale Insects

Gao (1987) reported *Dufouriella ater* (Dufour) to be a potential predator of Japanese pine bast scale *Matsucoccus matsumarae* (Nuwana) in China. The authors recorded *C. affinis* preying on the latania or palm scale *Hemiberlesia lataniae* infesting agave.

3.6.6 Predators of Mites

Blaptostethus pallescens Poppius (originally described as *Blaptostethus piceus* Fieber var. *pallescens* Poppius from Celebes) was assessed as a potential predator of pests in the maize ecosystem in Egypt (Tawfik and El-Sherif, 1969; Tawfik and El-Husseini, 1971; Tawfik et al., 1974). In India, the authors have conducted several studies on *B. pallescens*, which have indicated that this predator can be successfully mass multiplied and utilized primarily against spider mites and broad mites. Ballal et al. (2009) reported that six releases of *B. pallescens* reduced 78% of spider mite population infesting bhendi in a net house study. This predator was also successfully evaluated against *T. urticae* infesting okra, brinjal, rose, apple and chilli in the states of Orissa, Punjab, Tamil Nadu, Kerala and Himachal Pradesh under All India Coordinated Research Project on Biological Control. Ahmad et al. (2020) found both immature as well as adults of *B. pallescens* to be potential biocontrol agents of two-spotted spider mite infesting different host plants as well as European red mite infesting apple.

3.6.7 Predators of Storage Pests

In storehouses, *Xylocoris flavipes* is known to suppress the populations of small, externally developing prey, particularly readily accessible eggs and early instars that are neither heavily sclerotized nor overly hirsute (Sing and Arbogast, 2008). *Lyctocoris campestris* (F.) (Parajulee et al., 1994) and *X. flavipes* (LeCato, 1976; LeCato et al., 1977; Brower and Press, 1992; Brower et al., 1996; Donnelly and Phillips, 2001) have proved their biocontrol efficacy against stored-product pests. *Xylocoris flavipes* could be effectively utilized for the suppression of rice moth infesting rice and wheat in the states of Kerala, Assam, Tamil Nadu, Andhra Pradesh and Himachal Pradesh based on the trials conducted by All India Coordinated Research Project (AICRP) on biological control. Releases of *X. flavipes* against *S. cerealella* infesting paddy resulted in an 89% reduction in pest emergence in the treatment batch compared with control. However, further research is warranted to standardize dose, timings and method of release to deploy *X. flavipes* in various stored commodities and seeds to manage stored product pests. Sing and Arbogast (2008) suggested that bruchids could be effectively managed if *X. flavipes* is introduced into the storage system as soon as legumes are stored.

3.6.8 Factors Affecting the Success of Anthocorid Predators as Bioagents

Different species of anthocorid predators, especially *Orius* spp., are reported to prey effectively on sucking pests, such as thrips, mites, whiteflies, mealybugs and lepidopteran eggs and neonates, hence widely used in biocontrol programmes in some countries, where they are commercially available. The success however depends on various factors. An optimal release rate is important when predators are used as biological control agents, and it would not be economically feasible to increase the release rate above the optimal rate. The control effect of *Orius* spp. against *Thrips palmi* Karny infesting greenhouse eggplant was evaluated by releasing varying numbers of predator nymphs, and it was found that when two, three and five last instar nymphs of the predator were released per plant, *T. palmi* number per leaf decreased to 0.6 several weeks after the release (regardless of the number of predators released). However, it was not possible to further lower the *T. palmi* density by increasing predator density (Kawai, 1995). Besides the release rate, the timing and method of biological control applications are also significant factors. Additional factors that would impact the success of a biocontrol programme include the fecundity, establishment rates, dispersal and cannibalism of the predator and prey availability. Hence, laboratory studies are important to understand the preference of a specific prey to be used as a host for rearing anthocorids but also to decide on the ideal predator to target a specific pest, while planning biological control field applications. The predatory bug, *Orius sauteri* (Poppius), is an important biocontrol agent in Asia that is often released on field crops and in greenhouses to control pests on vegetable and ornamental plants. Laboratory studies conducted by Wang et al. (2014) indicated that *Frankliniella occidentalis* was the optimal prey type based on the fastest development, largest adult body size, shortest pre-oviposition period and highest fecundity and longevity of *O. sauteri*. However, other prey species, like *T. urticae*, *M. persicae*, *A. gossypii*, *A. craccivora* and *Megoura japonica* (Matsumara), did not seem to impact the numerical response of the predator. The biocontrol efficiency of an anthocorid depends not only on the prey but also the crop. Osekre et al. (2008) based on a study in Florida reported that *O. insidiosus* (an efficient predator of thrips in vegetable ecosystems) was not an effective predator of *Frankliniella tritici* (Fitch), the most dominant pest in field cotton during the reproductive stage. Periodic inundative releases of *O. laevigatus* at three bugs per cucumber plant every 2 weeks could reduce *Thrips tabaci* population, but it failed to provide sustainable control of greenhouse cucumber thrips in south of Iran (Rajabpour et al., 2011). Silveira et al. (2004) reported effective results through releases of 9.5 *O. insidiosus*/m^2 for the control of thrips infesting chrysanthemum in tropical Brazilian greenhouses. However, he further mentioned that this control method would be successful only if applied under practical conditions of a complete and well planned IPM programme (including integration with chemical pesticides that are safe to natural enemies).

3.6.9 Compatibility of Anthocorid Predators with Microbial Bioagents

Arthropod natural enemies help in disseminating fungal pathogens if used together to target a pest. Besides predation, *O. laevigatus* when used with an entomopathogenic fungus (*Lecanicillium longisporum* or *L. muscarium*) served a dual function by disseminating the fungal pathogen, thus improving the control of the pest species (*Myzus persicae*, *Bemisia tabaci* and *Frankliniella occidentalis*) (Down et al., 2009). Otieno et al. (2017) reported a synergistic interaction when *O. laevigatus*+Neem and *O. laevigatus*+*Metarhizium anisopliae* ICIPE-69+*Steinernema carpocapsae* combinations were used to manage western flower thrips. However, some reports presented contrasting results concerning the compatibility of azadirachtin and *O. laevigatus* (Biondi et al., 2012; Bonsignore and Vacante, 2012). Shipp et al. (2003) cautioned that *O. insidiosus* and *B. bassiana* strain GHA should not be used together because of high mortality of the adult bug caused due to direct application of the fungus. On the contrary, Gao et al. (2012) monitored *O. sauteri* over their entire life span and found no significant mortality of the anthocorid caused by *B. bassiana* (strain Bb-RSB), which was highly virulent to *F. occidentalis*. Hence, both biocontrol agents were recommended for managing *F. occidentalis* in an IPM programme. Similarly, Pourian et al. (2011) reported that there were no or only weak side effects of entomofungal pathogens on *Orius* spp.

3.7 REARING OF ANTHOCORID PREDATORS

Anthocorid predators are amenable to rearing and hence have been mass reared and utilized for field releases. Some species of anthocoridae are mass reared in commercial insectaries for shipment and release, which include *Orius insidiosus*, *Orius laevigatus* (Fieber), *Orius majusculus* (Reuter) and *Anthocoris nemoralis*. The successful rearing of anthocorid bugs depends on various factors, viz., the type and stage of host, oviposition substrate and founder population size.

Eggs of lepidopteran storage pests are generally used for rearing anthocorid predators. *Lyctocoris campestris*, which is a predator of stored product pests, could be reared on *Plodia interpunctella* (Hübner). The first two nymphal instars survived well on eggs or small host larvae, whereas the later instars fed on active larvae of any size (Parajulee and Phillips, 1992). Majority of the rearing systems developed for *Orius* spp. utilized UV-irradiated or frozen eggs of storage pests, which are amenable to rearing. Eggs of *C. cephalonica* were observed to be the ideal host for rearing several Indian anthocorid bugs including *Cardiastethus exiguus, Cardiastethus affinis, Buchananiella indica, Blaptostethus pallescens, Xylocoris flavipes, Xylocoris afer, Amphiareus constrictus, Montandoniola indica* and *Anthocoris muraleedharani* (Ballal and Yamada, 2016). Saini et al. (2003) reported on the usefulness of *S. cerealella* eggs as effective laboratory hosts for the multiplication of *O. insidiosus*. Eggs of *E. kuehniella, P. interpunctella* and *S. cerealella* were found suitable for rearing *O. tantillus* by Nagai et al. (1998), Kawamoto et al. (1999) and Gupta and Ballal (2009), respectively. Devi and Gupta (2010) successfully reared *O. bifilarus* and *O. niger* on eggs of *C. cephalonica*. Eggs of the lepidopteran *Heliothis virescens* were utilized for rearing *O. insidiosus* (Bush et al., 1993) and sorghum aphid, and *Melanaphis sacchari* Zehntner for rearing *O. sauteri* (Wang et al., 1996). Toft et al. (2020) attempted to rear *O. majusculus* on the aphid *Rhopalosiphum padi* (Linnaeus) and the results confirmed that *R. padi* is a low-quality food for rearing *O. majusculus*, though *O. majusculus* could significantly reduce the population of the aphid in field situations.

In India, protocols have been developed at the ICAR-National Bureau of Agricultural Insect Resources (NBAIR) for rearing some of the potential anthocorid predators, viz., *O. tantillus, B. pallescens, C. exiguus, C. affinis, X. flavipes, M. indica* and *A. muraleedharani* (Ballal and Yamada, 2016). *Cardiastethus exiguus* could be reared on UV-irradiated eggs of *C. cephalonica*. Utilizing 1 cc of *C. cephalonica* eggs, 450 *C. exiguus* adults could be produced (Ballal et al., 2003b). Tawfik and El-Husseini (1971) attempted to rear *B. pallescens* on lepidopteran larvae, aphids and mites, and reported that the nymphs were unable to moult when the food was restricted to aphids or plant sap. Ballal et al. (2003a) developed a protocol for rearing *B. pallescens* on UV-irradiated *C. cephalonica* eggs and French bean pods were used as oviposition substrates. Problems of cannibalism in the rearing containers and excessive handling could be avoided through the adoption of this protocol and a 1,000 *B. pallescens* could be produced utilizing 9 cc of *C. cephalonica* eggs. *Blaptostethus pallescens* is an ideal candidate for targeting red spider mites infesting polyhouse crops. Though *Orius tantillus* was commonly recorded in the field conditions compared to *B. pallescens,* it was less amenable to rearing as its survival and fecundity were significantly lower during laboratory rearing.

Pollen is used in many of the rearing systems for anthocorids. In China, *O. sauteri* was reared on a wide range of arthropods, pollen or an artificial diet (Zhou and Wang, 1989). The survival rate was recorded to be significantly higher on pollen diet in comparison to artificial diet. However, Funao and Yoshiyasu (1995) reported that maize pollen alone was not sufficient for rearing *O. sauteri*, whereas a combination of pollen and *Aphis gossypii* was a suitable diet. Calixto et al. (2013) also reported that pollen did not improve the performance parameters of *O. insidiosus* when offered exclusively or as a complementary food.

In anthocorid rearing, selection of an ideal oviposition substrate is one of the most important steps. Some of the species, which do not require a plant substrate for egg laying and can oviposit on cotton strands, include *C. exiguus, C. affinis, X. flavipes, Buchananiella indica* Muraleedharan and *Amphiareus constrictus* Stal. (Ballal and Yamada, 2016; Ballal et al., 2016, 2019). Hence, these species can be easily mass reared continuously in insectaries as they do not require a green house facility for continuous production of host plants. *Amphiareus constrictus* could be reared continuously for more than 20 generations on *C. cephalonica* eggs in the laboratory (Ballal et al., 2019). Plate 3.1 presents the images of some of the Indian anthocorid predators amenable to rearing.

For a species that require a plant substrate for egg laying, the plant species is known to affect the oviposition behaviour as the adult females prefer to oviposit on those plants and plant parts, where the newly hatched nymphs can survive and perform the best. Several of the anthocorids can lay eggs on French bean pods. Common bean, *Phaseolus vulgaris* L. (Richards and Schmidt, 1996; Seagraves and Lundgren, 2010) and inflorescence of Farmer's friend *Bidens pilosa* Linnaeus (Bueno et al., 2006) were recorded as ideal oviposition substrates for *O. insidiosus*; soybean for *O. strigicollis, O. laevigatus* (Ito, 2007) and *O. sauteri* (Zhou et al., 1991); bean, cucumber and apple for *O. niger* (Ahmadi, 2010) and leaves of geranium, *Pelargonium peltatum* (L.) L'Hér for *O. majusculus* (Alauzet et al., 1992).

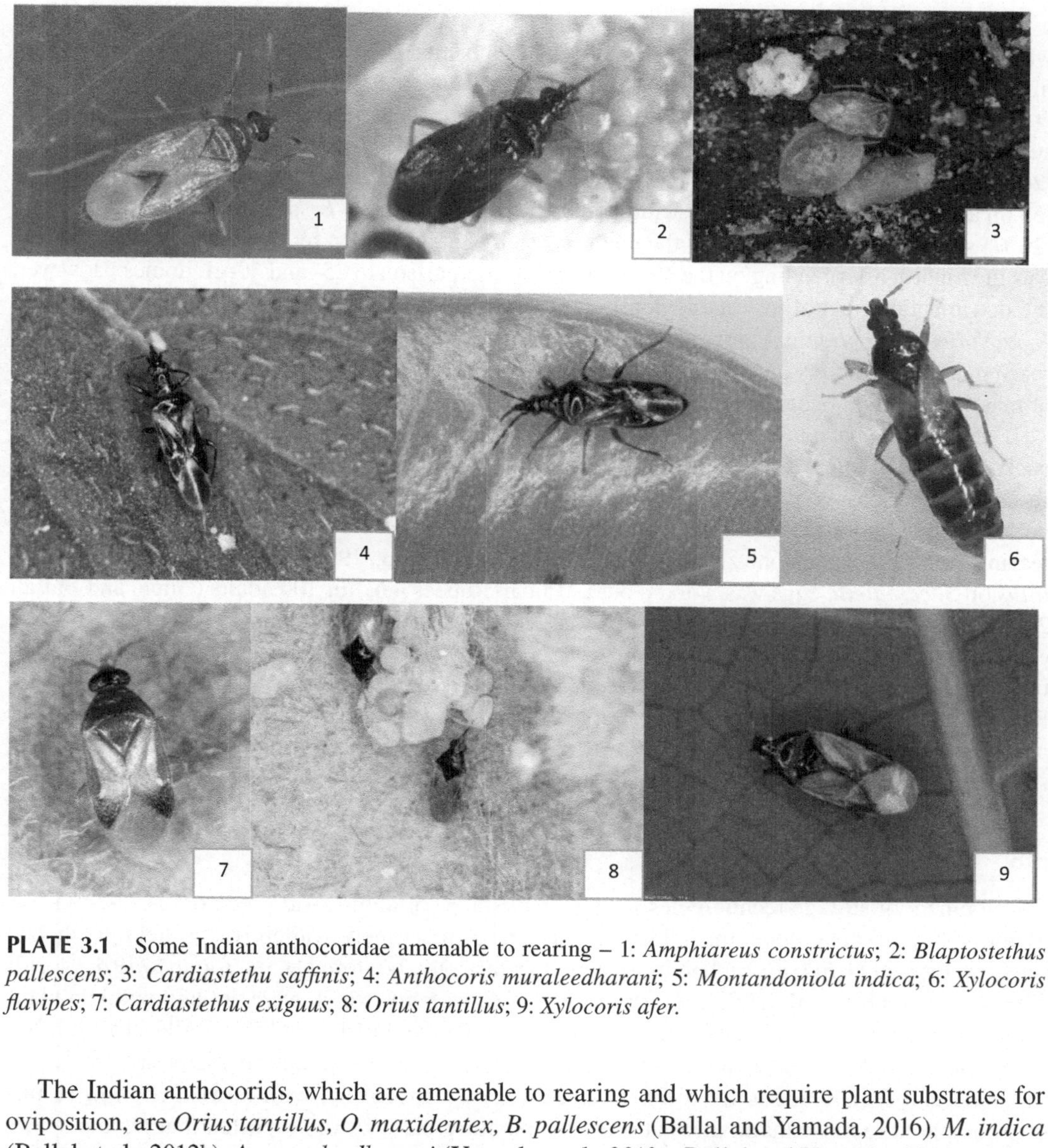

PLATE 3.1 Some Indian anthocoridae amenable to rearing – 1: *Amphiareus constrictus*; 2: *Blaptostethus pallescens*; 3: *Cardiastethu saffinis*; 4: *Anthocoris muraleedharani*; 5: *Montandoniola indica*; 6: *Xylocoris flavipes*; 7: *Cardiastethus exiguus*; 8: *Orius tantillus*; 9: *Xylocoris afer*.

The Indian anthocorids, which are amenable to rearing and which require plant substrates for oviposition, are *Orius tantillus, O. maxidentex, B. pallescens* (Ballal and Yamada, 2016), *M. indica* (Ballal et al., 2012b), *A. muraleedharani* (Yamada et al., 2010a; Ballal and Yamada, 2016), *X. afer* and *Blaptostethoides pacificus* Herring (Chandish R. Ballal, unpublished). *Anthocoris muraleedharani* was reared on eggs of *C. cephalonica* and crawlers of *Ferrisia virgata* and bean pods and potato sprouts were used as oviposition substrates. *Montandoniola indica* can be efficiently reared on *C. cephalonica* eggs, with bean pods used as oviposition substrates. In a recent study by Díaz et al. (2020), it was reported that by using a mixed diet of *F. occidentalis* and eggs of *S. cerealella*, maintaining constant conditions of the laboratory for rearing and following a 48-hour feeding frequency, a batch production process could be standardized for continuous and mass scale production of *O. insidiosus*.

Researchers have attempted to rear anthocorids on artificial medium or using artificial oviposition substrates, with limited success, either because of the high cost involved or due to the low-quality individuals produced when reared on artificial diets. The initial attempts to rear anthocorid bugs on artificial diets were made by Mukherjee et al. (1971) for the rearing of *X. flavipes*; by Martin et al. (1978) for *O. insidiosus* and by Bronnimann (1964) for *Anthocoris nemorum* L. and *A. nemoralis*. *Orius sauteri* was reared on an artificial diet by Zhou and Wang (1989) and *O. strigicollis* by Lee and Lee (2007). For *O. insidiosus*, Castane and Zalom (1994) developed an artificial oviposition medium made of carrageenan salt of potassium chloride covered with paraffin wax, while Shapiro and Ferkovich (2006) devised an oviposition substrate for *O. insidiosus* by making water-filled domes with parafilm-M. A microencapsulated artificial diet was developed by Tan et al. (2013), using which *O. sauteri* with increased biological and physiological characters could be successfully reared and evaluated in greenhouses. Arijs and De Clercq (2004) and Rajabpour et al. (2018) observed lower fecundity and survival rate of *Orius laevigatus* and *Orius albidipennis* when fed on artificial diet in comparison with factitious food, i.e., eggs of *E. kuehniella* . Ferkovich et al. (2007) studied the effect of size of the diet domes on egg production by *O. insidiosus.* Oviposition was the highest after feeding on the 10-µL domes in comparison to 25- and 50-µL domes. However, the detrimental effect of unsuitable dome size was negated when the diet was supplemented with *E. kuehniella* egg protein. Future studies have to focus on the shape of the diet domes, the effect of rearing anthocorid bugs on artificial diets for continuous generations and on the cost-effectiveness of mass rearing anthocorid bugs on artificial diets.

If anthocorids have to be effectively utilized in biocontrol or IPM systems, storage technology, packaging methodology for shipping and cost of production are important aspects to be considered. In Ontario, the cost of production of *O. insidiosus* could be reduced from 0.73 to 0.03 Canadian dollar per bug and up to 100,000 bugs could be reared per week. A method was standardized for rearing 200–600 bugs in one zip-lock plastic bag (Schmidt et al., 1995). The cost of rearing *O. tantillus* on *S. cerealella* eggs was worked out as Indian Rupees 4.67 for 100 adults (Gupta and Ballal, 2009). However, there is a need to improve the rearing efficiency of *O. tantillus* as problems were encountered with respect to continuous rearing of quality culture and scaling up the production. To improve the cost effectiveness of *O. laevigatus* rearing, a plantless rearing system was developed by Puysseleyr et al. (2014), wherein wax paper was used as a walking substrate, water encapsulated in parafilm domes as water source, a substrate made of parafilm and moist cotton wool as oviposition substrate and *E. kuehniella* eggs as food source.

3.8 STORAGE STUDIES

Development of storage technologies for different stages of anthocorid predators is essential for planning large-scale production and storage, which ensures timely availability for field/polyhouse releases. Eggs of *Orius sauteri* laid in soybean sprouts could be stored at 5°C for up to 19 days with 75% hatching (Guo et al., 2002). However, eggs of *C. exiguus* (which were not laid within a plant substrate) could be safely stored for only up to 5 days at 10°C and 10 days at 15°C (Ballal et al., 2012c). Guo et al. (2002) reared the adults of *O. sauteri* at 15°C for 4 days and then stored at 5°C, and the male and female adults lived for 98 and 29 days, respectively, with no reduction in fecundity compared to those maintained at room temperature. Kim (2009) reported that *O. laevigatus* adults could be stored at 10°C for up to 36 days, with 70% survival. *Cardiastethus exiguus* adult females (which were not acclimatized to the low-temperature regime) could be stored for only up to 15 days at 15°C (Ballal et al., 2012c). After removal from storage for 15 days, the adult female could survive for more than a month and progeny production was not adversely affected. In the case of *Orius majusculus*, fecundity was substantially reduced after 20 days of storage (Rudolf et al., 1993).

Diapause induction in laboratory-reared anthocorids or collection of diapausing individuals from the natural field conditions are two options to manage long-term storage. Naturally overwintered adults of *O. sauteri* were successfully stored for 2 months recording a survival rate of 55% (Wang et al., 1994). Parker (1981) collected *A. nemorum* as adults from fields in early autumn and artificially induced diapause. At the termination of diapause during early spring, they were reared to reach the desired stage and then cold stored till the time of requirement.

It is important to maintain the quality of mass-reared anthocorid predators. Ballal et al. (2003b) reported that for *C. exiguus*, in order to maintain the quality attributes, rejuvenation of the culture with wild population may be required after rearing continuously in the laboratory for more than a year. Carvalho et al. (2011) evaluated the performance of the wild and laboratory-reared populations of two anthocorids, *O. laevigatus* and *O. insidiosus*, based on their searching behaviour and orientation towards prey. Though the searching behaviour of the laboratory-reared *O. laevigatus* seemed to be inferior to the wild population, such a difference was not observed in *O. insidiosus*. This indicates that some of the species could be more amenable to continuous rearing and maintain their quality attributes, like searching and predation efficiency, through generations. Such species are potential candidates for the biocontrol of pests.

3.9 CONCLUSIONS

Globally, systematic surveys in unexplored regions to search for and document the anthocorid fauna can result in identifying novel and potential bioagents. Anthocorid taxonomy (including immature taxonomy) along with molecular taxonomy and phylogenetic studies will need more attention to clear the existing lack of clarity in the systematic position. The important role played by species belonging to Anthocoridae as biological control agents has been substantiated by several researchers. The potential research areas to strengthen our knowledge on this family for their improved utilization as bioagents include seasonal population dynamics, tritrophic interactions, interactions with other macrobials and microbials, standardization of production protocols, which can be up-scaled, long-term storage studies and extensive evaluation studies, especially in greenhouses and storage facilities. Studies have revealed that in certain situations, some of the naturally occurring indigenous species of Anthocoridae can provide significant reduction in the pest population. However, in majority of the cases, it is necessary to mass rear and release the predators in order to augment their populations and thus manage the pests. Globally, maximum successes have been recorded where anthocorid predators have been released against thrips and other sucking pests infesting polyhouse crops. This can be attributed to the commercial availability of anthocorid predators in the United States, Europe, Australia and Middle east. In India, thrips and mites infesting ornamentals and polyhouse crops and stored grain pests in granaries are major issues, and chemical insecticide treatments are not providing a tangible solution. The solution can only be through biological control and commercial availability of bioagents including anthocorid predators. The promising anthocorid predators, which can be reared utilizing simple production protocols and whose performance has been validated in small-scale trials can be further evaluated through large-scale field trials in collaboration with commercial entrepreneurs so that they are motivated to adopt the technology for mass rearing of these predators and further supply to the farmers.

3.10 POINTS TO REMEMBER

1. Anthocorids bugs are effective predators and play an important role in the natural control of many insect pests in different parts of the world. Emphasis should be on searching for and documenting the anthocorid fauna of the unexplored regions.
2. There should be major research focus on taxonomy of Anthocoridae, which, besides traditional taxonomy, should include immature taxonomy, molecular taxonomy and phylogenetic studies
3. Techniques have been evolved in several countries to mass rear anthocorid predators and release them in the field or glasshouses for the management of several pests, especially sucking pests like thrips and mites infesting polyhouses.
4. In India, though several species of anthocorid predators have been identified as promising bioagents, and production protocols have been standardized, they are not available commercially. The available standardized mass-rearing technologies for anthocorid predators have to be disseminated to commercial insectaries.
5. There is a need to conduct tritrophic interaction studies and to evaluate the available Indian anthocorid predators systematically against the pests infesting different crop ecosystems.

6. An understanding of the response of anthocorid predators to semiochemicals can aid in manipulating their behaviour and improve their performance as biocontrol agents.
7. The success of a biocontrol programme utilizing anthocorid predators depends on identifying the optimal release rate, the timing and method of applications, the fecundity and establishment rates, cannibalism of the predator, prey and host plant preferences, prey availability and dispersal. Hence, pre-release studies are extremely important to decide on the ideal anthocorid species to target a specific pest.
8. Globally, stored grain pests cause very serious problems in granaries and since chemical treatments are not providing a solution to the problem, specific species of anthocorid predators (e.g. *Xylocoris flavipes*; *Lyctocoris campestris*) can be evaluated against storage pests either alone or in combination with other parasitoids, microbial biopesticides or botanicals.
9. Compatibility studies have to be taken up between anthocorid predators and other potential bioagents (viz. insect pathogens and parasitoids).
10. Temperate species of anthocoridae enter into diapause and hence have limited use in polyhouses. The Indian anthocorid species do not enter into diapause and hence are ideal candidates for bilateral exchange of bioagents between countries for global biocontrol programmes.

REFERENCES

Abdurahiman, U. C., Mohammmed, U. V. K., and Remadevi, O. K. (1982) Studies on the biology of a predator, *Cardiastethus* sp. (Hemiptera: Anthocoridae) found in the galleries of *Nephantis serinopa* Meyr. (Lepidoptera: Xyloricticlae). *Current Science* 51: 574–576.

Ahmad, M.J., Mohiudin, S., Manzar, A.B.U., and Sherwani, A. (2020) Laboratory evaluation of anthocorid bug, *Blaptostethus pallescens* Poppius (Heteroptera: Anthocoridae) against European red mite, *Panonychus ulmi* (Koch) and two spotted spider mite, *Tetranychus urticae* Koch infesting apple in Kashmir. *Journal of Entomology and Zoology Studies* 8: 1750–1755.

Ahmadi, K. (2010) Egg-laying preferences of the predatory flower bug *Orius niger* (Wolff) to different substrates of oviposition. *Julius-Kühn-Archiv* No. 428, pp. 444

Alauzet, C. Dargagnon, D., and Hatte, M. (1992) Production of the heteropteran predator: *Orius majusculus* (Het.: Anthocoridae). *Entomophaga* 37: 249–252.

Aldrich, J.R., Oliver, J.L., Shifflet, T., Smith, C.L., and Dively, G.P. (2007) Semiochemical investigations of the insidious flower bug, *Orius insidiosus* (Say). *Journal of Chemical Ecology* 33: 1477–1493.

Al-Jboory, I. J., Katbeh-Bader, A., and Shakir, A. (2012) First observation and identification of some natural enemies collected from heavily infested tomato by *Tuta absoluta* (Meyrick) (Lepidoptera: Gelechiidae) in Jordan. *Middle-East Journal of Scientific Research* 2012(11): 787–790.

Ananthakrishnan, T. N. and Thangavelu, K. (1976) The cereal thrips *Haplothrips ganglbaueri* Schmutz with particular reference to the trends of infestation on *Oryza sativa* and the weed *Echinochloa crusgalli*. *Proceedings of the Indian Academy of Sciences - B* 83: 196–201.

Ananthakrishnan, T. N. and Varadarasan, S. (1977) *Androthrips flavipes* Schmutz (Insecta: Thysanoptera), a predatory inquiline in thrips galls. *Entomon* 2: 105–107.

Arab, A., Trigo, J.R., Lourenção, A.L., Peixoto, A.M., Ramos, F., and Bento, J.M.S. (2007) Differential attractiveness of potato tuber volatiles to *Phthorimaea operculella* (Gelechiidae) and the predator *Orius insidiosus* (Anthocoridae). *Journal of Chemical Ecology* 33: 1845–1855.

Ardanuy, A., Albajes, R., and Turlings, T.C. (2016) Innate and learned prey searching behavior in a generalist predator. *Journal of Chemical Ecology* 42: 497–507.

Arijs, Y. and De Clercq, P. (2004) Liver-based artificial diets for the production of *Orius laevigatus*.*BioControl* 49(5): 505–516.

Arnó, J., Roig, J., and Riudavets, J. (2008) Evaluation of *Orius majusculus* and *O. laevigatus* as predators of *Bemisia tabaci* and estimation of their prey preference. *Biological Control* 44: 1–6.

Aukema, B., Rieger, C. H., and Rabitsch, W. (2013) *Catalogue of the Heteroptera of the Palaearctic Region. VI. Supplement*. The Netherlands Entomological Society, Amsterdam, xxiii+629 pp.

Ballal, C. R., Akbar, S. A., Yamada, K., Wachkoo, A. A., and Varshney, R. (2018a) Annotated catalogue of the flower bugs from India (Heteroptera: Anthocoridae, Lasiochilidae). *Acta Entomologica Musei Nationalis Pragae* 58(1): 207–226.

Ballal, C. R., Gupta, T., and Joshi, S. (2012a) Effect of different laboratory hosts on the fertility table parameters and continuous rearing of an anthocorid predator *Orius tantillus* (Motsch.). *Pest Management in Horticultural Ecosystems* 18(1): 24–28.

Ballal, C. R., Gupta, T., and Joshi, S. (2012b) Morphometry and biology of a new anthocorid *Montandoniola indica*, a potential predator of *Gynaikothrips uzeli*. *Integrated Control in Protected Crops, Mediterranean Climate, IOBC - WPRS Bulletin*, 80: 79–84.

Ballal, C. R., Gupta, T., and Joshi, S (2012c) Production protocol for and storage efficacy of an Anthocorid predator *Cardiastethus exiguus* Poppius. *Journal of Environmental Entomology* 34(1): 50–56.

Ballal, C. R., Gupta, T., and Joshi, S. (2018b) Effect of constant temperature regimes on the biological parameters of an anthocorid predator *Orius tantillus* (Motsch.). *Journal of Biological Control* 31: 146–158.

Ballal, C. R., Gupta, T., Joshi, S., and Chandrashekhar, K. (2009) Evaluation of an anthocorid predator *Blaptostethus pallescens* against two-spotted spider mite *Tetranychus urticae*. In: Castane, C., Perdikis, D. (eds.). *Integrated Control in Protected Crops. IOBC/WPRS Bulletin* 49: 127–132.

Ballal, C. R., Singh, S. P., Poorani, J., and Gupta, T. (2003a) Biology and rearing requirements of an anthocorid predator, *Blaptostethus pallescens* Poppius (Hetroptera: Anthocoridae). *Journal of Biological Control*, 17(1): 29–33.

Ballal, C. R., Singh, S. P., Poorani, J. and Gupta, T. (2003b) Feasibility of mass multiplication and utilization of *Cardiastethus exiguus* Poppius, a potential anthocorid predator of *Opisina arenosella* Walker (Lepidoptera: Oecophoridae) (pp. 29–33). In: Tandon P. L., Ballal, C. R., Jalali S. K., and Rabindra, R. J. (eds.), *Biological Control of Lepidopteran Pests*. Society for Biocontrol Advancement, Bangalore, India, 354 pp.

Ballal, C. R., Varshney, R., and Joshi, S (2019) Morphology, biology and predation capacity of *Amphiareus constrictus* (Stål) (Hemiptera: Anthocoridae). *Neotropical Entomology* 48: 668–677.

Ballal, C. R. and Yamada, K. (2014) Diversity of Indian Anthocoridae (p. 89). In *World Biodiversity Congress – 2014*, November 24th to 27th 2014, Colombo, Sri Lanka, 212 pp.

Ballal, C. R. and Yamada, K. (2016) Anthocorid predators (pp. 329–366). In: Omkar s.), *Ecofriendly Pest Management for Food Security*. Elsevier, London, 727 pp.

Ballal, C. R., Yamada, K., and Joshi, S. (2016) Morphology and biology of the litter-inhabiting *Buchananiella indica* Muraleedharan (Hemiptera: Anthocoridae). *Entomon* 41: 11–20.

Banihashemi, A.S., Seraj, A.A., Yarahmadi, F., and Rajabpour, A. (2017) Effect of host plants on predation, prey preference and switching behaviour of *Orius albidipennis* on *Bemisia tabaci* and *Tetranychus turkestani*. *International Journal of Tropical Insect Science* 37: 176–182.

Battu, G. S., Bains, S. S., and Atwal, A. S. (1975) Natural enemies of *Trogoderma granarium* Everts infesting wheat in the rural stores in the Punjab. *Bulletin of Grain Technology* 13: 52–55.

Bhagat, R. C. (2015) Diversity and updated systematic checklist of Cimicomorpha bugs (Heteroptera: Hemiptera) of Jammu, Kashmir and Ladakh Himalaya (India). *Indian Journal of Fundamental and Applied Life Sciences* 5(1): 275–279.

Biondi, A., Desneux, N., Siscaro, G., and Zappala' L. (2012) Using organic-certified rather than synthetic pesticides may not be safer for biological control agents: Selectivity and side effects of 14 pesticides on the predator *Orius laevigatus*. *Chemosphere* 87: 803–812.

Blaeser, P., Sengonca, C., and Zegula, T. (2004) The potential use of different predatory bug species in the biological control of *Frankliniella occidentalis* Pergande (Thysanoptera: Thripidae). *Journal of Pest Science* 77: 211–219.

Bonsignore, C.P. and Vacante, V. (2012) Influences of botanical pesticides and biological agents on *Orius laevigatus - Frankliniella occidentalis* dynamics under greenhouse conditions. *Journal of Plant Protection Research* 52: 15–23.

Bonte, J., Musolin, D. L., Conlong, D., and De Clercq, P. (2016) Diapause and winter survival of two *Orius* species from southernAfrica. *BioControl* 61(5): 519–532.

Bouagga, S., Urbaneja, A., Rambla, J.L., Granell, A., and Pérez-Hedo, M. (2018) *Orius laevigatus* strengthens its role as a biological control agent by inducing plant defenses. *Journal of Pest Science* 91: 55–64.

Brødsgaard, H. F. and Enkegaard, A. (1995) Interactions among polyphagous Anthocorid bugs used for thrips control and other beneficials in multi-species biological pest management system. *Mededelingen Faculteit Landbouwkundigeen Toegepaste Biologische Wetenschappen Universiteit Gent* 60(3a): 893–900.

Brødsgaard, H. F. and Enkegaard, A. (2005) Intraguild predation between *Orius majusculus* (Reuter) (Hemiptera: Anthocoridae) and *Iphiseius degenerans* Berlese (Acarina: Phytoseiidae). *IOBC/WPRS Bulletin* 28: 19–22.

Bronnimann, H. (1964) Rearing anthocorids on an artificial medium. *CIBC Technical Bulletin* 4: 147–150.

Brower, J. H. and Press, J. W. (1992) Suppression of residual populations of stored-product pests in empty corn bins by releasing the predator *Xylocoris flavipes*. *Biological Control* 2: 66–72.

Brower, J. H., Smith, L., Vail, P. V., and Flinn, P. W. (1996) Biological control (pp. 223–286). In: B. Subramanyam and Hagstrum, D. W. (eds.), *Integrated Management of Insects in Stored Products*. Marcel Dekker, New York.

Bu, W. J. and Zheng, L. Y. (2001) *Hemiptera Lasiochilidae Lyctocoridae Anthocoridae Fauna Sinica, Insecta.* Vol. 24. Science Press, Beijing, vii+267 pp (in Chinese, with English summary).

Bueno, V. H. P., Mendes, S. M., and Carvalho, L. M. (2006) Evaluation of a rearing method for the predator *Orius insidiosus*. *Bulletin of Insectology* 59: 1–6.

Bush, L., Kring, T. J., and Ruberson, J. R. (1993) Suitability of greenbugs, cotton aphids and *Heliothis virescens* eggs for development and reproduction of *Orius insidiosus*. *Entomologia Experimentalis et Applicata* 67: 217–222.

Butt, A. and Xaaceph, M. (2015) Functional response of *Oxyopes javanus* (Araneidae: Oxyopidae) to *Sogatella furcifera* (Hemiptera: Delphacidae) in laboratory and mesocosm. *Pakistan Journal of Zoology* 47: 89–95.

Calixto, A. M., Bueno, V. H. P., Montes, F. C., Silva, A. C., and van Lenteren, J. C. (2013) Effect of different diets on reproduction, longevity and predation capacity of *Orius insidiosus* (Say) (Hemiptera: Anthocoridae). *Biocontrol Science and Technology* 23: 1245–1255.

Carayon, J. (1972) Caractèressystématiques et classifi cation des Anthocoridae (Hemipt.). *Annales de la Société Entomologique de France, Nouvelle Série* 8: 309–349.

Carpintero, D. L. (2002) Catalogue of the neotropical Anthocoridae. *Revista de la Sociedad Entomológica Argentina* 61: 25–44.

Carpintero, D. L. (2015) Minute pirate bugs (Anthocoridae and Lyctocoridae) (pp. 217–236). In: Panizzi, A. R. and Grazia, J. (eds.), *True Bugs (Heteroptera) of the Neotropics*. Springer, Doderecht, Heidelberg, New York, London.

Carvalho, L.M., Bueno, V.H.P. and Castane, C. (2011) Olfactory response towards its prey *Frankliniella occidentalis* of wild and laboratory-reared *Orius insidiosus* and *Orius laevigatus*. *Journal of Applied Entomology* 135: 177–183.

Cassis, G. and Gross, G. F. (1995) Hemiptera: Heteroptera (Coleorrhyncha to Cimicormorpha). Anthocoridae (pp. 23–42). In: Houston, W. K. and Maynard, G. V. (eds.), *Zoological Catalogue of Australia*. Vol. 27.3A. CSIRO Publishing, Melbourne, xv+506 pp.

Castane, C. and Zalom, F. G. (1994) Artificial oviposition substrate for rearing *Orius insidiosus* (Hemiptera: Anthocoridae). *Biological Control* 4: 88–91.

Chacko, M. J. (1973) Observations on some natural enemmies of *Pineus* sp. (Hem.: Adelgidae) at Shillong (Meghalaya), India, with special reference to *Tetraphleps raoi* Ghauri (Hem.: Anthocoridae). *Technical Bulletin of the Commonwealth Institute of Biological Control* 16: 41–46.

Chow, A., Chau, A., and Heinz, K.M. (2008) Compatibility of *Orius insidiosus* (Hemiptera: Anthocoridae) with *Amblyseius degenerans* (Acari: Phytoseiidae) for control of *Frankliniella occidentalis* (Thysanoptera: Thripidae) on cut roses. *Biological Control* 44: 259–270.

Chow, A., Chau, A., and Heinz, K.M. (2010) Compatibility of *Amblyseius (Typhlodromips) swirskii* (Athias-Henriot) (Acari: Phytoseiidae) and *Orius insidiosus* (Hemiptera: Anthocoridae) for biological control of *Frankliniella occidentalis* (Thysanoptera: Thripidae) on roses. *Biological Control* 53: 188–196.

Coll, M. and Izraylevich, S. (1997) When predators also feed on plants: Effects of competition and plant quality on omnivore-prey population dynamics. *Annals of the Entomological Society of America* 90: 155–161.

Coll, M. and Ridgway, R.L. (1995) Functional and numerical responses of *Orius insidiosus* (Heteroptera: Anthocoridae) to its prey in different vegetable crops. *Annals of the Entomological Society of America* 88: 732–738.

Coll, M., Smith, L. A., and Ridgway, R. L. (1997) Effects of plants on the searching efficiency of a generalist predator: The importance of predator-prey spatial association. *Entomologia Experimentalis et Applicata* 83: 1–10.

Devasahayam, S. (2000) *Bioecology of Leaf Gall Thrips, Liothrips karnyi Bagnall Infesting Black Pepper*. Ph.D. thesis. Department of Zoology, University of Calicut, Kerala, India, 187 pp.

Devasahayam, S. and Koya, A. (1994) Natural enemies of major insect pests of black pepper (*Piper nigrum* L.) in India. *Journal of Spices and Aromatic Crops* 3: 50–55.

Devi, N. and Gupta, P. R. (2010) Anthocorid bugs encountered on cultivated crops and ornamentals, and an attempt to rear *Orius niger* Wolff under laboratory conditions. *Pest Management and Economic Zoology* 18: 313–320.

Devi, W. Sangeeta, Singh, O. D., and Varatharajan, R. (2007) Predatory efficiency and biology of two species of anthocorid bugs feeding on *Megalurothrips distalis* (Karny) (Thysanoptera:Thripidae). *Journal of Biological Control* 21 (Special Issue): 193–196.

Dhaliwal, G. S. (1976) Intensity of insect infestation under rural storage conditions in the Punjab. *Entomologists' Newsletter* 6: 8–9.

Dhaliwal, G. S. (1977) Incidence of storage insect pests in rural areas. *Indian Journal of Entomology* 39: 114–116.

Díaz, M. A., Avellaneda, J., and Rodríguez, D. (2020) Effect of diet, maintenance frequency, and environmental conditions on the rearing of *Orius insidiosus* (Hemiptera: Anthocoridae) in Neotropical highlands. *International Journal of Tropical Insect Science* 40: 503–512.

Distant, W. L. (1904) Rhynchotal notes, XXI. *Annals and Magazine of Natural History, Series 7* 14: 219–222.

Distant, W. L. (1906) Rhynchota. – Vol. III. (Heteroptera–Homoptera). In: Bingham, C. T. (ed.), *The Fauna of British India Including Ceylon and Burma*. Taylor and Francis, London, xiv+503 pp.

Distant, W. L. (1910) Rhynchota. – Vol. V. Heteroptera: Appendix. In: Shipley, A. E. and Marshall, G. A. K. (eds.), *The Fauna of British India Including Ceylon and Burma*. Taylor and Francis, London, xii+362 pp.

Donnelly, B. E. and Phillips. T. W. (2001) Functional response of *Xylocoris flavipes* (Hemiptera: Anthocoridae): Effect of prey species and habitat. *Environmental Entomology* 30: 617–624.

Down, R. E., Cuthbertson, A. G. S., Mathers, J. J., and Walters, K. F. A. (2009) Dissemination of the entomopathogenic fungi, *Lecanicillium longisporum* and *L. muscarium*, by the predatory bug, *Orius laevigatus*, to provide concurrent control of *Myzus persicae*, *Frankliniella occidentalis* and *Bemisia tabaci*. *Biological Control* 50: 172–178.

Drukker, B., Bruin, J., and Sabelis, M. W. (2000) Anthocorid predators learn to associate herbivore induced plant volatiles with presence or absence of prey. *Physiological Entomolgy* 25: 260–265.

Drukker, B., Scutareanu, P., and Sabelis, M. W. (1995) Do anthocorid predators respond to synomones from psylla-infested pear trees under field conditions? *Entomologia Experimentalis et Applicata* 77(2): 193–203.

El-Basha, N. A., Salman, M. S., and Osman, M. A. (2012) Functional response of *Orius albidipennis* (Hemiptera: Anthocoridae) to the two-spotted spider mite *Tetranychus urticae* (Acari: Tetranychidae). *Journal of Entomology* 9: 248–256.

Falamarzhi, S. H., Asad, G. H., and Hosseini, R. (2000) Species inventory, preys and host plants of Anthocoridae *sensu lato* (Hemiptera: Heteroptera) in Shiraz and its environs (Iran, Fars Province). *Acta Entomologica Musei Nationalis Pragae* 49: 33–42.

Fathi, S.A.A., Asghari, A., and Sedghi, M. (2008) Interaction of *Aeolothrips intermedius* and *Orius niger* in controlling *Thrips tabaci* on potato. *International Journal of Agriculture & Biology* 10: 521–525.

Ferkovich, S. M., Venkatesan, T., Shapiro, J. P., and Carpenter, J. E. (2007) Effects of composition and size of prey and diet domes on egg production by *Orius insidiosus* (Heteroptera: Anthocoridae). *Florida Entomologist* 90(3): 502–508.

Finke, D. L. and Denno, R. F. (2003) Intraguild predation relaxes natural enemy impacts on herbivore populations. *Ecological Entomology* 28: 67–73.

Fischer, S., Linder, C., and Freuler, J. (1992) Biology and utilization of *Orius majusculus* Reuter (Heteroptera: Anthocoridae) for the control of thrips *Frankliniella occidentalis* Perg. and *Thrips tabaci* Lind., in greenhouses. *Rev Suisse Viticult Arboricult Horticult* 24: 119–127.

Fransen, J.J. and Tolsma, J. (1992) Releases of the minute pirate bug, *Orius insidiosus* (Say) (Hemiptera: Anthocoridae), against western flower thrips, *Frankliniella occidentalis* (Pergande), on chrysanthemum. *Mededelingen Faculteit Landbouwwetenschappen Rijksuniversiteit Gent* 57: 479–489.

Funao, T. and Yoshiyasu, Y. (1995) Development and fecundity of *Orius sauteri* (Poppius) (Hemiptera: Anthocoridae) reared on *Aphis gossypii* Glover and corn pollen. *Japanese Journal of Applied Entomology and Zoology* 39: 84–88.

Gao, W. C. (1987) The anthocorid *Duforiella ater* (Dufour) - a new natural enemy of Japanese pine bast scale *Matsucoccus matsumarae* (Nuwana). *Acta Entomologica Sinica* 30: 271–276.

Gao, Y., Reitz, S. R., Wang, J., Tamez-Guerra, P., Wang, E., Xu, X., and Lei, Z. (2012) Potential use of the fungus *Beauveria bassiana* against the western flower thrips *Frankliniella occidentalis* without reducing the effectiveness of its natural predator *Orius sauteri* (Hemiptera:Anthocoridae). *Biocontrol Science and Technology* 22: 803–812.

Gardiner, M.M. and Landis, D.A. (2007) Impact of intraguild predation by adult *Harmonia axyridis* (Coleoptera: Coccinellidae) on *Aphis glycines* (Hemiptera: Aphididae) biological control in cage studies. *Biological Control* 40: 386–395.

Gentleman, W., Leising, A., Frost, B., Strom, S., and Murray, J. (2003) Functional responses for zooplankton feeding on multiple resources: A review of assumptions and biological dynamics. *Deep-Sea Research II: Topical Studies in Oceanography* 50: 2847–2875.

Ghahari, H., Carpentiro, D. L., and Ostovan, H. (2009) An annotated catalogue of the Iranian anthocoridae (Hemiptera: Heteroptera: Cimicomorpha). *Acta Entomologica Musei Nationalis Pragae* 49: 43–58.

Ghauri, M. S. K. (1964) Notes on the Hemiptera from Pakistan and adjoining areas. *Annals and Magazine of Natural History* 6: 409–421.

Ghauri, M. S. K. (1965) Notes on the Hemiptera from Pakistan and adjoining areas. *Annals and Magazine of Natural History, Series* 13: 673–688.

Ghauri, M. S. K. (1972) The identity of *Orius tantillus* (Motschulsky) and notes on Oriental Anthocoridae (Hemiptera, Heteroptera). *Journal of Natural History* 6: 409–421.

Ghosh, D., Poddar, S., and Raychaudhuri, D. N. (1981) Natural enemy complex of *Aphis craccivora* Koch. And *Lipaphis erysimi* (Kalt.) in and around Calcutta, West Bengal. *Science and Culture* 47: 58–60.

Gillespie, D. R. and Quiring, D. J. M. (1992) Competition between *Orius tristicolor* (White) (Hemiptera: Anthocoridae) and *Amblyseius cucumeris* (Oudemans) (Acari: Phytoseiidae) feeding on *Frankliniella occidentalis* (Pergande) (Thysanoptera: Thripidae). *The Canadian Entomologist* 124: 1123–1128.

Gitonga, L. M., Overhalt, W. A., Lohr, B., Magambo, J. K., and Mueke, J. M. (2002) Functional response of *Orius albidipennis* (Hemiptera: Anthocoridae) to *Megalurothrips sjostedti* (Thysanoptera: Thripidae). *Biological Control* 24: 1–6. https://doi.org/10.1016/S1049-9644(02)00001-4.

Gómez-Polo, P., Alomar, O., Castañé, C., and Agustí, N. (2012) Analysing predation of *Orius majusculus* (Heteroptera: Anthocoridae) in lettuce crops by PCR. *IOBC/WPRS Bulletin* 80: 103–108.

Gómez-Polo, P., Alomar, O., Castañé, C., Riudavets, J., and Agustí, N. (2013) Identification of *Orius* spp. (Hemiptera: Anthocoridae) in vegetable crops using molecular techniques. *Biological Control* 67: 440–445.

Guo, J. Y., Wu, M., and Wan, F. H. (2002) Effects of cold storage on the adults and eggs of *Orius sauteri*. *Chinese Journal of Biological Control* 18: 10–12.

Gupta, T. (2009) *Studies on Production and Utility of Anthocorid predators – With Special Reference to Orius spp. and Blaptostethus spp.* Ph.D. thesis. Submitted to Mysore University, vi+207 pp.

Gupta, T. and Ballal, C. R. (2015) Biology and feeding potential of *Xylocoris flavipes* (Reuter) predating on eggs of different moth pests infesting stored grains. In: *Proceedings of International Conference on Innovative Insect Management Approaches for Sustainable Agro Ecosystem*, 27–30 January, 2015, 348 pp.

Gupta, T. and Ballal, C. R. (2006) Biology and feeding potential of an Anthocorid predator, *Orius tantillus* (Motsch) (Heteroptera: Anthocoridae) on *Sitotroga cerealella*. *Indian Journal of Plant Protection* 34(2): 168–172.

Gupta, T. and Ballal, C. R. (2007) Feeding preference of anthocorid predators for parasitised and un-parasitised eggs. *Journal of Biological Control* 21: 73–78.

Gupta, T. and Ballal, C. R. (2009) Protocols for commercial production of *Orius tantillus* (Motschulsky) (Hemiptera: Anthocoridae). *Journal of Biological Control* 23(3): 385–391.

Gupta, T., Ballal, C. R. and Kadam, Shashikala S. (2018) Morphology and performance specifications of *Blaptosthethus pallescens* Poppius (Heteroptera: Anthocoridae) when reared on two alternate laboratory hosts. *Journal of Biological Control* 32(3): 172–178.

Hagler, J. R., Jackson, C. G., Isaacs, R., and Machtley, S. A. (2004) Foraging behavior and prey interactions by a guild of predators on various lifestages of *Bemisia tabaci*. *Journal of Insect Science* 4: 1–13. www/insectscience.org/14.11.

Hagler, J. R. and Naranjo, S. E. (1994) Determining the frequency of heteropteran predation on sweet potato whitefly and pink bollworm using multiple ELISAs. *Entomologia Experimentalis et Applicata* 72: 59–66.

Hassanzadeh-Avval, M., Sadeghi-Namaghi, H., and Fekrat, L. (2018) Prey preference and prey switching in *Anthocoris minki* Dohrn (Hemiptera: Anthocoridae). *Journal of Asia-Pacific Entomology* 21: 1116–1121. https://doi.tps://doi.org/10.1016/j.aspen.2018.08.007.

Hassanzadeh-Avval, M., Sadeghi-Namaghi, H., and Fekrat, L. (2020) Molecular and morphological identification of *Anthocoris* spp. (Hemiptera: Anthocoridae) predators of three economically important psyllid species in Razavi Khorasan province, Northeastern Iran. *Biologia* 75: 1667–1677.

Henry, T. J. (1988) Family Anthocoridae (pp. 12–28). In: Henry, T. J. and Froeschner, R. C. (eds.), *Catalog of the Heteroptera, or True Bugs, of Canada and the Continental United States*. E. J. Brill, Leiden, New York, København, Köln, xix+958 pp.

Hodgson, C. and Aveling, C. (1988) Anthocoridae (pp. 279–292). In: Minks, A. K. and Harrewijin (eds.), *Aphids, Their Biology, Natural Enemies and Control*. Elsevier Science Publishers, Amsterdam, The Netherlands.

Horton, D. R. (2008) Minute Pirate Bugs (Hemiptera: Anthocoridae). In: Capinera, J. L. (eds.), *Encyclopedia of Entomology*. Springer, Dordrecht. https://doi.org/10.1007/978-1-4020-6359-6_4633

Horton, D. R., Lewis, T. M., and Dobbs, T. T. (2013) Interceptions of Anthocoridae, Lasiochilidae, and Lyctocoridae at the Miami Plant Inspection Station (Hemiptera: Heteroptera). *Florida Entomologist* 96(2): 482–497.

Horton, D. R., Lewis, T. M., Garczynski, S. F., Thomsen-Archer, K., and Unruh, T. R. (2016) Morphological and genetic re-appraisal of the *Orius* Fauna of the Western United States (Hemiptera: Heteroptera: Anthocoridae). *Annals of the Entomological Society of America* 109(2): 319–334.

Hulshof, J. and Linnamaki, M. (2002) Predation and oviposition rate of the predatory bug *Orius laevigatus* in the presence of alternative food. *IOBC/WPRS Bulletin* 25: 107–110.

Isenhour, D. J. and Yeargan, K. V. (1981) Interactive behavior of *Orius insidiosus* [Hem.: Anthocoridae] and *Sericothrips variabilis* [Thys.: Thripidae]: Predator searching strategies and prey escape tactics. *Entomophaga* 26: 213–219.

Ito, K. (2007) A simple mass rearing method for predaceous *Orius* bugs in the laboratory. *Applied Entomology and Zoology* 42: 573–577.

Ito, K. and Nakata, T. (1998a) Diapause and survival in winter in two species of predatory bug, *Orius sauteri* and *O. minutus*. *Entomologia Experimentalis et Applicata*, 89: 271–276.

Ito, K. and Nakata, T. (1998b) Effect of photoperiod on reproductive diapause in the predatory bugs, *Orius sauteri* (Poppius) and *O. minutus* (Linnaeus) (Heteroptera: Anthocoridae). *Applied Entomology and Zoology*, 33(1): 115–120.

Jalalizand, A., Modaresi, M., Tabeidian, S. A., and Karimy, A. (2011) Functional response of *Orius niger niger* (Hemiptera: Anthocoridae) to *Tetranychus urticae* (Acari: Tetranychidae): Effect of host plant morphological feature. In: *International Conference on Food Engineering and Biotechnology: IPCBEE 9*, IACSIT Press, Singapore, 2011.

Janssen, A., Pallini, A., Venzon, M., and Sabelis, M. W. (1998) Behavior and indirect interactions in food webs of plant inhabiting arthropods.*Experimental and Applied Acarology* 22: 497–521.

Jung, S. and Lee, S. (2011) Systematic and phylogenetic study of the Korean *Orius* species based on morphological and molecular data (Heteroptera: Anthocoridae). *Applied Entomology and Zoology* 46(2): 153–164.

Jung, S., Yamada, K., and Lee, S. (2013) Annotated catalog, biological notes and diagnoses of the flower bugs (Heteroptera: Anthocoridae *sensu lato*) in the Korean Peninsula. *Journal of Asia-Pacific Entomology* 16: 421–427.

Kawai, A. (1995) Control of *Thrips palmi* Karny (Thysanoptera, Thripidae) by *Orius* spp. (Heteroptera, Anthocoridae) on greenhouse eggplant. *Applied Entomology and Zoology* 30: 1–7.

Kawamoto, K., Sasaki, M., Nakaima, H., Onuma, M., and Uchida, N. (1999) Development and reproductive ability of *Orius tantillus* (Motschulky) (Hemiptera: Anthocoridae) reared on egg of *Plodia interpunctella* (Hubner). *Research Bulletin of the Plant Protection Service, Japan* 35: 87–91.

Ke, Y. and Bu, W. (2007) Female copulatory tubes and the subdivision of the genus *Anthocoris* (Heteroptera: Anthocoridae: Anthocorini). *European Journal of Entomology* 104: 89–98. http://www.eje.cz/scripts/viewabstract.php?abstract=1202.

Kemasa, O., Kaewpradit, A., and Suasaard, W. (2008) Biology of *Orius maxidentex* Ghauri (Hemiptera: Anthocordidae) and feeding capacity against *Bemisia tabaci* (Gennadius) (Homoptera: Aleyrodidae). In: *Proceedings of the 46th Kasetsart University Annual Conference*, pp. 52–57.

Kheradmand, K., Pourali, Z., and Jamshidnia, A. (2017) Influence of temperature on the functional response of the predatory bug, *Anthocoris minki pistaceae* (Hemiptera: Anthocoridae), a predator of *Agonoscena pistaciae* (Hemiptera: Psyllidae). *Zoology and Ecology* 26: 64–68.

Kim, J. H. (2009) Cold storage effect on the biological characteristics of *Orius laevigatus* (Fieber) (Hemiptera: Anthocoridae) and *Phytoseiulus persimilis* Athias-Henriot (Acari: Phytoseiidae). *Korean Journal of Applied Entomology* 48: 361–368.

Kiman, Z. B. and Yeargan, K. V. (1985) Development and reproduction of the predator *Orius insidiosus* (Hemiptera: Anthocoridae) reared on diets of selected plant material and arthropod prey. *Annals of the Entomological Society of America* 78: 464–467.

Kobayashi, T. and Osakabe, M. (2009) Pre-winter copulation enhances overwintering success of *Orius* females (Heteroptera: Anthocoridae). *Applied Entomology and Zoology* 44 (1): 47–52.

Krishnaswamy, S. and Yaseen, M. I. (1972) Hemiptera: Anthocoridae. In: Rao, P. and Ghani, M. A. (eds.), *Studies on Predators of Adelges spp. in the Himalayas*. Miscellaneous Publication, Commonwealth Institute of Biological Control, No. 3, pp. 4–14.

Kumar, N. S. and Ananthakrishnan, T. N. (1984) Predator-thrips interactions with reference to *Orius maxidentex* Ghauri and *Carayanocoris indicus* Muraleedharan (Anthocoridae: Heteroptera). *Proceedings of the Indian National Science Academy, B* 50: 139–145.

Lattin, J. D. (1999) Bionomics of the Anthocoridae. *Annual Review of Entomology* 44: 207–231.

LeCato, G. L. (1976) Predation by *Xylocoris flavipes*: Influence of stage, species and density of prey and of starvation and density of predator. *Entomophaga* 21: 217–221.

LeCato, G. L., Collins, J. M., and Arbogast, R. T. (1977) Reduction of residual populations of stored product insects by *Xylocoris flavipes* (Hemiptera: Anthocoridae). *Journal of the Kansas Entomological Society* 50: 84–88.

Lee, K. S. and Lee, J. H. (2007) Rearing of *Orius strigicollis* (Heteroptera: Anthocoridae) on artificial diet. *Entomological Research* 34: 299–303.

Lehrman, A., Boddum, T., Stenberg, J. A., Orians, C. M., and Björkman, C. (2013) Constitutive and herbivore-induced systemic volatiles differentially attract an omnivorous biocontrol agent to contrasting Salix clones. *AoB Plants* 5: plt005. doi:10.1093/aobpla/plt005.

Lingren, P. D. (1977) *Campoletis sonorensis*: Maintenance of a population on tobacco budworms in a field cage. *Environmmental Entomology* 6: 72–76.

Lingren, P. D., Lukefahr, M. J., Diaz, M. Jr. and Harstack, A. Jr. (1978) Tobacco budworm control in caged cotton with a resistant variety, augmentative releases of *Campoletis sonorensis*, and natural control by other beneficial species. *Journal of Economic Entomology* 71: 739–745.

Lundgren, J. G., Wyckhuys, K. A. G., and Desneux, N. (2009) Population responses by *Orius insidiosus* to vegetational diversity. *Biocontrol* 54: 135–142.

Lyla, K. R., Beevi, P., and Ballal, C. R. (2006) Field evaluation of anthocorid predator, *Cardiastethus exiguus* Poppius against *Opisina arenosella* Walker (Lepidoptera: Oecophoridae) in Kerala. *Journal of Biological Control* 20: 229–232.

Madadi, H., Enkegaard, A., Brødsgaard, H.F., Kharrazi-Pakdel, A., Ashouri, A., and Mohaghegh-Neishabouri, J. (2008) *Orius albidipennis*: Intraguild predation of and prey preference for *Neoseiulus cucumeris* on different host plants. *Entomologica Fennica* 19: 32–40.

Madadi, H., Enkegaard, A., Brødsgaard, H.F., Kharrazi-Pakdel, A., Ashouri, A., and Mohaghegh-Neishabouri, J. (2009) Interactions between *Orius albidipennis* (Heteroptera: Anthocoridae) and *Neoseiulus cucumeris* (Acari: Phytoseiidae): Effects of host plants under microcosm condition. *Biological Control* 50: 137–142.

Magalhães, S., Tudorache, C., Montserrat, M., van Maanen, R., Sabelis, M.W., and Janssen, A. (2004) Diet of intraguild predators affects antipredator behavior in intraguild prey. *Behavioral Ecology* 16: 364–370.

Martin, P. B., Ridgeway, R. L., and Schuetze, C. E. (1978) Physical and biological evaluations of an encapsulated diet for rearing *Chrysopa carnea*. *Florida Entomologist* 61: 145–152.

McCaffrey, J. P. and Horsburgh, R. L. (1986) Functional response of *Orius insidiosus* (Hemiptera: Anthocoridae) to the European red mite *Panonychus ulmi* (Acari: Tetrnychidae), at different constant temperatures. *Environmental Entomol*ogy 15: 532–535.

McDaniel, S. G., Sterling, W. L., and Dean, D. A. (1981) Predators of tobacco budworm larvae in Texas cotton. *Southwestern Entomologist* 6: 102–108.

Mendes, S. M., Bueno, H., Argolo, V. M., and Silvera, L. C. (2002) Type of prey influences biology and consumption rate of *Orius insidiosus* (Say) (Hemiptera, Anthocoridae). *Revista Brasileira de Entomologia* 46(1): 99–103.

Messelink, G. J. and Janssen, A. (2014) Increased control of thrips and aphids in greenhouses with two species of generalist predatory bugs involved in intraguild predation. *Biological Control* 79: 1–7.

Meyling, N. and Pell, J. K. (2006) Detection and avoidance of an entomopathogenic fungus by a generalist insect predator. *Ecological Entomology* 31: 162–171.

Meyling, N. V., Enkegaard, A., and Brodsgaard, H. (2003) Two *Anthocoris* bugs as predators of glasshouse aphids – voracity and prey preference. *Entomologia Experimentalis et Applicata* 108: 59–70.

Mochizuki, M. and Yano, E. (2007) Olfactory response of the anthocorid predatory bug *Orius sauteri* to thrips-infested eggplants. *Entomologia Experimentalis et Applicata* 123: 57–62.

Montserrat, M., Albajes, R., and Castane, C. (2000) Functional response of four heteropteran predators preying on greenhouse whitefly (Homoptera: Aleyrodidae) and western flower thrips (Thysanoptera: Thripidae). *Environmental Entomology* 29: 1075–1082.

Mukherjee, A. B., Chaudhury, A. K. S., Bronniman, H., and Chu, Y. (1971) Artificial diets for rearing of *Xylocoris flavipes* (Reuter), a predator of some pests of stored cereals. *Indian Journal of Entomology* 33: 356–358.

Muraleedharan, N. (1975) New record of *Cardiastethus pygmaeus pygmaeus pauliani* (Lansbury) from India. *Newsletter, Zoological Survey of India* 1: 67.

Muraleedharan, N. (1977a) Notes on a collection of Lyctocorinae (Hemiptera: Anthocoridae) from Garo Hills (Meghalaya). *Records of the Zoological Survey of India* 73: 239–246.

Muraleedharan, N. (1977b) Some genera of anthocorinae (Heteroptera: Anthocoridae) from south India. *Entomon* 2: 231–236.

Muraleedharan, N. (1977c) A new genus of Anthocoridae (Heteroptera) from South India. *Oriental Insects* 11: 463–466.

Muraleedharan, N. and Ananthakrishnan, T. N. (1971) Bionomics of *Montandoniola moraguesi* (Puton) (Heteroptera, Anthocoridae) a predator on gall thrips. *Bulletin of Entomology* 12: 4–10.

Muraleedharan, N. and Ananthakrishnan, T. N. (1974a) New and little-known species of *Orius* Wolff from South India. *Oriental Insects* 8(1): 37–41.

Muraleedharan, N. and Ananthakrishnan, T. N. (1974b) The genus *Buchananiella* Reuter from India with description of a new species (Heteroptera: Anthocoridae).*Oriental Insects* 8: 33–35.

Muraleedharan, N. and Ananthakrishnan, T. N. (1974c) A new species of the genus *Scoloposcelis* Fieber (Hem. Anthocoridae) from South India. *Journal of Natural History* 8: 511–512.

Muraleedharan, N. and Ananthakrishnan, T. N. (1978a) On the tribes Cardiastethini and Almeidini from South India (Hemiptera: Anthocoridae). *Bulletin of the Zoological Survey of India* 1: 65–70.

Muraleedharan, N. and Ananthakrishnan, T. N. (1978b) Bioecology of four species of *Anthocoridae* (Hemiptera: Insecta) predaceous on thrips, with key to genera of anthocorids from India. *Occasional Paper, Records of the Zoological Survey of India* 11: 1–32.

Murdoch, W. W. and Oaten, A. (1975) Predation and population stability. *Advances in Ecological Research* 9: 1–131.

Musolin, D. L. and Ito, K. (2008) Photoperiodic and temperature control of nymphal development and induction of reproductive diapause in two predatory *Orius* bugs: Interspecific and geographic differences. *Physiological Entomology* 33(4): 291–301.

Nagai, K., Takagi, M., Nakashima, Y., and Hiramatsu, T. (1998) Selection of alternative prey for rearing *Orius tantillus* (Motschulsky). *Japanese Journal of Applied Entomology and Zoology* 42: 85–87

Naranjo, S. E. (2007) Intraguild predation on *Eretmocerus* sp. nr. *emiratus*, a parasitoid of *Bemisia tabaci*, by three generalist predators with implications for estimating the level and impact of parasitism. *Biocontrol Science and Technology* 17(5/6): 605–622.

Nasser, M. and Abdurahman, U. C. (1990) Reproductive biology and predatory behaviour of the anthocorid bugs (Anthocoridae: Hemiptera) associated with the coconut caterpillar, *Opisina arenosella* (Walker). *Entomon* 15: 3–4.

Önder, F. (1982) Contribution to the study of Turkish Anthocoridae (Heteroptera). *Eğe Üniversitesi Ziraat Fakultesi Yayınları* 459: 1–159.

Osekre, E. A., Wright, D. L., Marois, J. J., and Mailhot, D. J. (2008) Predator-prey interactions between *Orius insidiosus* (Heteroptera: Anthocoridae) and *Frankliniella tritici* (Thysanoptera: Thripidae) in cotton blooms. *The Journal of Cotton Science* 12: 195–201.

Ostovan, H., Ghahari, H., and Moulet, P. (2017) Updated catalogue of Iranian Anthocoridae (Hemiptera: Heteroptera: Cimicomorpha). *Zootaxa* 4311: 451–479.

Otieno, J. A., Pallmann, P., and Poehling, H. M. (2017) Additive and synergistic interactions amongst *Orius laevigatus* (Heteroptera: Anthocoridae), entomopathogens and azadirachtin for controlling western flower thrips (Thysanoptera: Thripidae). *BioControl* 62: 85–95. https://doi.org/10.1007/s10526-016-9767-7.

Parajulee, M. N. and Phillips, T. W. (1992) Laboratory rearing and field observations of *Lyctocoris campestris* (Heteroptera: Anthocoridae), a predator of stored product insects. *Annals of the Entomological Society of America* 85: 736–743.

Parajulee, M. N., Phillips, T. W., and Hogg, D. B. (1994) Functional response of *Lyctocoris campestris* (F.) adults: Effects of predator sex, prey species, and experimental habitat. *Biological Control* 4: 80–87.

Parker, N. J. B. (1981) A method for mass rearing the aphid predator *Anthocoris nemorum. Annals of Applied Biology* 99(3): 217–223.

Parrella, M. P. and Murphy, B. (1996) Western flower thrips: Identification, biology and research on the development of control strategies. *IOBC/WPRS Bulletin* 19(1): 115–118.

Patel, H. K., Patel, J. R., and Patel, S. N. (1975) Records of predators and their parasites from Gujarat. *Entomologists' Newsletter* 5: 8–9.

Péricart, J. (1972) *Faune de l' Europe et du BassinMéditerrèen*. No. 7. *Hémiptères Anthocoridae. Cimicidae et Microphysidae de l' Ouest- Paléarctique*. Masson et Cie, Paris, 402pp.

Péricart, J. (1996) Family Anthocoridae Fieber, 1836 – Flower bugs, minute pirate bugs (pp. 108–140). In: Aukema, B. and Riegler, Ch. (eds.), *Catalogue of the Heteroptera of the Palaearctic Region. Vol. 2. Cimicomorpha I.* The Netherlands Entomological Society, Amsterdam, xiv+361 pp.

Pillai, G. B. and Nair, K. R. (1993) A checklist of parasitoids and predators of *Opisina arenosella* Wlk. on coconut. *Indian Coconut Journal, Cochin* 23: 2–9.

Pirzadfard, S., Zandi-Sohani, N., Sohrabi, F., and Rajabpour, A. (2020) Intraguild interactions of a generalist predator, *Orius albidipennis*, with two *Bemisia tabaci* parasitoids. *International Journal of Tropical Insect Science* 40: 259–265. https://doi.org/10.1007/s42690-019-00075-w

Poppius, B. (1909) BeiträgezurKenntnis der Anthocoriden. *Acta Societatis Scientiarum Fennicae* 37(9): 1–43.

Poppius, B. (1913) ZurKenntnis des Miriden, Isometopiden, Anthocoriden, Nabiden und Schizopteriden Ceylons. *Entomologisk Tidskrift* 34: 239–260.

Pourali, Z., Kheradmand, K., and Mehrnejad, M. R. (2012) Effect of temperature on biological parameters of *Anthocoris minki pistaceae* (Hemipetera: Anthocoridae). *Zoology and Ecology* 22: 240–245.

Pourian, H. R., Talaei-Hassanloui, R., Kosari, A. A., and Ashouri, A. (2011) Effects of *Metarhizium anisopliae* on searching, feeding and predation by *Orius albidipennis* (Hem., Anthocoridae) on *Thrips tabaci* (Thy., Thripidae) larvae. *Biocontrol Science and Technology* 21: 15–21. https:/doi.org/10.1080/09583157.2010.517603

Pumariño, L., Alomar, O., and Lundgren, J. G. (2011) The influence of intraguild competitors on reproductive decisions by two predatory Heteroptera, *Orius insidiosus* (Anthocoridae) and *Nabis americoferus* (Nabidae). *Biocontrol Science and Technology* 21: 1321–1330.

Puysseleyr, V. D., Höfte, M., and De Clercq, P. (2014) Continuous rearing of the predatory anthocorid *Orius laevigatus* without plant materials. *Journal of Applied Entomology* 138: 45–51.

Qin, S. H. (1985) Mass rearing of predatory insects in the field for control of injurious mites in apple orchards. *Natural Enemies of Insects KunchongTiandi* 7: 117–124.

Queiroz, O. S., Ramos, R. S., Gontijo, L. M., and Picanco, M. C. (2015) Functional response of three species of predatory pirate bugs attacking eggs of *Tuta absoluta* (Lepidoptera: Gelechiidae). *Environmental Entomology* 44: 246–251.

Rajabpour, A., Seraj, A. A., Allahayari, H., and Shishehbor, P. (2011) Evaluation of *Orius laevigatus* (Heteroptera: Anthocoridae) for biological control of *Thrips tabaci* Lindeman (Thysanoptera: Thripidae) on greenhouse cucumber in south of Iran. *Asian Journal of Biological Sciences* 4: 257–267.

Rajabpour, A., Seraj, A. A., and Mehrnia, A. (2018) Developing an artificial diet for rearing *Orius albidipennis* Reuter (Het., Anthocoridae). *Journal of Plant Protection Research* 58(3): 276–281.

Rajasekhara, K. (1973) A new species of *Blaptostethus* (Hemiptera: Anthocoridae) from Mysore, India. *Annals of Entomological Society of America* 66: 86–87.

Rajasekhara, K. and Chatterji, S. (1970) Biology of *Orius indicus* a predator of *Taeniothrips nigricornis*. *Annals of Entomological Society of America* 63: 364–367.

Rao, S. V. R. (1976) Studies on the biology, bionomics and chemical control of the sorghum midge, *Contarinia sorghicola* (Coquillett). *Entomologists' Newsletter* 6: 14–15.

Rehman, S. U., Zhou, X., Ali, S., Asim Rasheed M., Islam, Y., Hafeez, M., Aamir Sohail M., and Khurram, H. (2020) Predatory functional response and fitness parameters of *Orius strigicollis* Poppius when fed *Bemisia tabaci* and *Trialeurodes vaporariorum* as determined by age-stage, two-sex life table. *Peer Journal* 8: e9540. https://doi.org/10.7717/peerj.9540

Reid, C. D. (1991) Ability of *Orius insidiosus* (Hemiptera: Anthocoridae) to search for, find and attack European corn borer and corn earworm eggs on corn. *Journal of Economic Entomology* 84: 83–86.

Reuter, O. M. (1884) *Monographia Anthocoridarum Orbis terrestris*. Helsingforsiae, 204 pp. [Also published in *Acta Societatis Scientiarum Fennicae*14 (1885): 555–758.]

Richards, P. C. and Schmidt, J. M. (1996) The suitability of some natural and artificial substrates as oviposition sites for the insidious flower bug, *Orius insidiosus*. *Entomologia Experimentalis et Applicata* 80: 325–333.

Rosenheim, J. A. (2005) Intraguild predation of *Orius tristicolor* by *Geocoris* spp. and the paradox of irruptive spider mite dynamics in California cotton. *Biological Control* 32: 172–179.

Rosenheim, J. A., Limburg, D. D., and Colfer, R. G. (1999) Impact of generalist predators on a biological control agent, *Chrysoperla carnea*: Direct observations. *Ecological Applications* 9: 409–417.

Rosenheim, J. A., Wilhoit, R., and Armer, C. A. (1993) Influence of intraguild predation among generalist insect predators on the suppression of an herbivore population. *Oecologia* 96: 439–449.

Rudolf, E., Malausa, J. C., Millot, P., and Pralavorio, R. (1993) Influence of cold temperature on biological characteristics of *Orius laevigatus* and *Orius majusculus* (Het.: Anthocoridae) [Influence des basses temperatures sur lespotentialitesbiologiquesd'Orius laevigatus et d'Oriusmajusculus (Het.: Anthocoridae)]. *Entomophaga* 38(3): 317–325.

Saini, E., Cervantes, V., and Alvarado, L. (2003) Efecto de la dieta, temperatura y hacinamiento, sobre la fecundidad, fertilidad y longevidad de Oriusinsidiosus (Say) (Heteroptera: Anthocoridae). *Revista de Investigación Agropecuarias* 32(2): 21–32.

Sanchez, J. and Lacasa, A. (2002) Modelling population dynamics of *Orius laevigatus* and *O. albidipennis* (Hemiptera: Anthocoridae) to optimize their use as biological control agents of *Frankliniella occidentalis* (Thysanoptera: Thripidae). *Bulletin of Entomological Research* 92: 77–88. https://doi.org/10.1079/BER2001136

Sanchez, J. A., Alcazar, A., Llamas, A., Lacasa, A., and Bielza, P. (2000) Integrated pest management strategies in sweet pepper plastic houses in the Southeast of Spain.*Bulletin OILB SROP* 23: 21–30.

Sanderson, J. P., Brødsgaard, H. F., and Enkegaard, A. (2005) Preference assessment of two *Orius* spp. for *Neoseiulus cucumeris* vs. *Frankliniella occidentalis*. *IOBC/WPRS Bulletin* 28: 221–224.

Sands, W. A. (1957) The immature stages of some British Anthocoridae (Hemiptera). *Transactions of Royal Entomological Society, London* 109: 295–310.

Sarker, A. C., Islam, W., and Parween, S. (2016) Host-stage specific effects on the biological parameters of *Xylocoris flavipes* (Reuter) preyed on *Cryptolestes pussilus* (Schon.). *Tropical Agricultural Research & Extension* 19(3–4): 305–312.

Sarmento, R. A., Pallini, A., Venzon, M., Souza, F. F., Molina-Rugama, A. J., and Oliveira, C. L. (2007) Functional response of the predator *Eriopis connexa* (Coleoptera: Coccinellidae) to different prey types. *Brazilian Archives of Biology and Technology* 50: 121–126.

Saulich, A. Kh. and Musolin, D. (2009) Seasonal development and ecology of anthocorids (Heteroptera, Anthocoridae). *Entomological Review* 89(5): 501–528.

Saxena, R. C. (1975) Integrated approach for the control of *Thrips tabaci* Lind. *Indian Journal of Agricultural Sciences* 45(9): 434–436.

Saxena, R. C. (1981a) Natural enemy complex of *Aphis craccivora* Koch. and *Lipaphis erisymi* (Kalt.) in and around Calcutta, West Bengal. *Science and Culture* 47: 58–60.

Saxena, R. C. (1981b) Observations on some predators and parasites of *Thrips tabaci* Lind. *Bulletin of Entomology* 22(1/2): 97–100.

Schausberger, P. and Walzer, A. (2001) Combined versus single species release of predaceous mites: Predator–predator interactions and pest suppression. *Biological Control* 20: 269–278.

Schmidt, J., Richards, P., Nadel, H., and Ferguson, G. (1995) A rearing method for the production of large numbers of insidiosus flower bug *Orius insidiosus* (Say) (Hemiptera: Anthocoridae). *The Canadian Entomologist* 127(3): 445–447. https://doi.org/10.4039/Ent127445-3

Schuh, R. T. and Slater, J. A. (1995) *True Bugs of the World (Hemiptera:Heteroptera). Classification and Natural History*. Cornell University Press, Ithaca and London, xii+337 pp.

Schuh, R. T. and Štys, P. (1991) Phylogenetic analysis of cimicomorphan family relationships (Heteroptera). *Journal of the New York Entomological Society* 99: 298–350.

Schuldiner-Harpaz, T. and Coll, M. (2012) Identification of *Orius* (Heteroptera: Anthocoridae) females based on egg operculum structure. *Journal of Economic Entomology* 105: 1520–1523.

Scutareanu, P., Drukker, B., Bruin, J., Posthumus, M. A., and Sabelis, M. W. (1997) Volatiles from Psylla-infested pear trees and their possible involvement in attraction of anthocorid predators. *Journal of Chemical Ecology* 23: 2241–2260.

Scutareanu, P., Lingeman, R., Drukker, B., and Sabelis, M. W. (1999) Cross-correlation analysis of fluctuations in local populations of pear psyllids and anthocorid bugs. *Ecological Entomology* 24: 354–362.

Seagraves, M. P. and Lundgren, J. G. (2010) Oviposition response by *Orius insidiosus* (Hemiptera: Anthocoridae) to plant quality and prey availability. *Biological Control* 55: 174–177.

Sengonca, C., Ahmadi, K., and Blaeser, P. (2008) Biological characteristics of *Orius similis* Zheng (Heteroptera, Anthocoridae) by feeding on different aphid species as prey. *Journal of Plant Diseases and Protection* 115: 32–38.

Shahpouri, A., Yarahmadi, F., and Sohani, N. Z. (2019) Functional response of the predatory species *Orius albidipennis* Reuter (Hemiptera: Anthocoridae) to two life stages of *Bemisia tabaci* (Genn.) (Hemiptera: Aleyrodidae). *Egyptian Journal of Biological Pest Control* 29: 14. https://doi.org/10.1186/s41938-019-0119-7.

Shakya, S., Weintraub, P.G. and Coll, M. (2009) Effect of pollen supplement on intraguild predatory interactions between two omnivores: The importance of spatial dynamics. *Biological Control* 50: 281–287.

Shapiro J. P. and Ferkovich S. M. (2006) Oviposition and isolation of viable eggs from *Orius insidiosus* in a parafilm and water substrate: Comparison with green beans and use in enzyme-linked immunosorbent assay. *Annals of the Entomological Society of America* 99: 586–591.

Shapiro, J. P., Shirk, P. D., Kelley, K., Lewis, T. M., and Horton, D. R. (2010) Identity of two sympatric species of *Orius* (Hemiptera: Heteroptera: Anthocoridae). *Journal of Insect Science* 10: 189.

Shapiro, J. P., Shirk, P. D., Reitz, S., and Koenig, R. (2009) Sympatry of *Orius insidiosus* and *O. pumilio* (Hemiptera: Anthocoridae) in North Central Florida. *Florida Entomologist* 92: 362–366.

Shimizu, T. and Kawasaki, K. (2001) Geographic variability in diapause response of Japanese *Orius* species. *Entomologia Experimentalis et Applicata* 98: 303–316.

Shipp, J. L., Zhang, Y., Hunt, D. W. A., and Ferguson, G. (2003) Influence of humidity and greenhouse microclimate on the efficacy of *Beauveria bassiana* (Balsamo) for control of greenhouse arthropod pests. *Environmental Entomology* 32: 1154–1163.

Sigsgaard, L. (2010) Habitat and prey preferences of the two predatory bugs *Anthocoris nemorum* (L.) and *A. nemoralis* (Fabricius) (Anthocoridae: Hemiptera-Heteroptera). *Biological Control* 53: 46–54.

Silveira, L. C. P., Bueno, V. H. P., and Van Lenteren, J. C. (2004) *Orius insidiosus* as biological control agent of thrips in greenhouse chrysanthemums in the tropics. *Bulletin of Insectology* 57: 103–109.

Sing, S. E. and Arbogast, R. T. (2008) Optimal *Xylocoris flavipes* (Reuter) (Hemiptera: Anthocoridae) density and time of introduction for suppression of bruchid progeny in stored legumes. *Environmental Entomology* 37: 131–142.

Singh, T. V. K., Singh, K. M., and Singh, R. N. (1991) Influence of inter-cropping: III. Natural enemy complex in groundnut. *Indian Journal of Entomology* 53: 333–368.

Skirvin, D. J., Kravar-Garde, L., Reynolds, K., Jones, J., de Courcy Williams, M. E. (2006) The influence of pollen on combining predators to control *Frankliniella occidentalis* in ornamental chrysanthemum crops. *Biocontrol Science and Technology* 16: 99–105.

Snyder, W. E., Ballard, S. N., Yang, S., Clevenger, G. M., Miller, T. D., Ahn, J. J., Hatten, T. D., and Berryman, A. A. (2004) Complementary biological control of aphids by the ladybird beetle *Harmonia axyridis* and the parasitoid *Aphelinus asychis* on greenhouse roses. *Biological Control* 30: 229–235.

Snyder, W. E. and Ives, A. R. (2001) Generalist predators disrupt biological control by a specialist parasitoid. *Ecology* 82: 705–716.

Sobhy, I. S., Sarhan, A. A., Shoukry, A. A., El-Kady, G. A., Mandour, N. S., and Reitz, S. R. (2010) Development, consumption rates and reproductive biology of *Orius albidipennis* reared on various prey. *Biological Control* 55(6): 753–765.

Sohrabi, F., Enkegaard, A., Shishehbor, P., Moosa Saber, M., and Mosaddegh, M. S. (2013) Intraguild predation by the generalist predator *Orius majusculus* on the parasitoid *Encarsia formosa*. *BioControl* 58: 65–72.

Southwood, T. R. E. (1956) The structure of the eggs of Terrestrial Heteroptera and its relationship to the classification of the group. *Transactions of the Royal Entomological Society, London* 108(6): 163–221.

Southwood, T. R. E. (1960) The flight activity of Heteroptera. *Transactions of Royal Entomological Society, London* 112: 173–220.

Stenberg, J. A., Lehrman, A., and Björkman, C. (2010) Uncoupling direct and indirect plant defences: Novel opportunities for improving crop security in willow plantations. *Agriculture, Ecosystems & Environment* 139: 528–533.

Štys, P. and Vilimová, J. (1989) Unusual numbers of instars in Heteroptera: A review. *Acta entomologica Bohemoslovaca* 86: 1–32.

Sun, M., Lü, B., Lyu, Z.-y., Zhang, Y., Zhai, Y.-f., Chen, H., Zheng, L., and Yu, Y. (2017) Comparison of diapause and non-diapause induced *Orius sauteri* (Hemiptera: Anthocoridae) based on a two sex life table in the laboratory. *Journal of Asia-Pacific Entomology* 20: 1301–1306.

Tan, X. L., Wang, S., and Zhang, F. (2013) Optimization an optimal artificial diet for the predatory bug *Orius sauteri* (Hemiptera: Anthocoridae). *PLoS One* 8(4): e61129. https://doi.org/10.1371/journal.pone.0061129

Tawfik, M. F. S. and El-Husseini, M. M. (1971) The life history of the anthocorid predator, *Blaptostethus piceus* Fieber var. *pallescens* Poppius (Hemiptera: Anthocoridae). *Bulletin de la Societe Entomologique d' Egypte* 55: 239–252.

Tawfik, M. F. S. and El-Sherif, S. I. (1969) The biology of *Pyroderces simplex* Wlsm., an important pest on maize plants in UAR (Lepidoptera: Lavernidae). *Bulletin de la SociteEntomologique d' Egypte* 53: 615–628.

Tawfik, M. F. S., Kira, M. T., and Metwally, S. M. I. (1974) On the abundance of major pests and their associated predators in corn plantations. *Bulletin de la SocieteEntomologique d' Egypte* 58: 168–177.

Thontadarya, T. S. and Rao, K. J. (1987) Biology of *Orius maxidentex* Ghauri (Hemiptera: Anthocoridae), a predator of the sorghum earhead midge, *Contarinia sorghicola* (Coquillet). *Mysore Journal of Agricultural Sciences* 21(1): 27–31.

Tillman, P. G. (2006) Relative attractiveness of developmental stages of sorghum panicles to predator, *Orius insidiosus* (Say), and prey, *Helicoverpa zea* (Boddie). *Journal of Entomological Science* 41: 248–252.

Toft, S., Jensen, K., Sørensen, J. G., Sigsgaard, L., and Holmstrup, M. (2020) Food quality of *Ephestia* eggs, the aphid *Rhopalosiphum padi* and mixed diet for *Orius majusculus*. *Journal of Applied Entomology* https://doi.org/10.1111/jen.12739

Traczyk, E., Funderburk, J., and Martini, X. (2020) Foraging behavior responses of *Orius insidiosus* to thrips cues. *Entomologia Experimentalis et Applicata* 168: 716–722.

Urbaneja, A., Leon, F. J., Gimenez, A., Aran, E., and van der Blom, J. (2003) Interaccion de *Neoseiulus (Amblyseius) cucumeris* (Oudemans) (Aca.: Phytoseiidae) en la instalacion de *Orius laevigatus* (Fieber) (Hem.: Anthocoridae) eninvernaderos de pimiento. *Boletin de Sanidad Vegetal Plagas* 29: 347–357.

van den Bosch, R., Reynolds, H. T., and Dietrick, E. J. (1956) Toxicity of widely used insecticides to beneficial insects in California cotton and alfalfa fields. *Journal of Economic Entomology* 49: 359–363.

van den Meiracker R. A. F. and Sabelis, M. W. (1999) Do functional responses of predatory arthropods reach a plateau? A case study of *Orius insidiosus* with western flower thrips as prey. *Entomologia Experimentalis et Applicata* 90: 323–329.

van Laerhoven, S., Gillespie, D. R., and Mcgregor, R. R. (2000) Leaf damage and prey type determine search effort in *Orius tristicolor*. *Entomologia Experimentalis et Applicata* 97: 167–174.

Varshney, R., Budhlakoti, N., and Ballal, C. R. (2020) Functional response of three species of predatory pirate bugs to different densities of blossom thrips, *Frankliniella schultzei* Trybom (Thysanoptera: Thripidae). *Current Science* 118: 827–833.

Veer, V. (1984) On some new records of predators of *Thrips flavus* Schrank and *Thrips hawaiiensis* Morgan (Thysanoptera: Thripidae) from Dehradun (India). *Indian Journal of Forestry* 7: 245–246.

Venkatesan, T., Ballal, C. R., and Rabindra, R. J. (2008) Biological control of coconut black-headed caterpillar *Opisina arenosella* using *Goniozus nephantidis* and *Cardiastethus exiguus*. *Technical Bulletin* No. 39. Project Directorate of Biological Control, Bangalore, 14 pp.

Venzon, M., Janssen, A., and Sabelis, M. W. (1999) Attraction of a generalist predator towards herbivore-infested plants. *Entomologia Experimentalis et Applicata* 93: 305–314.

Vieira, N. F., Truzi, C. C., Veiga, A. C. P., Sipriano-Nascimento, T. P., Vacari, A. M., and De Bortoli, S. A. (2018) Life table of *Xylocoris afer* (Hemiptera: Anthocoridae) feeding on eggs of *Corcyra cephalonica* (Lepidoptera: Pyralidae) and *Plutella xylostella* (Lepidoptera: Plutellidae). *Journal of Asia-Pacific Entomology* 21: 1379–1383. https://doi.org/10.1016/j.aspen.2018.10.016.

Viswanathan, T. R. and Ananthakrishnan, T. N. (1973). Observations on the biology. ecology. and behaviour of *Ecacantkothrips sanquineus* Bagnall (Tubulifera. Thysanoptera) *Current Science* 42: 727–728.

Viswanathan, T. R. and Ananthakrishnan, T. N. (1974) Population fluctuations of three species of anthophilous Thysanoptera in relation to the numerical response of their predator, *Orius minutus* L. (Anthocoridae: Hemiptera). *Current Science* 43: 19–20.

Wang, C. L., Lee, P. C., and Wu, Y. J. (2001) Field augmentation of *Orius strigicollis* (Heteroptera: Anthocoridae) for the control of thrips in Taiwan. In: *Extension Bulletin*, Vol. 500, Food and Fertilizer Technology Center, Taipei, ROC, pp. 1–9.

Wang, F. H., Zhou, W. R., and Wang, R., (1994) Studies on the overwintering survival rate of *Orius sauteri* Hem.: Anthocoridae in Beijing. *Chinese Journal of Biological Control* 10: 100–102.

Wang, F. H., Zhou, W. R., and Wang, R. (1996) Studies on the method of rearing *Orius sauteri*. *Chinese Journal of Biological Control* 12: 49–51.

Wang, S., Michaud, J. P., Tan, X. L., and Zangi, F. (2014) Comparative suitability of aphids, thrips and mites as prey for the flower bug *Orius sauteri* (Hemiptera: Anthocoridae) *European Journal of Entomology* 111(2): 221–226.

Wang, Sx., Di, N., Chen, X., Zhang, F., Biondi, A., Desneux, N., and Wang, S. (2019) Life history and functional response to prey density of the flower bug *Orius sauteri* attacking the fungivorous sciarid fly *Lycoriella pleuroti*. *Jornal of Pest Science* 92: 715–722.

Wang, T., Zhang, P., Ma, C., Yasir Ali, M., Gao, G., Lu, Z., and Zalucki, M. P. (2021) Is *Orius sauteri* Poppius a promising biological control agent for walnut aphids? An assessment from the laboratory to field. *Insects* 12: 25. https://doi.org/10.3390/insects12010025

Whitcomb, W. H. and Bell, K. (1964) Predaceous insects, spiders, and mites of Arkansas cotton fields. *Bulletin 690*, Agricultural Experiment Station, Division of Agriculture, University of Arkansas, Fayetteville.

Wittmann, E. J. and Leather, S. R. (1997) Compatibility of *Orius laevigatus* Fieber (Hemiptera: Anthocoridae) with *Neoseiulus* (*Amblyseius*) *cucumeris* Oudemans (Acari: Phytoseiidae) and *Iphiseius* (*Amblyseius*) *degenerans* Berlese (Acari: Phytoseiidae) in the biocontrol of *Frankliniella occidentalis* Pergande (Thysanoptera: Thripidae). *Experimental and Applied Acarology* 21: 523–538.

Yamada, K., Ballal, C. R., Gupta, T., and Poorani, J. (2010a) Description of a new species of *Anthocoris* (Hemiptera: Heteroptera: Anthocoridae) from southern India, associated with striped mealybug on purple orchid tree. *Acta Entomologica Musei Nationalis Pragae* 50: 415–424.

Yamada, K., Bindu, K., and Nasser M. (2008) Taxonomic and biological notes on *Cardiastethus affinis* and *C. pseudococci pseudococci* (Hemiptera: Heteroptera: Anthocoridae) in India. *Zootaxa* 1910: 59–68.

Yamada, K., Bindu, K., and Nasser, M. (2010b) The second species of the genus *Rajburicoris* Carpintero and Dellapé (Hemiptera: Heteroptera: Anthocoridae) from southern India, with reference to autapomorphies and systematic position of the genus. *Proceedings of the Entomological Society, Washington* 112: 464–472.

Yamada, K., Bindu, K., Nasreem, A., Nasser, M., Ballal, C. R., and Poorani, J. (2010c) The flower bugs found in agro-ecosystems of Southern India (Heteroptera:Anthocoridae). *Presented at the Fourth Quinquennial Meeting of the International Heteropterists Society*, July 12–12, 2010, Tianjin, China.

Yamada, K., Bindu, K., Nasreem, A., and Nasser, M. (2011) A new flower bug of the genus *Montandoniola* (Hemiptera: Heteroptera: Anthocoridae), a predator of gall-forming thrips on black pepper in southern India. *Acta Entomologica Musei Nationalis Pragae* 51: 1–10.

Yamada, K., Yasunaga, T., and Ishikawa, T. (2016) Family Anthocoridae (pp. 422–428). In: Hayashi, M., Tomokuni, M., Yoshizawa, K. and Ishikawa, T. (eds.), *Catalogue of the Insects of Japan. Vol. 4. Paraneoptera*. The Entomological Society of Japan &Toukashobo, Tokyo, 629 pp.

Yano, E. and van Lenteren, J. C. (1996) Biology of *Orius sauteri* (Poppius) and its potential as a biocontrol agent for *Thrips palmi* Karny. *Bulletin OILB SROP* 19: 203–206.

Yasunaga, T. (1997) The flower bug genus *Orius* Wolff (Heteroptera, Anthocoridae) from Japan and Taiwan, Part II. *Applied Entomology and Zoology* 32: 379–386.

Zambrano, J. A. (2009) Evaluación de cuatroraciones de huevos de *Sitotroga cerealella* comoalimento de ninfas de *Orius insidiosus* (Say) (Hemiptera: Anthocoridae) y dos sustratosvegetales (*Ipomoea batata* y *Bidens pilosa*) para la oviposición de adultosencondiciones de laboratorio. Trabajo de gradoIngeniero Agrónomo. Universidad de Zamorano, Honduras, 20pp.

Zeng, G., Zhi, J. R., Zhang, C. R., Zhang, T., Ye, J. Q., Zhou, L., Hu, C. X., and Ye, M. (2021) *Orius similis* (Hemiptera: Anthocoridae), a promising Candidate Predator of *Spodoptera frugiperda* (Lepidoptera: Noctuidae). *Journal of Economic Entomology* https://doi.org/10.1093/jee/toaa318

Zhang S. C., Zhu F., Zheng X. L., Lei C. L., and Zhou X. M. (2012) Survival and developmental characteristics of the predatory bug *Orius similis* (Hemiptera: Anthocoridae) fed on *Tetranychus cinnabarinus* (Acari: Tetranychidae) at three constant temperatures. *European Journal of Entomology* 109(4): 503–508.

Zhou, W. and Wang, R. (1989) Rearing of *Orius sauteri* (Hem.: Anthocoridae) with natural and artificial diets. *Chinese Journal of Biological Control* 5: 9–12.

Zhou, W., Wang, R., and Qiu, S. (1991) Use of soybean sprouts as the oviposition material in mass rearing of *Orius sauteri* (Het:.Anthocoridae). *Chinese Journal of Biological Control* 7: 7–9.

4 Mirid and Geocorid Predators

Richa Varshney
ICAR-National Bureau of Agricultural Insect Resources

CONTENTS

LEARNING OBJECTIVES

1. To know the systematics and diagnosis of predatory mirid and geocorid bugs.
2. To learn the diversity and target prey species.
3. To understand the basic biology, behavior and optimum conditions for rearing.
4. To know the competition and intraguild relationships.
5. To know about functional response.
6. To learn their role as biocontrol agents.

4.1 INTRODUCTION

Predaceous mirids (Hemiptera: Miridae) and geocorids (Hemiptera: Geocoridae) belong to suborder Heteroptera (true bugs) of the order Hemiptera. All species belonging to this suborder have a hemimetabolous life cycle. Juvenile stages (nymphs) resemble the adults but do not have genitalia

DOI: 10.1201/9781003370864-4

or wings. Heteroptera typically possess two pairs of wings: forewings (hemelytra) are partly membranous, whereas hindwings are fully membranous. Most species have thoracic scent glands, which can be used for defense. Heteropteran bugs possess piercing-sucking mouthparts to pierce plant or animal tissues (Wheeler, 2001; Gullan and Cranston, 2010; Schaefer, 2009).

4.2 PREDACEOUS MIRIDS

4.2.1 Systematics

The Miridae, or plants bugs, are the largest family of true bugs, containing about 1,300 genera and approximately 11,139 described species distributed worldwide (On-line Systematic Catalogue of Plant Bugs (http://research.amnh.org/pbi/catalog)). The Miridae are one of the most diverse families of class Insecta and second largest after Cicadellidae among Hemimetabola (Konstantinov and Namyatova, 2019). Mirid bugs also use plants as an oviposition substrate. These bugs exhibit a wide range of food preferences, like phytophagy, carnivory and omnivory (Cassis and Schuh, 2012). In Europe, several omnivorous mirids are used in augmentative biocontrol programs and are in high demand. The evolutionary success of the group was evidently determined by close trophic associations and parallel evolution with seed plants (Konstantinov and Namyatova, 2019). The suprageneric phylogeny developed by Schuh (1976) serves as a baseline for mirid relationships in the literature with eight recognized subfamilies. They are Bryocorinae, Cylapinae, Deraeocorinae, Psallopinae, Isometopinae, Mirinae, Phylinae and Orthotylinae. Among them Isometopinae, Cylapinae and Deraeocorinae are sometimes considered strict or obligate predators and other families show mixed feeding preferences (Wheeler, 2001). Some mirid species exhibit significant economic impacts; some are pests of food and fiber crops (Wheeler, 2000), whereas others are beneficial species (Wheeler, 2001) used as biocontrol agents (Sanchez et al., 2004; Wheeler, 2001).

4.2.1.1 Family Diagnosis

Mirids are fragile and delicate bugs, varying in size, and are elongate to ovoid in shape. The head is triangular and prognathous but it can be hypognathous. The eyes are usually large and ocelli are absent except in Isometopinae, which have two ocelli and greatly enlarged eyes. The antennae and labium are four segmented. The forewings cover the abdomen and are commonly deflexed at the costal fracture. The hemelytra are characterized by a cuneus at the apex of the corium and one or two closed cells in the forewing membrane. Wing polymorphism is observed (aptery, coleoptery, brachyptery, microptery and macroptery).

Species of *Dicyphus* (*Dicyphus*) (Heteroptera: Miridae: Bryocorinae: Dicyphini) are routinely used as biocontrol agents of vegetable crop pests in the Northern Hemisphere. Their identification is difficult. Sanchez and Cassis (2018) employed an integrative approach, combining molecular, morphometrics and morphological data, for the delimitation of species. They have synonymized a few species of *Dicyphus*. *Dicyphus marocannus* Wagner and *Dicyphus bolivari atlanticus* Wagner are synonymized with *D. bolivari* Lindberg; *D. umbertae* Sanchez and Cassis are synonymized with *D. cerastii* Wagner and *D. baezi* Ribes is synonymized with *D. rubicundus* Blote.

4.2.2 Diversity and Prey

Plant bug family is a hyperdiverse family and is found in all major biogeographic regions of the world (Schuh, 1995, 2013), and it occupies all the layers of plant communities. This family includes mainly plant feeders and some of them are serious agricultural and forest pests (Wheeler, 2001). However, this family also contains a wide range of omnivorous bugs showing behavior ranging from phytozoophagy to zoophytophagy (Alomar and Wiedenmann, 1996; Coll and Guershon, 2002; Eubanks et al., 2003). Distant (1904, 1910) was one of the earliest to work on the Indian fauna of this group. In the Fauna of British India series, Distant provided descriptions of 251 species of

Miridae and this even today serves as an important work for the study of this group in the Indian subcontinent. Ballard (1927) described 15 species of mirids from south India, which forms another important work from southern India (Yeshwant, 2013).

Varshney et al. (2018a) reported *Dortus primarius* Distant as a predator of thrips, aphids and other soft-bodied insects in the field. It feeds on eggs and first instar larvae of *Tuta absoluta* in laboratory (Personal observation). *Termatophylum orientale* Poppius was reported for the first time from India. It was collected from mango, papaya and *Peltophorum pterocarpum* where it was found sharing niche along with other predators, like anthocorids and geocorids, and pests, like thrips, mites and lepidopteran larvae (Varshney and Yeshwanth, 2018). Yasunaga et al. (2015) discovered *Chimairacoris lakshimiae* Yasunaga, Schuh and Cassis, which inhabit the leaf-margin roll galls produced by *Trioza* sp. on the leaves of Indian banyan tree. This morphologically and ecologically specialized mirid belongs to the subfamily Phylinae on the basis of genitalic structures and is currently considered best placed in the subtribe Oncotylina of the tribe Phylini. The gall-producing psyllid, *Trioza* sp., and a mealybug, *Phenacoccus parvus* Morrison, are confirmed as prey of this mirid (Yasunaga et al., 2015).

Mirids are generalist omnivorous predators and are able to use various food resources, and thereby establish prior to the pest infestation (Lu et al., 2012; Messelink et al., 2012; Pina et al., 2012). The role of zoophytophagous bugs, such as *Nesidiocoris tenuis* (Reuter) and *M. pygmaeus*, in controlling thrips, leaf miners, aphids, spider mites, *Tuta absoluta* Meyrick and lepidopterans have been very well proven by different researchers (Calvo et al., 2009a, 2012; Perdikis and Lykouressis, 2002; van Lenteren, 2012; Arnó et al., 2010; Messelink et al., 2011; Urbaneja et al., 2012; Pérez-Hedo and Urbaneja, 2015; Mollá et al., 2011; Biondi et al., 2018). Other dicyphine predatory bugs, such as *Dicyphus cerastii* Wagner, *Dicyphus errans* (Wolff), *Dicyphus tamaninii* Wagner, *Macrolophus costalis* Fieber, are distributed in the Mediterranean region on tomato. *Dicyphus hesperus* Knight is native to North America and has been reported to be the most promising agent for pest control on the tomato crop (McGregor et al., 1999; Sanchez et al., 2003a; Shipp and Wang, 2006; Ingegno et al., 2013; Smith and Krey, 2019). *Deraeocoris nebulosus* (Uhler) occurs throughout the United States and Canada (Carvalho, 1957; Henry and Wheeler, 1988) and is recorded as a good predator of sweet potato whitefly (Jones and Snodgrass, 1998). The predatory plant bug, *Stethoconus praefectus* (Distant), belonging to subfamily Deraeocorinae was found feeding on avocado lace bug, *Pseudacysta perseae* (Heidemann), in Florida (Henry et al., 2009), coconut and banana lace bug (*Stephanitis typica* (Distant)) in India (Mathen and Kurian, 1972; Poorani et al., 2019), lantana tingid, *Teleonemia scrupulosa* (Stal) in India (Ganga Visalakshy and Jayanth, 1994), and *Stephanitis subfasciata* in Japan (Yasunaga et al., 1997). Species of *Stethoconus* are considered obligate lace bug predators (Mathen and Kurian, 1972; Neal et al., 1991; Wheeler, 2001). Mirid predators can be found on many different types of host plants, *viz*., Solanaceae, Asteraceae, Cucurbitaceae, Fabaceae and Pedaliaceae (Naselli et al., 2017; Biondi et al., 2016; Sánchez et al., 2003c).

These predatory mirids are being used in several biocontrol programs in the tomato-producing areas of southern Europe, North Africa and the Middle East (Urbaneja et al., 2012; Zappalà et al., 2013). However, *N. tenuis* and *M. pygmaeus* are produced in large numbers by different private companies in Europe, particularly in Mediterranean regions. *Deraeocoris lutescens* Schilling is a predatory bug found across Middle East and Europe, and feeds on a wide range of arthropod pests, such as aphids, small caterpillars, mites and insect eggs, on a wide variety of plants. *Nesidiocoris tenuis* was reported to feed on whitefly in tomato ecosystem in India (Sridhar et al., 2012a, b) and on *T. absoluta* (Sridhar et al., 2014; Ballal et al., 2016; Varshney and Ballal, 2017c).

4.2.3 Biology and Life History

The life cycle of mirids comprises an egg, five nymphal instars and adults. Eggs are laid interstitially within stems and petioles of plants. Research on the biology of mirids focuses typically on their food preferences and host and habitat affiliations. Mirids are phytophagous (Bryocorinae,

Orthotylinae, Phylinae and Deraeocorinae), mycetophagous (Cylapinae), carnivorous (Isometopinae and Deraeocorinae) and zoophytophagous (Bryocorinae: Dicyphinae), but the details of their life history need further investigation. For example, predaceous genera, such as *Phytocoris* Fallén show very strong associations with given plant species, but whether this occurs because they require some plant matter for completion of development or because they feed exclusively on herbivores associated with the plant is known only for very few species.

Biological traits of *N. tenuis* and *M. pygmaeus* have been studied on both *T. absoluta* and factitious prey *Ephestia kuehniella* Zeller (Sanchez et al., 2009; Urbaneja et al., 2005; Vandekerkhove and De Clercq, 2010; Mollá et al., 2014). Both the predators consume more numbers of *T. absoluta* eggs compared to the factitious host. However, *M. pygmaeus* could produce only less number of offsprings when it consumed only *T. absoluta* eggs (Mollá et al., 2014). The other predatory mirid, *D. errans,* was reported to prey actively on *T. absoluta* eggs and first-instar larvae to complete its life cycle on this pest. However, a high mortality of the predator was recorded when only first-instar larvae of tomato pinworm were provided (Ingegno et al., 2013). Rearing of different predatory mirid bugs on different diets (artificial food, bee pollen, various preys and plants) is responsible for differences in developmental period (Gillespie and McGregor, 2000; Perdikis and Lykouressis, 2000; Castañé et al., 2002; Castañé and Zapata, 2005). Generally, fast development was recorded on *E. kuehniella* eggs compared to natural prey, such as *Myzus persicae* Sulzer, *Trialeurodes vaporariorum* Westwood and *Tetranychus urticae* Koch (Fauvel et al., 1987; Tedeschi et al., 1999). Similarly, *D. bolivari* (=*D. maroccanus*) could complete its development in 19 days when *E. kuehniella* eggs were provided. Developmental period of *D. bolivari* and *D. errans* was approximately 6 and 4 days longer than that reported for *N. tenuis* and *M. pygmaeus,* respectively (Abbas et al., 2014). The number of *E. kuehniella* eggs consumed by *D. bolivari* was approximately three times more than *N. tenuis* and *M. pygmaeus,* probably because of longer nymphal duration of *D. bolivari* (Abbas et al., 2014). *Deraeocoris lutescens* Schillin was able to successfully feed, develop and reproduce when fed on eggs of *Sitotroga cerealella* (Olivier) as food on bean plants at an optimal temperature (Zadeh and Parvar, 2014). Similarly, *D. nebulosus* can survive, develop and reproduce normally using sweet potato whitefly immatures as prey (Jones and Snodgrass, 1998). Biology of this predatory bug has also been studied on oak lace bug *Corythucha arcuata* (Say) (Wheeler et al., 1975) and the cotton aphid *Aphis gossypii* Glover (Snodgrass, 1991) as prey. *Stethoconus praefectus* completed developmental period from egg to adult in 16 days when fed on avocado lacebug and consumed two to seven lacebugs per day (Holguin et al., 2009).

Biology and rearing protocol were developed for *D. primarius* (Figure 4.1). This predatory mirid can be easily reared on Ultraviolet (UV)-irradiated *Corcyra cephalonica* (Stainton). Total nymphal duration was 17 days and fecundity was 127 eggs/female (Varshney et al., 2018a). Until now, its role to damage plant has not been recorded. Its biology has been studied on *Frankliniella schultzei* Trybom and *T. absoluta* (Varshney and Ballal, 2019). Similarly, rearing protocol has been developed for *T. orientale*. Total developmental period from egg to adult was 16–21 days and adult longevity was 15–20 days. A female lays 45–50 eggs during its life span (Varshney and Yeshwanth, 2018).

Zoophytophagy is common in dicyphines. Several researchers assume that zoophytophagous bugs feed on plants to obtain water for maintaining their physiological status during extraoral digestion (Gillespie and McGregor, 2000). However, research on this aspect showed that besides water, several other nutrients can be taken up from plant materials and could improve life history traits of zoophytophagous bugs. Plant bugs, like *D. hesperus* and *M. pygmaeus* nymphs, when fed on *E. kuehniella* eggs with plants (tomato and tobacco, respectively), exhibited a higher survival and a shorter development than their counterparts provided with insect eggs and supplementary water (Gillespie and McGregor, 2000; De Puysseleyr, 2007).

4.2.4 Ecology and Behavior

Establishment and efficiency of generalist zoophytophagous predators depend on alternative prey, plant material during prey scarcity, competition, intraguild interactions and abiotic factors.

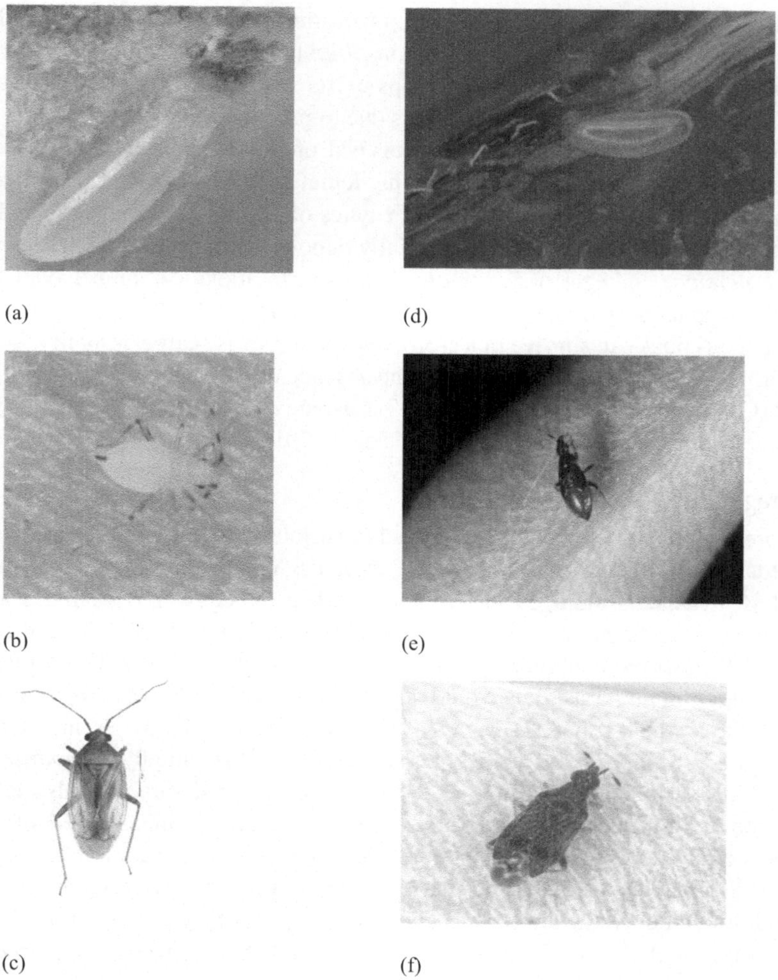

FIGURE 4.1 Life stages of *Dortus primarius*: A) Egg; B) Nymph; C) Adult and *Termatophylum orientale*: D) Egg; E) Nymph; F) Adult.

These factors may have an impact on population dynamics. Therefore, the interactions of these factors on biological parameters of zoophytophagous predators are utterly important to interpret the population dynamics and behavior to further exploit these predators in a biocontrol program. Sanchez et al. (2009) reported that temperature from 20°C to 30°C was optimum for *N. tenuis* development. Nymphal mortality was high at 35°C while low reproduction rate and high nymphal mortality were observed at 15°C. This indicates low adaptability of *N. tenuis* at low temperature (15°C), but it is better adapted to high temperatures (35°C) than other species, such as *M. caliginosus, M. pygmaeus* and *D. tamaninii*, and linked to its highest abundance in the summer months (Tavella and Goula, 2001; Sanchez et al., 2003b, 2006; Arnó et al., 2006). However, other species, *viz.*, *M. pygmaeus* and *M. caliginosus*, proved better than *N. tenuis* at a low temperature because their temperature threshold was below 10°C (Sanchez et al., 2009). *Nesidiocoris tenuis* is able to develop on a range of plant species including tomato (*Solanum lycopersicum*), pepper (*Capsicum annuum* L.), aubergine (*Solanum melongena*), cucumber (*Cucumis sativus* L.) and beans (Fabaceae spp.) (Calvo and Urbaneja, 2004). However, it is unable to complete development to adulthood on a strictly phytophagous diet (Urbaneja et al., 2005). Other plant bugs, *viz.*, *D. tamaninii*, *M. costalis* and *M. pygmaeus,* are able to develop without preying on plants. In India, *N. tenuis* is reported as a pest on sesame (Ahirwar et al., 2009), tobacco (Patel, 1980), bottle gourds, tomatoes and cucurbits

(Patel, 1980; Reddy and Kumar, 2004). However, much attention has been given toward the predator side of this bug and it is evaluated against *Tuta absoluta* in tomato in India (Varshney and Ballal, 2017c). The period of higher risk in tomato crops exists when *N. tenuis* reaches high population peaks and its prey decreases to very low numbers due to predation (Sanchez, 2009).

Oviposition behavior of bugs can increase survival of progeny and performance. Along with ensuring the availability of prey before egg laying, females also assess the physical and chemical structure of plant to identify high quality host. Females of *D. lutescens*, have the ability to distinguish different nutritional sources and preferentially deposit the eggs onto the leaves infested with *A. gossypii, M. persicae* and eggs of *S. cerealella*. This form of oviposition behavior appears to be correlated with nymphal development success (Azimizadeh et al., 2012).

Zoophytophagous bugs can survive in a crop even when prey is scarce or totally absent (Sanchez et al., 2004; Urbaneja et al., 2005). However, in the absence of prey, they may cause injury to plants and yield loss (Calvo et al., 2009a; Sanchez, 2009; Castañé et al., 2011; Biondi et al., 2016) and the level of damage varies species to species (Castañé et al., 2011).

4.2.4.1 Intraguild Predation

The use of more than one natural enemy may lead to various interactions, such as competition and intraguild predation (IGP), and may impact the efficiency of biocontrol agents. Adult females of *N. tenuis* and *M. pygmaeus* indulged in IGP and attacked the eggs of *T. absoluta* parasitized by *Trichogramma achaeae* Nagaraja and Nagarkatti. However, IGP did not reduce the overall level of biocontrol of *T. absoluta* (Chailleux et al., 2013; Cabello et al., 2015). These findings showed that adding *Trichogramma* parasitoids would significantly increase the level of control of the pest over what could be attained using *M. pygmaeus* or *N. tenuis* alone. However, interactions between *M. caliginosus* and *D. tamaninii* did not decrease the control of the whitefly *T. vaporariorum* (Lucas and Alomar, 2001; Castañé et al., 2004). Coexistence of two predators may also lead to exploitative competition and displacement. In the absence of prey, approximately 25% of *M. pygmaeus* nymphs did not complete their development in the presence of *N. tenuis* adult females (Perdikis and Lykouressis, 2000). Studies also depicted that *M. pygmaeus* nymphs were vulnerable to predation by *N. tenuis* adult females in the absence of alternative prey (Perdikis et al., 2009; Moreno-Ripoll et al., 2012; Perdikis et al., 2014). Furthermore, *N. tenuis* and *D. bolivari* (=*D. maroccanus*) avoided preying on each other in the presence of alternative prey though *N. tenuis* has a greater ability to outcompete (Gervassio et al., 2017).

4.2.4.2 Functional Response

Functional response study is necessary to find out the potential and capability of predator to find, kill and consume the prey (Nachman, 2006) and is one of the fundamental aspects to understand prey–predator dynamics to assess its effectiveness in controlling pest populations (Fernández-Arhex and Corley, 2003). Functional response is the change in an individual predator's rate of prey consumption with the change in prey density. Together, the numerical response (a reproduction response) and the functional response (a predation response) determine the pest population regulation capacity of a predator (Coll and Ridgway, 1995; Van Den Meiracker and Sabelis, 1999).

Natural enemies exhibiting Type III response could be considered as an efficient biocontrol agent to manage the prey population (Fernández-Arhex and Corley, 2003). Type III functional response was observed in *M. caliginosus* at varying density of *T. vaporariorum* (Enkegaard et al., 2001; Hamdan 2006); *N. tenuis* against *B. tabaci* pupae (Madbouni et al., 2017); *Macrolophus basicornis* (Stal) and *Engytatus varians* (Distant) at varying density of *T. absoluta* eggs (Van Lenteren, et al., 2016). The Type III response was originally considered to be characteristic of vertebrate predators and associated with learning behavior (Holling, 1959). Later, it was reported in invertebrate predators and parasitoids. Several studies indicated a Type II response for most of the predatory mirid bugs (Foglar et al., 1990; Song and Heong, 1997; Montserrat et al., 2000; Fantinou et al., 2008, 2009; Alvarado et al., 1997; Maselou et al., 2014, 2015; Madbouni et al., 2017; Ingegno et al., 2019). Types of response depends on various factors, *viz*., prey species and size (Hassell et al., 1977; Streams, 1994), predator's hunger level, size, and stage (Holling, 1959; Varshney et al., 2018b), temperature and other abiotic factors (Mohaghegh et al., 2001; Madbouni et al., 2017), experimental arena (Hemerik and Yano, 2011); and types of analysis used for the assessment of functional response (Van Lenteren et al., 2016).

4.2.4.3 Chemical Ecology

Numerous studies demonstrated that the herbivory of insects induces the production of herbivore-induced plant volatiles (HIPVs) (Leitner et al., 2005; Shiojiri et al., 2006) and HIPVs may be attractive to natural enemies of pests. Pérez-Hedo et al. (2018) demonstrated that HIPVs produced by the plant feeding of *N. tenuis* and *M. pygmaeus* were found as repellent for *Bemisia tabaci* (Gennadius) and attractive to *Encarsia formosa* Gahan, whereas they showed no effect on *T. absoluta*. Olfactory response study of *N. tenuis* exhibited that volatiles from eggplant infested with *Spodoptera litura* Fabricius were more attractive than eggplant infested with red spider mite (Rim et al., 2015).

4.2.5 As Biocontrol Agents

Predaceous mirids are receiving increasing attention as biocontrol agents of different pests (Thrips, whitefly, lepidopteran insects, etc.) in protected cultivation in various parts of Europe and North America. Role of the mirid bug as a biocontrol agent against the insect pest has been given in Table 4.1.

Macrolophus pygmaeus has been found in the northeast of Spain (Castañé et al., 2004; Gabarra et al., 2004) colonizing the tomato crop. It has been commercially produced and used as a generalist predator in European tomato greenhouses (van Lenteren, 2003; Arnó et al., 2009a). This is a well-known predator and is used for controlling several vegetable crop pests, including thrips, leafminers, aphids, mites and the eggs of lepidopteran pests. It has been shown to be a predator of the South American tomato pinworm, *T. absoluta* (Arnó et al., 2009b; Urbaneja et al., 2009).

Laboratory studies showed the efficacy of *D. tamaninii* and *M. caliginosus* against western flower thrips (WFT) (Riudavets and Castañé, 1998). *D. tamaninii* was very active in reducing and maintaining low *Frankliniella occidentalis* (Pergande) populations in cucumber (Gabarra et al., 1995; Castañé et al., 1996). Similarly, *M. caliginosus* was also recorded preying on *F. occidentalis* in a tomato greenhouse (Trottin-Caudal and Millot, 1994). Laboratory experiments have been conducted to evaluate the efficacy of *M. pygmaeus* against *T. absoluta*. An adult can feed on 100 or more than 100 *T. absoluta* eggs daily (Arnó et al., 2009b; Chailleux et al., 2013) and predation rate for *T. absoluta* is higher than other lepidopteran pests (Urbaneja et al., 2009). However, when both *B. tabaci* and *T. absoluta* were offered to this predator, it did not show any preference between whitefly nymph and *T. absoluta* eggs, but exhibited lower preference toward larvae of tomato pinworm (Jaworski et al., 2013). In greenhouse, it reduced *B. tabaci* population by 30 fold and pinworm population by 3.7 fold (Bompard et al., 2013). Similarly, Mollá et al. (2009) found that the release of this predator reduced leaf and fruit infestations by pinworm to the tune of 76% and 56%, respectively (Nannini et al., 2012; Urbaneja et al., 2012).

An Integrated Pest Management (IPM) program in Spain based on the augmentation and conservation of *M. pygmaeus* and *N. tenuis* resulted in 80% and 95% reduction in leaves and fruit damages in tomato due to pinworm attack compared to conventional approach including insecticides (Arnó et al., 2009a). Furthermore, *N. tenuis* along with *Bacillus thuringiensis* has resulted in 97% reduction in leaf damage by *T. absoluta* and a significantly high yield was obtained with this treatment compared to control (Mollá et al., 2011). Predatory mirids take some time to reach a level to be effective against the pest. Hence, focus should be more for early establishment of these bugs in order to achieve success. There are different strategies that help in early establishment of mirids on crop like release of predators in nursery and use of banker plants (Arnó et al., 2000; Lenfant et al., 2000). A strategy like release of 0.5 *N. tenuis* per plant with *E. kuehniella* eggs as an alternative prey in nursery stage proved to be successful for controlling *T. absoluta* and was implemented in more

TABLE 4.1
Predatory Mirids and Geocorids as Biocontrol Agents against Different Pests

Species	Target (s)	Crop (s)	Region of Use	Reference
		Miridae		
***Nesidiocoris tenuis* Reuter**	Whitefly (*Bemisia tabaci*)	Tomato	Spain	Calvo et al. (2009a, 2012)
		Sweet Pepper	Spain	Calvo et al. (2009b)
		Tomato	Spain	Sanchez (2008)
	Western flower thrips (WFT) and Whitefly	Sweet pepper	Spain	Bouagga et al. (2018)
	Tuta absoluta	Tomato	Algeria	Urbaneja et al. (2012)
		Tomato	Spain	Arnó et al. (2009a)
		Tomato	Spain	Belda et al. (2010)
		Tomato	India	Varshney and Ballal (2017c)
***Macrolophus caliginosus* Wagner**	Western flower thrips and Whitefly	Tomato	France	Trottin-Caudal and Millot (1994)
		Sweet pepper	Spain	Bouagga et al. (2018)
***Macrolophus pygmaeus (Rambur)*: Aphids**	Western flower thrips	*Euphorbia milli* and *Serissa foetida* flowers	Germany	Blaeser et al. (2004)
	Whitefly and *Tuta absoluta*	Tomato	France	Bompard et al. (2013)
	Tuta absoluta	Tomato	France	Biondi et al. (2018)
		Tomato	Spain	Arnó et al. (2009b); Guenaoui et al. (2011)
***Macrolophus caliginosus* Wagner**		Tomato	Iran	Baniameri and Cheraghian (2011)
		Tomato	Algeria	Urbaneja et al. (2012)
	Whitefly (*Bemisia tabaci*)	Tomato	Spain	Barnadas et al. (1998)
	Greenhouse Whitefly	Tomato	France	Fauvel et al. (1987)
	Leafminer	Tomato	Spain	Arnó et al. (2003)
***Dicyphus tamaninii* Wagner**	Western flower thrips and Whitefly	Cucumber	Spain	Gabarra et al. (1995); Castañé et al. (1996); Riudavets and Castañé (1998)
	Whitefly	Tomato	Spain	Barnadas et al. (1998)
	Aphid (*Myzus persicae*)	Sweet Pepper	The Netherlands	Messelink et al. (2015)
		Cucumber	Jordan	Ghabeish et al. (2008)
	Western flower thrips	*Euphorbia milli* and *Serissa foetida* flowers	Germany	Blaeser et al. (2004)
***Dicyphus hesperus* Knight**	Whitefly and Thrips	Tomato	Canada	McGregor et al. (1999); Gillespie et al. (2007)

(*Continued*)

TABLE 4.1 (*Continued*)
Predatory Mirids and Geocorids as Biocontrol Agents against Different Pests

Species	Target (s)	Crop (s)	Region of Use	Reference
***Dicyphus bolivari* (Lindberg) (=*D. maroccanus* Wagner)**	Western flower thrips and Whitefly	Tomato	Spain	Abbas et al. (2014)
***Dortus primarius* Distant**	Thrips	Capsicum	India	Varshney et al. (2019)
	Tuta absoluta	Tomato	India	Varshney and Ballal (2019)
Geocoridae				
***Geocoris proteus* Distant**	Red spider mites	Strawberry and Watermelon	Japan	Oida and Kadeno (2007)
	Thrips (WFT)	Sweet Pepper	Japan	Oida et al. (2007)
***Geocoris punctipes* (Say)**	*Helicoverpa (Heliothis)* spp.	Cotton	Texas	Lingren et al. (1968)
	Pectinophora gossypiella (Saunders)	Cotton	Southern California	Orphanides et al. (1971)
	Helicoverpa (Heliothis) zea	Cotton	USA	Bell and Whitcomb (1964)
	Pea aphid	Lima beans	USA	Eubanks and Denno (2000)
***Geocoris pallens* Stal**	*Manduca sexta* (Linnaeus)	*Nicotiana attenuata*	South western Utah	Schuman et al. (2013)
***Geocoris uliginosus* (Say)**	*Spodoptera frugiperda* (J.E. Smith)	Turfgrass	USA	Braman et al. (2003)
***Geocoris ochropterus* Fieber**	Whitefly	Cotton	India	Kapadia and Puri (1991)
	Thrips	Tomato	India	Varshney et al. (2018c)
	Red spider mites	Tomato	India	Varshney et al. (2018c)
	Helicoverpa armigera eggs	Tomato	India	Varshney and Ballal (2018b)

than 3,500 ha of commercial greenhouses in Spain (Urbaneja et al., 2012). In open-field processing tomatoes, release of *N. tenuis* resulted in the reduction of pinworm damage in tomato in Israel (Harpaz et al., 2016).

Nesidiocoris tenuis when released at the rate of one and four *N. tenuis* per plant was able to reduce 90% of the whitefly population in cages of 8 m^2 (Calvo et al., 2009a). *Nesidiocoris tenuis* and *M. pygmaeus* could establish on sweet pepper plant following the pre-plant release strategy before the infestation of thrips and whiteflies and result in continuous suppression of both the pests (Bouagga et al., 2018). Similarly, the combination of *M. pygmaeus* and *O. laevigatus* improved the biocontrol of *F. occidentalis* and *M. persicae* in sweet pepper crops (Messelink and Janssen, 2014). Conversely, the combined release of *N. tenuis* with *Amblyseius swirskii* Athias-Henriot did not improve the biocontrol of *B. tabaci* on sweet pepper (Calvo et al., 2009b). It could be due to their involvement in IGP. Messelink et al. (2015) demonstrated that *M. pygmaeus* and *D. tamaninii* could successfully reduce aphid populations when released prior to an artificially introduced aphid infestation on sweet pepper.

Several laboratory studies were conducted to study the predatory potential of *D. tamaninii* on different insect pests attacking vegetable crops. The role of this bug as predator of some key pests that affect vegetable crops was documented in laboratory studies that showed its voracious capacity for consuming whiteflies, thrips (Barnadas et al., 1998; Montserrat et al., 2000; Blaeser et al., 2004) and aphids (Sengonça and Saleh, 2002). Its efficacy to control whiteflies, thrips and aphids on tomato and cucumber crops was tested in cage studies (Gabarra et al., 1995; Castañé et al., 1996, 2009; Blaeser et al., 2004; Ghabeish et al., 2008). Plant damage was recorded in the form of punctures when pest density was low, and it decreased when pest density increased (Castañé et al., 2003). Another mirid bug, *Dicyphus hesperus,* was commercially produced and released in greenhouse-grown tomatoes for the biocontrol of whiteflies or thrips in both Canada and northern Europe (Gillespie et al., 2007). Releases in experimental tomato greenhouses resulted in the effective control of greenhouse whitefly, spider mites and *F. occidentalis* (McGregor et al., 1999; Shipp and Wang, 2006). Control of thrips and whitefly was achieved after introducing *D. hesperus*, but it could not control spider mites. Introduction of *D. hesperus* also proved effective in the control of gerbera pests and of aphids in pepper crops (Gillespie et al., 2007). In India, *N. tenuis* was evaluated against *T. absoluta*. Three releases of *N. tenuis* (1 pair/plant) resulted in the reduction of number of eggs and % mined area by 83% and 89%, respectively. Furthermore, number of necrotic rings per plant increased with decreasing *T. absoluta* population but flower abortion was not observed (Varshney and Ballal, 2017c).

4.3 GEOCORID BUGS

4.3.1 Systematics

Geocoridae, or big-eyed bugs, are a worldwide distributed, moderately species-rich, morphologically heterogenous group of the superfamily Lygaeoidea (Malipatil, 1994; Kóbor, 2019) comprising 30 genera and more than 290 described species (Henry, 2009, 2013; Malipatil, 2012; Rengifo-Correa et al., 2013; Dellapé et al., 2015). Geocoridae is represented by five subfamilies: the Australocorinae Malipatil, Bledionotinae Reuter, Geocorinae Stål, Henestarinae Douglas and Scott, and Pamphantinae Barber and Bruner (Slater, 1999; Malipatil, 2012; Henry et al., 2015). Among all the subfamilies, Geocorinae occurs in all the zoographical regions and includes more than 220 described species of 16 genera (Brailovsky, 2016; Kóbor, 2019). The Australocorinae are restricted to Australia (Malipatil, 2012), the Bledionotinae to the Palearctic (Slater, 1964; Péricart, 2001), the Except one species from Argentina, The Henestarinae belong to old world (Henry et al., 2015) and the Pamphantinae from the New World tropics (Henry, 2013; Rengifo-Correa et al., 2013), with the exception of one genus and species described from Queensland, Australia (Slater, 1981; Henry et al., 2015).

In India, most of the geocorids described so far belong to subfamily Geocorinae. From India, so far ten species of *Geocoris* Fallen have been reported from West Bengal and 23 species from the rest of India (Mukhopadhyay, 1988). Recently, Malipatil (2017) described a new tribe Indopamphantini of subfamily Pamphantinae of family Geocoridae, with new genus *Indopamphantus* containing one new species, *I. makutaensis,* as the first member of this subfamily from the Oriental Region. This myrmecomorphic species was collected from the Makuta range area within Coorg (Kodagu) district at altitudes of up to 909 m in the canopy of *Vateria indica* L. (Dipterocarpaceae), a tree indigenous to the Western Ghats in south India. Taxonomy and systematics of big eyed bugs are poorly studied and needs revision.

4.3.1.1 Family Diagnosis

The members of the family Geocoridae are recognized by their large reniform eyes that extend laterally beyond the anterior margin of the pronotum, the broad heads, the posteriorly curved abdominal sutures between terga IV–V and V–VI, abdominal spiracles on segments II, III and IV dorsal and those on V, VI and VII usually ventral (Henry, 1997), except for the members of the subfamily Australocorinae, which has all abdominal spiracles II–VII dorsal (Malipatil, 2012).

4.3.2 Diversity and Prey

Representatives of the Geocoridae inhabit most of the biomes with warm and moderate climates and can be found even in extreme biotopes, e.g., deserts and high mountains. Big-eyed bugs are generalist predators. However, their feeding on plants and seeds was also reported by researchers (Tamaki and Weeks, 1972; Eubanks and Denno, 1999). Studies also showed that despite plant feeding, the net effect of *Geocoris* spp. and plant interactions is usually beneficial to plants and it can be effective biocontrol agents in many agricultural systems (Hunter, 2009). They are reported to prey on different insects, including prey from at least 3 classes, 10 orders and 30 families of arthropods (Tamaki and Weeks, 1972; Crocker and Whitcomb, 1980). Along with prey, they have also been reported to feed on seeds and plant parts (Eubanks and Denno, 1999; Sweet, 2000). Both adults and nymphs are found to prey on various life stages of a wide range of arthropod pests. They are most known as predators of aphids and thrips (Kumar and Ananthakrishnan, 1985), representatives of lepidopteran families of twirler moths (Gelechiidae) (Bueno et al., 2016), owlet moths or armyworms (Noctuidae) (Torres et al., 2004), snout moths (Pyralidae) and hawkmoths or hornworms (Sphingidae) (Schuman et al., 2013), Colorado beetle (*Leptinotarsa decemlineata*) (Koss et al., 2004), phytophagous plant bugs, viz., *Adelphocoris* sp., *Lygus* sp. (Tong et al., 2011; Tamaki et al., 2019), spider mites (Oida and Kadono, 2007; Krey et al., 2017; Varshney et al., 2018c) and mealybugs (Varshney and Ballal, 2017b). *Geocoris lubra* Kirkaldy is one of the seven Australian *Geocoris* species (Malipatil, 1994) that is found in crops, such as cotton, lucerne and soybean (Evans, 1985; Mensah, 1999; Johnson et al., 2000). *Geocoris floridanus* Blatchley occurs in the southern United States, ranging from East Texas to Florida and north to Washington DC, although Georgia and Florida appear to be the center of distribution (Readio and Sweet, 1982). *Geocoris punctipes* (Say) is frequently recorded on strawberries and vegetables grown in plastic tunnels in the USA (Hagler and Sanchez, 2011) and is an effective predator of various pests of greenhouse crops (Tamaki and Weeks, 1972; Pendleton, 2002). During the 1990s, this species was commercially produced and sold to manage lepidopteran insect pests in North America (van Lenteren, 2012). In India, *Geocoris ochropterus* Fieber is a very common species found on cotton (*Gossypium* spp.), Lucerne (*Medicago sativa* L.), maize (*Zea mays* L.), sunflower (*Helianthus annuus* L.), tea (*Camellia sinensis* (L.) Kuntze), fruit, vegetable and flowering plants. It has been reported as a predator of thrips (Thysanoptera) (Kumar and Ananthakrishnan, 1985), cotton whitefly (Hemiptera) (Kapadia and Puri, 1991), tea spider mites (Trombidiformes) (Sannigrahi and Mukhopadhyay, 1992) and *Helicoverpa armigera* (Hubner) eggs (Varshney et al., 2018b).

4.3.3 Biology and Life History

Life history traits have been characterized in laboratory for *G. atricolor* Montandon (Dunbar and Bacon, 1972a), *G. bullatus* (Tamaki and Weeks, 1972), *G. lubra* Kirkaldy (Mansfield et al., 2007), *G. pallens* (Tamaki and Weeks, 1972; Dunbar and Bacon, 1972a), *G. punctipes* (Dunbar and Bacon, 1972a; Champlain and Sholdt, 1967a), *G. ochropterus* Fieber (Varshney and Ballal, 2017a) and *G. superbus* Montandon (Varshney and Ballal, 2017b). Under natural circumstances, big-eyed bugs overwinter as adults showing reproductive diapause (Brannon et al., 2006). Upon reaching adulthood, female starts egg laying. Mean pre-oviposition period recorded in studies is 6–10 days. Females lay their eggs on leaves or stems of plants (Hagler, 2020; Lundgren, 2011). Among plants, soybean is a very suitable and preferred oviposition site for *G. punctipes* (Geocoridae) (Naranjo and Stimac, 1985, 1987). In laboratory, *Geocoris* species was observed to lay eggs on both plant and artificial surfaces (Naranjo and Stimac, 1987; Varshney and Ballal, 2017a, b). Fecundity also varies according to the physiological state of females and 10–178 eggs laid by a single female was reported (Sweet, 2000; Varshney and Ballal, 2017a, b; Rajan et al., 2018). Davis (1981) reported that *G. uliginosus, Geocoris lividipennis* Stål, *Geocoris pallens* (Say) and *G. punctipes* fed with mealworm moth eggs produced 96–347 eggs, with the lowest and the highest fecundities found for *G. uliginosus* and *G. punctipes*, respectively, at 26.7°C.

Female *G. bullatus* lays 75 eggs over her life span (Tamaki and Weeks, 1972). A single female *G. punctipes* can lay up to 20 eggs on a single day (Dunbar and Bacon, 1972a). Hatching occurs in 7–10 days at normal temperature and is influenced by temperature (Calixto et al., 2014). There are five nymphal instars in different *Geocoris* species. Rearing protocol of *Geocoris* spp. has been developed on *E. kuehniella* (*Geocoris proteus* Distant and *Geocoris varius* (Uhler)) (Oida et al., 2008); on *C. cephalonica* eggs (*Geocoris erythrocephalus*) (Lepeletier & Serville) (Rajan et al., 2018); and on *S. cerealella* (*Geocoris ochropterus* Fieber and *G. superbus*) (Varshney and Ballal, 2017a, b) (Figures 4.2 and 4.3). Development and reproduction also varies according to prey species. *Geocoris punctipes* exhibited variability in development, reproduction and survival when fed with different insects (Dunbar and Bacon, 1972b). *Geocoris floridanus* Blatchley could develop and reproduce on both corn earworm eggs and beet armyworm larvae. However, better performance was recorded on corn earworm eggs (Torres et al., 2004). Similarly, nymphal survival of *G. lubra* was low on *Aphis gossypii* Glover compared to *Helicoverpa armigera* (Hu¨bner) eggs (Mansfield et al., 2007). Based on nutrient profiles of tobacco budworm eggs (*Heliothis virescens* F.; Cohen and Patana, 1985), an artificial diet was developed for *G. punctipes* (Cohen, 1985). *Geocoris varius* can be reared on both fresh- and lyophilized artificial feeds (Igarashi and Nomura, 2013).

However, development time of *G. varius* fed with artificial diets was significantly delayed compared with *G. varius* fed on *E. kuehniella* eggs. Supplementing a prey diet with plant food improved development and increased survival and reproduction of *G. punctipes* (Naranjo and Stimac, 1985). More information on biology of geocorids is required to mass produce them and to make their practical use.

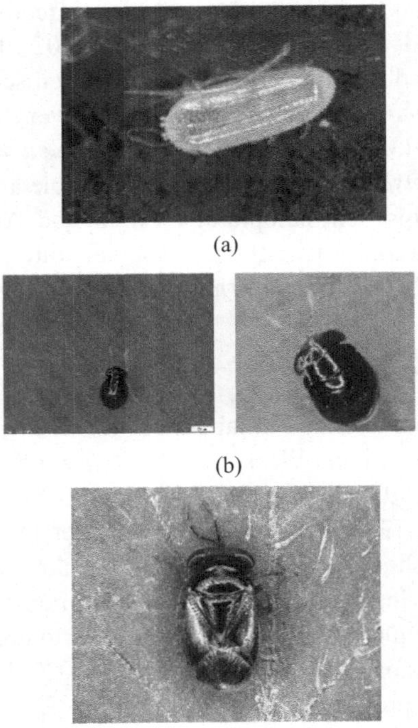

FIGURE 4.2 Life stages of *Geocoris ochropterus* a) Egg; B) Nymph; C) Adult.

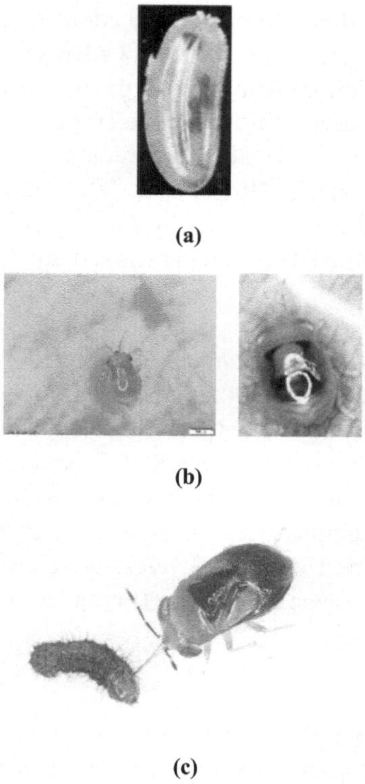

FIGURE 4.3 Life stages of *Geocoris superbus* a) Egg; B) Nymph; C) Adult.

4.3.4 Ecology and Behavior

Abiotic factors like temperature and relative humidity influence reproduction and other biological parameters of geocorids. This knowledge is required to develop effective rearing methodology. Data on development and reproduction at different physical conditions, prey selection, population dynamics, interspecies interactions and chemical-mediated interactions can provide valuable insights into ecological attributes and biocontrol potential of these bugs. Oida and Kadono (2012) found that the optimal rearing temperatures for *G. varius* is 24°C–30°C and for *G. proteus* is 26°C–33°C. Calixto et al. (2014) studied the effect of constant and alternating temperatures on survival and development of *G. punctipes*. The optimal temperature range for development and survival of *G. punctipes* was 24.5°C–30°C. The developmental time of the various instars of this bug recorded at different temperatures are similar to those reported for *G. punctipes* reared at 25°C and fed on *Spodoptera frugiperda* (J.E. Smith) (Champlain and Sholdt, 1967a) and at 30°C and fed on *Phthorimaea operculella* (Zeller) (Dunbar and Bacon, 1972b). Fast development and nymphal survival of *G. lubra* Kirkaldy was observed at 27°C compared with 25°C. Photoperiod also affected development time but not survival of the nymph of geocorids (Mansfield et al., 2007). Populations of geocorids are likely to develop more rapidly in warm than in cold spring (Tamaki and Weeks, 1972; Champlain and Sholdt, 1967a, b). The development rate from egg to adult of *Geocoris* spp. correlates positively with temperatures between 21°C and 37°C but may differ considerably among species (Schuman et al., 2013) and with the quality of the prey.

4.3.4.1 Intraguild Predation

In some of the studies, *Geocoris* bugs preferred unparasitized eggs or nymphs compared to parasitized eggs and nymphs. Velasco-Hernández et al. (2013) reported a preference of *G. punctipes* for nonparasitized *T. vaporariorum* nymphs in choice tests. Bao-Fundora et al. (2016) demonstrated that though *G. punctipes* engages in IGP on immature and adult stages of *Eretmocerus eremicus* Rose and Zolnerowich under semi-field conditions, it does not adversely affect whitefly control. Similarly, *G. ochropterus* preferred to prey on unparasitized eggs of *H. armigera* and *C. cephalonica* over parasitized eggs (Varshney and Ballal, 2018). Nonpreference to parasitized host could be due to hardening of the cuticle (Chailleux et al., 2013), physiological/chemical changes (Gelman et al., 2002) in parasitized eggs and prey species (Roger et al., 2000). Conversely, Naranjo (2007) reported that *G. punctipes* preferred parasitized *B. tabaci* by *Eretmocerus* sp. nr. *emiratus* (Hymenoptera: Aphelinidae) or could not discriminate between parasitized and nonparasitized nymphs.

4.3.4.2 Functional Response

The nymphs and the adult females of *G. pallidipennis* exhibited Type II functional response to both *M. persicae* reflected and artificial diet. Adult female *G. pallidipennis* reared on either *M. persicae* or artificial diet produced a significantly better performance than the juvenile stages tested (Liu and Zeng, 2014). Third, fourth and fifth instars, and adult of *G. ochropterus* exhibited Type II functional response to *H. armigera* eggs. Adult *G. ochropterus* followed by fifth instar exhibited the highest egg consumption and attack rate compared to other stages. Adult female could consume 56 eggs in 24 hours (Varshney et al., 2018b). However, *G. punctipes* consumed 44 eggs of *Helicoverpa zea* (Boddie) in 24 hours and exhibited Type II response (Parajulee et al., 2006).

4.3.5 Chemical Ecology

Female of *G. puntipes* produces sex pheromone ((E)-2-octenyl acetate), which triggers very active male searching behavior directed particularly toward any moving object. It has implications for attempting to use the pheromone for monitoring purposes (Marques et al., 2000). *Geocoris* spp. responds to volatiles emitted from *Nicotiana attenuata* after herbivory exhibited increased predation (Kessler and Baldwin, 2001; Halitschke et al., 2008; Skibbe et al., 2008; Meldau et al., 2009; Allmann and Baldwin, 2010; Schuman et al., 2012), resulting in a fitness benefit for plants (Schuman et al., 2012). Big-eyed bug, *Geocoris pallens*, responded to methyl salicylate, indole and trans-2-hexen-1-al (James 2005). However, Yu et al. (2017) evaluated nine HIPVs with one methyl jasmonate in field and found that *G. pallidipennis*, did not respond to any of the tested HIPVs.

4.3.5.1 As Biocontrol Agents

Geocorids are general predators and feed on lepidopteran and soft-bodied insect pests. Their role as predators of some insect pests has been given in Table 4.1. Bueno et al. (2016) reported low predation of *T. absoluta* eggs (0.5 eggs per day) by adults of *G. punctipes. Geocoris punctipes* and other *Geocoris* spp. are reported to attack larvae of *T. absoluta, H. armigera, Helicoverpa* (=*Heliothis*) *zea* (Boddie) and *Heliothis virescens* (F.), nymphs of *B. tabaci, A. gossypii, M. persicae, F. occidentalis, Phenacoccus solenopsis* Tinsley and spider mites (Lingren et al., 1968; Joseph and Braman, 2009; Oida and Kadono, 2012; Rondon et al., 2004; Torres et al., 2004; Velasco-Hernández et al., 2013; Varshney et al., 2018b, c; Varshney and Ballal, 2017b). *Geocoris punctipes* nymphs were also reported to prey on *S. frugiperda* larvae (Joseph and Braman, 2009). Female predators consumed more numbers of *T. absoluta* (Bueno et al., 2016) and *S. frugiperda* larvae than males (Joseph and Braman, 2009). Fifth-instar nymph and adult *G. ochropterus* consumed maximum number of *Frankliniella schultzei* Trybom at 1:60 (Varshney et al., 2018c). In cage study, three releases of *G. ochropterus* at 20 nymphs/plant reduced red spider mites' population by 79% in tomato (Varshney et al., 2018c).

During the sixties and seventies, *G. punctipes* had been successfully applied in inoculative releases in cotton crops (Lingren et al., 1968) and in open field strawberries in the USA (Wood, 1998). In laboratory studies, among six turfgrass cultivars tested, *G. uliginosus* nymphs varied in ability to reduce numbers of *S. frugiperda* larvae. Field experiments depicted the reduction in *S. frugiperda* larvae by a low density of *G. uliginosus* on Sea Isle 1 and 561–79 seashore paspalum grass (Braman et al., 2003).

4.4 CONCLUSIONS

Predaceous mirids and geocorids are neglected as biocontrol agents because basic biological and other ecological studies have been conducted for only a few species. Most of the mirid bugs are zoophytophagous in nature and they exhibit greater fecundity, survival and lifespan. Thus, conservational strategies should be adopted for early establishment of these bugs in fields to achieve better results. Plant and prey diets vary in chemical composition and zoophytophagous bugs require specific digestive enzymes. Studies with enzymatic profiles of zoophytophagous bugs are scarce. This area needs further deep research, which would certainly help to understand the omnivorous behavior of these bugs and its implication for pest management. Although zoophytophagous bugs are efficient biocontrol agents, their feeding on plants in the absence of prey impede their optimum use in pest management. This could be improved through artificial selection on behavioral traits, *viz.*, beneficial zoophagy, diet specialization and detrimental phytophagy of candidate zoophytophagous predators. Interaction studies between zoophytophagous bugs and other agents, such as parasitoids, entomopathogenic fungi, bacteria and virus, are warranted to harness maximum benefits of these biocontrol agents. Research efforts should be made toward exploration and identification of nonhost plants for these zoophytophagous bugs as conservation strategies. Studies showed that plant-feeding activities of these bugs could differentially induce HIPVs. These HIPV might have a role as repellents and attractants for pest and natural enemies, respectively. Further research in this area might open a new avenue to exploit these volatiles as pest management strategies. In European countries, these bugs are used in greenhouse crops against many sucking and lepidopteran pests. More studies are required to understand basic biology and rearing protocol of potential generalist zoophytophagous bugs so that they could be produced in large numbers and used to manage insect pests in greenhouses. Emphasis should be given to provide funding for basic studies, ecology and rearing techniques for continuous mass production and conservation strategies for these neglected yet potent predaceous bugs.

4.5 POINTS TO REMEMBER

1. The Miridae or plants bugs are the largest family of true bugs comprising about 1,300 genera and approximately 11,139 described species distributed worldwide.
2. These bugs exhibit a wide range of food preferences, like phytophagy, carnivory and omnivory.
3. Geocoridae, or big-eyed bugs, comprise 30 genera and more than 290 described species worldwide.
4. These predatory mirids are being used in several biocontrol programs in the tomato-producing areas of southern Europe, North Africa and the Middle East.
5. Both predatory mirids and geocorids are amenable to rearing in the laboratory on eggs of *Ephestia kuehniella*, *Corcyra cephalonica* and *Sitotroga cerealella*.
6. Populations of geocorids are likely to develop more rapidly in warm than in cold spring.
7. The herbivory of bugs induces the production of HIPVs and might act as repellents and attractants to other pests and natural enemies, respectively.
8. Many studies exhibited positive interactions between zoophytophagous bugs and parasitoids. Addition of *Trichogramma* parasitoids significantly increased the level of control of *Tuta absoluta* over what could be attained using *M. pygmaeus* or *N. tenuis* alone.
9. *Geocoris* bugs prefer unparasitized eggs or nymphs compared to parasitized eggs and nymphs in many studies.
10. Predatory mirids, viz., *Macrolophus pygmaeus*, *D. tamaninii*, and *Nesidiocoris tenuis*, are effective predators of whiteflies, thrips, *T. absoluta*, etc. In India, *N. tenuis* and *D. primarius* are amenable to rearing in the laboratory and have potential to control insect pests.

ACKNOWLEDGEMENTS

The authors are grateful to the Indian Council of Agricultural Research and the Director, ICAR-National Bureau of Agricultural Insect Resources for the facilities provided.

REFERENCES

Abbas, S., Pérez-Hedo, M., Colazza, S. & Urbaneja, A. (2014). The predatory mirid *Dicyphus maroccanus* as a new potential biological control agent in tomato crops. *BioControl*, 59: 565–574.

Ahirwar, R. M., Banerjee, S. & Gupta, M. P. (2009). Seasonal incidence of Insect pests of Sesame in relation to Abiotic factors. *Annals of Plant Protection Sciences*, 17(2): 351–356.

Allmann, S. & Baldwin, I.T. (2010). Insects betray themselves in nature to predators by rapid isomerization of green leaf volatiles. *Science*, 329: 1075–1078.

Alomar, O. & Wiedenmann, R.N. (Eds.). (1996). *Zoophytophagous Heteroptera: Implications for Life History and Integrated Pest Management*. Thomas Say Publications in Entomology, Entomological Society of America, Landham, MD.

Alvarado, P., Baltà, O. & Alomar, O. (1997). Efficiency of four Heteroptera as predators of *Aphis gossypii* and *Macrosiphum euphorbiae* (Hom.: Aphididae). *Entomophaga*, 42: 215–226. doi:10.1007/BF02769899.

Arnó, J., Alonso, E. & Gabarra, R. (2003). Role of the parasitoid *Diglyphus isaea* (Walker) and the predator *Macrolophus caliginosus* Wagner in the control of leafminers. *IOBC/WPRS Bulletin*, 26(10): 79–84.

Arnó, J., Arino, J., Espanol, R., Martí, M. & Alomar O. (2000). Conservation of *Macrolophus caliginosus* Wagner (Het. Miridae) in commercial greenhouses during tomato crop-free periods. *IOBC/WPRS Bulletin*, 23: 241–246.

Arnó, J., Castañé, C., Riudavets, J., Roig, J. & Gabarra, R. (2006). Characterization of damage to tomato plants produced by the zoophytophagous predator *Nesidiocoris tenuis*. *IOBC/WPRS Bulletin*, 29: 249–254.

Arnó, J., Gabarra, R., Estopà, M., Gorman, K., Peterschmitt, M., Bonato, O., Vosman, B., Hommes, M. & Albajes, R. (2009a). Implementation of IPM programs in European greenhouse tomato production areas. Tools and constraints. Edicions de la Universitat de Lleida.

Arnó, J., Sorribas, R., Prat, M., Matas, M., Pozo, C., Rodriguez, D., Garreta, A., Gomez, A. & Gabarra, R. (2009b). *Tuta absoluta* a new pest in IPM tomatoes in the north east of Spain. *IOBC/WPRS Bulletin*, 49: 203–208.

Arnó, J., Gabarra, R., Liu, T.X., Simmons, A.M. & Gerling, D, (2010). Natural enemies of *Bemisia tabaci*: Predators and parasitoids. In: *Bemisia: Bionomics and Management of a Global Pest* (Eds. Stansly, P.A. & Naranjo S.E.). Springer, Dordrecht, The Netherlands, pp. 385–421.

Azimizadeh, N., Ahmadi, K., Imani, S., Takalloozadeh, H. & Sarafrazi, A. (2012). Evaluation of oviposition-site preference behavior in predatory bug *Deraeocoris lutescens* Schilling (Hemiptera: Miridae). *Munis Entomology & Zoology*, 7(1): 506–515.

Ballal, C.R., Gupta, A., Mohan, M., Lalutha, Y. & Verghese, A. (2016). The new invasive pest *Tuta absoluta* (Meyrick) (Lepidoptera: Gelechiidae) in India and its natural enemies along with evaluation of Trichogrammatids for its biological control. *Current Science*, 110(11): 2155–2159.

Ballard, E. (1927). Some new Indian Miridae (Capsidae). Memoirs of the Department of Agriculture in India. *Botanical series*, 10(4): 61–68.

Baniameri, V. & Cheraghian, A. (2011). The current status of *Tuta absoluta* in Iran. International Symposium on Management of *Tuta absoluta* (tomato leafminer), Agadir, Morocco, November 16–18, 2011.

Bao-Fundora, L., Ramirez-Romero, R., Sánchez Hernández, C.V., Sánchez Martínez, J., Desneux, N. (2016). Intraguild predation of *Geocoris punctipes* on *Eretmocerus eremicus* and its influence on the control of the whitefly *Trialeurodes vaporariorum*. *Pest Management Science*, 72(6): 1110–1116.

Barnadas, I., Gabarra, R. & Albajes, R. (1998). Predatory capacity of two mirid bugs preying on *Bemisia tabaci*. *Entomologia Experimentalis et Applicata*, 86: 215– 219.

Belda, J.E., Calvo, F.J. & Giménez, A. (2010). Estrategia para el control biológico de Tuta absoluta en tomate mediante sueltas de *Nesidiocoris tenuis* en pre-trasplante. *Phytoma-España*, 217: 48–52.

Bell, K.O. & Whitcomb, W.H. (1964). "Field studies on egg predators of the bollworm, Heliothis zea (Boddie)." *Florida Entomologist*, 47: 171–180.

Biondi, A., Guedes, R.N.C., Wan, F. H. & Desneux, N. (2018). Ecology, worldwide spread, and management of the invasive South American tomato pinworm, *Tuta absoluta*: Past, present, and future. *Annual Review of Entomology*, 63: 239–258.

Biondi, A., Zappalà, L., Di Mauro, A., Garzia, G.T., Russo, A., Desneux, N. & Siscaro, G. (2016). Can alternative host plant and prey affect phytophagy and biological control by the zoophytophagous mirid *Nesidiocoris tenuis*? *Biocontrol*, 61(1): 79–90.

Blaeser, P., Sengonça, C. & Zegula, T. (2004). The potential use of different predatory bug species in the biological control of *Frankliniella occidentalis* (Pergande) (Thysanoptera: Thripidae). *Journal of Pest Science*, 77: 211–219.

Bompard, A., Jaworski, C.C., Bearez, P. & Desneuz, N. (2013). Sharing a predator: Can an invasive alien pest affect the predation on a local pest? *Population Ecology*, 55(3): 433–440.

Bouagga, S., Urbaneja, A. & Pérez-Hedo, M. (2018). Comparative biocontrol potential of three predatory mirids when preying on sweet pepper key pest. *Biological Control*, 121: 168–174.

Brailovsky, H. (2016). A review of the Geocoridae of Mexico (Hemiptera: Heteroptera: Lygaeoidea), with descriptions of four new species, new distributional records, and a key to the known subfamilies, tribes, genera and species. *Zootaxa*, 4173(5): 401–448.

Braman, S.K., Duncan, R.R., Hanna, W.W. & Engelke, M.C. (2003). Arthropod predator occurrence and performance of *Geocoris uliginosus* (Say) on pest-resistant and susceptible turfgrasses. *Environmental Entomology*, 32(4): 907–914.

Brannon, S.L., Decker, K. B., and Yeargan, K.V. (2006). Photoperiodic induction of reproductive diapause in the predator *Geocoris uliginosus* (Hemiptera: Geocoridae). *Annals of the Entomological Society of America*, 99(2): 300–304.

Bueno, V.H.P., Silva, D.B., Calixto, A.M., Montes, F.C. & Van Lenteren, J. C. (2016). *Geocoris punctipes* nymphs and adults easily prey on leaf-mining larvae of *Tuta absoluta* on tomato. *Bulletin of Insectology*, 69(2): 271–276.

Cabello, T. Bonfil, F., Gallego, J. R., Fernandez, F.J., Gamez, M. & Garay, J. (2015). Can Interactions between an omnivorous hemipteran and an egg parasitoid limit the level of biological control for the tomato pinworm? *Environmental Entomology*, 44(1): 12–26. doi: 10.1093/ee/nvu017.

Calixto, A.M., Bueno, V.H., Montes, F.C. & Van Lenteren, J.C. (2014). Development and thermal requirements of the Nearctic predator *Geocoris punctipes* (Hemiptera: Geocoridae) reared at constant and alternating temperatures and fed on *Anagasta kuehniella* (Lepidoptera: Pyralidae) eggs. *European Journal of Entomology*, 111(4): 521–528.

Calvo, F.J., Bolckmans, K. & Belda J.E. (2009b). Development of a biological control-based Integrated Pest Management method for *Bemisia tabaci* for protected sweet pepper crops. *Entomologia Experimentalis et Applicata*, 133: 9–18.

Calvo, F.J., Bolckmans, K., Stansly, P.A. & Urbaneja, A. (2009a). Predation by *Nesidiocoris tenuis* on Bemisia tabaci and injury to tomato. *BioControl*, 54: 237–246.

Calvo, F.J., Lorente, M.J., Stansly, P.A. & Belda, J.E. (2012). Preplant release of *Nesidiocoris tenuis* and supplementary tactics for control of *Tuta absoluta* and *Bemisa tabaci* in greenhouse tomato. *Entomologia Experimentalis et Applicata*, 143: 111–119.

Calvo, J. & Urbaneja, A. (2004). *Nesidiocoris tenuis*, un aliado para el control biologico de la mosca blanca. *Horticultura Internacional*, 44: 20–25.

Carvalho, J. C. M. (1957). Catalogue of the Miridae of the world. Part 1. *Archivos do Museu Nacional do Rio de Janeiro*, 44: 1–158.

Cassis, G. & Schuh, R.T. (2012). Systematics, biodiversity, biogeography, and host associations of the Miridae (Insecta: Hemiptera: Heteroptera: Cimicomorpha). *Annual Review of Entomology*, 57: 377–404.

Castañé, C., Alomar, O., Goula, M. & Gabarra, R. (2004). Colonization of tomato greenhouses by the predatory mirid bugs *Macrolophus caliginosus* and *Dicyphus tamaninii*. *Biological Control*, 30: 591–597.

Castañé, C., Alomar, O. & Riudavets, J. (1996). Management of western flower thrips on cucumber with *Dicyphus tamaninii* (Heteroptera: Miridae). *Biological Control*, 7: 114–120.

Castañe, C., Alomar, O. & Riudavets, J. (2003). Potential risk of damage to zucchinis caused by mirid bugs. *IOBC/WPRS Bulletin*, 26(10): 135–138.

Castañé, C., Arnó, J., Gabarra R. & Alomar, O. (2011). Plant damage to vegetable crops by zoophytophagous mirid predators. *Biological Control*, 59:22–29.

Castañé, C., Iriarte, J. & Lucas, E. (2002). Comparison of prey consumption by *Dicyphus tamaninii* reared conventionally, and on a meat-based diet. *Biocontrol*, 47: 657– 666.

Castañé, C., Riudavets, J., Alomar, O. (2009). El depredador generalista *Dicyphus tamaninii* en el control de poblaciones mixtas de mosca blanca y de trips en pepino de invernadero. *Boletín de Sanidad Vegetal, Plagas*, 35: 29–37

Castañé, C. & Zapata, R. (2005). Rearing the predatory bug *Macrolophus caliginosus* on a meat-based diet. *Biological Control*, 34: 66–72.

Chailleux, A., Bearez, P., Pizzol, J., Amiens-Desneux, E., Ramirez-Romero, R. & Desneux, N. (2013). Potential for combined use of parasitoids and generalist predators for biological control of the key invasive tomato pest, *Tuta absoluta*. *Journal of Pest Science*, 86: 533–541.

Champlain, R.A. & Sholdt, L.L. (1967a). Life history of *Geocoris punctipes* (Hemiptera: Lygaeidae) in the laboratory. *Annals of Entomological Society of America*, 60: 881–883.

Champlain, R.A. & Sholdt, L.L. (1967b). Temperature range for development immature stages of *Geocoris punctipes* (Hemiptera: Lygaeidae). *Annals of Entomological Society of America*, 60: 883–885.

Cohen, A. C. (1985). Simple method for rearing the insect predator *Geocoris punctipes* (Heteroptera: Lygaeidae) on a meat diet. *Journal of Economic Entomology*, 78: 1173–1175.

Cohen, A.C. & Patana, R. (1985). Chemical composition of tobacco budworm eggs during development. *Comparative Biochemistry and Physiology Part B: Comparative Biochemistry*, 81: 165–169.

Coll, M. & Guershon, M (2002). Omnivory in terrestrial arthropods: Mixing plant and prey diets. *Annual Review of Entomology*, 47: 267–297.

Coll, M. & Ridgway, R.L. (1995). Functional and numerical responses of *Orius insidiosus* (Heteroptera: Anthocoridae) to its prey in different vegetable crops. *Annals of Entomological Society of America*, 88: 732–738.

Crocker, R.L. & Whitcomb, W.H. (1980). Feeding niches of the big-eyed bugs *Geocoris bullatus*, *G. punctipes*, and *G. uliginosus* (Hemiptera: Lygaeidae: Geocorinae). *Environmental Entomology*, 9: 508–513.

Davis, D. L. (1981). Population dynamics of four species of *Geocoris* in the peanut agroecosystem. Ph.D. dissertation, Texas A&M University, College Station.

De Puysseleyr, V. (2007). Rol van de plant voor de zo¨ofytofage wants *Macrolophus pygmaeus*. Master thesis, Ghent, Belgium.

Dellapé, P.M., Dellapé, G. & Martínez, F. (2015). The rediscovery of Stenogeocoris horvathi Montandon (Heteroptera, Geocoridae) in Argentina. *Iheringia, Série Zoologia, Porto Alegre*, 105(2):245–251.

Distant, W. L. (1904). Rhynchota II, Heteroptera. In: *The Fauna of British India, Including Ceylon and Burma* (Ed. Blanford, W.T.). Taylor & Francis, London, vii+503 pp.

Distant, W.L. (1910). Descriptions of Oriental Capsidae. *The Annals and Magazine of Natural History; Zoology, Botany, and Geology*, 5(8): 10–22.

Dunbar, D.M. & Bacon, O.G. (1972a). Influence of temperature on development and reproduction of *Geocoris atricolor*, *G. pallens* and *G. punctipes* (Heteroptera: Lygaeidae) from California. *Environmental Entomology*, 1: 596–599.

Dunbar, D.M. & Bacon, O.G. (1972b). Feeding, development, and reproduction of *Geocoris punctipes* (Heteroptera: Lygaeidae) on eight diets. *Annals of Entomological Society of America*, 65: 892–895.

Enkegaard, A., Brødsgaard, H.F. & Hansen, D.L. (2001). *Macrolophus caliginosus*: Functional response to whiteflies and preference and switching capacity between whiteflies and spider mites. *Entomologia Experimentalis et Applicata*, 101: 81–88. doi: 10.1046/j.1570-7458.2001.00893.x.

Eubanks, M.D. & Denno, R.F. (1999). The ecological consequences of variation in plants and prey for an omnivorous insect. *Ecology*, 80: 1253–1266.

Eubanks, M.D. & Denno, R. F. (2000). Host plants mediate omnivore-herbivore interactions and influence prey suppression. *Ecology*, 81(4): 936–947.

Eubanks, M.D., Styrsky, J. D., Robert, F. & Denno. R. F. (2003). The evolution of omnivory in heteropteran insects. *Ecology*, 84(10): 2549–2556.

Evans, M. L. (1985). Arthropod species in soybeans in southeast Queensland. *Journal of the Australian Entomological Society*, 24: 169–177.

Fantinou, A. A., Perdikis, D. C., Maselou, D. A. & Lambropoulos, P. D. (2008). Prey killing without consumption: Does *Macrolophus pygmaeus* show adaptive foraging behavior? *Biological Control*, 47: 187–193.

Fantinou, A.A., Perdikis, D.C., Maselou, D.A. & Lambropoulos, P.D. (2009). Preference and consumption of *Macrolophus pygmaeus* preying on mixed instar assemblages of *Myzus persicae*. *Biological Control*, 51: 76–80.

Fauvel, G., Malausa, J.C. & Kaspar, B. (1987). Etude en laboratoire des principales caracteristiques biologiques de *Macrolophus caliginosus* (Heteroptera: Miridae). *Entomophaga*, 32: 529–543.

Fernández-arhex, V. & Corley, J.C. (2003). The functional response of parasitoids and its implications for biological control. *Biocontrol Science and Technology*, 13: 403–413. doi: 10.1080/0958315031000104523.

Foglar, H., Malausa, J. & Wajnberg, E. (1990). The functional response and preference of *Macrolophus caliginosus* [Heteroptera: Miridae] for two of its prey: *Myzus persicae* and *Tetranychus urticae*. *Entomophaga*, 35: 465–474.

Gabarra, R., Alomar, O., Castañé, C., Goula, M. & Albajes, R. (2004). Movement of the greenhouse whitefly and its predators between in- and outside of Mediterranean greenhouses. *Agriculture, Ecosystems and Environment*, 102: 341–348.

Gabarra, R., Castañé, C. & Albajes, R. 1995. The mirid bug *Dicyphus tamaninii* as a greenhouse whitefly and western flower thrips predator on cucumber. *Biocontrol Science & Technology*, 5: 475–488.

Ganga Visalakshy, P. N. & Jayanth, K. P. (1994). *Stethoconus praefectus* (Distant), a predator of Teleonemia scrupulosa Stål in Bangalore, India. *Entomon*, 19: 177–178.

Gelman, D.B., Blackburn, M.B. & Hu, J.S. (2002). Timing and ecdysteroid regulation of the molt in last instar greenhouse whiteflies (*Trialeurodes vaporariorum*). *Journal of Insect Physiology*, 48: 63–73.

Gervassio, N.G.S., Pérez-Hedo, M., Luna, M.G. & Urbaneja, A. (2017). Intraguild predation and competitive displacement between *Nesidiocoris tenuis* and *Dicyphus maroccanus*, two biological control agents in tomato pests. *Insect Science*, 24: 809–817. doi: 10.1111/1744-7917.12361.

Ghabeish, I., Saleh, A. & Al-Zyoud, F. (2008). *Dicyphus tamaninii*: Establishment and efficiency in the control of Aphis gossypii on greenhouse cucumber. *Journal of Food Agriculture and Environment*, 6(3–4): 346–349.

Gillespie, D.R. & McGregor, R.R., (2000). The functions of plant feeding in the omnivorous predator *Dicyphus hesperus*: Water places limits on predation. *Ecological Entomology*, 25: 380–386.

Gillespie, D.R., McGregor, R., Sanchez, J.A., Van Laerhoven, S.L., Quiring, D.M.J., Roitberg, B.D., Foottit, R.G., Schwartz, M.D. & Shipp, J.L. (2007). An endemic omnivorous predator for control of greenhouse pests. In: *Biological Control: A Global Perspective* (Ed. Vincent, C., et al.). CABI McGregor Publishing, Wallingford, UK, pp. 128–135.

Guenaoui, Y., Bensaad, R. & Ouezzani, K. (2011). Importance of native polyphagous predators able to prey on Tuta absoluta Meyrick (Lep: Gelechiidae) on tomato crops. EPPO/IOBC/FAO/NEPPO Joint International Symposium on Management of *Tuta absoluta*, Agadir, Marocco. Book of Abstract, 38, November 16–18, 2011.

Gullan, P.J & Cranston, P.S. (2010). *The Insects: An Outline of Entomology*, 4th edition. Wiley Blackwell, New York.

Hagler, J. (2020). *Geocoris* spp. (Heteroptera: Lygaeidae), bigeyed bug. Biological control: A guide to natural enemies in North America. Cornell University. Available at: http://nysaes.cornell.edu/ent/biocontrol/predators/geocoris.html. Retrieved on February 4, 2021.

Hagler, J. & Sanchez, N. (2011). *Geocoris* spp. (Heteroptera: Lygaeidae). Available at https://biocontrol.entomology.cornell.edu/predators/Geocoris.php. Accessed on February 3, 2014.

Halitschke, R., Stenberg, J. A., Kessler, D., Kessler, A. & Baldwin, I. T. (2008). Shared signals: 'alarm calls' from plants increase apparency to herbivores and their enemies in nature. *Ecology Letters*, 11: 24–34.

Hamdan, A-J.S. (2006). Functional and numerical responses of the predatory bug *Macrolophus caliginosus* Wagner fed on different densities of eggs of the greenhouse whitefly, *Trialeurodes vaporariorum* (Westwood). *Journal of Biological Research*, 6: 147–154.

Harpaz, L.S., Gerling, D., Graph, S., Kedoshim, H., Azolay, L., Rozenberg, T., Nachache, Y. et al. (2016). Control of the Tomato Leafminer, *Tuta absoluta* (Lepidoptera: Gelechiidae), in open-field tomatoes by indigenous natural enemies occurring in Israel. *Journal of Economic Entomology*, 109(1): 120–131. doi: 10.1093/jee/tov309.

Hassell, M., Lawton, J. & Beddington, J. (1977). Sigmoid functional responses by invertebrate predators and parasitoids. *Journal of Animal Ecology*, 46: 249–262.

Hemerik, L. & Yano, E. (2011). Scaling up from individual behaviour of *Orius sauteri* foraging on Thrips palmi to its daily functional response. *Population Ecology*, 53: 563–572.

Henry, T.J. (1997). Phylogenetic analysis of family groups within the infraorder Pentatomomorpha (Hemiptera: Heteroptera), with emphasis on the Lygaeoidea. *Annals of the Entomological Society of America*, 90(3): 275–301. https://doi.org/10.1093/aesa/90.3.275.

Henry, T.J. (2009). Biodiversity of Heteroptera. In: *Insect biodiversity: Science and Society*. (Eds. Foottit, R. & Adler, P.). Oxford, Blackwell, pp. 223–263.

Henry, T.J. (2013). *Cymapamphantus valentineorum*, a new genus and species of Pamphantinae (Heteroptera: Lygaeoidea: Geocoridae) from the British Virgin Islands, with a checklist of the species and keys to the tribes and genera of the subfamily. *Proceedings of the Entomological Society of Washington*, 115: 392–401.

Henry, T.J., Dellape, P.M. & de Paula, A.S. (2015). The Big-Eyed Bugs, Chinch Bugs, and Seed Bugs (Lygaeoidea). In: *True Bugs (Heteroptera) of the Neotropics, Entomology in Focus 2* (Eds. Panizzi, A.R. & Grazia, J.). Springer Science+Business Media Dordrecht (outside the USA), pp. 459–514. doi: 10.1007/978-94-017-9861-7_16.

Henry, T.J., Peña, J.E., Long, D. & Acevedo, F. (2009). *Stethoconus praefectus* (Hemiptera: Miridae): First North American records of an Old World plant bug predacious on avocado lace bug, *Pseudacysta perseae* (Hemiptera: Tingidae), in Florida. *Proceedings of the Entomological Society of Washington*, 111(1): 98–105.

Henry, T.J. & Wheeler, A.G., Jr. (1988). Family Miridae Hahn, 1833. In: *Catalog of the Heteroptera, or True Bugs, of Canada and the Continental United States* (Eds. Henry, T. J. & Froeschner, R.C.). E. J. Brill Publishers, Leiden, Netherlands, pp. 251–507.

Holguin, C. M., Pena, J. E., Henry, T. J. & Acevedo, F. E. (2009). Biology of *Stethoconus praefectus* (Heteroptera: Miridae), a newly established predator of the avocado lace bug, *Pseudacysta perseae* (Heteroptera: Tingidae). *Florida Entomologist*, 92: 54–57.

Holling, C.S. (1959). Some characteristics of simple types of predation and parasitism. *Canadian Entomologist*, 91: 385–398.

Hunter, M. D. (2009). Trophic promiscuity, intraguild predation and the problem of omnivores. *Agricultural and Forest Entomology*, 11: 125–131.

Igarashi, K. & Nomura, M. (2013). Development and reproduction of *Geocoris varius* (Hemiptera: Geocoridae) on two types of artificial diet. *Applied Entomology and Zoology*, 48: 403–407.

Ingegno, B.L., Ferracini, C., Gallinotti, D., Alma, A. & Tavella, L. (2013). Evaluation of the effectiveness of *Dicyphus errans* (Wolff) as predator of *Tuta absoluta* (Meyrick). *Biological Control*, 67: 246–252.

Ingegno, B.L., Messelink, G.J., Bodino, N., Iliadou, A., Driss, L., Woelke, J.B., Leman, A., Tavella L (2019) Functional response of the mirid predators *Dicyphus bolivari* and Dicyphus *errans* and their efficacy as biological control agents of *Tuta absoluta* on tomato. *Journal of Pest Science*, 92: 1457–1466. doi: 10.1007/s10340-019-01079-8.

James, D. G. (2005). Further field evaluation of synthetic herbivore-induced plant volatiles as attractants for beneficial insects. *Journal of Chemical Ecology*, 31: 481–495.

Jaworski, C.C., Bompard, A., Genies, L., Amiens-Desneux, E. & Desneux, N. (2013). Preference and prey switching in a generalist predator attacking local and invasive alien pests. *PLoS One*, 8(12): e82231.

Johnson, M. L., Pearce, S., Wade, M., Davies, A., Silberbauer, L., Gregg, P. & Zalucki, M. (2000*). Review of Beneficials in Cotton Farming Systems*. Cotton Research and Development Corporation, Narrabri, NSW.

Jones, W.A. & Snodgrass, G.L. (1998). Development and fecundity of *Deraeocoris nebulosus* (Heteroptera: Miridae) on *Bemisia argentifolii* (Homoptera: Aleyrodidae). *Florida Entomologist*, 81(3): 345–350.

Joseph, S. V. & Braman, S. K. (2009). Predatory potential of *Geocoris* spp. and *Orius insidiosus* on fall armyworm in resistant and susceptible turf. *Journal of Economic Entomology*, 3: 1151–1156.

Kapadia, M. N. & Puri, S. N. (1991). Biology and comparative predation efficacy of three heteropteran species recorded as predators of *Bamisia tabaci* in Maharashtra. *Entomophaga*, 36: 555–559.

Kessler, A. & Baldwin, I T. (2001). Defensive function of herbivore induced plant volatile emissions in nature. *Science*, 291: 2141–2144.

Kóbor, P. (2019). Umbrageocoris kondorosyi: A new genus and species of big-eyed bugs from New Guinea (Heteroptera: Lygaeoidea: Geocoridae). *Acta Zoologica Academiae Scientiarum Hungaricae*, 65(1): 1–8.

Konstantinov, F.V. & Namyatova, A.A. (2019). Taxonomic revisions and specimen databases in the internet age: Dealing with a species rich insect taxon. *Entomological Review*, 99: 340–361.

Koss, A. M., Chang, G. C. & Snyder, W. E. (2004). Predation of green peach aphids by generalist predators in the presence of alternative, Colorado potato beetle egg prey. *Biological Control*, 31(2): 237244.

Krey, K.L., Blubaugh, C.K., Chapman, E.G., Lynch, C.A., Snyder, G.B., Jensen, A.S. & Snyder, W.E. (2017). Generalist predators consume spider mites despite the presence of alternative prey. *Biological Control*, 115: 157–164.

Kumar, N.S. & Ananthakrishnan, T.N. (1985). *Geocoris ochropterus* Fabr. as a predator of some thrips. *Proceedings of the National Academy of Sciences, India Section B: Biological Sciences*, 51: 185–193.

Leitner, M., Boland, W., Mithofer, A. (2005). Direct and indirect defences induced by piercing-sucking and chewing herbivores in *Medicago truncatula*. *New Phytologist*, 167:597–606.

Lenfant, C., Ridray, G. & Schoen, L. (2000). Biopropagation of *Macrolophus caliginosus* (Wagner) for a quicker establishment in southern tomato greenhouses. *IOBC/WPRS Bulletin*, 23: 241–246.

Lingren P. D., Ridgway R. L. & Jones S. L. (1968). Consumption by several common arthropod predators of eggs and larvae of two *Heliothis* species that attack cotton. *Annals of Entomological Society*, 61: 613–618.

Liu, F. & Zeng, F. (2014). The influence of nutritional history on the functional response of *Geocoris pallidipennis* to its prey, *Myzus persicae*. *Bulletin of Entomological Research*, 10: 702–706.

Lu, Y.H., Wu, K.M., Jiang, Y.Y., Guo, Y.Y. & Desneux, N. (2012). Widespread adoption of Bt cotton and insecticide decrease promotes biocontrol services. *Nature*, 487:362–365.

Lucas, E. & Alomar, O. (2001). *Macrolophus caliginosus* (Wagner) as an intraguild prey for the zoophytophagous *Dicyphus tamaninii* Wagner (Heteroptera: Miridae). *Biological Control*, 20: 147–152.

Lundgren, J.G. (2011). Reproductive ecology of predaceous Heteroptera. *Biological Control*, 59(1): 37–52.

Madbouni, M.A.Z., Samih, M.A., Namvar, P. & Biondi, A. (2017). Temperature-dependent functional response of *Nesidiocoris tenuis* (Hemiptera: Miridae) to different densities of pupae of cotton whitefly, *Bemisia tabaci* (Hemiptera: Aleyrodidae). *European Journal of Entomology*, 114: 325–331. doi: 10.14411/eje.2017.040.

Malipatil, M.B. (2012). Australocorinae, a new subfamily of Geocoridae (Hemiptera: Heteroptera, Lygaeoidea) from Australia, with descriptions of a new genus and two new species. *Zootaxa*, 3554: 7588.

Malipatil, M.B. (1994). Revision of Australian *Geocoris* Fallen and *Stylogeocoris* Montandon (Heteroptera: Lygaeidae: Geocorinae). *Invertebrate Systematics*, 8(2): 299–327.

Malipatil, M.B. (2017). *Indopamphantus makutaensis*, a new genus and species, and Indopamphantini, a new tribe of Pamphantinae (Hemiptera: Heteroptera: Geocoridae), as the first representative of the subfamily from the Oriental Region. *Zootaxa*, 4242: 281–298.

Mansfield, S., Scholz, B., Armitage S. & Johnson M.L. (2007). Effects of diet, temperature and photoperiod on development and survival of the bigeyed bug, *Geocoris lubra*. *BioControl*, 52: 63–74.

Marques, F. de Assis, Mcelfresh, J.S. & Millar, J.G. (2000). Female-produced sex pheromone of the predatory bug *Geocoris punctipes*. *Journal of Chemical Ecology*, 26: 2843–2855.

Martínez-Cascales, J.I., Cenis, J.L., Cassis, G. & Sanchez, J.A. (2006). Species identity of *Macrolophus melanotoma* (Costa 1853) and *Macrolophus pygmaeus* (Rambur 1839) (Insecta: Heteroptera: Miridae) based on morphological and molecular data and bionomic implications. *Insect Systematics & Evolution*, 37:385–404.

Maselou, D.A., Perdikis, D.C. & Fantinou, A. (2015). Effect of hunger level on prey consumption and functional response of the predator *Macrolophus pygmaeus*. *Bulletin of Insectology*, 68:211–218.

Maselou, D.A., Perdikis, D.C., Sabelis, M.W. & Fantinou, A.A. (2014). Use of plant resources by an omnivorous predator and the consequences, for effective predation. *Biological Control*, 79: 92–100. doi: 10.1016/j.biocontrol.2014.09.002

Mathen, K. & Kurian, C. (1972). Description, life-history and habits of *Stethoconus praefectus* (Distant), predacious on Stephanitis typicus Distant (Heteroptera: Tingidae), a pest of coconut palm. *Indian Journal of Agricultural Sciences*, 42: 255–262.

McGregor, R.R., Gillespie, D.R., Park, C.G., Quiring, D.M.J. & Foisy, M.R.J. (1999).Potential use of *Dicyphus hesperus* Knight (Heteroptera: Miridae) for biological control of pests of greenhouse tomatoes. *Biological Control*, 16: 104–110.

Meldau, S., Wu, J. & Baldwin, I. T. (2009). Silencing two herbivory activated MAP kinases, SIPK and WIPK, does not increase *Nicotiana attenuata's* susceptibility to herbivores in the glasshouse and in nature. *New Phytologist*, 181: 161–173.

Mensah, R.K. (1999). Habitat diversity: Implications for the conservation and use of predatory insects of *Helicoverpa* spp. in cotton systems in Australia. *International Journal of Pest Management*, 45: 91–100.

Messelink, G.J., Bloemhard, C.M.J., Hoogerbrugge, H., van Schelt, J., Ingegno, B. L. & Tavella, L. (2015). Evaluation of mirid predatory bugs and release strategy for aphid control in sweet pepper. *Journal of Applied Entomology*, 139: 333–341.

Messelink, G.J., Bloemhard, C.M.J., Kok, L. & Janssen, A. (2011). Generalist predatory bugs control aphids in sweet pepper. *IOBC/WPRS Bulletin*, 68:115–118.

Messelink, G.J. & Janssen, A. (2014). Increased control of thrips and aphids in greenhouses with two species of generalist predatory bugs involved in intraguild predation. *Biological Control*, 79: 1–7.

Messelink, G.J., Sabelis, M.W. & Janssen, A. (2012).Generalist predators, food web complexities and biological pest control. In: *Greenhouse Crops Integrated Pest Management and Pest Control—Current and Future Tactics* (Eds. Larramendy, M.L. & Soloneski, S.). In Tech, Rijeka, Croatia, pp. 191–214.

Mohaghegh, J., De Clercq, P., Tirry, L. (2001). Functional response of the predators *Podisus maculiventris* (Say) and *Podisus nigrispinus* (Dallas) (Heteroptera: Pentatomidae) to the beet armyworm, *Spodoptera exigua* (Hübner) (Lep., Noctuidae): Effect of temperature. *Journal of Applied Entomology*, 125: 131–134.

Mollá, O., Biondi, A., Alonso-Valiente, M. & Urbaneja, A. (2014). A comparative life history study of two mirid bugs preying on *Tuta absoluta* and *Ephestia kuehniella* eggs on tomato crops: Implications for biological control. *BioControl*, 59:175–183.

Mollá, O., Gonzalez-Cabrera, J. & Urbaneja, A. (2011). The combined use of *Bacillus thuringiensis* and *Nesidiocoris tenuis* against the tomato borer *Tuta absoluta*. *BioControl*, 56: 883–891.

Mollá, Ó., Montón, H., Vanaclocha, P., Francisco, B. & Urbaneja, A. (2009). Predation by the mirids *Nesidiocoris tenuis* and *Macrolophus pygmaeus* on the tomato borer *Tuta absoluta*. *IOBC/WPRS Bulletin*, 49: 209–214.

Montserrat, M., Albajes, R. & Castañé, C. (2000). Functional response of four Heteroptera predators preying on greenhouse whitefly and western flower thrips. *Environmental Entomology*, 29: 1075–1082.

Moreno-Ripoll, R., Agusti, N., Berruezo, R. & Gabarra, R. (2012). Conspecific and heterospecific interactions between two omnivorous predators on tomato. *Biological Control*, 62: 189–196.

Mukhopadhyay, A. (1988). Biological notes on two species of big-eyed bugs (Insecta: Hemiptera: Lygaeidae: Geocorine). *Journal of Bombay Natural History Society*, 85: 298–310.

Nachman, G. (2006). A functional response model of a predator population foraging in a patchy habitat: Functional response in a patchy environment. *Journal of Animal Ecology*, 75: 948–958. doi: 10.1111/j.1365-2656.2006.01114.x

Nannini, M., Atzori, F., Murgia, G., Pisci, R. & Sanna, F. (2012). Use of predatory mirids for control of the tomato borer Tuta absoluta (Meyrick) in Sardinian greenhouse tomatoes. *EPPO/OEPP Bulletin*, 42: 255–259. doi: 10.1111/epp.2563.

Naranjo, S.E. (2007). Intraguild predation on *Eretmocerus* sp. nr. *emiratus*, a parasitoid of *Bemisia tabaci*, by three generalist predators with implications for estimating the level and impact of parasitism. *Biocontrol Science & Technology*, 17: 605–622.

Naranjo, S.E. & Stimac, J. L., (1985). Development, survival, and reproduction of *Geocoris punctipes* (Hemiptera: Lygaeidae): Effects of plant feeding on soybean and associated weeds. *Environmental Entomology*, 14: 523–530.

Naranjo, S.E. & Stimac, J. L. (1987). Plant influence on predation and oviposition by *Geocoris punctipes* (Hemiptera: Lygaeidae) in soybeans. *Environmental Entomology*, 16: 182–189.

Naselli, M., Zappalà, L., Gugliuzzo, A., Garzia, G.T., Biondi, A., Rapisarda, C., Cincotta, F., Condurso, C., Verzera, A. & Siscaro, G. (2017). Olfactory response of the zoophytophagous mirid *Nesidiocoris tenuis* to tomato and alternative host plants. *Arthropod Plant Interactions*, 11(2): 121–131.

Neal, J. W., Haldemann, R. H. & Henry, T. J. (1991). Biological control potential of a Japanese plant bug *Stethoconus japonicus* (Heteroptera: Miridae), an adventive predator of the azalea lace bug (Heteroptera: Tingidae). *Annals of Entomological Society of America*, 84(3): 287–293.

Oida, H. & Kadono, F. (2007). Biological control of twospotted spider mite of strawberry and cotton aphid of watermelon by two big-eyed bugs, *Piocoris varius* and *Geocoris proteus* (Heteroptera: Geocoridae) in greenhouses. *Annual Report of Kanto Plant Protection Society*, 54: 133–138.

Oida, H. & Kadono, F. (2012). Development of *Geocoris varius* and *G. proteus* (Hemiptera: Geocoridae) provided with *Ephestia kuehniella* (Lepidoptera: Pyralidae) eggs. *Applied Entomology and Zoology*, 47(4): 365–372.

Oida, H., Kadono, F., Goto, C., Mukawa, S. & Kobayashi, N. (2007). Biological control of *Frankliniella intonsa* and *F. occidentalis* by the polyphagous predator, *Piocoris varius* (Heteroptera: Geocoridae) on sweet pepper in greenhouses. *Annual Report of the Kanto-Tosan Plant Protection Society*, 54: 139–142 (in Japanese with English summary).

Oida, H., Kadono, F., Shimizu, K. & Yoshii, S. (2008). Oviposition site preference and rearing for successive generations of the polyphagous big-eyed bugs *Piocoris varius* (Uhler) and *Geocoris proteus* Distant (Heteroptera: Geocoridae). *Bulletin of the Chiba Prefectural Agriculture Research Center (Japan)*, 7: 53–61 (in Japanese with English summary).

Orphanides, G.M., Gonzalez, D. & Bartlett, B.R. (1971). Identification and evaluation of pink bollworm predators in southern California. *Journal of Economic Entomology*, 64: 421–424.

Parajulee, M. N., Shrestha, R. B., Leser, J. F., Wester, D. B. & Blanco, C. A. (2006). Evaluation of the functional response of selected arthropod predators on bollworm eggs in the laboratory and effect of temperature on their predation efficiency. *Environmental Entomology*, 35: 379–386.

Patel, N.G. (1980). The bionomics and control measures of tobacco bug, *Nesidiocoris tenuis* Reuter (Miridae: Hemiptera). *Gujarat Agricultural University Research Journal*, 5(2): 60.

Pendleton, N.D. (2002). Development and impact of *Geocoris punctipes* (Say) (Hemiptera; Lygaeidae) on selected pests of greenhouse ornamentals. Master thesis, University of Tennessee, Knoxville, 93 pp.

Perdikis, D. & Lykouressis, D. (2000). Effects of various items, host plants and temperatures on the development and survival of *Macrolophus pygmaeus* Rambur (Hemiptera: Miridae). *Biological Control*, 17: 55–60.

Perdikis, D., Lucas, E., Garantonakis, N., Giatropoulos, A., Kitsis, P., Maselou, D., Panagakis, S., Paraskevopoulos, A., Lykouressis, D. & Fantinou, A. (2009). Intraguild predation between *Macrolophus pygmaeus* and *Nesidiocoris tenuis*. *IOBC/WPRS Bulletin*, 49: 301–305.

Perdikis, D., Lucas, E., Garantonakis, N., Giatropoulos, A., Kitsis, P., Maselou, D., Panagakis, S., Lampropoulos, P., Paraskevopoulos, A., Lykouressis, D. & Fantinou, A. (2014). Intraguild predation and sublethal interactions between two zoophytophagous mirids, *Macrolophus pygmaeus* and *Nesidiocoris tenuis*. *Biological Control*, 70: 35–41.

Perdikis, D. & Lykouressis, D.P. (2002). Life table and biological characteristics of *Macrolophus pygmaeus* when feeding on *Myzus persicae* and *Trialeurodes vaporariorum*. *Entomologia Experimentalis et Applicata*, 102: 261–272.

Pérez-Hedo, M., Rambla, J.L., Granell, A. & Urbaneja, A. (2018). Biological activity and specificity of Miridae-induced plant volatiles. *BioControl*, 63: 203–213. doi: 10.1007/s10526-017-9854-4.

Pérez-Hedo, M. & Urbaneja, A. (2015). Prospects for predatory mirid bugs as biocontrol agents of aphids in sweet peppers. *Journal of Pest Science*, 88: 65–73. doi: 10.1007/s10340-014-0587-1.

Péricart, J. (2001). Family Lygaeidae Schilling, 1829 seed-bugs. In: *Catalogue of the Heteroptera of the Palearctic Region. Pentatomomorpha I.* Vol 4. (Eds. Aukema, B. & Rieger, C.). The Netherlands Entomological Society, Amsterdam, pp. 35–220.

Pina, T., Argolo, P.S., Urbaneja, A. & Jacas, J.A. (2012). Effect of pollen quality on the efficacy of two different life-style predatory mites against *Tetranychus urticae* in citrus. *Biological Control*, 61: 176–183.

Poorani, J., Padmanaban, B. & Thanigairaj, R. (2019). Natural enemies of banana lacewing bug, *Stephanitis typica* (Distant) in India, including first report of *Anagrus* sp. (Hymenoptera: Mymaridae) as its egg parasitoid. *Munis Entomology & Zoology*, 14 (1): 83–87.

Rajan, S.J., Latha, E.S. & Sathish, R. (2018). Biology of Big-eyed bug, *Geocoris erythrocephalus* (Lepeletier & Serville) on cabbage aphid, *Brevicoryne brassicae* (L.). *International Journal of Current Microbiology and Applied Sciences*, 7(7): 3301–3305.

Readio, J. & Sweet, M.H. (1982). Are view of the Geocorine of the United States east of the 100th meridian (Hemiptera: Lygaeidae). *Miscellaneous publications of the Entomological Society of America*, 12: 1–91.

Reddy, N.A. & Kumar, A. (2004). Studies on the seasonal incidence of insect pests of tomato in Karnataka. *Pest Management in Horticultural Ecosystems*, 10(2):113121.

Rengifo-Correa, L.; Brailovsky, H.; Henry, T. J. & Morrone, J. J. (2013). Phylogenetics and evolutionary morphology of the Neotropical true bug genus *Epipolops* (Hemiptera: Heteroptera: Geocoridae). *Systematic Entomology*, 39:127–140.

Rim, H., Uefune, M., Ozawa, R. & Takabayashi, J. (2015). Olfactory response of the omnivorous mirid bug *Nesidiocoris tenuis* to eggplants infested by prey: Specificity in prey developmental stages and prey species. *Biological Control*, 91:47–54.

Riudavets, J. & Castañé, C. (1998). Identification and Evaluation of Native Predators of *Frankliniella occidentalis* (Thysanoptera: Thripidae) in the Mediterranean. *Environmental Entomology*, 27(1): 86–93.

Roger, C., Coderre, D. & Boivin, G. (2000). Differential prey utilization by the generalist predator *Coleomegilla maculata lengi* according to prey size and species. *Entomologia Experimentalis et Applicata*, 94: 3 –13.

Rondon S. I., Cantliffe D. J., Price J. F., (2004). The feeding preferences of the bigeyed bug, minute pirate bug, and pink spotted lady beetle relative to main strawberry pests. *Environmental Entomology*, 33: 1014–1019.

Sanchez, J.A. (2008). Zoophytophagy in the plantbug *Nesidiocoris tenuis*. *Agricultural and Forest Entomology*, 10: 75–80.

Sanchez, J.A. (2009). Density thresholds for *Nesidiocoris tenuis* (Heteroptera: Miridae) in tomato crops. *Biological Control*, 51:493–498.

Sanchez, J.A. & Cassis, G. (2018). Towards solving the taxonomic impasse of the biocontrol plant bug subgenus *Dicyphus (Dicyphus)* (Insecta: Heteroptera: Miridae) using molecular, morphometric and morphological partitions. *Zoological Journal of the Linnean Society*, 184: 330–406.

Sanchez, J.A., del Pino-Pérez, M., Davo, M.M., Martinez-Cascales, J.I. & Lacasa, A. (2006). Zoophytophagy of the plantbug *Nesidiocoris tenuis* in tomato crops in southeast Spain. *IOBC/WPRS Bulletin*, 29: 243–248.

Sanchez, J.A., Gillespie, D.R. & Gillespie, R.R. (2004). Plant preference in relation to life history traits in the zoophytophagous predator *Dicyphus hesperus*. *Entomologia Experimentalis et Applicata*, 112:7–19.

Sanchez, J.A., Gillespie, D.R. & McGregor, R.R. (2003a). The effects of mullein plants (Verbascum thapsus) on the population dynamics of *Dicyphus hesperus* (Heteroptera: Miridae) in tomato greenhouses. *Biological Control*, 28: 313–319.

Sanchez, J.A., Gillespie, D.R. & McGregor, R.R. (2004). Plant preference in relation to life history traits in the zoophytophagous predator *Dicyphus hesperus*. *Entomologia Experimentalis et Applicata*, 112:7–19.

Sanchez, J.A., Lacasa, A., Arnó, J., Castañé, C. & Alomar, O. (2009). Life history parameters for *Nesidiocoris tenuis* (Reuter) (Het., Miridae) under different temperature regimes. *Journal of Applied Entomology*, 133: 125–132.

Sanchez, J.A., Martinez-Cascales, J.I. & Lacasa, A. (2003b). Abundance and wild host plants of predator mirids (Heteroptera: Miridae) in horticultural crops in the Southeast of Spain. *IOBC/WPRS Bulletin*, 26: 147–151.

Sánchez, J.A., Martinez-Cascales, J.I. & Lacasa, A. (2003c). Abundance and wild host plants of predator mirids (Heteroptera: Miridae) in horticultural crops in the Southeast of Spain. *IOBC/WPRS Bulletin*, 26(10):147–152.

Sannigrahi, S. & Mukhopadhyay, A. (1992). Laboratory evaluation of predatory efficiency of *Geocoris ochropterus* Fieber (Hemiptera: Lygaeidae) on some common tea pests. *Sri Lanka Journal of Tea Science*, 61: 39–44.

Schaefer, C.W. (2009). Prosorrhyncha (Heteroptera and Coleorrhyncha). In: *Encyclopedia of Insects*, 2nd Edition, (Eds. Resh, V. & Carde, R.). Academic Press, Burlington, NJ. pp.839–855.

Schuh, R.T. (1976). Pretarsal structure in the Miridae (Hemiptera) with a cladistics analysis of relationships within the family. *American Museum Novitates*, 2601:391–39.

Schuh, R.T (1995). *Plant Bugs of the World (Insecta: Heteroptera: Miridae)*. Systematic Catalog, Distributions, Host List, and Bibliography, New York, 1329 pp.

Schuh, R.T. (2013). On-line systematic catalog of plant bugs (Insecta: Heteroptera: Miridae). Available at: http://research.amnh.org/pbi/catalog/.

Schuman, K.B., Kessler, D. & Baldwin I.T. (2013). Ecological observations of native *Geocoris pallens* and *G. punctipes* populations in the Great Basin Desert southwestern Utah. *Psyche*, 2013: 1–11.

Schuman, M.C., Barthel, K. & Baldwin, I.T. (2012). Herbivory induced volatiles function as defenses increasing fitness of the native plant *Nicotiana attenuata* in nature. *eLife*, 1: Article ID e00007.

Sengonça, C. & Saleh, A. (2002). Prey consumption of the predatory bug *Dicyphus tamaninii* Wagner (Heteroptera: Miridae) during nymphal and adult stages by feeding on *Aphis gossypii* Glover (Homoptera: Aphididae) as prey. *Journal of Plant Diseases and Protection*, 109: 430–439.

Shiojiri, K., Kishimoto, K., Ozawa, R., Kugimiya, S., Urashimo, S., Arimura, G., Horiuchi, J., Nishioka, T., Matsui, K. & Takabayashi, J. (2006). Changing green leaf volatile biosynthesis in plants: An approach for improving plant resistance against both herbivores and pathogens. *Proceedings of the National Academy of Sciences of the United States of America*, 103:16672–16676.

Shipp, J.L. & Wang, K. (2006). Evaluation of *Dicyphus hesperus* (Heteroptera: Miridae) for biological control of *Frankliniella occidentalis* (Thysanoptera: Thripidae) on greenhouse tomato. *Journal of Economic Entomology*, 99: 414–420.

Skibbe, M., Qu, N., Galis, I. &. Baldwin, I.T. (2008). Induced plant defenses in the natural environment: *Nicotiana attenuata* WRKY3 and WRKY6 coordinate responses to herbivory. *Plant Cell*, 20: 1984–2000.

Slater, J.A. (1964). *A Catalogue of the Lygaeidae of the World*. University of Connecticut, Storrs, 1668 pp.

Slater, J.A. (1981). Two new genera of Lygaeidae from northern Australia including the first member of the Pamphantini from the Eastern Hemisphere (Hemiptera: Heteroptera). *Journal of Australian Entomological Society*, 20: 111–118. doi: 10.1111/j.1440-6055.1981.tb01010.x.

Slater, J.A. (1999). The systematic position of the Pamphantinae with the description of two new tribes and a new species of *Cattarus* (Hemiptera: Lygaeoidea: Geocoridae). *Acta Societatis Zoologicae Bohemicae*, 63: 199–208.

Smith, H.A. & Krey, K.L. (2019). Three Release Rates of *Dicyphus hesperus* (Hemiptera: Miridae) for Management of *Bemisia tabaci* (Hemiptera: Aleyrodidae) on Greenhouse Tomato. *Insects*, 10: 213. doi: 10.3390/insects10070213.

Snodgrass, G. L. (1991). *Deraecoris* (sic) *nebulosus* (Heteroptera: Miridae): Little known predator in cotton in the Mississippi Delta. *Florida Entomologist*, 74: 340344.

Song, Y.H. & Heong, K.L. (1997). Changes in searching response with temperature of *Cyrtorhinus lividipennis* Reuter (Hemiptera: Miridae) on the eggs of the brown planthopper, *Nilaparvata lugens* (Stal.) (Homoptera: Delphacidae). *Researches on Population Ecology*, 39: 201–206.

Sridhar, V., Chakravarthy, A. K., Asokan, R, S., Vinesh, K., Rebijith B. & Vennila, S. (2014). New record of the invasive South American tomato leaf miner, *Tuta absoluta* (Meyrick) (Lepidoptera: Gelechiidae) in India. Pest Management in Horticultural Ecosystems, 20: 148–154.

Sridhar, V., Jayashankar, M. & Vinesh, L. S. (2012a). Zoophytophagous mirid bug, *Nesidiocoris tenuis* (Reuter) (Heteroptera: Miridae) on tomato (*Lycopersicon esculentum*) Pest or Beneficial. *IV* National Symposium on Plant Protection in Horticultural Crops, Bengaluru, India, April 25–28, p. 28.

Sridhar, V., Jayashankar, M. & Vinesh, L. S. (2012b). Population dynamics of zoophytophagous mirid bug, *Nesidiocoris tenuis* (Reuter) (Heteroptera: Miridae) and its prey, *Bemisia tabaci* Genn. (Homoptera: Aleyrodidae) on tomato (*Solanum lycopersicum* Mill). *Pest Management in Horticultural Ecosystems*, 18 (1): 35–38.

Streams, F. (1994). Effect of prey size on attack components of the functional response by *Notonecta undulata*. *Oecologia*, 98: 57–63.

Sweet, M.H. (2000). Economic importance of predation by big-eyed bugs (Geocoridae). In: *Heteroptera of Economic Importance* (Eds. Schaefer, C.W. & Panizzi, A.R.). CRC Press, Boca Raton, FL, pp. 713–724.

Tamaki, G., Olsen, D. P. & Gupta, R. K. (2019). Laboratory evaluation of *Geocoris bullatus* and *Nabis alternatus* as predators of *Lygus*. *Journal of the Entomological Society of British Columbia*, 75: 35–37.

Tamaki, G. & Weeks, R.E. (1972). Biology and Ecology of Two Predators, *Geocoris pallens* Stal and *G. bullatus* (Say) (Technical Bulletin no. 1446). US Department of Agriculture, Washington, DC.

Tavella, L. & Goula, M. (2001). Dicyphini collected in horticultural areas of north-western Italy (Heteroptera Miridae). *Bollettino di Zoologia Agraria e di Bachicoltura*, 33: 93–102.

Tedeschi, R., Clercq, P. de, Veire, M. van de & Tirry, L. (1999). Development and predation of *Macrolophus caliginosus* (Heteroptera: Miridae) on different prey. Proceedings of the 51st International Symposium on Crop Protection, Gent, Belgium, May 4, 1999. Part I, 64 (3a):pp. 235–240.

Tong, Y., Lu, Y. & Wu, K. (2011). Predation of *Geocoris* pallidipennis on Adelphocoris *lineolatus*. Chinese *Journal of Applied Entomology*, 48(1): 136–140.

Torres, J.B., Silva-Torres, C.S. & Ruberson, J. R. (2004). Effect of two prey types on life-history characteristics and predation rate of *Geocoris floridanus* (Heteroptera: Geocoridae). *Environmental Entomology*, 33(4): 964974.

Trottin-Caudal, Y. & Millot, P. (1994). Lutte integree contre les ravageurs sur tomate sous abri. Situation et perspectives en France. Bull. International Organization for Biological and Integrated Control of Noxious Animals and Plants, *West Paleartic Regional Section*, 17(5): 5–13.

Urbaneja, A., González-Cabrera, J., Arnó, J. & Gabarra, R. (2012). Prospects for the biological control of *Tuta absoluta* in tomatoes of the Mediterranean Basin. *Pest Management Science*, 68: 1215–1222.

Urbaneja, A., Montón, H. & Mollá, O. (2009). Suitability of the tomato borer *Tuta absoluta* as prey for *Macrolophus pygmaeus* and *Nesidiocoris tenuis*. *Journal of Applied Entomology*, 133: 292–296.

Urbaneja, A., Tapia, G. & Stansly, P. (2005). Influence of host plant and prey availability on developmental time and survivorship of *Nesidiocoris tenius* (Het.: Miridae). *Biocontrol Science & Technology*, 15: 513–518.

Van Den Meiracker, R.A.F. & Sabelis, M.W. (1999) Do functional responses of predatory arthropods reach a plateau? A case study of *Orius insidiosus* with western flower thrips as prey. *Entomologia Experimentalis et Applicata*, 90: 323–329. doi: 10.1046/j.1570-7458.1999.00452.x.

van Lenteren, J.C. (2003). Commercial availability of biological control agents. In: *Quality Control and Production of Biological Control Agents: Theory and Testing Procedures* (Ed. van Lenteren, J.C.). CABI Publishing, London, UK, pp. 167–179.

van Lenteren, J.C. (2012). The state of commercial augmentative biological control: Plenty of natural enemies, but a frustrating lack of uptake. *BioControl*, 57:1–20.

van Lenteren, J.H., Hemerik, L., Lins, J.C. Jr. & Bueno, V.H.P. (2016). Functional responses of three neotropical mirid predators to eggs of *Tuta absoluta* on tomato. *Insects*, 7: 34. doi: 10.3390/ insects703 0034.

Vandekerkhove, B. & De Clercq, P. (2010). Pollen as an alternative or supplementary food for the mirid predator *Macrolophus pygmaeus*. *Biological Control*, 53:238–242.

Varshney, R. & Ballal, C.R. (2017b). Biology and feeding potential of *Geocoris superbus* Montandon (Heteroptera: Geocoridae), a predator of mealybug. *Journal of Entomology and Zoology Studies*, 5: 520524.

Varshney, R. & Ballal, C.R. (2017a). Biological, morphological and life table parameters of *Geocoris ochropterus* Fieber. (Hemiptera: Geocoridae) fed on *Sitotroga cerealella* eggs. *Egyptian Journal of Biological Pest Control*, 27(2): 189–194.

Varshney, R. & Ballal, C.R. (2018). Intraguild predation on *Trichogramma chilonis* Ishii (Hymenoptera: Trichogrammatidae) by the generalist predator Geocoris ochropterus Fieber (Hemiptera: Geocoridae). *Egyptian Journal of Biological Pest Control*, 28(1): 1–6. doi: 10.1186/s41938-017-0006-z.

Varshney R. & Ballal, C.R. (2019). Prospects for predatory mirids in biological control. In *XIX International Plant Protection Congress IPPC2019*, at Hyderabad, Telangana, India from 10–14 November 2019.

Varshney, R. & Ballal, C.R. (2017c). Studies on evaluation of *Nesidiocoris tenuis* (Reuter) (Hemiptera: Miridae) preying on invasive insect pest Tuta absoluta (Meyrick) (Lepidoptera: Gelechiidae) and its damage to tomato plant. *Journal of Biological Control*, 31(2): 69–73.

Varshney, R., Budhlakoti, N. & Ballal, C.R. (2018b). Functional response of *Geocoris ochropterus* Fieber (Hemiptera: Geocoridae) to different egg densities of *Helicoverpa armigera* (Hübner) (Lepidoptera: Noctuidae). *Phytoparasitica*, 46: 451–458.

Varshney, R., Budhlakoti, N. & Ballal, C.R. (2018c). *Geocoris ochropterus* Fieber: A potential predator. In *First International Conference on Biological Control: Approaches and Applications*, September 27th–29th 2018, NBAIR, Bengaluru, India, 267 pp.

Varshney, R., Budhlakoti, N. & Ballal, C.R. (2019). Biology and functional response of *Dortus primarius* Distant (Hemiptera: Miridae) to common blossom thrips, *Frankliniella schultzei* (Trybom) (Thysanoptera: Thripidae). In *International conference on Plant protection in Horticulture- Advances and Challenges at ICAR-IIHR*, IIHR, Bengaluru, India, July 24–27, 2019.

Varshney, R. & Yeshwanth H. M. (2018). First Record of *Termatophylum orientale* Poppius (Hemiptera: Miridae: Deraeocorinae) from India with Biological Note. *Journal of Entomological Research Society*, 20(3):67–73.

Varshney, R., Yeshwanth H. M. & Ballal, C.R. (2018a). Biology and rearing protocol for *Dortus primarius* Distant, a predatory mirid (Hemiptera: Miridae). *Journal of Biological Control*, 32(4): 224–229. doi: 10.18311/jbc/2018/22438.

Velasco-Hernández, M.C., Ramirez-Romero, R., Cicero, L., Michel-Rios, C. & Desneux, N. (2013) Intraguild predation on the whitefly parasitoid *Eretmocerus eremicus* by the generalist predator *Geocoris punctipes*: A behavioral approach. *PLoS One*, 11: e80679.

Wheeler, A. G. Jr. (2000). Plant bugs (Miridae) as plant pests. In: *Heteroptera of Economic Importance* (Eds. Schaefer, C, W. & Panizzi, A.R.). CRC Press, Boca Raton, FL, pp. 37–83.

Wheeler, A. G. Jr. (2001). *Biology of the Plant Bugs (Hemiptera:Miridae). Pests, Predators, Opportunists.* Cornell University Press, Ithaca, NY, 507 pp.

Wheeler, A. G. Jr., Stinner, B. R. & Henry, T.J. (1975). Biology and nymphal stages of *Deraeocoris nebulosus* (Hemiptera; Miridae), a predator of arthropod pests on ornamentals. *Annals of Entomological Society of America*, 68: 1063–1068.

Wood, M. (1998). Bugslinger! *Agricultural Research*, 46(8): 12–14.

Yasunaga, T., Schuh, R.T., Poorani, J. & Cassis, G. A. (2015). Remarkable new genus and new species of the plant bug (Heteroptera: Miridae: Phylinae), inhabiting psyllid leaf margin roll gall on Indian banyan, Ficus benghalensis. *American Museum Novitates*, 3839: 1–16.

Yasunaga, T., Takai, M. & Nakatani, Y. (1997). Species of the genus *Stethoconus* of Japan (Heteroptera, Miridae): Predaceous deraeocorine plant bugs associated with lace bugs (Tingidae). *Applied Entomology and Zoology*, 32: 261–264.

Yeshwant, H.M. (2013). Taxonomy of mirid bugs (Hemiptera: Miridae) of South India. Ph.D. Thesis. University of Agricultural Sciences, GKVK, Bangalore, India.

Yu, H., Khashaveh, A., Li, Y., Li, X. & Zhang, Y. (2017). Field trapping of predaceous insects with synthetic herbivore-induced plant volatiles in cotton fields. *Environmental Entomology*, 47(1): 114–120. doi: 10.1093/ee/nvx201.

Zadeh, N.A. & Parvar, A. (2014). Study of mass production of predatory bugs *Deraeocoris* sp. for use in biological control against sucking pests in different crops. *Annals of Biological Research*, 5(1): 100–105.

Zappalà, L., Biondi, A., Alma, A., Al-Jboory, I.J, Arnó, J., Bayram, A., Chailleux, A., El-Arnaouty, A., Gerling, D., Guenaoui, Y., Shaltiel-Harpaz, L.,.Siscaro, G., Stavrinides, M., Tavella, L., Vercher Aznar, R., Urbaneja, A. & Desneux, N. (2013). Natural enemies of the South American moth, *Tuta absoluta*, in Europe, North Africa and Middle-East, and their potential use in pest control strategies. *Journal of Pest Science*, 86: 635–647.

5 Asopinae (Pentatomidae) Predators

S. Salini and K. J. David
National Bureau of Agricultural Insect Resources

CONTENTS

LEARNING OBJECTIVES

1. To understand the diversity of the species of Asopinae and their prey,
2. To know the status of commercialization and utilization of potential Asopinae species across the world,
3. To investigate suitable biocontrol agents and their prey amenable for mass multiplication.

5.1 INTRODUCTION

Pentatomidae is the third largest family of Hemiptera, with majority of the species being phytophagous and a small group specialized in predatory habits (Schwertner et al., 2021). Of the ten subfamilies, only four subfamilies, viz., Asopinae, Pentatominae, Phyllocephalinae and Podopinae are reported from India (Salini and Viraktamath, 2015). Among these, Asopinae are exclusively predatory, and they comprise 303 species belonging to 63 genera (Rider et al., 2018). In India, this subfamily is represented by 30 species in 17 genera (Salini, 2019). An autapomorphic character

DOI: 10.1201/9781003370864-5

defined by Gaupd (1991) was the presence of a crassate labium, an adaptation for the predatory habit. A few of them are known for their dark red or metallic blue sheen, but a few have bizarre appearance with robust body.

First instars of Asopinae do not attack prey and require plant juices and moisture for their development and from second instars onwards, they depend on animal food for their development (De Clercq, 2000). Asopinae are recognized by the crassate labium; the first segment of labium is markedly thickened and free except the base, which is embedded between bacculae. Besides, the bacculae are united behind the oral groove, which is considered to be plesiomorphic, as it is found in other pentatomid groups as well (Rider et al., 2018). A few genera possess subapical spines on forefemora with foliate expansion on foretibiae. In order to minimize the problem of scent giving away their presence to potential prey, many asopines have a reduced evaporatorium. Asopines are considered to be a natural group (monophyletic group) mainly based on the synapomorphies of the genitalia – the presence of a combination of thecal shield and genital plates (Thomas, 1992). While evaluating selected foreleg characters with respect to their phylogenetic relevance, Barão et al. (2013) opined that though Asopinae are considered to be a monophyletic group, only further phylogenetic analysis using additional characters can establish the monophyly of this group.

The Asopinae were first recognized as a group by Amyot and Serville (1843) as "Spissirostres" under the tribe "Coniscutes." The first comprehensive revision of this group was done by Schouteden (1907), which listed 152 species in 21 genera from the western hemisphere. Later, Thomas (1992) revised Asopinae for the same region with 110 species in 26 genera and provided a key to the new world Asopinae. Moreover, Thomas (1992) also provided an excellent review of various nomenclatural problems associated with the subfamily Asopinae and undertook a preliminary study of different characters and phylogenetic relationships of the world asopine genera. Based on this preliminary study, he pointed out that the various tribes erected in Asopinae could not be justified due to the lack of a congruent set of characters and tribal relationships. Thomas (1994) published taxonomic revision of 187 Old World species along with a key to genera and various nomenclatural changes. A note on nomenclatural changes and checklist of Chinese species were provided by Rider and Zheng (2002). The Euro-Mediterranean taxa were treated by Péricart (2010). Khuong (2004) recorded *Blachia ducalis* Walker, *Cazira bhoutanica* Schout., *C. breddini* Schout., *C. similis* Distant, *C. ulcerata* (H.-Sch.) and *C. friwaldskyi* Horv. from Vietnam. Roell et al. (2019) revised the African shield bug, *Afrius* Stål from Africa. Brugnera et al. (2020a) resolved the taxonomic confusion associated with a few species of *Podisus* (H.-Sch.). The new world genus, *Tylospilus* Stål, is studied by Brugnera et al. (2020b), who provided a key to the included species, redescribed *T. nigrobinotatus* (Berg.) and described *T. armatusimilis* Brugnera, Paim & Grazia from Uruguay. The egg structure and external morphology of immature stages of *Tynacantha marginata* Dallas, a Neotropican species, were studied by Brugnera and Grazia (2018).

The tribal classification in Asopinae is unresolved and Rider et al. (2018) discussed about various characters, such as the presence of a pair of pilose glands on the abdominal sternum of males, extension of seventh abdominal tergites, pedunculate eyes distant from the cervix (a membranous area between the head and thorax), a bilobed thorax and spinous humeri used for the proposition of Asopini, Armini, Jallini and Stilbotini by their respective authors. They also found that the three groups recognized in Asopinae by Gapon and Konstantinov (2006), based on aedeagal characters, did not correspond with the groups possessing or lacking male abdominal glands. Therefore, they suggested that this lack of congruence reveal that either the male genital characters or the abdominal glands or both are subject of convergent or homoplastic evolution and could not support the present tribal classification.

5.1.1 *Amyotea* Ellenrieder

Chinese species of *Amyotea* Ellenrieder was reviewed by Zhao et al. (2011) who described *A. lata* from Xizang, Yunnan, China. Ahmad and Rana (1992) redescribed *Amyotea malabrica* (Fab.)

based on specimens from Bihar, India and discussed the relationships with a few related genera. Azim and Shafee (1982) described *Asopus rufus* from a single female specimen collected from Agra, Uttar Pradesh, India, but later, this species was synonymized with *A. malabarica* by Rider and Zheng (2002). Ahmad and Kamaluddin (1996) described *A. maliki* from Pakistan. Salini (2016) redescribed *Amyotea malabrica* (Fab.) from India and provided short notes on the colour variability, bionomics and distribution.

5.1.2 *Anasida* Karsch

Ahmad and Kamaluddin (1983) described *Anasida ikrami* from CIBC (Commonwealth Institute of Biological Control) collection in India and reported the prey insect as caterpillar of *Porthetia* sp. (Lepidoptera). Ahmad and Rana (1991) reviewed *Anasida* and provided a key to the four species.

5.1.3 *Andrallus* (Bergroth)

Shanker et al. (2017) reported *Andrallus spinidens* (F.) as a predator on lepidopteran larvae in rice fields in India. They also studied the basic biology and predatory potential of this species under laboratory conditions and tested its suitability as a possible candidate for biological control of armyworm management in rice. Lefroy and Howlett (1909) reported that *A. spinidens* preys upon larvae of *Thermesia rubricans* and various other caterpillars. Fletcher (1914) opined that *Andrallus spinidens* is distributed throughout south India up to about 4,000 ft. Chopra and Sucheta (1986) redescribed *Andrallus* and reported *A. spinidens* from Hisar, Haryana, India.

5.1.4 *Arma* Hahn

Zheng (1981) studied the fauna of *Arma* from China. This genus is revised with description of *A. neocustos* and *A. neoinsperata* from Turkey by Ahmad and Önder (1990). Zhao et al. (2018) synonymized *Arma chinensis* with *A. custos* based on morphological, molecular and geographical data.

5.1.5 *Blachia* Walker

This genus was revised by Ahmad and Rana (1994) and later Sucheta and Khokhar (2006) redescribed the genus *Blachia* Walker along with its type species, *Blachia ducalis* Walker.

5.1.6 *Picromerus* Amyot & Serville

This genus was revised by Ahmad and Önder (1990) from the western Palaearctic region with description of *P. brachypterus* and *P. pseudobidens* from Turkey. They dealt with *P. nigridens* (Fabr.), *P. conformis* (H. Sch.) and *P. nigridens*, and provided a key to the western Palaearctic species. Zhao et al. (2013) reviewed *Picromerus* Amyot & Serville, described *P. elevatus* and provided a key to the species from China.

5.1.7 *Cecyrina* Walker

A review of *Cecyrina* Walker along with description of *C. flata* and *C. rubra* was done by Zhao et al. (2016).

5.1.8 *Cazira* Amyot & Serville

C. schwarzi was described by Abbasi and Rishi (1974) from Pakistan.

5.1.9 *Eocanthecona* Bergroth

Ahmad and Rana (1988) revised this genus from Pakistan, provided a key to the included species and described *E. neotibialis* (as *Canthecona neotibialis*) and *E. populusi* (as *Canthecona populusi*). They found *E. parva* feeding on caterpillars of *Utetheisa pulchella* (Linn.). Sridhar and Jhansirani (2003) reported *E. parva* as a predator of *Hypena quadralis* (Walker), a pest on the medicinal plant, *Tylophora indica* (Burm. f) in India. Fletcher (1914) found that both nymphs and adults of *Eocanthecona furcellata* were reported to feed on *Athalia proxima* and on various small caterpillars and gave its distribution as throughout plains of south India. A preliminary phylogenetic analysis of the included species of this genus was done by Ahmad et al. (1988). Shophiya and Sahayaraj (2014) studied the predatory potential of *E. furcellata* against larvae of *Pericallia ricini* (Fab.).

5.1.10 *Zicrona* Amyot & Serville

Chopra and Sucheta (1984) described *Z. hisarensis* from Hisar, Haryana, India along with the illustrations of male and female genitalia. Recently, Zhao et al. (2020) sequenced the complete mitochondrial genome of *Z. caerulea* (Linnaeus).

5.1.11 Species list of Asopinae

See Table 5.1.

5.2 DIVERSITY OF PREDATORS AND PREY

Andrallus spinidens (F.) is the most important predatory pentatomid on which several works related to various aspects, like biology (Bhojendra et al., 2019), predatory potential (Mohaghegh and Amir-Maafi, 2007; Bhojendra et al., 2019), climatic suitability (Bhatnagar et al., 2019; Bhojendra et al., 2019), suitability for mass rearing (Bhojendra et al., 2019), feeding behaviour and types of prey (Distant, 1902; Lefroy and Howlett, 1909; Pawar, 1976; Bhojendra et al., 2019), were carried out by different workers. This is a ubiquitous generalist predator.

Pawar (1976) recorded the nymphs and adults of *Andrallus spinidens* (F.) feeding on larvae of *Pelopidas mathias* (F.), *Melanitis leda ismene* (Cram.), *Spodoptera litura* (F.), *S. mauritia* (Boisd.), *Eurema hecabe* (L.) (=*Terias hecabe*) and *Mythimna separata* (Wlk.) in rice fields in the Kangra Valley, Himachal Pradesh, India. Mohaghegh and Amir-Maafi (2007) reported that both nymphs and adults of this species can prey on several caterpillars (e.g., *Naranga aenescens* Moore, *Helicoverpa armigera* (Hübner) and *Chilo suppressalis* (Walker)), in different field crops in Northern Iran. Bal and Biswas (2013) found that *A. spinidens* is predaceous on the Yellow Stem Borer, *Pamara mathias* (Fabr.), *Sesamia inferens* (Walk.), *Spodoptera litura* (Fabr.) and *Melanitis leda ismene* (Cramer).

Another important predator on which a lot of information is generated on various aspects of life history parameters (Kalaiyarasi et al., 2020), like predatory potential, feeding behaviour, etc. is *E. furcellata*. *E. furcellata* (Wolff) feeds on limacodids (Lepidoptera: Limacodidae) (Ghorpade, 1973; Mallik et al., 1996), sawflies (Hymenoptera: Tenthredinidae), *Oreta* Walker, *Setora* Walker, *Catopsilia* Huebner, *Bombyx* Linnaeus (all Lepidoptera) and also larvae of *Calopepla leayana* Latreille (Coleoptera: Chrysomelidae), *Henosepilachna vigintioctopunctata* (Fabr.) (Coleoptera: Coccinellidae) (Kalaiyarasi et al., 2020). This species was also considered as a serious pest of Tasar silkworms, like *Antheraea mylitta* Drury cultivated in the open (Kalshoven, 1981; Lefroy and Howlett, 1909; Fletcher, 1914; Singh et al., 2019). Shylesha and Sravika (2018) reported *E. furcellata* as a potential biocontrol agent for the control of fall armyworm, *Spodoptera frugiperda* (J. E. Smith). They also reported *A. spinidens* as a predator of *S. frugiperda*. *Amyotea malabrica* was found feeding on *Melanitis leda ismene* and was found to be a potential predator of rice bug, *Leptocorisa acuta* (Thunb.). *Cazira chiroptera* (Herrich-Schaffer) feeds on larvae of *Epilachna* Dejean (Coleoptera: Coccinellidae). Vanitha et al. (2018) found that this species prey

TABLE 5.1
List of Species of Asopinae Reported from India

Sl. No	Species Reported from India
1.	*Amyotea malabarica* (Fabr.)
2.	*Anasida orientalis* Dist.
3.	*Andrallus spinidens* (Fabr.)
4.	*Arma velata* Walk.
5.	*Blachia ducalis* Walk.
6.	*Cantheconidea humeralis* (Dist.)
7.	*Cantheconidea javana* (Dallas)
8.	*Cazira frivaldskyi* Horv.
9.	*Cazira insignis* (Schoutenden)
10.	*Cazira internexa* Walker
11.	*Cazira similis* Dist.
12.	*Cazira vegeta* Kirkaldy
13.	*Cazira verrucosa* (Westw.)
14.	*Cecyrina platyrhinoides* Walk.
15.	*Eocanthecona binotata* (Dist.)
16.	*Eocanthecona concinna (Walker)*
17.	*Eocanthecona furcellata* (Wolff.)
18.	*Eocanthecona ornatula* (Dist.)
19.	*Eocanthecona parva* (Dist.)
20.	*Eocanthecona robusta* (Dist.)
21.	*Eocanthecona tibialis* (Dist.)
22.	*Glypsus fuscispinus* Stål
23.	*Martinina prima* (Dist.)
24.	*Perillus bioculatus* (Fabr.)
25.	*Picromerus griseus* (Dall.)
26.	*Platynopus (Platynopus) indicus*
27.	*Pseudanasida fallax* Schout.
28.	*Troilus luridus* (Fabr.)
29.	*Zicrona caerulea* (Linn.)
30.	*Zicrona hisarensis* Chopra and Sucheta

Source: Mainly compiled from Walker (1867); Distant (1902, 1908, 1918); Kirkaldy (1909); Chatterjee (1934) and Thomas (1994).

upon a wide variety of leaf-eating caterpillars and flower pests of cashew. Singh et al. (2019) studied the feeding behaviour of this species on the muga silkworm (*Antheraea assamensis* (Helfer)) and reported that *E. furcellata* is an important predator contributing to muga crop loss. *Zicrona caerulea* (Linn.) was found feeding on various adults of chrysomelid beetle (*Altica* sp.) (Figures 5.2, 5.3) (Salini, person. observation), especially on those feeding on rose bushes in Himachal Pradesh. Bal and Biswas (2013) reported that both adults and nymphs of *Z. caerulea* are predaceous on larvae of *Spodoptera litura, Semiothisa pervolgata, Catopsilia pyranthe, Eurema hecabe, Anticarsia irrorata, Helicoverpa armigera* and *Pelopidas mathias.*

5.3 CHARACTERISTICS OF PENTATOMIDS

In general, Asopinae are have markedly thickened labium, which helps the predator to be active by feeding on the prey. Biocontrol agents are of generalist and specialist types. Generalist predators eat on more than one prey species and they often migrate from field to field if the prey density is less. Majority of Asopines are generalist predators and are suitable for augmentative biological control programmes. The predators should have good predatory potential, high fecundity, short life cycle, high survival rate, adaptability to various climatic conditions, be tolerant to commonly used pesticides, less mortality rate, high percentage of egg hatchability, survival, etc. Vanitha et al. (2018) reported the suitable attributes of *E. furcellata* as successful biological control agents. The attributes, like short life cycle, higher fecundity, high survival rate, feasibility of mass rearing of *E. furcellata*, make this species an ideal candidate for biological control programmes.

One of the important attributes required for a successful biological control agent is the adaptability to various climatic conditions and their compatibility with commonly used insecticides. Bhatnagar et al. (2019) and Bhojendra et al. (2019) studied the variability among various populations of *E. canthecona* in different agroecosystems of north western Himalayas. Their results showed that *E. furcellata* is adapted to different climatic conditions to survive with good predatory efficiency. Hence, this predator could be ideal in biological control programme for effective management of insect pests in various regions of western Himalayas. Zibaee et al. (2020) found that *A. spinidens* have shown compatibility with some insecticides and also the entomopathogenic fungus, *Beauveria bassiana*.

There are several values, which need to be monitored throughout a mass rearing process, in order to ensure the maintenance of quality production of biological control agents. They are egg viability; density and yield of immature forms; survival, size and yield of adults; adult density, sex ratio and age; life history measurements; diet texture, composition and microbial load; and environmental conditions (Chambers, 1977). But these attributes are understudied in almost all species of predatory pentatomids.

5.4 COMMERCIALIZATION

Presently, over 150 species of arthropods are commercially available worldwide for augmentation programmes targeting a wide array of mite and insect pests. About half of the species commercialized for biological control in European protected cultivation are exotic. This is related to the fact that most European countries impose no restrictions on distributors of beneficials for the importation and use of non-native species (De Clercq, 2002). Though some species have received considerable attention throughout the globe, only 10% have been studied in more or less detail out of the 300 plus species of Asopinae reported from the world. They attack slow-moving, soft-bodied insects, primarily larval forms of Lepidoptera, Coleoptera and Hymenoptera, though they are known to feed on prey from several other insect orders. The predatory potential of these insects were assessed on a worldwide scale. However, use of these bugs in agricultural practice is still negligible except a few (e.g. The spined soldier bug, *Podisus maculiventris* (Say) on Mexican bean beetle, *Epilachna varivestis* (Mulsant), *Perillus bioculatus* (F) on Colorado potato beetle (*Leptinotarsa decemlineata*), but within the context of future integrated pest management, several predatory pentatomids are believed to have a future for augmentation programmes to suppress economically important crop pests (De Clercq, 2000).

None of the Asopinae is truly host specific although there appears to be some degree of specialization in a number of species (De Clercq, 2002). The spined soldier bug, *Podisus maculiventris* (Say), is one of the best known predatory species and is commercialized in North America and Europe mainly for the biological control of Colorado potato beetle in Potatoes and Noctuid caterpillars in vegetables (De Clercq, 2002).

5.5 EFFECTIVE BIOCONTROL AGENTS

Among the various species of Asopinae known in India, *Amyotea malabarica* (Fabr.) (Figure 1a), *Andrallus spinidens* (Fabr.) (Figure 1b), *Blachia ducalis* Walk., *Cazira breddini* Schoutenden,

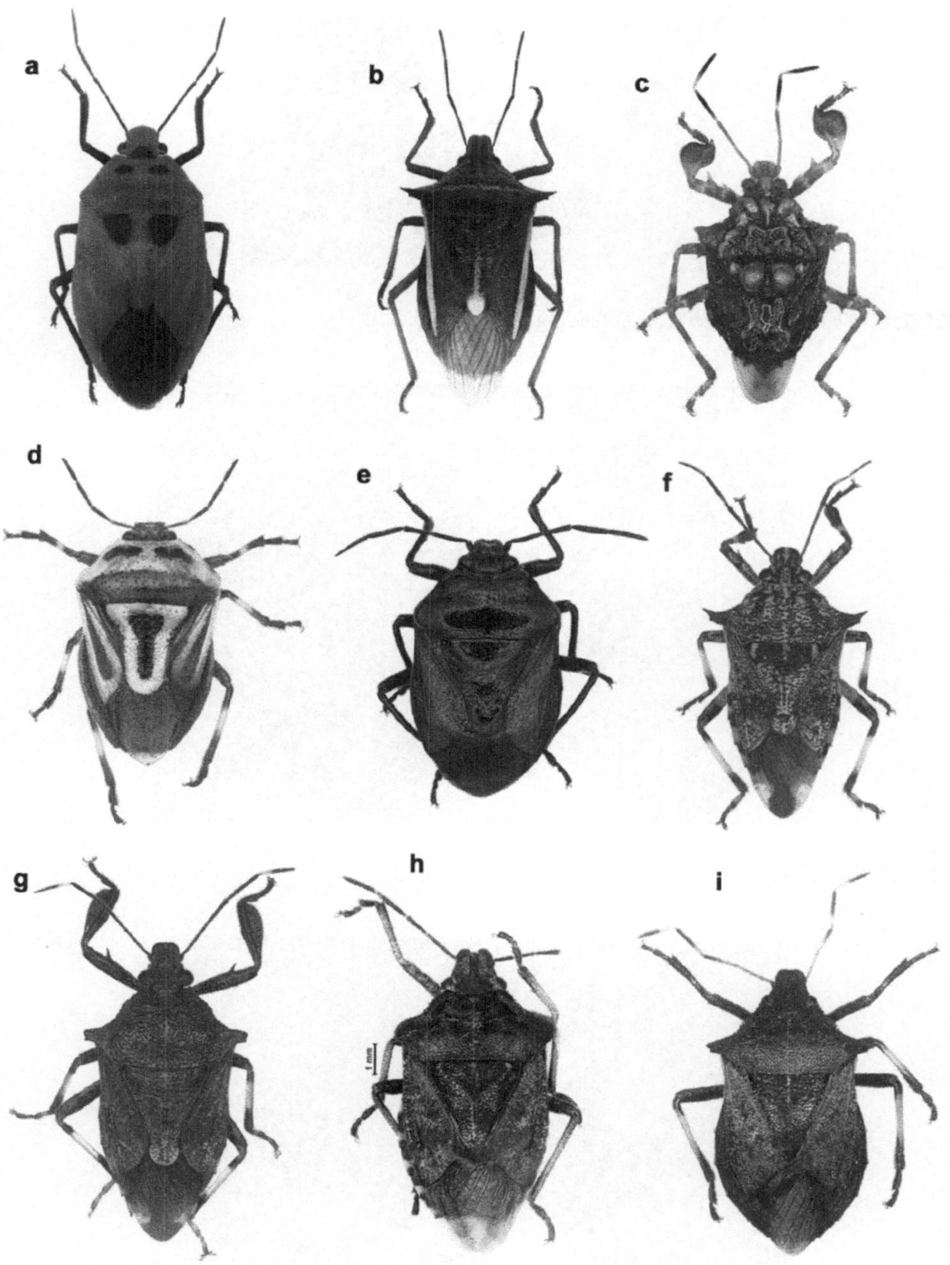

FIGURE 5.1 Habitus of Asopinae. a, *Amyotea malabarica* (Fabr.); b, *Andrallus spinidens* (Fabr.); c, *Cazira verrucosa* (Westw.); d, *Perillus bioculatus* (Fabr.); e, *Zicrona caerulea* (Linn.); f, *Eocanthecona furcellata* (Wolff.); g, *Eocanthecona concinna* (Walker); h, *Troilus luridus* (Fabr.); i, *Picromerus griseus* (Dallas).

Cazira frivaldskyi Horv., *Cazira insignis* (Schoutenden), *Cazira vegeta* Kirkaldy, *Cazira verrucosa* (Westw.) (Figure 1c); *Cecyrina platyrhinoides* Walk., *Eocanthecona furcellata* (Wolff.) (Figure 1f), *Eocanthecona parva, E. concinna* (Figure 1g), *Perillus bioculatus* (Fabr.) (Figure 1d), *Picromerus griseus* (Dall.) (Figure 1i); and *Troilus luridus* (Fabr.) (Figure 1h), *Zicrona caerulea* (Linn.) (Figure 1e) are the commonly occurring predatory pentatomids.

Ray and Khan (2011) studied the biology of *E. furcellata* and they found that this bug is good for minimizing the population of poplar defoliator, *Clostera fulgurita* Walker. Tiwari et al. (2017)

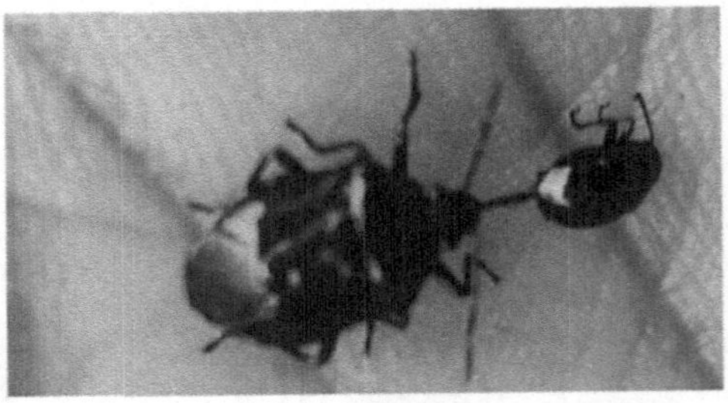

FIGURE 5.2 *Zicrona caerulea* (Linn.) feeding on *Altica* sp.

FIGURE 5.3 *Altica* sp.

FIGURE 5.4 *Eocanthecona furcellata* feeding on Citrus semi looper.

studied the impact of different lepidopteran hosts on the biology and predatory efficiency of *E. furcellata* under laboratory conditions. Their analysis showed that the maximum life period was on the laboratory host, *Corcyra cephalonica*. But in this host, the predators' egg laying was found to be the lowest. However, *E. furcellata* showed maximum predation on *Maruca vitrata*. Prasad and Pal (2015) reported *Perillus bioculatus* (Fabr.), commonly called as the two-spotted stink bug, for the first time from India (Meerut, Uttar Pradesh). They studied the biology of this bug along with its predatory potential. *Podisus maculiventris* (Say) is a generalist predator of over 100 species of insects, including approximately 50 different crop pests; its primary prey items are larval Coleoptera and Lepidoptera (Herrick and Reitz, 2009). But, at present, the distribution of *P. maculiventris* is restricted to Asia, Europe and North America, and is identified as a natural enemy of 40 species of insects (CABI, 2020).

5.6 MASS REARING OF PREDATORY PENTATOMID BUGS

This is very essential for a successful biological control programme to ensure the accessibility of the desirable numbers of predatory species to be used in agroecosystems. But as far as Asopinae are concerned, perusal of literature reveals that majority of the research on this group is centred around only a few species, like *A. spinidens*, *Eocanthecona furcellata* and *Amyotea malabarica* in India and not much information is available on the remaining species.

Successful mass rearing of *A. spinidens* is carried out on several hosts including *Corcyra cephalonica*, *Ephestia kuehniella*, *Galleria mellonella* and *S. litura* by various researchers (Mohaghegh and Amir-Maafi, 2007; Shanker et al., 2017; Bhojendra et al., 2019; Zibaee et al., 2020). Bhojendra et al. (2019) mass reared *A. spinidens* on second instar nymphs of *S. litura* under laboratory conditions (16 hours light: 8 hours darkness, 32°C ± 2°C, 75% ± 5% relative humidity). Zibaee et al. (2020) successfully reared this species on *Galleria mellonella* larvae.

Srikumar et al. (2018) developed a mass rearing procedure of *E. concinna* on *Corcyra cephalonica* and found that it is a potential predator of various lepidopteran pests of tea plantations. Vanitha et al. (2018) reported that *E. furcellata* can be successfully reared on *Galleria mellonella* and this laboratory host is suitable for its large-scale mass multiplication. Tiwari et al. (2017) found that *E. furcellata* can be successfully reared on *Spodoptera litura* and *Corcyra cephalonica*. Tuan et al. (2016) found that *Plutella xylostella* is a much superior host for rearing *E. furcellata* as compared to *S. litura*.

There are also studies using frozen insects as prey to mass multiply the predatory bugs. The predatory stinkbug, *Eocanthecona furcellata* (Wolff), using live larvae of *Spodoptera litura* (Fabr.) was heavier and had also more fertile eggs than that fed on frozen larvae (Yasuda and Wakamura, 1992). Use of frozen prey for mass rearing can ensure its availability and also storage possibility. Mohaghegh and Amir-Maafi (2007) studied the effect of live and frozen larvae of *Galleria mellonella* and *Ephestia kuehniella* and observed five nymphal instars of *A. spinidens*. They found that predators feeding on live larvae of *E. kuehniella* weighed more than those fed on corresponding frozen prey. They also found that the fecundity and longevity of the predators were affected neither by prey species nor by prey freezing and mentioned that it is a cost-effective method requiring less expense and labour in comparison to live prey.

Coudron and Kim (2004) attempted mass rearing of *Perillus bioculatus* on a zoophytogenous artificial or insect-free diet. They found that both in nymphs and adults, when fed with an artificial diet, their developmental times were prolonged, and their net reproductive rate and the cost of mass rearing decreased as compared to when fed with insects like *Trichoplusia ni* (Hübner). Another similar study is on record by Zou et al. (2015) on *Arma chinensis* (Fallou). They studied the developmental rate, life history parameters and fertility of *A. chinensis* on a zoophytogenous insect-free artificial diet and a secondary prey, pupae of Chinese oak silk moth (*Antheraea pernyi* (Guérin-Méneville)) (Lepidoptera: Saturniidae). They found that the developmental times were prolonged on the artificial diet and fecundity, egg viability and net reproductive rates were declined.

5.7 CONCLUSIONS

Asopinae are one of the most underexplored groups of predatory bugs. Quite a few species are commercially available for mass multiplication, especially in countries abroad. But the mass rearing and commercialization of predatory Pentatomidae is still in its infancy in India. They can also be successfully used under protected cultivation. By considering the basic characteristics of predatory Pentatomidae, it may be noted that they could be ideal candidates for augmentative biological control.

5.8 POINTS TO REMEMBER

1. Asopinae are generalist predators and are ideal candidates for augmentative biological control programmes.
2. Of the 300 plus species of Asopinae reported from the world, only 10% have been studied in more or less detail. The use of frozen prey for mass rearing will ensure the availability of biocontrol agents when required.
3. The spined soldier bug, *Podisus maculiventris* (Say), is one of the best known predatory species and is commercialized in North America and Europe. This species is identified as a natural enemy of 40 species of insects.
4. There are reports that *Galleria mellonella*, *Spodoptera litura* and *Corcyra cephalonica* are ideal laboratory hosts for the large-scale mass multiplication of *E. furcellata.*
5. *E. concinna* could be successfully mass multiplied on *Corcyra cephalonica* and it is a potential predator of various lepidopteran pests of tea plantations.

ACKNOWLEDGEMENTS

The authors are grateful to Dr. S. N. Sushil, Director, ICAR-NBAIR, Bengaluru, India for the support and facilities extended for this work.

REFERENCES

Abbasi, Q.A. & Rishi, Z. (1974). Eine neuepaläarktische Wanzen-Art der Gattung *Cazira* aus Pakistan (Heteroptera, Pentatomidae). *Entomologische Zeitschrift*, 84(5): 41–44.

Ahmad, I. & Kamaluddin, S. (1983). A new species of *Anasida* Karsch (Pentatomidae: Pentatominae: Asopini) with the first record of the genus from Pakistan. *Pakistan Journal of Zoology*, 15(1): 75–78.

Ahmad, I. & Kamaluddin, S. (1996). New record and new species of a predaceous asopine stink bug genus *Amyotea* Ellenrieder (Hemiptera: Pentatomidae: Pentatominae) from pulses field of Sialkot district, Pakistan. *Pakistan Journal of Zoology*, 28(4): 311–313.

Ahmad, I. & Önder, F. (1990). Revision of the genus *Arma* Hahn (Hemiptera: Pentatomidae: Pentatominae: Asopini) with description of two new species from Turkey. *Türkiye Entomologii Dergisi*, 14(1): 3–12.

Ahmad, I. & Rana, N.A. (1988). A revision of the genus *Canthecona* Amyot et Serville (Hemiptera: Pentatomidae: Pentatominae: Asopini) from Indo-Pakistan subcontinent with description of two new species from Pakistan. *Türkiye bitki korumadergisi*, 12(2): 75–84.

Ahmad, I. & Rana, N.A. (1991). A review and cladistic analysis of the genus *Anasida* Karsch (Pentatomidae: Pentatominae: Asopini). *Proceedings of the Pakistan Congress of Zoology*, 10: 271–285.

Ahmad, I. & Rana, N.A. (1992). Redescription of *Amyotea malabarica* (F.) (Hemiptera: Pentatomidae: Pentatominae: Asopini) and its cladistic relationships. *Proceedings of the Pakistan Congress of Zoology*, 12: 237–241.

Ahmad, I. and Rana, N.A. 1994. A revision of asopine stink bug genera *Blachia* Walker and *Breddiniella* Schouteden (Hemiptera: Pentatomidae). *Proceedings of the Pakistan Congress of Zoology,* 14: 137–146.

Ahmad, I., Rana, N.A. & Zaidi, R.H. (1988). A cladistic analysis of the genus *Canthecona* Amyot et Serville (Hemiptera: Pentatomidae: Pentatominae: Asopini) from Indo Pakistan subcontinent. *Proceedings of the Pakistan Congress of Zoology*, 8: 147–153.

Amyot, C.J.B. & Serville, A. (1843). *Histoire Naturelle des Insectes. Hémipteres.* Librairie Encyclopedique de Roret ed., Paris. xxvi+675 p.

Azim, M. N. & Shafee, S.A. (1982). New descriptions, a new species of the genus *Asopus* (Heteroptera: Pentatomidae). *Journal of Bombay Natural History Society*, 79(2): 361–362.

Bal, A. & Biswas, B. (2013). *Andrallus spinidens* (Fabricius), *Amyotea malabaricus* (Fabricius), *Eocanthecona furcellata* (Wolff.), *Zicrona caerulea* (Linnaeus).-24 In: Handbook on Major Hemipteran Predators of India (Ed. The Director, Zoological Survey of India, Kolkata). Zoological Survey of India, Kolkata, pp.21–24.

Barão, K.R., Ferrari, A. & Grazia, J. (2013). Comparative morphology of selected characters of the Pentatomidae foreleg (Hemiptera: Heteroptera). Arthropod Structure and Development, 42: 425–435.

Bhatnagar, V.R., Maurya, R.P., Brijwal, L., Dobhal, P., Suyal, P. & Singh, N.K. (2019). Study on variability among predatory stink bug, *Eocanthecona furcellata* (Wolff) (Hemiptera: Pentatomidae) population thriving in different agro-ecosystems of north-western Himalayas, India. *Journal of Entomology and Zoology Studies*, 7(3): 726–736.

Bhojendra, P., Maurya, R., Brijwal, L., Patwal, H. & Suyal, P. (2019). Effect of diverse locations on the biological attributes of predatory bug, *Andrallus spinidens* (Fabricius) (Hemiptera: Pentatomidae) thriving in North-Western Himalayan region. *Journal of Biological Control*, 33(1): 18–26.

Brugnera, R. & Grazia, J. (2018). External morphology of immature of *Tynacantha marginata* Dallas, 1851 (Hemiptera, Heteroptera, Pentatomidae). *Zootaxa*, 4378(1): 121–128.

Brugnera, R., M.R. Paim, T. Roell, G. Dellapé, & Grazia, J. (2020b). Taxonomic modifications in *Tylospilus* Stål (Hemiptera: Pentatomidae: Asopinae): redescription of *T. nigrobinotatus* (Berg), description of a new species and an updated identification key for the genus. *Zootaxa*, 4766(1): 128–138.

Brugnera, R.T., Roell, T., Campos, L. A., & Grazia, J. (2020a). Taxonomy of widespread Neotropical species of *Podisus* Herrich-Schäffer (Hemiptera: Pentatomidae: Asopinae): Redescription of *P. distinctus* (Stål, 1860) and revalidation of *P. fuscescens* (Dallas, 1851). *Zootaxa*, 4751(3): 546–562.

CABI, (2020). Invasive Species Compendium. CAB International. Wallingford, UK. www.cabi.org/isc.

Chambers, D.L. (1977). Quality control in mass rearing. *Annual Review of Entomology*, 22: 289–308.

Chatterjee, N. C. (1934). Entomological investigations on the spike disease of sandal (24). Pentatomidae (Hemipt.). *Indian Forest Records*, 20: 1–31.

Chopra, N.P. & Sucheta. (1984). A new species of the genus *Zicrona* Amyot & Serville (Pentatomidae: Hemiptera). *Bulletin of Entomology*, 25(1): 38–41

Chopra, N.P & Sucheta. (1986). Taxonomic Studies on genus *Andrallus* Bergroth (Hemiptera: Pentatomidae: Asopinae). *Bulletin of Entomology*, 27(1): 37–40.

Coudron, T.A. & Kim, Y. (2004). Life history and cost analysis for continuous rearing of *Perillus bioculatus* (Heteroptera: Pentatomidae) on a zoophytogenous artificial diet. *Journal of Economic Entomology*, 97(7): 807–812.

De Clercq, P. (2000). Predaceous Bugs (Pentatomidae: Asopinae). In: *Heteroptera of Economic Importance* (Ed. C.W. Schaefer & A.R. Panizzi). CRC press, Boca Raton London, New York, Washington D C. p. 737–789.

De Clercq, P. (2002). Dark clouds and their silver linings: Exotic generalist predate ors in augmentative biological control. *Neotropical Entomology*, 31(2): 169–176.

Distant, W.L. (1902). Rhynchota Vol. I, Heteroptera. In: *The Fauna of British India Including Ceylon and Burma* (Ed. Blanford, W. T.). Taylor and Francis, London, p. xxxvii+438p.

Distant WL. (1908). Rhynchota Vol. IV. Homoptera and Appendix (Pt). In: *The Fauna of British India Including Ceylon and Burma* (Ed. Bingham). Taylor and Francis, London. xv+ 501p.

Distant WL. (1918). Rhynchota Vol.VII. Homoptera: Appendix. Heteroptera: Addenda. In: *The Fauna of British India Including Ceylon and Burma* (Eds. A. E Shipley and A. K. Marshall). Taylor and Francis, London. Vii+ 210p.

Fletcher, T.B. (1914). *Some South Indian Insects and Other Animals of Importance*. Publishers: Bishen Singh Mahendra Pal Singh, Dehra Dun. The Superintendent, Government Press, xxii+565 p.

Gapon, D.A., & Konstantinov, F.V. (2006). On the structure of the aedeagus of shield bugs (Heteroptera, Pentatomidae): III. Subfamily Asopinae. *Entomological Review,* 86: 491–507.

Gapud, V.P. (1991). A generic revision of the subfamily Asopinae, with consideration of its phylogenetic position in the family Pentatomidae and superfamily Pentatomoidea (Hemiptera-Heteroptera). Parts I and II. *Philippine Entomologist,* 8: 865–961.

Ghorpade, K.D. (1973). A predaceous pentatomid bug, *Cantheconidea furcellata* (Wolff) attacking *Latoia lepida* (Cramer) on mango near Bangalore. *Journal of Bombay Natural History Society*, 72(2): 596–599.

Herrick, N.J.& Reitz, S.R. (2009). Temporal occurrence of *Podisus maculiventris* (Hemiptera: Heteroptera: Pentatomidae) in North Florida. *Florida Entomologist,* 87(4): 587–590.

Kalaiyarasi, L., Livingstone, A.R. & Pavithra, S. (2020). Predation efficiency and life cycle of pentatomid bug, *Eocanthecona furcellata* (Wolff.) on the Brinjal pest, *Henosepilachna vigintioctopunctata. Journal of Biopesticides*, 13(1): 69–78.

Kalshoven, L.G.E. (1981). *Pests of Crops in Indonesia*. Revised and translated by P. A. Van Der Laan., P.T. Ichtiar Basu-Van Hoeve, Jakarta, xix+701p.

Khuong, Dang Duc. (2004). One new genus and six new species of the subfamily Asopinae (Pentatomida: Heteroptera) for Vietnam. *Academia Journal of Biology,* Flag counter, 27(1): 18-20. https://doi.org/10.15625/0866-7160/v27n1.5247.

Kirkaldy, G.W. (1909). *Catalogue of the Hemiptera (Heteroptera) with biological and anatomical references, lists of food plants and parasites etc. Vol. I Cimicidae (M). Berlin*, F.L. Dames, XI+392 p.

Lefroy, M.H. & Howlett, F.M. (1909). *Indian Insect Life, a manual of the insects of the plains (Tropical India)* [reprinted edition]. Today and Tomorrow's Printers and Publishers, New Delhi, 786p.

Mallik, B., Viraktamath, C.A. & Puttaswamy. (1996). Damage by slug caterpillar *Macroplectra nararia* (Lepidoptera: Limacodidae) to Coconut around Bangalore, South India. *Indian Coconut Journal*, 27(2): 2–4.

Mohaghegh J, Amir-Maafi M. (2007). Reproduction of the predatory stinkbug *Andrallus spinidens* (F.) (Heteroptera: Pentatomidae) on live and frozen prey. *Applied Entomology and Zoology*, 42(1): 15–20. https://doi. org/10.1303/aez.2007.15

Pawar, A.D. (1976), *Andrallus spinidens* (Fabricius) (Asopinae: Pentatomidae: Hemiptera) as a predator of insect pests of rice in Himachal Pradesh, India. *Rice Entomology Newsletter*, 4: 23–24.

Péricart, J. (2010). *Hémiptères Pentatomoidea Euro-Méditerranéens 3. Podopinae et Asopinae. In: Faune de France Volume 93.* Fédération Française des Sociétés de Sciences Naturelles, Paris. xxiv+291p.

Prasad, C.S. & Pal, R. (2015). First record of two spotted stink bug, *Perillus bioculatus* (Fab.) from Meerut (U.P.) North India. *International Journal of Environmental and Agriculture Research*, 1(3): 9–11.

Ray, S.N. & Khan, M.A. (2011). Biology of a predatory bug, *Canthecona furcellata* Wolff. (Hemiptera: Pentatomidae) on poplar defoliator, *Closterafulgurita* walker (Lepidoptera: Notodontidae). *Journal of Biopesticides*, 4(2): 109–111.

Rider, D.A., Schwertner, C.F., Vilímová, J., Rédei, D., Kment, P. & Thomas, D.B. (2018). *Higher Systematics of the Pentatomoidea. In: Invasive Stink Bugs and related species (Pentatomoidea) Biology, Higher Systematics, Semiochemistry and Management* (Ed. J.E. McPherson,). CRC Press, Tayler & Francis group, Boca Raton, London, New York, pp. 25–201.

Rider, D.A. & Zheng, L. (2002). Checklist and nomenclatural notes on the Chinese Pentatomidae (Heteroptera) I. Asopinae. *Entomotaxonomia*, 24(2): 107–115.

Roell, T., Lemaître, V. A. & Webb, M. (2019). Revision of the African Shield bug genus, *Afrius* Stål, 1870 (Hemiptera: Heteroptera: Pentatomidae: Asopinae). *European Journal of Taxonomy*, 520: 1–44. https://doi.org/10.5852/ejt.2019.520.

Salini, S. (2016). Redescription of predatory stink bug, *Amyotea malabarica* (Fabricius, 1775) (Hemiptera: Pentatomidae: Asopinae). *Journal of Biological Control*, 30(4): 240–247.

Salini, S. (2019). Pentatomidae (Hemiptera: Heteroptera: Pentatomidea) of India. *Indian Insects Diversity and Science, a festschrift for Professor C. A. Viraktamath's 75th birthday* (Ed. S. Ramani et al.). CRC Press, Taylor and Francis Group, Boca Raton, London, New York, pp. 121–146.

Salini, S. &Viraktamath, C.A. (2015). Genera of Pentatomidae (Hemiptera: Pentatomoidea) from south India - an illustrated key to genera and checklist of species. *Zootaxa*, 3924(1): 1–76.

Schouteden, H. (1907). Descriptions d'Asopiens nouveaux. *Annales de la Société Entomologique de Belgique*, 51: 36–51.

Schwertner, C.F., Carrenho, R., Moreira, F.F.F. & Cassis, G. (2021) Hemiptera Sampling Methods (In: *Measuring Arthropod Biodiversity* (Eds. J.C. Santos, and G.W. Fernandes). Springer, Cham. pp. 289-313. http://doi-org-443.webvpn.fjmu.edu.cn/10.1007/978-3-030-53226-0_12

Shanker, C., Sapna, P. Shabbir, K., Sunil, V., Jhansirani, B & Katti, Biology and functional Response of *Andrallus spinidens* (F) to the rice army worm *Spodoptera mauritia* (Boisduval). *Journal of Biological Control*, 31(4): 201–204.

Shophiya, N.J., & Sahayaraj, K. (2014). Biocontrol potential of entomophagous predator, *Eocanthecona furcellata* (Wolff.) against *Pericallia ricini* (Fab.) larvae. *International Journal of Current Research*, 6(10): 9052–9056.

Shylesha, A.N., & Sravika, A. (2018). Natural occurrence of predatory bugs, *Eocanthecona furcellata* (Wolff) and *Andrallus spinidens* (Fabr.) on *Spodoptera frugiperda* (Smith) (Hemiptera: Pentatomidae) in maize and their potential in management of fall army worm. *Journal of Biological Control*, 32(3): 209–211.

Singh, S., Jigyasu, D.K., Roy, D., Shabnam, A.A. & Das, R. (2019). Feeding behaviour of two important predator bugs *Eocanthecona furcellata* Wolff and *Sycanus collaris*Fabricius in Muga ecosystem. *Research Journal of Agricultural Sciences*, 10(1): 185–188.

Sridhar, V & Jhansirani, B. (2003). New Record of *Canthecona parva* (Distant) (Heteroptera: Pentatomidae) as a predator of *Hypena quadralis* (Walker), a pest of Antamul, *Tylophora indica* (Burm, f) Merr. *Journal of Biological Control*, 17(2): 179–180.

Srikumar, K.K., Salini, S., Smitha, S., Kumar, B.S. & Radhakrishnan, B. (2018). Taxonomy, bionomics and predatory potential of *Eocanthecona concinna* (Walker) (Hemiptera: Pentatomidae: Asopinae). *Journal of Biological Control*, 32(2): 81–86.

Sucheta & Khokhar, K.S. (2006). Taxonomic studies on the genus *Blachia* Walker and the type species (Pentatomidae: Asopinae). *Journal of Entomological Research*, 30(3): 267–271.

Thomas, D.B. (1994). Taxonomic synopsis of old the world Asopine genera (Heteroptera: Pentatomidae). *Insecta Mundi*, 8(3–4): 145–212.

Thomas, D.B. Jr. (1992). *Taxonomic synopsis of the Asopine Pentatomidae (Heteroptera) of the Western Hemisphere. Thomas Say Foundation Monographs No. 16.* Entomological Society of America, Lanham MD, 156p.

Tiwari, S., Maurya, R.P. & Pandey, A.K. (2017). Effect of different insect hosts on biology and predation efficiency of *Eocanthecona furcellata* Wolff (Hemiptera: Pentatomidae). *The Bioscan*, 12(1): 193–197.

Tuan, S.J., Yeh, C.C., Atlihan, R. & Chi, H. (2016). Linking life table and predation rate for biological control: A comparative study of *Eocantheconafurcellata*(Hemiptera: Pentatomidae) fed on *Spodoptera litura* (Lepidoptera: Noctuidae) and *Plutella xylostella* (Lepidoptera: Plutellidae). *Journal of Economic Entomology*, 109(1): 13–24.

Vanitha, K., Raviprasad, T.N. & Shwetha, V. (2018). Life cycle of *Eocantheconafurcellata* Wolff. (Hemiptera: Pentatomidae) a predatory bug in cashew plantations, upon rearing on waxmoth larvae. *Journal of Entomology and Zoology Studies*, 6(2): 3007–3010.

Walker, F. (1867). *Catalogue of the specimens of heteropterous Hemiptera in the collection of the British Museum.* Part II. *Scutata*. E. Newman, London, 177 pp. [pp. 241–417].

Yasuda, T. & Wakamura, S. (1992). Rearing of the predatory stink bug Eocanthecona furcellata (Wolff) (Heteroptera: Pentatomidae) on frozen larvae of *Spodoptera litura* (Fabricius) (Lepidoptera: Noctuidae). *Applied Entomology and Zoology*, 27(2): 303–305.

Zhao, Q., Bu, W-j. & Liu, G-q. (2011). The genus *Amyotea* Ellenreider from China (Heteroptera, Pentatomidae). *Acta Zootaxonomica Sinica*, 36(4): 950–955.

Zhao, Q., Bu, W-j. &Liu, G-q. (2016). The genus *Cecyrina* Walker, 1867, with the description of two new species (Hemiptera: Pentatomidae: Asopinae). *Zootaxa*, 4114(3): 309–319.

Zhao, Q., Cassis, G., Zhao, L., Yifan, H Zhang, H. & Wei, J. (2020): The complete mitochondrial genome of *Zicronacaerulea* (Linnaeus) (Hemiptera: Pentatomidae: Asopinae) and its phylogenetic implications. *Zootaxa*, 4747 (3): 547–561, https://doi.org/10.11646/zootaxa.4747.3.8

Zhao, Q., Liu, G-q. & Bu, W-j. (2013). A review of the Chinese species of the genus *Picromerus* Amyot and Serville, with description of a new species (Hemiptera: Heteroptera: Pentatomidae: Asopinae). *Zootaxa*, 3613(2):146–164.

Zhao, Q., Wei, J-f., Bu, W-j., Liu, G-q. &Zhang, H-f. (2018). Synonymize *Arma custos* based on morphological, molecular and geographical data. *Zootaxa*, 4455(1): 161–176.

Zheng, L-y. (1981). Chinese species of genus *Arma* Hahn (Hemiptera: Pentatomidae). *Natural Enemies of Insects*, 3(4): 28–32.

Zibaee, A., Ramzi, S. & Hoda, H. (2020). The stink bug, *Andrallusspinidens* (Hemiptera: Pentatomidae), a potential predator for effective biological control. *Journal of Plant Protection Research*, 60(4): 336–344.

Zou, D.Y., Coudron, T.A., Wu, H.H., Gu, X.S., Xu, W.H., Zhang, L.S. & Chen, H.Y. (2015). Performance and Cost comparisons for continuous rearing of *Arma chinensis* (Hemiptera: Pentatomidae: Asopinae) on a Zoophytogenous Artificial Diet and a Secondary Prey. *Journal of Economic Entomology*, 108(2): 454–461.

6 Chrysopids (Neuroptera: Chrysopidae)

N. Bakthavatsalam and Richa Varshney
ICAR-National Bureau of Agricultural Insect Resources

CONTENTS

LEARNING OBJECTIVES

1. To know the systematics and species diversity of chrysopids
2. To learn about the target prey species
3. To understand the basic biology
4. To identify the suitable prey for laboratory rearing
5. To study the rearing techniques, release methods and storage
6. To understand competition and intraguild relationships
7. To know about strain improvement
8. To understand chemical ecology
9. To learn about the compatibility between chrysopids and microbials.

6.1 SYSTEMATICS

Neuropterida, one of the most ancestral groups of Holometabola, diverged from a coleopteran/strepsipteran ancestor some 276 million years ago (Misof et al., 2014; Peters et al., 2014). Neuropterida include three orders: Raphidioptera (snakeflies), Megaloptera (dobsonflies, alderflies) and Neuroptera (lacewings, antlions, dustywings and allies) (Aspöck et al., 2012; Winterton et al., 2010). The order name Neuroptera is derived from Greek (neuro=nerve; pteron=wing) and refers to the reticulate nature of the wing venation. The Neuroptera are a relatively small, cosmopolitan order of endopterygote, neopterous insects that are best represented in tropical regions. Sometimes, Neuroptera is referred to as Planipennia (Tauber et al., 2009). The Neuroptera is one of the smaller insect orders with ~6,000 living species and three suborders: Nevrorthiformia, Myrmeleontiformia and Hemerobiiformia (Tauber et al., 2009;

DOI: 10.1201/9781003370864-6

Winterton et al., 2010) including 16 families (Winterton et al., 2010; Winterton and Makarkin, 2010). Nevrorthiformia is represented by just one family, Nevrorthidae, with 19 species, Myrmeleontiformia include the most specious family, antlions (Myrmeleontidae; ~2,100 species) (Stange, 2004; Tauber et al., 2009); and Hemerobiiformia include the second-largest and most agriculturally important family, green lacewings (Chrysopidae) (Tauber et al., 2009; Winterton and Freitas, 2006). Neuropterans are soft-bodied; the compound eyes are well developed, but ocelli are usually absent. The antennae are long, multisegmented and sometimes display a club-shaped enlargement at the apex. Mouthparts are of the mandibulate biting type. Neuroptera usually display four large, membranous wings that are subequal in size; venation is abundant and netlike. The larvae are distinguished by the suctorial jaws formed of mandibles and maxillae. They are generally very active and always carnivorous.

Chrysopidae are the second most species-rich family in the order, with ~1,416 species grouped in 82 genera (Garzón-Orduña et al., 2019; Oswald, 2020). The family is traditionally divided into three extant subfamilies – Apochrysinae, Nothochrysinae, Chrysopinae – and the extinct subfamily Limaiinae (Garzón-Orduña et al., 2019). Chrysopidae comprises adults that are distinguished by their lace green wings, prominent green eyes and large left mandible, and in some species the left mandible has a more prominent tooth-like projection on the inner margin. The chrysopids have been extensively used in the suppression of several pest species worldwide because of their biodiversity, amenability to laboratory rearing and easy availability due to commercial production. Extensive studies have been made on chrysopids by several researchers on the biology, rearing techniques, field release techniques conservation and utilization of chrysopids in biorational integrated pest management (BIPM). Detailed account of chrysopids has been given in *Biology of Chrysopidae* (Canard et al., 1984).

The first chrysopid to be known from the Indian subcontinent was *Ankylopteryx octopunctata* Fabricius, in 1798. Later, several researchers have contributed to the faunistic studies of Chrysopidae. Species *Ankylopteryx tesselatus* Needham was recorded in 1909 by Needham (1909). Brauer (1864) recorded species like *Ankylopteryx anomala* Brauer. *Chrysoperla carnea* is the common species encountered in several agroecosystems. The identification of this species based on the vibrational song revealed it to be *Chrysoperla zastrowi arabica* (Henry et al. 2006; Venkatesan et al., 2008). Similarly, the other species *Mallada boninensis* (Okamoto) has been identified as *M. desjardinsi* (Navás) and *M. astur* (Banks) as *Pseudomallada astur* (Banks) based on taxonomic characters and vibrational songs.

In India, before 2010, publications on the common green lacewings in agricultural crop lands have used a variety of names, including *Chrysopa carnea* (later *Chrysoperla carnea*), *Chrysopa scelestes* and *Brinckochrysa scelestes* (Duelli, 2001). Consequently, at present, there is no other *Chrysoperla* species than *C. zastrowi sillemi* known from India. Therefore, in this chapter, references to *Chrysopa carnea* (later *Chrysoperla carnea*) from India are qualified by *Chrysoperla zastrowi sillemi* (Figure 6.1) as true *C. carnea* does not occur in India.

6.2 CRYPTIC SONGS FOR THE IDENTIFICATION OF CHRYSOPIDS SPECIES

Green lacewings of the genus *Chrysoperla* Steinmann include at least one large clade of strikingly cryptic species, informally known as the "*carnea* group" (Duelli, 1996). Originally, this complex was considered to be a single Holarctic species, *Chrysoperla carnea* (Stephens), based upon many adult specimens of remarkably similar morphology found all across North America, Europe, northern Africa and Asia (Tjeder, 1960). Now this group has radiated rapidly and recently into an especially large and important complex of cryptic species; each species is identifiable by its own substrate-borne, vibrational mating signal produced by courting males and females in a precise duetting format (Henry, 1980, 2006). Venkatesan et al. (2008) through the cryptic song analysis proved that the so-called *C. carnea* is in fact *Chrysoperla zastrowi arabica* and now this is confirmed as a junior synonym of *C. zastrowi sillemi* (Esben-Petersen) and not *C. carnea* as claimed earlier. Such cryptic song-based taxonomic analysis will become a boon for the identification of several other species of chrysopids. Henry et al. (2013) further emphasized that incompatible duetting songs are powerful barriers to hybridization between and among sympatric species of lacewings in the wild.

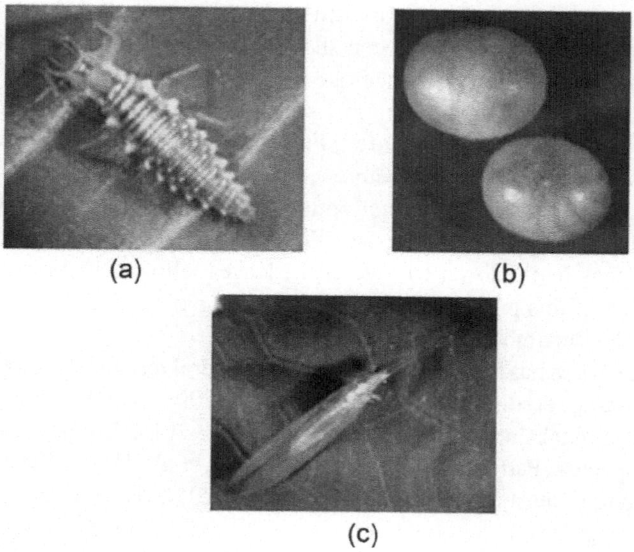

FIGURE 6.1 Chrysoperla *zastrowi sillemi*. (a) Larva, (b) pupae and (c) adult. (ICAR-NBAIR database.)

6.3 BIOLOGY AND LIFE HISTORY

All chrysopid larvae are characteristically voracious and they often have a broad prey range. They feed on small, comparatively soft-bodied arthropods, including aphids, leafhoppers, whiteflies and psyllids (Homoptera); thrips (Thysanoptera); and eggs of numerous insect pest species. lepidopteran larvae are also being attacked (Principi and Canard, 1984). Some adult chrysopids feed as carnivores, as do the larvae, and are also glyco-pollinivores (Principi and Canard, 1984). Adults of the genus *Chrysoperla* are glyco-pollinivorous (Chen et al., 2006).

The biology of several species of chrysopids and a few species of hemerobiid was studied in detail. However, species belonging to other families were seldom studied.

Chrysopids lay stalked oval, white, yellow and green eggs on the peduncle of the leaves, buds, stems and even fruiting bodies, sometimes with droplets of defensive fluid on the stalks (Eisner et al., 1996). Stalked eggs provide protection against predation and cannibalism (Rüzicka, 1997). The size of the stalk depends on mother's body size and environmental factors (Tauber et al., 2003). However, there are some platygastrid wasps, which parasitize chrysopid eggs (Ruberson et al., 1995; Alasady et al., 2011). There are few reports, which mentioned that some of the green lacewings (Chrysopidae) (*Anomalochrysa* spp.) have lost the stalked egg characteristic (Albuquerque et al., 2012; Canard et al., 1984; Tauber et al., 2009) in the Hawaiian Islands. Generally, most of the species lay the eggs singly. Species like *Apertochrysa* laid the eggs in groups that varied in numbers from 9 to 23 eggs (Bakthavatsalam et al., 1996). There are differences in the length of the eggstalk and posture of the eggs. Even though the egg stalk of *C. carnea* and *P. astur* looks alike, the eggs of *M. astur* remains erect while that of the *C. carnea* bends.

There are three larval instars. The larva of *C. carnea* is longer than broader while that of *M. astur* is broader than longer. There are two groups of larvae on the basis of their morphology and behaviour (Tauber et al., 2003). One group of larvae (*Apertochrysa*, *Dichochrysa*, *M. astur*, *M. boninensis*) collects trash, plant material, etc. with their mouthparts and place them on their dorsa and forms pocket or trash. The other group does not carry trash (*C. carnea*) on their back. Upon completion of larval development, the third instar larva spins a cocoon and remains inside this.

Males and females can be easily distinguished by their genitalia. Females are generally bigger than males. Mating takes place on the second day of adult emergence. The oviposition starts usually on the fifth day of emergence or third day of mating, which, however, varies with the species.

The biology of chrysopids is affected by different environmental factors. In the case of lacewing species, increased temperature has been shown to affect development as well as the rate of survival and reproductive potential (Albuquerque et al., 1994; Figueira et al., 2002; Fonseca et al., 2001; Maia et al., 2000). High temperature (35°C–37°C) was found to be detrimental to *C. carnea* (Pathan et al., 2016) and *Chrysoperla genanigra* Freitas (Bezerra et al., 2012). Diet and prey species also influence the biology of chrysopids. Kabissa et al. (1995) found that *H. armigera* eggs and *A. gossypii* nymphs were both adequate but not optimal diets for larval the development of *Mallada desjardinsi* (Navas) and *Chrysoperla congrua* (Walker). The adults of *Chrysopa scelestes* Banks preferred to lay the eggs on the lower surface of the leaves followed by branches (Patel and Vyas, 1985). Larval duration of this predator was 8.60 days when fed on *H. armigera* eggs and 11.70 days when fed on larvae (Krishnamoorthy and Mani, 1982).

The biology of *C. carnea* has been studied in detail by several researchers on *Corcyra cephalonica* Stainton (Bakthavatsalam et al., 1994; Saminathan et al., 2003; Khulbe et al., 2005; Chakraborty and Korat, 2010; Geethalakshmi et al., 2000; Shah et al., 2013; Kumar et al., 2014; Khanzada et al., 2018), aphid species (Patro and Behera, 2002; Sattar et al., 2011; Mhaske et al., 2017; Panth et al., 2017), mealybug (Kaur et al., 2008; Mhaske et al., 2017; Sattar et al., 2011) and other lepidopteran hosts. The egg, larval and pupal periods are generally 4–5, 8–10, 8–10 days, respectively. However, variation was observed between different strains/ecotypes from different geographical locations, ecosystems and the prey (Varma and Shenhmar, 1983; Bakthavatsalam et al., 1994; Mani and Krishnamoorthy, 1990; Balakrishnan et al., 2005; Patil and Hallolli, 2007; Bezerra et al., 2012; Panth et al., 2017). *Chrysoperla externa* (Hagen) successfully completed its development when fed on *Sipha flava* and can be reared at temperatures ranging from 16°C to 28°C, but not at 12°C and 32°C (Oliveira et al., 2010).

Apertochrysa astur (Banks, 1937) (=Pseudomallada astur) (=*Dichochrysa astur*) (Figure 6.2a) was recorded from several crops, like guava, mango and coffee (Singh and Narasimham, 1992). It feeds on leafhoppers, spiralling whitefly (*Aleurodicus dispersus* Russel) (Mani and Krishnamoorthy, 1999) and *Chloropulvinaria polygonata* (Ckll) (Mani and Krishnamoorthy, 1998). Patnaik and Bhagat (1989) studied the biology of *Chrysopa orestes* Banks on *Melanaphis sacchari* (Zehntner) ((*Longiunguis sacchari* (Zehntner)), *Ferrisia virgata*, *Thrips tabaci* Lindemann and *Tetranychus urticae* Koch. Around 86.33 eggs were laid by a single female and the larvae consumed 25.5 aphids/day.

Mallada desjardinsi (=*M. boninensis*) (Figure 6.2b) is a generalist predator and has been recorded to feed on pests, such as mealybugs (Mani and Krishnamoorthy, 1989), *Aphis gossypii*, eggs of *H. armigera* and *C. cephalonica* (Bakthavatsalam et al., 1996), whiteflies (Joshi and Yadav, 1990; Selvakumaran et al., 1996), aphids (Kabissa et al., 1996; Gade et al., 2011), neonates of *Helicoverpa armigera* (Hubner), *Spodoptera litura* F. and *Earias vittella* (F.) (Kabissa et al., 1996; Nagamallikadevi et al., 2013), red spider mite (Babu et al., 2011) and spiralling whitefly,

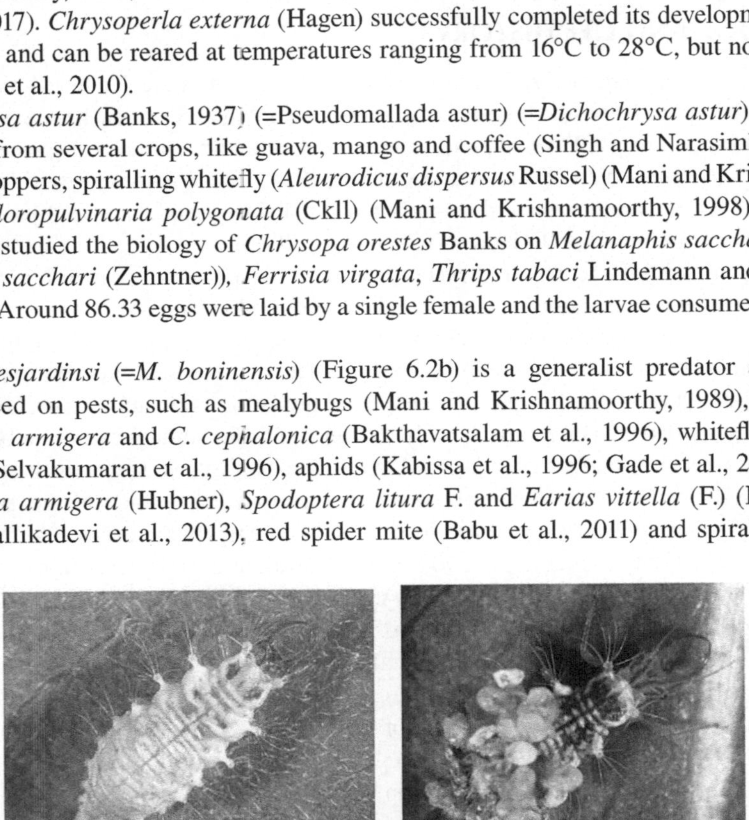

(a) (b)

FIGURE 6.2 (a) *Pseudomallada astur* (Banks) (=*Dichochrysa astur* (Banks)) and (b) *Mallada desjardinsi* (Navas) (=*M. boninensis* (Okamoto)). (ICAR-NBAIR database.)

Aleurodicus dispersus Russell (Boopathi et al., 2016). *Mallada desjardinsi* and *Chrysoperla congrua* (Walker) were found to be associated with *H. armigera* and *A. gossypii* and were able to complete their development on these pest species. (Kabissa, 1993). Joshi and Yadav (1990) reported that *Mallada desjardinsi* can consume an average of 628.75 eggs of *C. cephalonica* and 453 nymphs of whiteflies. The biology of this chrysopid was also studied on *Aleurocanthus woglumi* Ashby (Lavhe, 2000; Lavhe and Nimbalkar, 2000; Ramkumar et al., 2005a, b). Guntupalli and Kalyanasundaram (2016) found eggs of *C. cephalonica* to be superior for all the biological parameters tested followed by *Aleurodicus dispersus*, *Aphis gossypii* (nymphs/adults) and *Aphis craccivora* Koch. A list of chrysopids along with their prey is given in Table 6.1.

TABLE 6.1
Chryopids Associated with Different Insect Pests

Species	Prey	Host plant	References
Mallada desjardinsi (Navás) (=*Mallada boninensis* (Okamoto))	*Lipaleyrodes euphorbiae* David and Subramaniam	Star goose berry	Mani and Krishnamoorthy (1995)
	Kanakarajiella cardamomi (David and Subramaniam)	Cardamom	Selvakumaran et al. (1996)
	Aphis gossypii Glover	Cotton	Jalali and Singh (1994)
	Amrasca biguttula biguttula Ishida	Several plants	Singh et al. (1993)
	Idioscopus clypealis Lethierry	Mango	Fasih and Srivastava (1990)
	Maconellicoccus hirsutus Green	Grape vine	Krishnamoorthy and Mani (1989)
	Bemisia tabaci Gennadius	Cotton	Joshi and Yadav (1990)
	Lipaphis erysimi (Kaltenbach)	Mustard	Kalra (1988)
	Aleurocanthus woglumi Ashby		Satpute et al. (1986)
	Myzus persicae Sulzer	Tobacco	Rao and Chandra (1985)
	Spodoptera litura Fabricius	Tobacco	Rao and Satyanarayana (1984)
	Greenidea (Trichosiphum) formosana heeri Raychaudhuri, Ghosh, Banerjee and Ghosh	*Psidium guajava*	Debnath et al. (1988)
	Planococcus citri Risso *Ferrisia virgata* Cockerell *P. lilacinus* (Cockerell)	Vineyards	Mani and Krishnamoorthy (1990)
Pseudomallada astur (Banks) (=*M.* astur)	*A. disperses*	Guava	Mani and Krishnamoorthy (2007)
Mallada sp.	*Ferrisia virgata* Cockerell	Coffee	Balakrishnan et al. (1991)
Mallada obvious Hölzel (*Anisochrysa obvia* Hölzel)	*Rhopalosiphum* sp.	Grass	Debnath et al. (1988)
Chrysoperla zastrowi sillemi (=*Chrysoperla carnea*)	*Lipaphis erysimi*	Mustard	Kalra (1988); Akhtar et al. (2006)
	Hyadaphis coriandri Das	Coriander	Sagar and Kumar (1996)
	P. citri	Citrus	Gautam (1994) Krishnamoorthy and Mani (1989)
	B. tabaci	Cotton	Rao et al. (1989)
	A. craccivora	Groundnut	Patel and Vyas (1984)
	Bemisia tabaci	Cotton	Rao et al. (1989)
	Aphids	Pigeon pea	Sharma et al. (2007)
	Sucking insects	Linseed	Patel and Thakur (2005)
	Thrips tabaci Lind	Onion	Kadri and Goud (2005)
	Nausinoe geometralis	Jasmine	Rajabaskar (2007)
	Unidentified	Linseed	Patel and Thakur (2005)
Mallada basalis (*Anisochrysa basalis*)	*P. citri*	Citrus	Krishnamoorthy and Mani (1989)

(*Continued*)

TABLE 6.1 (*Continued*)
Chryopids Associated with Different Insect Pests

Species	Prey	Host plant	References
Plesiochrysa lacciperda	*Idioscopus clypealis* *I. nitidulus* *Adisura atkinsoni* *P. citri*	Mango	Fasih and Srivastava (1990), Krishnamoorthy and Mani (1989), and Mani and Krishnamoorthy (1990)
Micromus igorotus	*Ceratovacuna lanigera*	Sugarcane	Chakravarthy and Thyagaraj (2005) and Patil and Bhosale (2007)
Micromus australis Hagen	*Melanaspis sacchari* (Zehntner)	Sorghum	Vidya et al. (2008a, b)

Source: Modified after Bakthavatsalam (2011).

Many researchers have studied the effects of several abiotic factors, such as temperature (Tauber et al., 2006; Albuquerque et al., 1994; Volkovich and Arapov, 1996; Lopez-Arroyo et al., 1999; Nakahira et al., 2005; Mantoanelli et al., 2006; Silva et al., 2007; Pappas et al., 2008), photoperiod (Canard, 1997, 2005; Grimal, 1988; Canard and Grimal, 1988; Chang et al., 1995; Volkovich and Arapov, 1996; Volkovich and Blumental, 1997; Fujiwara and Nomura, 1999; Macedo et al., 2003; Nakahira and Arakawa, 2005), relative humidity (Tauber and Tauber, 1983; Pappas et al., 2008) and CO_2 (Gao et al., 2010) on certain life-history traits and predation capacity of chrysopids.

6.4 REARING AND STORAGE

Generally, species that have shorter life duration and low mortally rate are considered amenable to laboratory rearing. Accordingly, *C. carnea* and *M. desjardinsi* were considered to be highly amenable for laboratory rearing and field release (Bakthavatsalam et al., 1994, 1996). Most of the species could complete their life cycle in the laboratory conditions. However, species like *Plesiochrysa* and *Ankylopteryx* could not be reared in the laboratory on the prey. Chrysopids are reared on the eggs of *Sitotroga*, *Ephestia* (*Anagasta*) and *Corcyra* that are easily culturable and nutritionally superior than artificial diets tested so far, but expensive (Tauber et al., 2000). Several methods for the mass rearing of chrysopid larvae on an artificial diet (Cohen and Smith, 1998; Vasquez et al., 1998; Zaki and Gesraha, 2001; Sattar et al., 2007; Sattar and Abro, 2009) or a factitious food (Singh and Varma, 1989; Pappas et al., 2007, 2008; Matos and Obrycki, 2006; Sattar and Abro, 2009) have been tested. *Mallada basalis* (Walker) can be mass produced on chicken egg yolk–based diet with vitamin C and sucrose (Ye et al., 2017). In India, Singh et al. (1994b) and Jalali et al. (2003) developed rearing procedures for the chrysopids on *C. cephalonica* eggs, which are widely practiced even in the insectaries. Around 4.25 mL of *C. cephalonica* eggs are required for feeding 100 larvae of *C. carnea* during their whole life time (Bakthavatsalam et al., 1995). Higher quantity of the eggs (around 6 mL) is required for culturing *M. desjardinsi*.

Several diet combinations were attempted for rearing of the adult chrysopids. The diet with 40% honey and the castor pollen was found to record highest fecundity followed by protinex mixture with fructose and castor pollen. The protinex diet is still followed for culturing the adults (Krishnamoorthy, 1984). However, time to time several substitutes of protinex were tested. The powdered blackgram when mixed with other ingredients like honey, sugar and yeast produced the same fecundity, such as that of the protinex (Singh et al., 1994a). Depending on the facilities, demand and requirements, commercial insectaries have successfully reared *C. carnea* with slight modifications. In North America and Europe, for many years, *C. carnea* and *C. rufilabris* have been mass reared.

Several researchers have worked on long-term storage of adults of *C. externa*, *C. nipponensis* and *C. carnea* (Tauber et al., 1993; Wang and Nordlund, 1994; Chang et al., 1995; Tauber et al., 1997). The eggs may be stored conveniently at 10°C for 15 days without any deterioration in the quality of hatching (Bakthavatsalam et al., 1997a) or for 7 days at 4°C–5°C (Kaur et al., 2001). Newly laid and up to 3 days old eggs of *Ceraeochrysa cubana* (Hagen), *C. externa* and *Ceraeochrysa smithi* (Navas) stored at 15.6°C for up to 21 days had high rates of hatching (>70%). *Chrysoperla externa* eggs can be stored at 12.8°C for 21 days with more than 95% hatching (Lopez-Arroyo et al., 2000). Pupae could be stored for up to 15 days at 8°C with 74.8% adult emergence (Osman and Selman, 1993).

Effective storage techniques are very much warranted to improve production efficiency to meet seasonal demands during peak periods. It provides a stock of chrysopids available for timely field release, and physiological and genetic research (Tauber et al., 1993). Diapause is one of the major adaptations that insects have evolved to overcome adverse environmental conditions and to synchronize the life cycle with optimal resources availability (Denlinger et al., 2012; Kostal, 2006). The occurrence of diapause could be used for increasing shelf life and promoting long-distance shipment of biological control agents (Colinet and Boivin, 2011; Denlinger, 2008). Photoperiod is the most reliable and important signal used for the onset of diapause, and the dominant role of day length in regulating diapause is well-known for many green lacewings, including free-living third instar larvae of *Pseudomallada macleodi* (Adams and Garland), *Pseudomallada perfectus* Banks and *Pseudomallada sierra* Banks (Tauber and Tauber, 2015), prepupae of *Chrysopa pallens* (Rambur) (Nakahira and Arakawa, 2005; Volkovich, 1998), *Chrysopa perla* (L.) (Volkovich, 1998) and *Chrysopa dorsalis* Burmeister (Volkovich, 2006), as well as adults of *Chrysoperla carnea* (Stephens) (Volkovich, 1998) and *Chrysoperla sinica* (Tjeder) (Xu et al., 2004). Li et al. (2018) demonstrated that the induction of diapause in *Chrysopa formosa* Brauer is governed by photoperiod, and modulated by temperature to some degree. Prepupal diapause will not occur unless the free-living third instar larvae receive the correct short-day photoperiodic signals.

6.5 ECOLOGY AND INTERACTION STUDY

The seasonal incidence and population dynamics of the neuropterans is rarely studied except for a few species of the chrysopids. Population dynamics of the *C. carnea* was correlated with the prey, *Myzus persicae* on sunflower while no such relationship was observed with prey *Amrasca biguttula biguttua* and *Aphis gossypii* (Mannan et al., 1995). The period of activity was from September on sorghum for chrysopids (Duffield, 1995). Rao et al. (1989) observed that the population of predators decreased after rainfall. The abundance of *C. scelestes* was generally negatively correlated with maximum daily temperature, wind speed and sunshine, and positively correlated with minimum daily temperature, Relative humidity (RH) and rainfall (Faleiro et al., 1990). Population of chrysopids had a negative correlation with minimum and maximum temperature (Dhaka and Pareek, 2007). Population of aphids was positively correlated with the population of chrysopids on lucerne (Ingawale and Tambe, 2007). Activity of *Micromus* larvae was observed during November and thereafter it decreased (Sharanabasappa et al., 2008).

Functional response is an important component to assess the potential of biological pest control agents (Fernandez-Arhex and Corley, 2003). This is an important point, because the number of consumed prey is the basis for the development, survival and reproduction of predators and parasitoids (Oaten and Murdoch, 1975). A functional response is also helpful to determine voracity and efficiency of the biocontrol agent at varying prey density. Type II functional response was observed in the predatory response of *C. carnea* on *Aphis fabae* (Khan and Zaki, 2008). Functional response can also provide information on the effect of abiotic factors (i.e. temperature and interaction time) and biotic factors (i.e. hosts and prey) on the predator food-finding efficiency (De Clercq et al., 2000; Mohaghegh et al., 2001; Skirvin and Fenlon, 2003; Mahdian et al., 2006; Jalali et al., 2010; Fathipour et al., 2018; Chacon-Hernandez et al., 2019). Many studies show type II functional response for *Chrysoperla carnea* to *Brevicoryne brassicae* (Mushtaq and Khan, 2010), *Hyalopterus pruni* (Athhan et al., 2004), *Aphis craccivora* Koch, *Uroleucon cichorii* (Koch) and *Pterochloroides persicae* (Cholodkovsky) (Mirabzadeh et al., 1998); *Creontiades pallidus* Rambur (Fathipour and Jaafari, 2003); *Euonymus* aphid (Khan and Zaki, 2008; *Tetranychus urticae* Koch (Hassanpour et al., 2009); *Aphis gossypii* Glover (Montoya-Alvarez et al., 2010); *Helicoverpa armigera* eggs and first instar larvae (Hassanpour et al., 2011); *Tuta absoluta* (Al-Nadawi, 2016); and *Saissetia oleae* (Olivier) (Mahzoum et al., 2020). Similarly, *Chrysopa oculata* Say exhibited a type II response to eggs of *Helicoverpa zea* (Boddie) (Parajulee et al., 2006). Type II response was also observed in *C. nipponensis* to *Aphis craccivora*, *Paracoccus marginatus* and *Bemisia tabaci*. Type III response has also been reported in *Chrysoperla congrua* (Walker) to *H. armigera eggs* (Kabissa et al., 1996) and in *C. carnea* to *H. armigera* eggs (Hassanpour et al., 2011) and to *Myzus persicae* nymphs (Ail-Catzim et al., 2019). Similarly, functional response studied was also conducted for other chrysopids. Kumari et al. (2020) reported type II functional response for *Mallada desjardinsi* (Navas) on melon aphid, *Aphis gossypii*.

The physical characteristics of plant and its biochemical properties may directly or indirectly (Price et al., 1980; Kasaei-Faradonbeh et al., 2015) affect the biological characteristics of natural enemies. The host plant affects the competence and suitability of an herbivore for a natural enemy by changing both the quantity and quality of that herbivore. Natural enemies show different consumption of herbivores on different host plants due to change in the quality of food of prey (Zvereva and Rank, 2003). The highest predation rate and shortest developmental time of *C. carnea* were observed on the peach (Farrokhi et al., 2017).

Development and reproduction of *Dichochrysa prasina* Burmeister was better when it was fed on *Myzus persicae*, which resulted in high immature survival rate, short larval developmental time and increased adult longevity and fecundity (Pappas et al., 2007). Feeding on *A. craccivora* and *Lipaphis erysimi* (Kaltenbach) was found to increase the developmental time and reduce the survival rate of this predator (Chen and Liu, 2001; Khuhro et al., 2012). Aqueel et al. (2014) found that aphid consumption by green lacewings was reduced with the increase in nitrogen content in the host plants of aphids. Predation of both *Rhopalosiphum padi* (L.) and S*itobion avenae* (F.) by first, second and third instars larvae of *C. carnea* was the highest on aphids reared on plants with the lowest rate of fertilization, suggesting a compensatory consumption to overcome reduced biomass (lower aphid size). The larval duration, pupal weight, pupal duration, fecundity and male and female longevity were significantly affected by the level of nitrogen fertilization to the aphid's host plants. Trash-carrying behaviour plays an important role in the prevention of attack by predators, both by camouflaging the potential prey and by presenting a physical barrier to attack. Larvae of *Mallada signata* (Schneider) carrying trash packages exhibited lower cannibalism and higher activity rates, which may subsequently enhance foraging capacity (Anderson et al., 2003).

Competition and predation are two major interactions between animals occurring together in nature. Intraguild predation (IGP) occurs when two predators share the prey (the extraguild prey: EGP) (Polis et al., 1989). Zarei et al. (2020) demonstrated that IGP intensity between *C. carnea* and *Hippodamia variegata* (Goeze) decreased with increasing EGP (*Aphis fabae*) density rather than with increasing arena complexity. However, in this study, *C. carnea* always turned out as an intraguild predator. On the contrary, the EGP density did not affect lacewing survival when confronted with the invasive ladybird *Harmonia axyridis* (Pallas) (Wells et al., 2017).

6.5.1 Interaction with Microbials

It is necessary to understand interaction between microbials and chrysopids to develop successful integrated pest management module. The entomopathogenic fungi (EPF) used for managing insect pests should have selective properties and low virulence towards non-target insects (Portilla et al., 2017). There are some studies that have demonstrated the compatibility of fungal and lacewing species (Generoso, 2012; Pessoa et al., 2005; Cardoso et al., 2007; Schilick et al., 2015). Dias et al. (2019) reported that the interactions between *C. externa* and EPF are positive because the fungi (*B. bassiana*, *M. anisopliae* and *M. rileyi*) do not interfere with the ability of *C. externa* to consume *H. armigera* larvae. However, high concentrations of entomopathogens may have a greater negative effect than lower concentrations (Portilla et al., 2017). Similarly, endophytic *B. bassiana* application did not significantly influence the number of *Aphis gossypii* consumed by *C. carnea* or the time taken to consume each aphid (Gonzalez-Mas et al., 2019). But they observed avoidance of the infected parts of the aphid by the lacewing, which would increase lacewing safety. This has also been reported in *C. carnea* feeding on living *Metarhizium brunneum*–infected *Spodoptera littoralis* larvae (Ríos-Moreno et al., 2018). It is known that soil-borne non-pathogenic microbes can modulate plant–insect above-ground interactions via inducing systemic resistance by triggering biochemical changes in the primary plant metabolism. Benítez et al. (2017) found that differences in the composition of the rhizosphere active microbiome correlate with emission of the volatile compound by plants under aphid attack, and it was found to be attractive to *C. carnea*.

6.6 STRAIN IMPROVEMENT

Chrysoperla has an important role in the biological control of different insect pests. Emphasis was laid on selecting and utilizing the best strains of chrysopid for biological control programmes. Strains of *C. carnea* (*C. z. sillemi*) collected from different places were varied in biological parameters (oviposition, longevity, predatory potential, etc.) (Kumar et al., 2007a, b, 2008). All the strains collected from different regions developed resistance to insecticides in three to four generations after exposure to pesticides; however, withdrawal of insecticides brought back the resistance level drastically (Bakthavatsalam et al., 1997b; Singh et al., 1994a).

Efforts were made to develop strains with definite traits and increased levels of insecticide resistance. The developed strain showed manifold resistance to monocrotophos than other strains (Patel and Yadav, 1996). These strains also showed resistance to other organophosphorus compounds, like dimethoate, acephate, phosphomidon and methyl demeton (Patel and Yadav, 1997). The resistant strain showed more steryl esters and free fatty acids, and the activity of the acetylcholine esterase was double in the resistant strains than the susceptible strain (Patel et al., 1996). *Chrysoperla carnea* exposed to insecticides developed resistance to endosulfan and chlorpyrifos to many folds at 11–12 generations of exposure (Maurya et al., 2009). The strains that showed resistance to organophosphorous compounds also developed cross-resistance to synthetic pyrethroids on continuous exposure to the pesticides (Anonymous, 2000a). Populations of *C. zastrowi arabica* (=*C. zastrowi sillemi*) collected from geographical areas with higher sprays of insecticides showed higher LC_{50} than other populations, and populations of *C. z. arabica* (=*C. z. sillemi*) from cotton recorded more resistance than other crops probably due to the large quantity of pesticide sprays (Venkatesan et al., 2009). However, the inheritance of pesticide resistance was through the recessive multiple genes as reported in other species of chrysopids. Therefore, there are greater chances of losing the resistance in mating with the susceptible strain. The successive generations in the field will eventually have least resistance due to inter-breeding with the field populations. Recently, a strain of *Chrysoperla zastrowi sillemi* (PTS-8) having tolerance to different groups of pesticides, viz., organophosphate, organochlorine and synthetic pyrethroids has been developed (ICAR-NBAIR, 2019). PTS-8 recorded highest resistant factor (RF) for acephate (277.5), fenvalerate (66.11) and endosulfan (50.36). Biochemical assays revealed that the PTS-8 has got higher detoxifying enzymes activity, i.e., >0.04 µM/minutes/milligram protein for esterase and 21.1 µM/minutes/milligram protein for glutathione S-transferase as compared to susceptible population.

6.7 BIOINTENSIVE IPM

Chrysopids have great potential to be involved in various IPM programmes. Several studies were conducted to demonstrate the successful use of chrysopids and hemerobiids for the management of several pests. Chrysopids are being released as larvae in the field, as eggs are susceptible to parasitization. Usually, larvae are released manually mixed in saw dust or with paper fringes. Srivastava et al. (1996) devised an applicator for the release of chrysopids. The dosage of release varies depending upon the level of pest infestation.

Chrysoperla carnea along with *Cheilomenes sexmaculata* (Fabricius) could reduce the infestation of *A. craccivora* on groundnut (Patel et al., 1976). *Brinckochrysa scelestes* was found to be effective against *Myzus persicae*, *Bemisia tabaci* and apple woolly aphid (Rao and Chandra, 1985; Thakur et al., 1988; Dhandapani et al., 1992). Periodic release of *B. scelestes* along with other parasitoids, such as *Trichogramma chilonis* Ishida, *Tetrastichus howardi* (Olliff) and *Telenomus remus* Nixon, decreased the population of *Spodoptera litura* on potato (Ansari et al., 1992). *Chrysoperla carnea* along with bacterial formulations of *Bacillus thuringiensis kurstaki* (*Btk*) (1 kg/ha) was found to be effective to manage pink bollworm in India (Dhandapani et al., 1992). When *C. carnea* (1.0 lakh/ha) along with *Bt* and *Helicoverpa armigera* nucleopolyhedrosis virus (HaNPV) was released to manage pod borer in pigeon pea, it resulted in an increase in pod yield. IPM module consisting of six releases of *Trichogramma chilonis* (2.5 CC/acre) with two releases of *C. carnea* (200,000 eggs/acre), yellow sticky traps (25 traps/acre) and brinjal fruit and shoot borer pheromone traps (5/acre) was found to be effective in managing lepidopteran as well as sucking pests in brinjal (Elanchezhyan et al., 2009). Mani and Krishnamoorthy (1990) found that the population of *Planococcus citri* and *Ferrisia virgata* was suppressed by *Plesiochrysa lacciperda* in combination with *Spalgis epeus* (Westwood) and *Cryptolaemus montrouzieri* Mulsant.

Release of *C. carnea* at varying doses was found to effectively reduce the population of the pests, such as *H. armigera* on cotton; chickpea, cardamom, safflower, sunflower for various pests (Patel et al., 1976; Manjunath, 1993; Bharpoda et al., 2000; Rabindra et al., 1994; Hegde and Lingappa, 1998; Mathew et al., 1998; Venkatesan et al., 1997; Badgujar et al., 2000; Rajaram et al., 2006). Pod borer infestation in pigeon pea was reduced when the larvae of *C. carnea* were released @ 1,25,000/ha (Hegde and Lingappa, 1998). The *C. carnea* released at 1:10 ratio achieved good control over the population of *Pentalonia nigronervosa* (Coquerel) on cardamom (Mathew et al., 1998).

Release of *C. carnea* followed by spray of monocrotophos (0.04%) or endosulfan (0.07%) greatly reduced the incidence of castor mite and *Tetranychus dellaris* on castor (Rajasekhar et al., 1999). Release of *C. carnea* along with *T. chilonis* effectively controlled *H. armigera* and aphids on Hybrid-6 cotton in Gujarat (Bharpoda et al., 2000). Biorationale pest management, including release of *C. carnea* @ 2,500 eggs/ha was found to reduce the incidence of *Plutella xylostella* on cabbage (Reddy and Guerro, 2000). Release of two or three larvae of *C. carnea* per plant resulted in the reduction of the population of *Uroleucon carthami* on safflower (Badgujar et al., 2000). IPM module involving release of *T. chilonis*, *C. carnea*, Spray of Nuclear Polydrosis virus (NPV), nimbecidine, dipel and synthetic chemicals imposed at different intervals depending on the pheromone catches could effectively control the pest economically (Reddy and Manjunatha, 2000). Use of *C. carnea* along with the application of neem seed kernel extract (NSKE) reduced the population of *L. erysimi* on oilseed crops in terms of socioeconomic and environmental values (Pandey and Singh, 2008; Sanjeev et al., 2007). Farkade et al. (1999) studied the adoption of the various biological control practices and found that the use of *Chrysopa* was adopted by around 7% of the agriculturists and ranks fourth in the adoption of biological control.

Two releases of *Chrysopa scelestes* @ 10,000 eggs per ha was found effective in managing *Saccharicoccus sacchari* when the population is low (Chelvi and Kandasamy, 2009). Release of *M. desjardinsi* (= *M. boninensis*) @50 and 60 eggs per tree effectively controlled *Aleurocanthus woglumi* on citrus (Zade et al., 2007; Bhagat et al., 2007) and in combination with the release of parasitoids effectively controlled the incidence of citrus leafminer (Rao and Raguraman, 2005; Ingole et al., 2005).

Chrysoperla carnea @ 50,000/ha at 40 and 70 Days After Transplantation (DAT) in combination with *Beauveria* @1.5 kg/ha reduced the incidence of *S. litura* on cabbage (Das and Mandal, 2005). Combined release of *T. chilonis* (1.5 lakh parasitized eggs/ha) with different rates of *C. carnea* were comparable with endosulfan (0.07%) and reduced the infestation in squares, flowers and green bolls on cotton (Panchbhai et al., 2005). With the sudden outbreak of sugarcane woolly aphid, *Ceratovacuna lanigera* (Hemiptera: Aphididae) during 2002, the potential of *Micromus igorotus* (Figure 6.3) along with *Dipha aphidivora* as an effective predator was proved at field conditions (Patil et al., 2007). An IPM module consisting of soil application of neem cake, application of Nitrogen: Phosphorous: Potassium (NPK) at 250:75:185 kg/ha and release of *M. igorotus* at 2,600

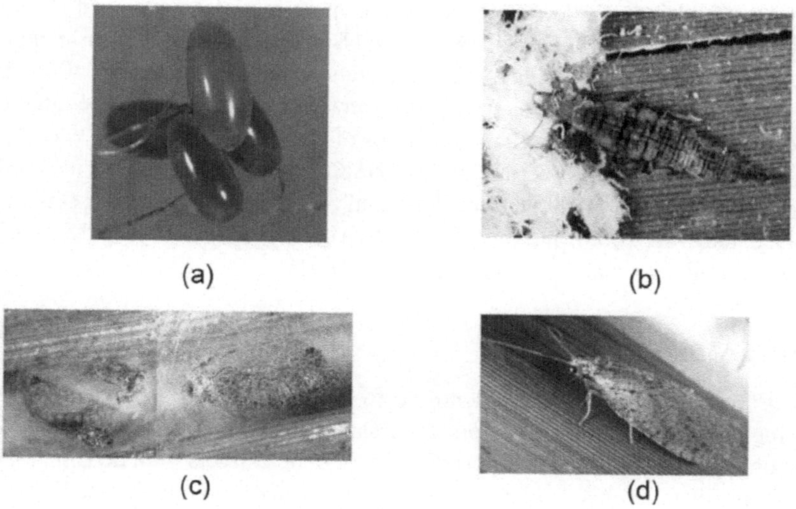

FIGURE 6.3 *Micromus igorotus* (Banks). (a) Eggs, (b) larva, (c) pupae and (d) adult. (ICAR-NBAIR database.)

grubs/ha and release of *D. aphidivora* at 1,000 larvae/ha recorded the highest reduction in sugarcane woolly aphid population (Mallapur et al., 2006).

In cotton sucking pests and *H. armigera* were effectively managed through the IPM package which includes intercrop with lucerne (1:1) population, Neem seed kernel extract (NSKE) (5%) two sprays and release of *C. carnea* 0.75 and 1.0 lakhs/ha (Hanumantharaya et al., 2008c). IPM module consisting of release of *T. chilonis*, *C. carnea* @ 1,00,00/ha and need-based application of insecticides can prove very useful in curtailing the excessive and indiscriminate use of insecticides and ensuring a sustainable yield of cotton (Brar et al., 2001).

Bio Intensive Pest Management (BIPM) practices including release of larvae of *C. carnea* on *Bt* cotton recorded efficient management of pests (Prasad et al., 2005; Shanmugam et al., 2006). Release of *Chrysoperla carnea* along with NSKE was effective against *Lipaphis erysimi* (Rai et al., 2007). *Chrysoperla carnea* after three to four releases at the rate of five to ten larvae per plant was found to be effective in reducing aphid and whitefly populations on sweet pepper and squash plants in open fields (Alghamdi et al., 2018). This was also found to be effective in greenhouse conditions on aphids and greenhouse whitefly, *Trialeurodes vaporariorum* West infesting different crops (El-Arnaouty et al., 2000; Ahmadzadeh and Hatami, 2006). This species was also tested against the aphid, *Chaetosiphon fragaefolii* on strawberry plant at the releasing rate of eight larvae/plant in open field and could significantly reduce the population of *C. fragaefolii* (Easterbrook et al., 2006). *Chrysoperla carnea* and *Chrysoperla rufilabris* (Burmeister) were also evaluated against tomato psyllids in greenhouse tomatoes (Al-Jabr, 2000). Nair et al. (2020) reported that a single release of *C. zastrowi sillemi* at four grubs per plant is effective against sucking pests in tomato grown under screen-house conditions. *Mallada desjardinsi* (=*M. boninensis*) had reduced the *Aleurocanthus woglumi* Ashby population in citrus plants by 34.6%–36.7% (Ingole et al., 2005). Similarly it was reported as the best biological control agent against *B. tabaci* and red spider mites (Babu et al., 2011). *Pseudomallada astur (=Dichochrysa astur*) is recorded as a very good predator of rugose spiralling whitefly (RSW) and Bondar's nesting whitefly (BNW), *Paraleyrodes bondari* Peracchi. Release of this predator at the rate of 100–150 eggs/palm during low incidence and 300 eggs/palm during medium incidence is recommended to tackle RSW problem in coconut (Chalapathi Rao et al., 2020). Methods and release of *Chrysoperla* spp. is also important. Biological control of vine leafhoppers, *Erythroneura variabilis* (Beamer) and *E. elegantula* (Osborn), by *Chrysoperla* spp. was affected by the method and timing of application (Daane and Yokota, 1997).

BIPM modules and insecticide resistance management modules developed by several authors resulted in the conservation of chrysopids (Ameta and Bunker, 2009; Suryawanshi et al., 2007). IPM modules encouraged *C. carnea* population on cotton and brinjal (Ameta et al., 2006; Sardana et al., 2006; Bhosle et al., 2007; Shanmugam et al., 2006; Tanwar et al., 2007). However, intercropping soyabean with cotton did not encourage the populations of chrysopids on cotton (Vennila et al., 2007). *Bt* and organic amendments, like vermicompost, NSKE, neem cakes, cow urine, and vermiwash, were safer to chrysopids in brinjal and okra (Jyoti and Goud, 2008; Jayakumar et al., 2008; Sharma et al., 2007; Deepika et al., 2009), and transgenic cotton conserved the chrysopids probably due to the reduced usage of pesticides than conventional cotton (Wadhwa and Gill, 2007).

6.8 CHEMICAL ECOLOGY

The 1*R*, 2*S*, 5*R* and 8*R*-iridodial were identified from abdominal extracts of males of a Nearctic green lacewing species, *Chrysopa oculata* Say, as a conspecific male attractant (Zhang et al., 2004). This compound has also been reported from thoracic extracts of male *Chrysopa nigricornis* Burmeister (Zhang et al., 2006a), and it was found to be attractive to other green lacewing species as well (Aldrich and Zhang, 2016). In *Chrysopa* species Skatole, the strong-smelling component of the secretion from the prothoracic glands is responsible for the defensive function (Blum et al., 1973). Colonization of chrysopids is influenced by the plants. Males of *C. carnea* had higher preference to sunflower and females had preference to sunflower and cotton while pigeon pea was least preferred. *Mallada desjardinsi (=M. boninensis)* males did not show any preference for any of the host plants (Ballal and Singh, 1999). *Chrysoperla carnea* was four times abundant on hybrids than open pollinated cultivars of cotton (Vennila, 1998). Mohite and Uthamasamy (1998) observed that predation of *H. armigera* by *C. carnea* was more on *Gossypium armourianum* (MCU-9) and *G. harkeressi.*

Some of the sunflower cultivars, like GKSFH 2002, recorded lower population of phytophagous insects and greater population of *C. carnea* and early sown sunflower crop recorded greater numbers of predators than late sown crop (Arora et al., 1993). Some reports exhibited more oviposition on cotton and sunflower by *C. carnea* but lesser on aster and coriander (Geetha and Swamiappan, 1998). Cotton cultivars, like MCU-10, MCU-9, NSPHH-5, Gujarat Hybrid 10, DHB290 and MCU-7, recorded more oviposition by this predator and among the sunflower cultivars, PCSH 243, KBSH-1 and SFL 103 recorded the highest oviposition under multiple choice conditions. The volatile profile of the varieties suggested the positive role of compounds, such as pinene, caryophyllene and other terpenoids (Anonymous, 2002; Bakthavatsalam et al., 2001b, 2002). Electrophysiological studies indicated that the cultivars of safflower, such as DSH 129, A-1, GMU 2518 and NARI–H15, elicited higher response than other cultivars (Hanumantharaya et al., 2008b). Higher oviposition was recorded on the LRA 5166 variety and higher predation by larvae on RCH-2, and trichome density showed negative relation with predation of *C. carnea* (Balakrishnan et al., 2006).

Singh et al. (1993) observed higher incidence of *C. carnea* on okra and sunflower, and suggested the possibility of exploiting these crops in cultural control of *H. armigera* when inter-cropped with cotton. Raising two rows of trap crop of tobacco or chickpea around flue-cured tobacco increased the population of chrysopids and decreased the incidence of *H. armigera* (Rao et al., 1994). Intercropping rice bean with redgram increased the population of chrysopids on rice bean (Satyanarayana et al., 1998). Duffield and Reddy (1997) found that in sorghum short duration pigeon pea intercrop system, egg laying by chrysopids was recorded only on sorghum and not on adjacent pigeon pea plants, and chrysopids did not migrate to pigeon pea despite the presence of potential prey. Spraying alternate rows (skip row coverage) with chemicals against *A. biguttula* and *Aphis gossypii* resulted in larger populations of *C. carnea* on cotton (Surulivelu and Kumaraswami, 1989). *Chrysoperla carnea* and *Coccinella septempunctata* were more on safflower intercropped with coriander and sorghum (Hanumantharaya et al., 2008a, b; Rao et al., 2007).

Kairomones are defined as chemicals released by an organism in nature, which evokes a behavioural response adaptively favourable to the receiver. The adult chrysopids were found to respond to the honey dew secretions from the aphids. On chemical analysis, it was found that L-tryptophan, the breakdown product of the honeydew, is the primary component attracting the adult chrysopids. Aphid sex pheromone components were found to attract males of *Chrysopa cognata* (Boo et al., 1998) and *Chrysopa* spp. (Koczor et al., 2010; Aldrich and Zhang, 2016). Jones et al. (2011, 2016) found attraction of a *Chrysopa nigricornis* Burmeister to traps baited with squalene, a plant-originated compound. A ternary blend, consisting of phenylacetaldehyde, methyl salicylate and acetic acid, was found to be more attractive to *Chrysoperla carnea* species complex (i.e. *C. carnea* complex) (Toth et al., 2009). Besides, attracted females laid eggs in the vicinity of the baits (Jaastad et al., 2010; Koczor et al., 2015a, 2017). Furthermore, Koczor et al. (2019) reported that traps baited with squalene and the ternary floral bait attracted adults of both *C. formosa* and *C. carnea* complex lacewings. Methyl salicylate was found to increase catches of *C. oculata*, *C. nigricornis* and *Chrysopa septempunctata* Wesmael (=*Chrysopa pallens* Rambur) males in combination with iridodial (Jones et al., 2011; Zhang et al., 2006b).

The adult chrysopids showed behavioural changes like reduced walking speed, more resting period and less speed on contact with the kairomones like tryptophan (Bakthavatsalam et al., 1997b). Electroantennogram studies showed that the fertile female *C. carnea* responded to the acid hydrolysed L-tryptophan (Bakthavatsalam et al., 2000a). In net house conditions, the L-tryptophan-treated plants recorded more oviposition compared to untreated plants (Bakthavatsalam and Singh, 1996). Further studies were conducted at field conditions, and it was found that oviposition was more in the cotton plants treated with L-tryptophan than untreated plants (Bakthavatsalam et al., 1999; Anonymous, 2000b). On the contrary, L-tryptophan was ineffective in increasing the oviposition on tomato (Bakthavatsalam et al., 2001a). Bakthavatsalam and Tandon (2007) recorded through electrophysiological studies that indole acetaldehyde and acid hydrolysed L-tryptophan elicited better response in *C. carnea*. The larvae of chrysopids responded to the scales of the hosts, such as *H. armigera*. The scale extract of *H. armigera* was found to attract the larval chrysopids in olfactometer studies (Bakthavatsalam and Singh, 1999). On chemical analysis, it was found that several compounds, like tricosane and pentacosane, were observed to be important in the attraction of chrysopids. Attempts were made to utilize the scale extract of *H. armigera*/*C. cephalonica* (laboratory waste) to increase the predatory potential of larval chrysopids. The predatory potential was more in the plants treated with the kairomonal formulation in cotton (Anonymous, 2002). However, these formulations did not increase the predatory potential of chrysopids on tomato, the least preferred host plants for chrysopids (Bakthavatsalam et al., 2001a).

Apart from the kairomones, synomones released from the cotton bolls, infested with *H. armigera*, also evoked behavioural responses in *C. carnea* (Bakthavatsalam et al., 2000b). Kairomone septa impregnated with n-tricosane along with the spraying of 3-day-old acid hydrolysed L-tryptophan increased the population of chrysopids on cotton (Bakthavatsalam et al., 2007). Acetone extracts of onion leaves and cluster bean enhanced the predatory activity of *C. carnea* (Elanchezhyan et al., 2009). A ternary floral bait was developed with the combination of phenylacetaldehyde, methyl salicylate and acetic acid (Tóth et al., 2009), and this bait was found to be a powerful attractant to adults of *Chrysoperla carnea* species complex, including *Chrysoperla carnea* (Stephens), *C. lucasina* (Lacroix), *C. pallida* Henry, Brooks, Duelli and Johnson (Tóth et al., 2009) and *Chrysoperla sinica* Tjeder (Deng et al., 2013). Females oviposit in the vicinity of the bait, and this bait can provide prospects for biological control to manipulate the oviposition site choice of females (Koczor et al., 2015b, 2017).

6.9 CONCLUSIONS

The taxonomy of chrysopids is poorly studied in India. A systematic approach is needed to record the various species along with their host plants and the pest species that will help to utilize them as a better predator. It is therefore essential to encourage researchers on the studies, like biosystematics and biodiversity of this group. The identification of species through the cryptic songs and molecular tools need to be undertaken for most of the species so that the identification becomes easier.

There is a need to address the population genetics of the species complex of chrysopids. The ecological studies with mass rearing technologies and storage of economically important chrysopids are very much essential to include them as one of the components of IPM.

Semiochemical data should be incorporated with morphological and genomic characters to strengthen phylogenetic inference, and chemical ecology knowledge should also be integrated with other management tactics to enhance its efficacy. *Chrysoperla externa* and *Chrysoperla* spp. from Peru are being produced in large numbers (1,00,000 per week) in Latin America to target lepidopteran pests and aphids, respectively (van Lenteren, 2012). Thus, in India, production of economically important chrysopids may be encouraged at village-level units with the help of trained personnel to meet the demand. Nowadays some of the private companies have started producing *C. z. sillemi* with back-up support of ICAR-NBAIR.

More than one type of interventions are required to tackle different kinds of insect pests of crops. Therefore, it is important that chrysopids should be compatible with other biopesticides. Emphasis should be given to conduct interaction studies to explore additive or synergistic effects.

Besides augmentation, studies should also be intensified on the conservation technologies using habitat manipulation techniques, safer pesticides, behavioural manipulation techniques, etc. Technology for the use of kairomones and supplementary food spraying has to be standardized to conserve the chrysopids. At the borders and surroundings of crops, the source of pollen and honeydew should be provided for common green lacewing adults. The vegetation should include plants that flower at different periods and should coincide with the flight activity of the chrysopids. Research grant is needed for basic studies, like biosystematics, ecology, standardization of rearing techniques and conservation technologies for chrysopids.

6.10 POINTS TO REMEMBER

1. Chrysopidae are the second most species-rich family in the order Neuroptera, with ~1,416 species grouped in 82 genera.
2. In India, *Chrysopa carnea* (later *Chrysoperla carnea*) does not exist; at present only *C. zastrowi sillemi* exists.
3. Cryptic song-based taxonomic analysis will become a boon for the identification of several other species of chrysopids.
4. Chrysopids lay stalked oval, white, yellow and green eggs on the peduncle of the leaves, buds, stems and even fruiting bodies.
5. Several methods for the mass rearing of chrysopid larvae on an artificial diet or a factitious food (*Corcyra cephalonica*) have been developed and tested.
6. Photoperiod is the most reliable and important signal used for the onset of diapause, for many green lacewings. The occurrence of diapause could be used for increasing the shelf life and promoting long-distance shipment of biological control agents.
7. Many studies exhibited positive interactions between chrysopids and EPF. Thus, both inputs can be used to manage insect pests in different ecosystems. However, high concentrations of entomopathogens may have a greater negative effect than lower concentrations.
8. A strain of *Chrysoperla zastrowi sillemi* (PTS-8) having tolerance to different groups of pesticides, *viz*., organophosphate, organochlorine and synthetic pyrethroid has been developed.
9. BIPM practices including release of larvae of *C. zastrowi sillemi* on *Bt* cotton recorded efficient management of pests. Besides this, it is compatible with other interventions and is recommended in crops like greenhouse tomato, squash, sweet pepper, citrus, etc.
10. The adult chrysopids respond to the honey dew secretions containing L-tryptophan from the aphids. A ternary blend, consisting of phenylacetaldehyde, methyl salicylate and acetic acid, was found to be more attractive to *Chrysoperla carnea* species complex (i.e. *C. carnea* complex). High oviposition was also recorded in the vicinity of the bait, which can provide prospects for biological control to manipulate the oviposition site choice of females.

ACKNOWLEDGEMENT

The authors are grateful to the Indian Council of Agricultural Research for the facilities provided.

REFERENCES

Ahmadzadeh, Z. & Hatami, B. (2006). Evaluation of integrated control of using *Chrysoperla carnea* (Steph.) and insecticide Confidor in greenhouse conditions. *Journal of Crop Production and Processing*, 9: 239–251.

Ail-Catzim, C.E., Rodriguez-Gonzalez, R.E., Hernandez-Juarez, A. & Chacon-Hernandez, J.C. (2019). Functional response of *Chrysoperla carnea* (Neuroptera: Chrysopidae) on *Myzus persicae* nymphs (Hemiptera: Aphididae). *Proceedings of the Entomological Society of Washington*, 121: 535–543.

Akhtar, M.S., Choudhury, R.A. & Shujauddin, S. (2006). Natural enemies of mustard aphid, *Lipaphis erysimi* (Kaltenbach) in Aligarh (India). *Environment and Ecology*, 24S: 781–784.

Alasady, M.A.A., Omar, D.B., Ibrahim, R.B. & Ibrahim, Y.B. (2011). The survey of green lacewings and occurrence of *Apertochrysa* sp. (Neuroptera: Chrysopidae) on various plants in Malaysia. *Journal of Entomological Science*, 8: 240–49.

Albuquerque, G.S., Tauber, C.A. & Tauber, M.J. (1994). *Chrysoperla externa* (Neuroptera: Chrysopidae): Life history and potential for biological control in Central and South America. *Biological Control*, 4: 8–13.

Albuquerque, G.S., Tauber, C.A. & Tauber, M.J. (2012). Green lacewing (Neuroptera: Chrysopidae): Predatory lifestyle. In: *Insect Bioecology and Nutrition for Integrated Pest Management,* (Eds. A.R. Panizzi & Parra, J.R.P.). Boca Raton, FL: CRC Press, pp. 593–631.

Aldrich, J. & Zhang, Q.H. (2016). Chemical ecology of Neuroptera. *Annual Review of Entomology*, 61: 197–218.

Alghamdi, A., Al-Otaibi, S. & Sayed, S.M. (2018). Field evaluation of indigenous predacious insect, *Chrysoperla carnea* (Steph.) (Neuroptera: Chrysopidae), fitness in controlling aphids and whiteflies in *two* vegetable crops. *Egyptian Journal of Biological Pest Control*, 28: 20. doi: 10.1186/s41938-018-0026-3.

Al-Jabr, A.M. (2000). Integrated pest management of tomato/potato psyllid, *Paratrioza cockerelli* (Sulc) (Homoptera: Psyllidae) with emphasis on its importance in greenhouse grown tomatoes. Dissertation, Colorado State University.

Al-Nadawi, F.A. (2016). Functional response of the local predator larvae to tomato moth eggs. *Food Science and Quality Management*, 51: 15–19.

Ameta, O.P. & Bunker, G.K. (2009). Validation and popularization of integrated pest management technology in cotton (*Gossypium hirsutum*). *Indian Journal of Agricultural Science*, 79: 527–530.

Ameta, O.P., Sharma, K.C., Rana, B.S. & Bambawale, O.M. (2006). Validation and popularization of IPM technology in cotton through farmers participatory approach in tribal area of Southern Rajasthan. *Annals of Agricultural Research*, 27: 162–166.

Anderson, K.L., Seymour, J.E. & Rowe, R. (2003). Influence of a dorsal trash-package on interactions between larvae of *Mallada signata* (Schneider) (Neuroptera: Chrysopidae). *Australian Journal of Entomology*, 42, 363–366.

Anonymous (2000a). National repository of natural enemies of crop pests and weeds. Annual report 1-4-1999 to 31-3-000, 23 pp.

Anonymous (2000b). Use of semiochemicals to increase the biocontrol potential of some important predators and parasitoids. Final technical report, Project Directorate of Biological Control, Bangalore.

Anonymous (2002). Implications of tritrophic interaction in the Integrated Pest Management of some important crop pests. Final technical report. Project Directorate of Biological Control, Bangalore, 221 pp, pp. 36–37.

Ansari, M.A., Pawar, A.D. & Kumar, D.A. (1992). Possibility for Biocontrol of tropical armyworm, *Spodoptera litura* (Lepidoptera: Noctuidae) on potato. *Plant Protection Bulletin Faridabad*, 44(4): 27–31.

Aqueel, M.A., Collins, C.M., Raza, A.M., Ahmad, S., Tariq, M., & Leather, S.R. (2014). Effect of plant nutrition on aphid size, prey consumption, and life history characteristics of green lacewing. *Insect Science*, 21: 74–82. doi: 10.1111/1744-7917.12019.

Arora, R., Gill, J.S. & Varma, G.C. (1993). Adverse effects of a chitin synthesis inhibitor, diflubenzuron on *Chrysoperla carnea* (Stephens) and *Trichogramma chilonis* Ishi. *Journal of Insect Science*, 6: 279–280.

Aspöck, U., Haring, E. & Aspöck, H. (2012). The phylogeny of the Neuropterida: Long lasting and current controversies and challenges (Insecta: Endopterygota). *Arthropod Systematics & Phylogeny* 70: 119–129.

Athhan, R., Kaydan, B. & Ozgokce, M.S. (2004). Feeding activity and life history characteristics of the generalist predator, *Chrysoperla carnea* (Neuroptera: Chrysopidae) at different prey densities. *Journal of Pest Science*, 77: 17–21.

Babu, A., Vasantha Kumar, D., Jasin Rahman, V.K., Roobak Kumar, A. & Sundaravadivelan, C. (2011). Potential of *Mallada boninensis* Okamoto (Neuroptera: Chrysophidae), as a biocontrol agent of *Oligonychus coffeae* Nietner (Acarina: Tetranychidae) infesting tea. *Journal of Plantation Crops,* 39: 193–195.

Badgujar, M.P., Deotale, V.Y., Sharnagat, B.K. & Nandanwar, V.N. (2000). Performance of *Chrysoperla carnea* against safflower aphid, *Dactynotus carthami* (HRL). *Journal of Soils and Crops*, 10: 125–127.

Bakthavatsalam, N. (2011). Neuroptera as biocontrol agent in India: Present status and future thrust. In: *Entomology, Ecology & Biodiversity*, (Eds. B.K. Tyagi & Vijay Veer). Jodhpur, India: Scientific Publishers, pp. 394–449.

Bakthavatsalam, N. & Singh, S.P. (1996). L-tryptophan as an ovipositional attractant for *Chrysoperla carnea* (Stephens) (Neuroptera: Chrysopidae). *Journal of Biological Control*, 10: 21–27.

Bakthavatsalam, N. & Singh, S.P. (1999). Behavioural responses of larvae of *Chrysoperla carnea* to kairomones. *Journal of Insect Science*, 12: 34–36.

Bakthavatsalam, N. & Tandon, P.L. (2007). Behavioural and electrophysiological responses of *Chrysoperla carnea* (Stephens) to kairomones: Acid hydrolyzed/oxidized L-tryptophan and its breakdown products. *Journal of Biological Control*, 21(Special): 85–87.

Bakthavatsalam, N., Singh, S.P., Pushpalatha, N.A. & Bhummannavar, B.S. (1994). Life tables of *four* species of chrysopids (Neuroptera: Chrysopidae). *Journal of Entomological Research*, 18: 357–360.

Bakthavatsalam, N., Singh, S.P., Pushpalatha, N.A. & Bhummannavar, B.S. (1995). Optimum dosage of eggs of *Corcyra cephalonica* Stainton for mass rearing *Chrysoperla carnea* (Stephens) (Neuroptera :Chrysopidae). *Journal of Insect Science*, 8: 90–92.

Bakthavatsalam, N., Singh, S.P., Pushpalatha, N.A. & Bhummannavar, B.S. (1996). Comparative Biology and feeding potential of four species of chrysopids (Neuroptera: Chrysopidae). *Hexapopoda,* 8: 45–53.

Bakthavatsalam, N., Singh, S.P., Pushpalatha, N.A. & Bhummannavar, B.S. (1997a). Optimum temperature for short term storage of *Chrysoperla carnea* (Stephens). *Journal of Biological Control*, 9: 45–46.

Bakthavatsalam, N., Singh, S.P., Pushpalatha, N.A. & Bhummannavar, B.S. (1997b). Evaluation of strains *of Chrysoperla carnea* (Stephens) (Neuroptera: Chrysopidae) resistant to monocrotophos. *National Symposium of Pest Management in Horticultural Ecosystems*, October 25–27, 1997, Bangalore.

Bakthavatsalam, N., Singh, S.P., Chaudhary, M. & Preethi, S. (1999). Kairomones for *Chrysoperla carnea* (Stephens) (Neuroptera: Chrysopidae). *Paper Presented at National Symposium on Emerging Trends in Biotechnological Approaches in the Integrated Pest Management*, February 26–27, 1999, Chennai.

Bakthavatsalam, N., Singh, S.P., Tandon, P.L., Chaudhary, M. & Preethi, S. (2000a). Electrophysiological responses of *Chrysoperla carnea* (Stephens) (Neuroptera: Chrysopidae) to potential kairomones. *Journal of Entomological Research*, 24: 109–114.

Bakthavatsalam, N., Singh, S.P., Tandon, P.L., Chaudhary, M. & Preethi, S. (2000b). Synomone mediated behavioural responses of *Chrysoperla carnea* (Stephens) (Neuroptera: Chrysopidae) to cotton infested by *Helicoverpa armigera* (Hubner) (Lepidoptera: Noctuidae). *Journal of Biological Control*, 14: 1–6.

Bakthavatsalam, N., Singh, S.P. & Tandon, P.L. (2001a). Attempts to increase the efficiency of *Chrysoperla carnea* (Stephens) (Neuroptera: Chrysopidae) on tomato (The Non Preferred host plant) through use of Kairomonal formulations. In *IPM in Horticultural Crops; Emerging Trends in the New Millennium*, October 17–19, 2001, Bangalore, pp. 188–189.

Bakthavatsalam, N., Singh, S.P., Tandon, P.L., Hanumantharaya, L. & Vellaikumar, S. (2001b). Influence of different cultivars of sunflower on the electroantennogram response and behaviour of *Chrysoperla carnea* (Stephens) (Neuroptera : Chrysopidae). In *Symposium on Biocontrol Based Pest Management for Quality Crop Protection in the Current Millennium*, July 13–19, 2001, Punjab Agricultural University, Ludhiana, pp. 52–53.

Bakthavatsalam, N., Singh, S.P., Tandon, P.L., Hanumantharaya, L. & Vellaikumar, S. (2002). Electroantennogram and ovipositional responses of *Helicoverpa armigera* (Hubner) and *Chrysoperla carnea* (Stephens) to volatiles of different cultivars of cotton in a tritrophic context. In *Symposium on Biological Control of Lepidopteran Pests*, July 17–18, 2002, Bangalore.

Bakthavatsalam, N., Tandon, P.L., Patil, S.B., Hugar, B. & Hosamani, A. (2007). Kairomone formulations as reinforcing agents for increasing abundance of *Chrysoperla carnea* (Stephens) in cotton ecosystem. *Journal of Biological. Control*, 21: 1–8.

Balakrishnan, M.M., Sreedharan, K., Venketesha, K. P. & Bhat, P.K. (1991). Observations on *Ferrisia virgata* (Ckll.) (Homoptera: Pseudococcidae) and its natural enemies of coffee, with new records of predators and host plants. *Journal of Coffee Research,* 21: 11–19.

Balakrishnan, N., Baskaran, R.K.M. & Mahadevan, N.R. (2005). Development and predatory potential of green lacewing *Chrysoperla carnea* (Stephens) (Neuroptera : Chrysophidae) on different prey insects. *Agricultural Science Digest*, 25: 194–197.

Balakrishnan, N., Baskaran, R.K.M. & Mahadevan, N.R. (2006). Influence of different cotton varieties/hybrid on the ovipositional preference and predation of *Chrysoperla carnea* Stephens and parasitization of *Trichogramma chilonis* Ishii. *Journal of Experimental Zoology*, 9: 93–96.

Ballal, C.R. & Singh, S.P. (1999). Host plant mediated orientational and ovipositional behaviour of three species of chrysopids (Neuroptera: Chrysopidae). *Biological Control*, 16: 47–53.

Benítez, E., Paredes, D., Rodríguez, E., Aldana, D., González, M., Nogales, R., Campos, M. & Moreno, B. (2017). Bottom-up effects on herbivore induced plant defences: A case study based on compositional patterns of rhizosphere microbial communities. *Scientific Reports*, 7: 6251. doi: 10.1038/s41598-017-06714-x.

Bezerra, C.E.S., Tavares, P.K.A., Nogueira, C.H.F., Macedo, L.P.M. & Araujo, E.L. (2012). Biology and thermal requirements of *Chrysoperla genanigra* (Neuroptera: Chrysopidae) reared on *Sitotroga cerealella* (Lepidoptera: Gelechiidae) eggs, *Biological Control*, 60: 113–118. doi: 10.1016/j.biocontrol.2011.11.010.

Bhagat, S.M., Deotale, V.Y., Zade, S.M., Makanwar, P.S. & Hiwase, A.B. (2007). Bioefficacy of *Mallada boninensis* (Okamoto) against citrus blackfly *Aleurocanthus woglumi* Ashby. *Journal of Soils and Crops*, 17: 125–127.

Bharpoda, T.M., Patel, H.P., Patel, H.P., Patel, G.P., Patel, J.J. & Patel, J.R. (2000). Integrated Pest Management (IPM) in cotton H-6 cultivated in middle Gujarat. *Indian Journal of Entomology*, 62: 327–331.

Bhosle, B.B., More, D.G., Bambawale, O.M., Sharma, O.P. & Patange, N.R. (2007). Effectiveness of cotton IPM module in rainfed Marathwada Region. *Annals of Plant Protection Sciences*, 15: 2125.

Blum, M.S., Wallace, J.B. & Fales, H.M. (1973). Skatole and tridecene: Identification and possible role in a chrysopid secretion. *Insect Biochemistry*, 3: 353–357.

Boo, K.S., Chung, I.B., Han, K.S., Pickett, J.A., & Wadhams, L.J. (1998). Response of the lacewing *Chrysopa cognata* to pheromones of its aphid prey. *Journal of Chemical Ecology*, 24: 631–643.

Boopathi, T., Singh, S.B., Ravi, M. & Manju, T. (2016). Distribution and Biology of *Mallada desjardinsi* (Neuroptera: Chrysopidae) in India and its predatory potential against *Aleurodicus dispersus* (Hemiptera: Aleyrodidae). *Journal of Economic Entomology*, 109(5), 1988–1994. doi: 10.1093/jee/tow154.

Brar, K.S., Sekhon, B.S., Singh, J., Shenhmar, M. & Bakhetia, D.R.C. (2001). Evaluation of biocontrol based IPM modules for the control of bollworm complex on cotton. *Journal of Insect Science*, 14: 19–22.

Brauer F (1864) Entomologische Beiträge. Verhandlungen der Kaiserlich-Königlichen Zoologisch-Botanischen Gesellschaft in Wien 14: 891–902

Brooks, S.J. & Barnard, P.C. (1990). The green lacewings of the World: A generic review (Neuroptera: Chrysopidae). Bulletin of the British Museum (Natural History). *Entomology*, 59: 117–286.

Canard, M. (1997). Can lacewings feed on pests in winter: (Neur.: Chrysopidae and hemerobiidae)?. *Entomophaga*, 42: 113–117.

Canard, M. (2005). Seasonal adaptations of green lacewings (Neuroptera: Chrysopidae). *European Journal of Entomology*, 102: 317–324.

Canard, M. & Grimal, A. (1988). Insect photoperiodism: Various ways of regulating univoltinism in lacewings (Planipennia: Chrysopidae). *Experientia*, 44: 523–525.

Canard, M., Semeria, Y. & New, T.R. (1984). *Biology of Chrysopidae*. The Hague, The Netherlands: Dr. W. Junk Publishers, 294pp.

Cardoso, E.R., Freitas, S., Nunes, H.T. & Pessoa, L.G.A. (2007). Selectivity of *Lecanicillium lecanii* and *Metarhizium anisopliae* for first instar larvae of *Ceraeochrysa cincta* (Neuroptera: Chrysopidae) in laboratory. *Acta Scientiarum Agronomy*, 29: 563–568.

Chacon-Hernandez, J.C., Cerna-Ch_Avez, E., Vanoye-Eligio, V., Ocho-Fuentes, Y.M., Ordaz-Silva, S., Soria-Dıaz, L. & Landeros-Flores, J. (2019). *Phytoseiulus persimilis* (Acari: Phytoseiidae) functional response in the biological control of *Tetranychus urticae* (Acari: Tetranychidae) eggs on rose foliage discs. *Journal of Entomological Science*, 54: 275–287.

Chakraborty, D. & Korat, D.M. (2010). Biology of green lacewing, *Chrysoperla carnea* (Stephens) in middle Gujarat conditions. *Karnataka Journal of Agricultural Sciences*, 23(3): 500–502.

Chakravarthy, A.K. & Thyagaraj, N.E. (2005). Role of predators in suppressing sugarcane woolly aphid, *Ceratovacuna lanigera* Zehntner (Homoptera: Aphididae) infestation in South Karnataka. *Insect Environment*, 11: 72–74.

Chalapathi Rao, N.B.V., Ramani, B.S., Roshan, D.R. & Bhagavan, B.V.K. (2020). Biocontrol management options for invasive whiteflies on coconut. *Indian Coconut Journal*, LXIII: 17–22.

Chang, Y.F., Tauber, M.J. & Tauber, C.A. (1995). Storage of the mass-produced predator *Chrysoperla carnea* (Neuroptera: Chrysopidae): Influence of photoperiod, temperature and diet. *Environmental Entomology*, 24: 1365–1374.

Chelvi, C.T. & Kandasamy, R. (2009). Field efficacy of *Chrysopa scelestis* Bank against sugarcane mealy bug *Saccaricoccus sacchari* Cockerell. *Cooperative Sugar*, 41: 37–39.

Chen, T.Y. & Liu, T.X. (2001). Relative consumption of three aphid species by the lacewing, *Chrysoperla rufilabris* and effects on its development and survival. *BioControl*, 46: 481–491.

Chen, T.-Y., Chu, C.-C., Hu, C., Mu, J.-Y., & Henneberry, T.J. (2006). Observations on midgut structure and content of *Chrysoperla carnea* (Neuroptera: Chrysopidae). *Annals of the Entomological Society of America*, 99(5): 917–919.

Cohen, A.C. & Smith, L.K. (1998). A new concept in artificial diets for *Chrysoperla rufilabris*: The efficacy of solid diets. *Biological Control*, 13: 49–54.

Colinet, H. & Boivin, G. (2011). Insect parasitoids cold storage: A comprehensive review of factors of variability and consequences. *Biological Control*, 58: 83–95.

Daane, K.M. & Yokota, G.Y. (1997). Release methods affect egg survival and distribution of augmentated green lacewings (Neuropteran: Chrysopidae). *Environmental Entomology*, 26: 455–464.

Das, T.K. & Mandal, S.M.A. (2005). Biological control of major lepidopteran pests in cabbage. *Journal of Plant Protection and Environment*, 2: 68–71.

De Clercq, P., Mohaghegh, J. & Tirry, L. (2000). Effect of host plant on the functional response of the predator *Podisus nigrispinus* (Heteroptera: Pentatomidae). *Biological Control*, 18: 65–70.

Debnath, N., Ghosh, D. & Chakrabarti, S. (1988). Predators and parasites of aphids *from* north west and western Himalaya. II. Records of eight aphidophaogus neuropterans (Insecta) *from* India. *Entomon*, 13: 137–139.

Deepika, K., Dahiya, K.K. & Lal, R. (2009). Impact analysis of genetically modified (Bt) cotton cultivars on natural enemies. *Journal of Cotton Research and Development*, 23: 111–116.

Deng, J.Y., Ye, B.B., Wang, F.M. & Zhang, Y. (2013). Attraction of *Chrysoperla sinica*, a natural enemy by synthetic attractants for controlling colorado potato beetle. *International Chemical Ecology Conference*, August 19–23, 2013, Melbourne, Australia.

Denlinger, D.L. (2008). Why study diapause? *Entomological Research*, 38: 1–9.

Denlinger, D.L., Yocum, G.D. & Rinehart, J.P. (2012). Hormonal control of diapause. In: *Insect Endocrinology* (Ed. Gilbert, L.I.). San Diego, CA: Academic Press, pp. 430–463.

Dhaka, S.R. & Pareek, B.L. (2007). Seasonal incidence of natural enemies of key insect pests of cotton and their relationship with weather parameters. *Journal of Plant Protection Research*, 47: 417–423.

Dhandapani, N., Kalyanasundaran, M., Swamiappan, M., Babu, P.C.S. & Jayaraj, S. (1992). Experiments on management of major pests of cotton with biocontrol agents in India. *Journal of Applied Entomology*, 114: 52–56.

Dias, P.M., de Souza Loureiro, E., Pessoa, L.G.A., de Oliveira Neto, F.M., de Souza Tosta, R.A. & Teodoro, P.E. (2019). Interactions *between* fungal-infected *Helicoverpa armigera* and the predator *Chrysoperla Externa*. *Insects*, 10: 309. doi:10.3390/insects10100309.

Duelli, P. (1996). The working group "*carnea*-complex:" Report on activities, results and cooperative projects. In: *Pure and Applied Research in Neuropterology*, (Eds. M. Canard, H. Aspock & M.W. Mansell), *Proceedings of the Fifth International Symposium on Neuropterology*, Cairo, Egypt, 1994. Toulouse, France: Sacco, pp. 307–311.

Duelli P. (2001). Lacewings in field crops. In *Lacewings in the Crop Environment*, (Eds. P.K. McEwen, T.R. New & A.E. Whittington). Cambridge University Press, Cambridge, UK, pp. 158–171.

Duffield, S.J. (1995). Crop specific differences in the seasonal abundance of *four* major predatory groups on sorghum and short duration pigeon pea. *International Chickpea and Pigeon Pea News Letter*, 2: 74–76.

Duffield, S.J. & Reddy, Y.V.R. (1997). Distribution and movement of predators of *Helicoverpa armigera* in intercropped sorghum and short duration pigeon pea. *Crop Research Hissar*, 14: 315–335.

Easterbrook, M.A., Fitzgerald, J.D. & Solomon, M.G. (2006). Suppression of aphids on strawberry by augmentative releases of larvae of the lacewing *Chrysoperla carnea* (Stephens). *Biocontrol Science and Technology*, 16(9): 893–900.

Eisner, T., Attygalle, A.B., Conner, W.E., Eisner, M., MacLeod, E. & Meinwald, J. (1996). Chemical egg defense in a green lacewing (*Ceraeochrysa smithi*). *PNAS*, 93: 3280–3283.

Elanchezhyan, K., Baskaran, R.K.M. & Rajavel, D.S. (2009). Kairomone effect of acetone extracts of intercrops on *Chrysoperla carnea* (Stephens) and *Trichogramma chilonis* Ishii. *Journal of Entomological Research*, 33: 51–54.

El-Arnaouty, S.A., Gaber, N. & Tawfik, M.FS. (2000). Biological control of the green peach aphid *Myzus persicae* by *Chrysoperla carnea* (Neuroptera: Chrysopidae). *Egyptian Journal of Biological Pest Control*, 10(12): 109–116.

Faleiro, J.R., Singh, K.M. & Singh, R.N. (1990). Influence of abiotic factors on the population build up of important insect pests of cowpea *Vigna unguiculata* (L) Walp and their biotic agents recorded at Delhi. *Indian Journal of Entomology*, 52: 675–680.

Farkade, B.C., Ahire, R.D., Patange, N.R. & Ahir, P.S. (1999). Extent of adoption of biological pest control in cotton. *Journal of Soils & Crops*, 9: 108–110.

Farrokhi, M., Gharekhani, G., Iranipour, S. & Hassanpour, M. (2017). Host plant–herbivore–predator interactions in *Chrysoperla carnea* (Neuroptera: Chrysopidae) and *Myzus persicae* (Homoptera: Aphididae) on *four* plant species under laboratory conditions. *Journal of Economic Entomology*, 110(6), 2342–2350. doi: 10.1093/jee/tox268.

Fasih, M. & Srivastava, R.P. (1990). Parasites and predators of insect pest of mango. *International Pest Control*, 32: 39–41.

Fathipour, Y. & Jaafari, A. (2003). Functional response of predators *Nabis capsiformis* and *Chrysoperla carnea* to different densities of *Creontiades pallidus* nymphs. *Journal of Agricultural Science and Natural Resources*, 10: 125–133.

Fathipour, Y., Karimi, M., Farazmand, A. & Talebi, A.A. (2018). Age-specific functional response and predation rate of *Amblyseius swirskii* (Phytoseiidae) on *two*-spotted spider mite. *Systematic and Applied Acarology*, 22: 159–169.

Fernandez-Arhex, V. & Corley, J.C. (2003). The functional response of parasitoids and its implications for biological control. *Biocontrol Science and Technology*, 13: 403–413.

Figueira, L.K., Lara, F.M. & Cruzi, I. (2002). Efeito de genótipos de sorgo sobre o predador *Chrysoperla externa* (Hagen, 1861) (Neuroptera: Chrysopidae) alimentado com *Schizaphis graminum* (Rondani, 1852) (Hemiptera: Aphididae). *Neotropical Entomology*, 31: 133–139.

Fonseca, A.R., Carvalho, C.F. & Souza, B. (2001). Capacidade predatória e aspectos biológicos das fases imaturas de *Chrysoperla externa* (Hagen, 1861) (Neuroptera: Chrysopidae) alimentada com *Schizaphis graminum* (Rondani, 1852) (Hemiptera: Aphididae) em diferentes temperaturas. *Ciêncea e Agrotecnologia*, 25: 251–263.

Fujiwara, C. & Nomura, M. (1999). Effects of photoperiod and temperature on larval development of *Chrysoperla carnea* Stephens (Neuroptera: Chrysopidae). *Japanese Journal of Applied Entomology and Zoology*, 43: 175–179.

Gade, R.S., Shetgar, S.S. & Chavan, S.L. (2011). Biology of *Mallada boninensis* (Okamoto) on three species of aphids. *Journal of Entomological Research*, 35: 139–141.

Gao, F., Chen, F. & Ge, F. (2010). Elevated CO_2 lessens predation of *Chrysopa sinica* on *Aphis gossypii*. *Entomologia Experimentalis et Applicata*, 135: 135–140.

Garzón-Orduña, I.J., Winterton, S.L., Jiang, Y. et al. (2019). The evolution of green lacewings (Neuroptera: Chrysopidae): A molecular super matrix approach. *Systematic Entomology*, 44: 499–513.

Gautam, R.D. (1994). Present status of rearing of chrysopids in India. *Bulletin of Entomological,* 35: 31–39.

Geetha, B. & Swamiappan, M. (1998). Influence of crop plants on adult biology of the predator *Chrysoperla carnea* (Chrysopidae: Neuroptera). *Insect Environment*, 3: 110.

Geethalakshmi, L., Muthukrishnan, N., Chandrasekharan, M. & Raghuraman, M. (2000). Chrysopids biology on *Corcyra cephalonica* and feeding potential on different host insects. *Annals of Plant Protection Sciences*, 8: 132–135.

Generoso, A.R. (2012). Compatibility of *Beauveria bassiana* and *Paecilomyces fumosoroseus* with *Chrysoperla externa* (Neuroptera: Chrysopidae) and Methodology for Evaluating Selectivity. Master's Thesis (Master in Entomology), Paulista State University, Jaboticabal-SP, Brazil, p. 63.

Gonzalez-Mas, N., Cuenca-Medina, M., Gutierrez-Sanchez, F. & Quesada-Moraga, E. (2019). Bottom-up effects of endophytic *Beauveria bassiana* on multitrophic interactions between the cotton aphid, *Aphis gossypii*, and its natural enemies in melon. *Journal of Pest Science*, 92: 1271–1281. doi: 10.1007/s10340-019-01098-5.

Grimal, A. (1988). Photoperiodic developmental requirements of the lacewing *Tjederina gracilis*. *Entomologia Experimentalis et Applicata*, 47: 189–194.

Guntupalli, S. & Kalyanasundaram, D.M. (2016). Biology and biological parameters of *Mallada boninensis* (Okamato) (Neuroptera: Chrysopidae) on different hosts. *The Bioscan*, 11(2): 695–698.

Hanumantharaya, L., Mallapur, C.P., Venkateshalu, Naik, V.R. & Kumar, C.J. (2008a). Pest status of safflower, *Carthamus tinctorius* L. in northern parts of Karnataka. In: *Safflower: Unexploited Potential and World Adaptability,* (Eds. S.E. Knights & T.D. Potter) *7th International Safflower Conference*, Wagga Wagga, Australia, November 3–6, 2008, pp. 1–3.

Hanumantharaya, L., Parameshwarappa, K.G., Suma, S.B. & Kumar, C.J. (2008b). Electroantennogram and ovipositional responses of *Helicoverpa armigera* (Hubner) and *Chrysoperla carnea* (Stephens) to volatiles of different cultivars of safflower, *Carthamus tinctorius*. In: *Safflower: Unexploited Potential and World Adaptability,* (Eds. S.E. Knights & T.D. Potter) *7th International Safflower Conference*, Wagga Wagga, Australia 3–6, November, 2008, pp. 1–6.

Hanumantharaya, L., Goud, K.B. & Naik, L.K. (2008c). Use of green lacewing, *Chrysoperla carnea* (Stephens) and neem seed kernel extract for management of insect pests on cotton. *Karnataka Journal of Agricultural Sciences*, 21: 41–44.

Hassanpour, M., Nouri-Ganbalani, G., Mohaghegh, J. & Enkegaard, A. (2009). Functional response of different larval instars of the green lacewing, *Chrysoperla carnea* (Neuroptera: Chrysopidae), to the two-spotted spider mite, *Tetranychus urticae* (Acari: Tetranychidae). *Journal of Food, Agriculture and Environment*, 7, 424–428.

Hassanpour, M., Mohaghegh, J., Iranipour, S., Nouri-Ganbalani, G. & Enkegaard, A. (2011). Functional response of *Chrysoperla carnea* (Neuroptera: Chrysopidae) to *Helicoverpa armigera* (Lepidoptera: Noctuidae): Effect of prey and predator stages. *Insect Science*, 18, 217–224.

Hegde, R. & Lingappa, S. (1998). Performance of *Chrysoperla carnea* (Stephens) against pigeon pea pod borer, *Helicoverpa armigera*. *Karnataka Journal of Agricultural Sciences*, 11: 658–661.

Henry, C.S. (1980). The courtship call of *Chrysopa downesi* Banks [(Neuroptera: Chrysopidae): Its evolutionary significance. *Psyche*, 86: 291–297.

Henry, C.S. (2006). Acoustic communication in neuropterid insects. In: *Insect Sounds and Communication: Physiology, Behaviour, Ecology and Evolution* (*Contemporary Topics in Entomology*) (Eds S. Drosopoulos & M. Claridge). Boca Raton, FL: CRC Press and Taylor & Francis Group, pp. 153–166.

Henry, C.S., Brooks, S.J., Peter Duelli, P., James B. Johnson, J.B., Marta M. Wells, M.M. & Atsushi Mochizuki, A. (2013). Obligatory duetting behaviour in the *Chrysoperla carnea*-group of cryptic species (Neuroptera: Chrysopidae): Its role in shaping evolutionary history. *Biological Reviews*, 88: 787–808. doi: 10.1111/brv.12027.

ICAR-NBAIR. (2019). Technologies ready for agribusiness (2019–2020). Technical bulletin No. 43, ICAR-National Bureau of Agricultural Insect Resources, Bengaluru-24, India.

Ingawale, U.V. & Tambe, A.B. (2007). Seasonal abundance of aphids and their natural enemies on lucerne. *Asian Journal of Biological Sciences*, 2: 8–10.

Ingole, D.A., Deotale, V.Y., Deotale, R.O. & Ramkumar, J. (2005). Preliminary study on the performance of *Mallada boninensis* (Okamoto) against *Aleurocanthus woglumi* Ashby on citrus. *Journal of Biological Control*, 19: 77–79.

Jaastad, G., Hatleli, L., Knudsen, G.K. & Toth, M. (2010). Volatiles initiate egg laying in common green lacewings. *International Organization for Biological and Integrated Control-West Palaearct Regional Section Bull*, 54: 77–82.

Jalali, S.K. & Singh, S.P. (1994). Effect of *Aphis gossypii* Glover number on *Mallada astur* (Banks) and *Cheilomenes sexmaculata* (Fabricius). *Biological Control*, 4: 45–47.

Jalali, S.K. Rabindra, R.J., Rao, N.S. & Dasan, C.B. (2003). Mass production of trichogrammatids and chrysopids. Technical Bulletin, 33, Project Directorate of Biological Control, Bangalore.

Jalali, M.A., Tirry, L. & De Clercq, P. (2010). Effect of temperature on the functional response of *Adalia bipunctata* to *Myzus persicae*. *BioControl*, 55: 261–269.

Jayakumar, R., Naganagoud, A. & Rajashekarappa, K. (2008). Influence of indigenous components on natural enemies in okra ecosystem. *Annals of Plant Protection Sciences*, 16(1): 102–106.

Jones, V.P., Steffan, S.A., Wiman, N.G., Horton, D.R., Miliczky, E., et al. (2011). Evaluation of herbivore-induced plant volatiles for monitoring green lacewings in Washington apple orchards. *Biological Control*, 56: 98–105.

Jones, V.P., Horton, D.R., Mills, N.J., Unruh, T.R., Miliczky, E. et al. (2016). Using plant volatile traps to develop phenology models for natural enemies: An example using *Chrysopa nigricornis* (Burmeister) (Neuroptera: Chrysopidae). *Biological Control*, 102: 77–84.

Joshi, B.C. & Yadav, D.N. (1990). Biology and feeding potential of *Mallada boninensis* (Okamoto), a chrysopid predator of whitefly *Bemisia tabaci* Gennadius. *Journal of Biological Control*, 4: 18–21.

Jyoti, D.P. & Goud, K.B. (2008). Safety of organic amendments and microbial pesticides to natural enemies in brinjal ecosystem. *Annals of Plant Protection Sciences*, 16: 111–114.

Kabissa, J.C.B. (1993). Studies on the biology and ecology of chrysopid predators (Neuroptera: Chrysopidae) of *Helicoverpa armigera* (Hubner) (Lepidoptera: Noctuidae) and *Aphis gossypii* (Glover) (Homoptera: Aphididae) on cotton, *Gossypium hirsutum* (L) in Eastern Tanzania. PhD thesis, University of Dar es Salaam.

Kabissa, J.C.B., Kayumbo, H.Y. & Yarro, J.G. (1995.) Comparative biology of *Mallada desjardinsi* (Navas) and *Chrysoperla congrua* (Walker) (Neuroptera: Chrysopidae), predators of *Helicoverpa armigera* (Hubner) (Lepidoptera: Noctuidae) and *Aphis gossypii* (glover) (Homoptera: Aphididae) on cotton in eastern Tanzania. *International Journal of Pest Management*, 41(4): 214–218. doi: 10.1080/09670879509371952.

Kabissa, J.C.B., Yarro, J.C., Kayumbo, H.Y. & Juliano, S.A. (1996). Functional responses of two chrysopid predators feeding on *Helicoverpa armigera* (Lep.: Noctuidae) and *Aphis gosssypii* (Hom.: Aphididae). *Entomophaga*, 41: 141–151.

Kadri, S. & Goud, B. (2005). Survey and seasonal incidence of onion thrips, *Thrips tabaci* (Lindmann). *Karnataka journal of Agricultural Sciences*, 18(2): 510–512.

Kalra, V.K. (1988). Population dynamics of various predators associated with mustard aphid, *Lipaphis erysimi* (Kalt). *Journal of Biological Control*, 2: 77–79.

Kasaei-Faradonbeh, S., Hassanpour, M., Razmjou, J., Golizadeh, A. & Esmaeilpour, B. (2015). Biological and population growth parameters of the melon aphid, *Aphis gossypii* Glover on cucumber plants grown at different vermicompost: soil ratios. *Plant Pests Research*, 4: 35–48.

Kaur, S., Shenhmar, M. & Brar, K.S. (2001). Effect of cold storage on eggs and pupae of *Chrysoperla carnea* (Stephens). *Journal of Insect Science*, 14: 58–59.

Kaur, R., Kaur, R. & Brar, K.S. (2008). Development and predation efficacy of *Chrysoperla carnea* (Stephens) on mealy bug, *Phenacoccus solenopsis* (Tinsley) under laboratory conditions. *Journal of Insect Science*, 21: 93–95.

Khan, A.A. & Zaki, F.A. (2008). Predatory response of *Chrysoperla carnea* (Stephens) (Neuroptera: Chrysopidae) feeding on the Euonymus aphid, *Aphis fabaesolanella* Theobald (Homoptera: Aphididae) in Kashmir. *Journal of Biological Control*, 22: 149–154.

Khanzada, K.K., Shaikh, H.M., Mastoi, S.M., Rajput, Z., Jat, M.I., Mastoi, P.M. & Kaleri, A.A. (2018). Some parameters of *Chrysoperla carnea* on floral feeds of castor and feeding potential on selective aphid species. *Journal of Entomology and Zoology Studies,* 6(2): 860–863.

Khuhro, N.H., Chen, H., Zhang, Y., Zhang, L. & Wang, M. (2012). Effect of different prey species on the life history parameters of *Chrysoperla sinica* (Neuroptera: Chrysopidae). *European Journal of Entomology*, 109: 175–180.

Khulbe, P., Maurya, R.P. & Khan, M.A. (2005). Biology of *Chrysoperla carnea* (Stephens) on different host insects. *Annals of Plant Protection Sciences*, 13(2): 351–354.

Koczor, S., Szentkir´alyi, F., Birkett, M.A., Pickett, J.A., Voigt, E., & Toth, M. (2010). Attraction of *Chrysoperla carnea* complex and *Chrysopa* spp. lacewings (Neuroptera: Chrysopidae) to aphid sex pheromone components and a synthetic blend of floral compounds in Hungary. *Pest Management Science*, 66: 1374–1379.

Koczor, S., Knudsen, G.K., Hatleli, L., Szentkirályi, F. & Tóth, M. (2015a). Manipulation of oviposition and overwintering site choice of common green lacewings with synthetic lure (Neuroptera: Chrysopidae). *Journal of Applied Entomology*, 139: 201–206.

Koczor, S., Szentkirályi, F., Pickett, J.A., Birkett, M.A. & Tóth, M. (2015b). Aphid sex pheromone compounds interfere with attraction of common green lacewings to floral bait. *Journal of Chemical Ecology*, 41: 550–556.

Koczor, S., Szentkirályi, F., Fekete, Z. & Tóth, M. (2017). Smells good, feels good: Oviposition of *Chrysoperla carnea*-complex lacewings can be concentrated locally in the field with a combination of appropriate olfactory and tactile stimuli. *Journal of Pest Science*, 90: 311–317.

Koczor, S., Szentkirályi, F. & Tóth, M. (2019). New perspectives for simultaneous attraction of *Chrysoperla* and *Chrysopa* lacewing species for enhanced biological control (Neuroptera: Chrysopidae). *Scientific Reports*, 9: 10303. doi: 10.1038/s41598-019-46621-x.

Kostal, V. (2006). Eco-physiological phases of insect diapause. *Journal of Insect Physiology*, 52: 113–127.

Krishnamoorthy, A. & Mani, M. (1982). Feeding potential and development *of Chrysopa scelestes* Banks on *Heliothis armigera* (Hubn.) under laboratory conditions. *Entomon*, 7: 385–388.

Krishnamoorthy, A. (1984). Influence of adult diet on the fecundity and survival of the predator *Chrysopa scelestes* (Neur.: Chrysopidae). *Entomophaga*, 29: 445–450.

Krishnamoorthy, A. & Mani, M. (1989). Records of lacewings preying on mealy bugs in India. *Current Science*, 58: 155–157.

Kumar, G.S., Thakare, D.R., Devi, A.R. & Devi, M.N. (2014). Bionomics of *Chrysoperla carnea* (Stephens) on eggs of *Corcyra cephalonica* (Stainton). *International Journal of Farm Sciences*, 4(3): 145–151.

Kumar, H.L.C., Chauhan, S.V., Ramkumar, J. & Irulandi, S. (2007a). Biology of *Chrysoperla carnea* (Stephens) strains collected from different agro climatic zones of Tamil Nadu. *Insect Environment*, 13: 42–44.

Kumar, H.L.C., Chauhan, S.V., Ramkumar, J. & Irulandi, S. (2007b). Predatory potential of different strains of *Chrysoperla carnea* larvae. *Insect Environment*, 13: 41–42.

Kumar, H.L.C., Swamiappan, M., Chauhan, S.V., Rajendran, R. & Thirumurugan, V. (2008). Ovipositional period and longevity of different strains of *Chrysoperla carnea* in Tamil Nadu. *Insect Environment*, 13: 147–148.

Kumari, D., Verma, S.C., Sharma, P.L. & Gaikwad, M.B. (2020). Biology, predatory potential and functional response of *Mallada desjardinsi* (Navas) on melon aphid, *Aphis gossypii* glover. *International Journal of Tropical Insect Science*. doi: 10.1007/s42690-020-00233-5.
Lavhe, N.V. (2000). Studies on the life cycle on chrysopid predator *Mallada boninensis* (Okamoto). *Insect Environment*, 5: 176.
Lavhe, N.V. & Nimbalkar, S.A. (2000). A record of predation of *Mallada boninensis* (Okamoto) on citrus blackfly. *Insect Environment*, 5: 172.
Li, Y., Wang, M., Gao, F., Zhang, H., Chen, H., Wang, M., Liu, C. & Zhang, L. (2018). Exploiting diapause and cold tolerance to enhance the use of the green lacewing *Chrysopa formosa* for biological control. *Biological Control*, 127: 116–126. doi: 10.1016/j.biocontrol.2018.08.024.
Lopez-Arroyo, J.I., Tauber, C.A. & Tauber, M.J. (1999). Comparative life histories of the predators *Ceraeochrysa cincta, C. cubana* and *C. smithi* (Neuroptera: Chrysopidae). *Annals of the Entomological Society of America*, 92: 208–217.
Lopez-Arroyo, J.I., Tauber, C.A. & Tauber, M.J. (2000). Storage of lacewing eggs: Post-storage hatching and quality of subsequent larvae and adults. *Biological Control*, 18: 165–171.
Macedo, L.P.M., Souza, B., Carvalho, C.F. & Ecole, C.C. (2003). Influence of the photoperiod on development and reproduction of *Chrysoperla externa* (Hagen) (Neuroptera: Chrysopidae). *Neotropical Entomology*, 32: 91–96.
Mahdian, K., Vantornhout, I., Tirry, L. & De Clercq P. (2006). Effects of temperature on predation by stink-bugs *Picromerus bidens* and *Podisus maculiventris* (Heteroptera: Pentatomidae) on noctuid caterpillars. *Bulletin of Entomological Research*, 96: 489–496.
Mahzoum, A.M., Villa, M., Benhadi-Marín, J. & José Alberto Pereira, J. (2020). Functional response of *Chrysoperla carnea* (Neuroptera: Chrysopidae) Larvae on *Saissetia oleae* (Olivier) (Hemiptera: Coccidae): Implications for biological control, *Agronomy*, 10: 1511; doi: 10.3390/agronomy10101511.
Maia, W.J.M.S., Carvalho C.F. & Souza, B. (2000). Exigências térmicas de *Chrysoperla externa* (Hagen, 1861) (Neuroptera: Chrysopidae) alimentada com *Schizaphis graminum* (Rondani, 1852) (Hemiptera: Aphididae) em condições de laboratório. *Ciêncea e Agrotecnologia*, 24: 81–86.
Mallapur, C.P., Sharanabasappa, Kambrekar, D.N., Gundannavar, K.P. & Kulkarni, K.A. (2006). Integrated management of sugarcane woolly aphid, *Ceratovacuna lanigera* (Zehntner). *Annals of Plant Protection Sciences*, 14: 337–339.
Mani, M. & Krishnamoorthy, A. (1989). Feeding potential and development of *Mallada boninensis* (Okamoto) on the grape mealy bug, *Maconellicoccus hirsutus* (Green). *Entomon*, 14: 19–20.
Mani, M. & Krishnamoorthy, A. (1990). Natural suppression of mealybugs in guava orchards. *Entomon*, 15: 245–247.
Mani, M. & Krishnamoorthy, A. (1995). Natural enemies of whitefly, *Lipaleyrodes euphorbiae* David and Subramaniam (Homoptera: Aleyrodidae). *Journal of Biological Control*, 9: 102–104.
Mani, M. & Krishnamoorthy, A. (1998). Biological control studies on the mango green scale *Chloropulvinaraia polygonata* (Ckll.) (Homoptera: Coccidae) in India. *Entomon*, 23: 105–110.
Mani, M. & Krishnamoorthy, A. (1999). Natural enemies of host plants of spiraling whitefly *Aleurodicus dispersus* Russel (Homoptera: Aleyrodidae) in Bangalore, Karnataka. *Entomon*, 24: 75–80.
Mani, M. & Krishnamoorthy, A. (2007). Recent trends in the biological suppression of guava pests in India. *Acta Horticulturae*, 735: 469–482. doi: 10.17660/ActaHortic.2007.735.64.
Manjunath, T.M. (1993). Biocontrol of *Helicoverpa armigera* and other cotton bollworms. *Indian Farming*, 42: 18–23.
Mannan, V.D., Varma, G.C. & Brar, K.S. (1995). Seasonal fluctuations and host predator releatiohsip of *Chrysoperla carnea* (Stephens) (Chrysopidae: Neuroptera). *Indian Journal of Ecology*, 22: 21–26.
Mantoanelli, E., Albuquerque, G.S., Tauber, C.A. & Tauber, M.J. (2006). *Leucochrysa* (*Leucochrysa*) *varia* (Neuroptera: Chrysopidae): Larval descriptions, developmental rates and adult color variation. *Annals of the Entomological Society of America*, 99: 7–18.
Mathew, M.J., Saju, K.A. & Venugopal, M.N. (1998). *Chrysoperla carnea* Stephens (Chrysopidae: Neuroptera): A potential biocontrol agent of cardamom aphid (*Pentalonia nigeronervosa* van der Goot). *Journal of Spices and Aromatic Crops*, 7: 155–156.
Matos, B. & Obrycki, J.J. (2006). Prey suitability of *Galerucella calmariensis* L. (Coleoptera: Chrysomelidae) and *Myzus lythri* (Schrank) (Homoptera: Aphididae) for development of *three* predatory species. *Environmental Entomology*, 35: 345–350.
Maurya, R.P., Khan, M.A., Kumar, S. & Kumar, A. (2009). Artificial selection of green lacewing, *Chrysoperla carnea* Stephens for tolerance to insecticides. *Annals of Plant Protection Sciences*, 17: 1–4.

Mhaske, S.H., Shetgar, S.S. & Shinde, P.R. (2017). Biology of *Chrysoperla zastrowi sillemi* (Esben-Petersen) on mealy bugs and aphids. *Journal of Biological Control*, 31(2): 110–113.

Mirabzadeh, A., Sahragard, A. & Azma, M. (1998). Functional response of *Chrysoperla carnea* Steph. larvae to *Pterochloroides persicae*, *Uroleucon cichorii* and *Aphis craccivora*. *Seed and Plant*, 14: 29–34.

Misof, B., Liu, S., Meusemann, K., Peters, R.S., Donath, A., et al. (2014). Phylogenomics resolves the timing and pattern of insect evolution. *Science*, 346: 763–767.

Mohaghegh, J., De Clercqm, P. & Tirry, L. (2001). Functional response of the predators *Podisus maculiventris* (Say) and *Podisus nigrispinus* (Dallas) (Het., Pentatomidae) to the beet armyworm, *Spodoptera exigua* (Hubner) (Lep., Noctuidae): Effect of temperature. *Journal of Applied Entomology*, 125: 131–134.

Mohite, P.B. & Uthamasamy, S. (1998). Host plant resistance and natural enemies interaction in the management of *Helicoverpa armigera* (Hubner) on cotton. *Indian Journal of Agricultural Sciences*, 32: 28–30.

Montoya-Alvarez, A.F., Ito, K., Nakahira, K. & Arakawa, R. (2010). Functional response of *Chrysoperla nipponensis* and *C. carnea* (Neuroptera: Chrysopidae) to the cotton aphid *Aphis gossypii* Glover (Homoptera: Aphididae) under laboratory conditions. *Applied Entomology and Zoology*, 45: 201–206.

Mushtaq, T. & Khan, A.A. (2010). Functional and aggregational response of *Chrysoperla* sp. (*carnea*-group) (Neuroptera: Chrysopidae) on *Brevicoryne brassicae* (Linnaeus) (Hemiptera: Aphididae). *Journal of Biological Control*, 24(1): 28–34.

Nagamallikadevi, M., Undirwade, D.B., Reddy, N.B., Ramadevi, A. & Srasvankumar, G. (2013). Biology of *Mallada boninensis* (Okamoto) [Chrysopidae: Neuroptera] on Aphids and Neonate Noctuids. *Trends in Biosciences*, 6: 827–830.

Nair, I.J., Sharma, S. & Kaur, R. (2020). Efficacy of the green lace wing, *Chrysoperla zastrowi sillemi* (Esben-Peterson) (Neuroptera: Chrysopidae), against sucking pests of tomato: An appraisal under protected conditions. *Egyptian Journal of Biological Pest Control*, 30: 74. doi: 10.1186/s41938-020-00277-2.

Nakahira, K. & Arakawa, R. (2005). Effect of photoperiod on the development and diapause of the green lacewing *Chrysopa pallens* (Neuroptera: Chrysopidae). *Entomological Science*, 8: 133–135.

Nakahira, K., Nakahara, R. & Arakawa, R. (2005). Effect of temperature on development, survival and adult body size of *two* green lacewings, *Mallada desjardinsi* and *Chrysoperla nipponensis* (Neuroptera: Chrysopidae). *Applied Entomology and Zoology*, 40: 615–620.

Needham, J. G. (1909). Notes on the Neuroptera in the collection of the Indian Museum. *Records of the Indian Museum, Calcutta*, 3: 185–210.

Oaten, A. & Murdoch, W.W. (1975). Functional response and stability in predator–prey systems. *The American Naturalist*, 109: 289–298.

Oliveira, S.A., Auad, A.M., Souza, B., Silva, D.M. & Carvalho, C.A. (2010). Effect of temperature on the interaction *between Chrysoperla externa* (Neuroptera: Chrysopidae) and *Sipha flava* (Hemiptera: Aphididae). *European Journal of Entomology*, 107: 183–188.

Osman, M.Z. & Selman, B.J. (1993). Storage of *Chrysoperla carnea* Steph. Chrysopidae: Neuroptera) eggs and pupae. *Journal of Applied Entomology*, 115: 420–424.

Oswald, J.D. (2020). Lacewing digital library [WWW document]. URL http://lacewing.tamu.edu/ [accessed on 22 November 2020].

Panchbhai, P.R., Sharnagat, B.K., Dhawad, C.S., Wadaskar, R.M. & Bagade, I.B. (2005). Effect of augmentation of *Trichogramma chilonis* Ishii and *Chrysoperla carnea* (Stephens); *two* natural enemies of *Helicoverpa armigera* (Hubner) on larval infestation and yield in cotton. *Pest Management and Economic Zoology*, 13: 77–81.

Pandey, R. & Singh, N.N. (2008). Effectiveness of biocontrol based IPM modules against *Lipaphis erysimi* Kaltenbach (Hemiptera: Aphididae). *Journal of Plant Protection Research*, 48: 1–7.

Panth, M., Giri, S. & Kafle, K. (2017). Biology and predatory potential of *Chrysoperla carnea* S. (Neuroptera: Chrysopidae) reared on different species of aphid under laboratory condition. *International Journal of Advanced Research in Science, Engineering and Technology*, 4: 3847–3851.

Pappas, M.L., Broufas, G.D. & Koveos, D.S. (2007). Effects of various prey species on development, survival and reproduction of the predatory lacewing *Dichochrysa prasina* (Neuroptera: Chrysopidae). *Biological Control*, 43: 163–170.

Pappas, M.L., Broufas, G.D. & Koveos, D.S. (2008). Effect of temperature on survival, development and reproduction of the predatory lacewing *Dichochrysa prasina* (Neuroptera: Chrysopidae) reared on *Ephestia kuehniella* eggs (Lepidoptera: Pyralidae). *Biological Control*, 45: 396–403.

Parajulee, M.N., Shrestha, R.B., Leser, J.F., Wester, D.B., & Blanco, C.A. (2006). Evaluation of the functional response of selected arthropod predators on bollworm eggs in the laboratory and effect of temperature on their predation efficiency. *Environmental Entomology*, 35, 379–386.

Patel, I.S. & Yadav, D.N. (1996). Studies on genetic improvement of green lacewing, *Chrysopa scelestes* Banks (Chrysopidae: Neuroptera) selection for monocrotophos resistance. *Gujarat Agricultural University Research Journal*, 21: 155.

Patel, R.K. & Thakur, B.S. (2005). Insect pest complex and seasonal incidence in linseed with particular reference to bud fly *(Dasineura lini* Barnes). *Journal of Plant Protection and Environment*, 2: 102–107.

Patel, K.G. & Vyas, H.N. (1984). Seasonal abundance of green lacewing, *Chrysopa* (=*Chrysoperla*) *scelestes* Banks on cotton, green gram and groundnut at Junagadh (Gujarat). *Agricultural Science Digest*, 4(1): 7–8.

Patel, K.G. & Vyas, H.N. (1985). Biology of green lacewing *Chrysosa* (*Chrysoperla*) *scelestes* Banks (Neuroptera: Chrysopidae) an important predator in Gujarat. *Gujarat Agricultural University Research Journal*, 11: 18–25.

Patel, I.S. & Yadav, D.N. (1997). Cross resistance of monocrotophos -resistant strain of green lace wing. *Indian Journal of Agricultural Sciences*, 67: 83–84.

Patel, R.C., Yadav, D.N. & Patel, J.R. (1976). Natural control of groundnut aphid, *Aphis craccivora* Koch in Central Gujarat. *Current Science*, 45: 34–35.

Patel, I.S., Yadav, D.N. & Shukla, Y.M. (1996). Biochemical basis of mechanisms in a monocrotophos resistant strain of the predator, green lace wing, *Chrysopa scelestes* Banks. *Annals of Plant Protection Sciences*, 4: 151–153.

Pathan, S.N., Bukero, A., Nizamani, I A., Lanjar, A.G. et al. (2016). Influence of varying temperature on life stages of *Chrysoperla carnea* (Stephens) under laboratory conditions. *Journal of Basic & Applied Sciences*, 12: 388–393.

Patil, N.G. & Bhosale, P.V. (2007). Impact of sugarcane woolly aphid, *Ceratovacuna lanigera* Zehntner (Aphididae: Homoptera), on production and recovery of sugarcane and their natural enemies distribution in Western Maharashtra. *Indian Sugarcane*, 57: 31–38.

Patil, R.H. & Hallolli, S.P. (2007). Biocontrol potential of *Chrysoperla carnea* (Stephens) on *Melanaphis sacchari* (Zehntner) and *Uroleucon compositae* (Theobald). *Journal of Biological Control*, 21(Special): 163–166.

Patil, A.S., Magar, S.B. & Shinde, V.D. (2007). Biological control of the sugarcane Woolly Aphid (*Ceratovacuna lanigera*) in Indian sugarcane through the release of predators. *XXVI Congress, International Society of Sugarcane Technologists*, ICC, Durban, South Africa, 29 July to 2 August, 2007, pp. 797–804.

Patnaik, N.C. & Bhagat, K.C. (1989). Life history of predator, *Chrysopa orestes* Banks and effect of an artificial diet on adult longevity and fecundity. *Bulletin of Entomology*, 30: 143–147.

Patro, B. & Behera, M.K. (2002). Biology and feeding potential of *Chrysoperla carnea* (Stephens) on bean aphid. *Journal of Biological Control*, 16: 77–79.

Pessoa, L.G.A., Cavalcanti, R.S., Moino Junior, A. & Souza, B. (2005). Compatibility *between Beauveria bassiana* and predator *Chrysoperla externa* in the laboratory. *Pesquisa Agropecuária Brasileira*, 40, 617–619.

Peters, R.S., Meusemann, K., Petersen, M., Mayer, C., Wilbrandt, J., et al. (2014). The evolutionary history of holometabolous insects inferred *from* transcriptome-based phylogeny and comprehensive morphological data. *BMC Evolutionary Biology*, 14: 1–16.

Polis, G.A., Myers, C.A. & Holt, R.D. (1989). The ecology and evolution of intraguild predation: Potential competitors that eat each other. *Annual Review of Ecology and Systematics*, 20: 297–330.

Portilla, M., Luttrell, R., Snodgrass, G., Zhu, Y.C. & Riddick, E. (2017). Lethality of the entomopathogenic fungus *Beauveria bassiana* NI8 strain in *Lygus lineolaris* (Hemiptera: Miridae) and its possible impact on beneficial arthropods. *Journal of Entomological Science*, 52: 352–369.

Prasad, T.M., Ghorpade, S.A., Pokharkar, D.S., Bade, B.A & Chandele, A.G. (2005). Effect of biointensive pest management on sucking pests, bollworms and natural enemies in MECH 162 Bt cotton in Maharashtra. *Insect Environment*, 11(2): 64–66.

Price, P.W., Bouton, C.E., Gross, P., Mcpheron, B.A., Thompson, J.N. & Weise, A.E. (1980). Interaction among three trophic levels: Influence of plants on Interactions between insect herbivores and natural enemies. *Annual Review of Ecology and Systematics*, 11: 41–65.

Principi, M.M. & Canard, M. (1984). Life histories and behavior: Feeding habits. In: *Biology of Chrysopidae* (Eds. M. Canard, et al.). The Hague, The Netherlands: Dr. W. Junk Publishers, pp. 76–92.

Rabindra, R.J., Muthuswami, M. & Jayaraj, S. (1994). Feasibility of nuclear polyhedrosis virus and *Chrysoperla carnea* (Stephens) use in *Helicoverpa armigera* management on chick pea. *Pest Management and Economic Zoology*, 2: 31–34.

Rai, S., Singh, N.N. & Monobrullah, M. (2007). Efficacy of management modules against *Lipaphis erysimi* on mustard. *Annals of Plant Protection Sciences*, 15: 251–253.

Rajabaskar, D. (2007). Stage preference and predatory potential of the bug *Chrysoperla carnea* against jasmine leaf webworm *Nausinoe geometralis*. *Journal of Ecobiology*, 19: 97–99.

Rajaram, V., Mathirajan, V.G. & Krishnasamy, S. (2006). IPM in cotton under dry farming condition. *International Journal of Agricultural Sciences*, 2: 557–558.

Rajasekhar, D.W., Rachappa, V.H. & Awaknavar, J.S. (1999). Role of *Chrysoperla carnea* Stephens and insecticides in suppression of castor mite. *Insect Environment*, 4: 151.

Ramkumar, J., More, G.D., Rajkumar, M.B. & Thiruvarasan, S. (2005b). Biology of *Mallada boninensis* (Okamoto) on different stages of *Aleurocanthus woglumi* Ashby. *Journal of Applied Zoological Research*, 16: 17–18.

Ramkumar, J., More, G.D., Rajkumar, M.B. & Thiruvarasan, S. (2005a). Predation potential of *Mallada boninensis* (Okamoto) on different stages of *Aleurocanthus woglumi* Ashby. *Journal of Applied Zoological Research*, 16: 19–20.

Rao, R.S.N. & Chandra, I.J. (1985). Biological control of *Myzus persicae* Sulz. *Indian Journal of Agricultural Sciences*, 55: 654–656.

Rao, N.S. & Raguraman, S. (2005). Influence of neem based insecticides on egg parasitoid, *Trichogramma chilonis* and green lacewing predator, *Chrysoperla carnea*. *Journal of Ecobiology*, 17: 437–443.

Rao, R.S.N. & Satyanarayana, S.V.V. (1984). Note on more additions to the natural enemy complex of *Spodoptera litura* F. and *Myzus persicae* Sulz. on tobacco. *Current Science*, 53: 201.

Rao, N.V., Reddy, A.S. & Reddy, D.D.R. (1989). Natural enemies of cotton whtiefly, *Bemisia tabaci* Gennadius in relation to hsot population and weather factors. *Journal of Biological Control*, 3: 10–12.

Rao, R.S.N., Sreedhar, U., Rao, S.G. & Satyavani, J.V.R. (1994). Bio environmental control of *Heliothis armigera* in FCV tobacco. *Tobacco Research*, 20: 36–39.

Rao, G.M.V.P., Venkateswarlu, P., Raju, K. & Sivaramakrishna, M. (2007). Role of tall growing millet crops in the biological suppression of aphid, *Myzus nicotianae* Blackman in bidi tobacco. *Journal of Biological Control*, 21(Special): 55–58.

Reddy, G.V.P. & Guerro, A. (2000). Pheromone based integrated pest management to control the diamond back moth *Plutella xylostella* in cabbage fields. *Pest Management Science*, 56: 882–888.

Reddy, G.V.P. & Manjunatha, M. (2000). Laboratory and field studies on the integrated pest management of *Helicoverpa armigera* (Hubner) in cotton, based on pheromone trap catch threshold level. *Journal of Applied Entomology*, 124: 213–221.

Ríos-Moreno, A., Quesada-Moraga, E. & Garrido-Jurado, I. (2018). Treatments with *Metarhizium brunneum* BIPESCO5 and EAMa 01/58- Su strains (Ascomycota: Hypocreales) are low risk for the generalist predator *Chrysoperla carnea*. *Journal of Pest Science*, 91: 385–394.

Ruberson, J.R., Tauber, C.A. & Tauber, M.J. (1995). Developmental effects of host and temperature on *Telenomus* spp. (Hymenoptera: Scelionidae) parasitizing chrysopid eggs. *Biological Control*, 5: 245–250.

Rüzicka, Z. (1997). Protective role of the egg stalk in Chrysopidae (Neuroptera). *European Journal of Entomology*, 94: 111–114.

Sagar, P. & Kumar, N. (1996). Natural enemies of *Hyadaphis corandri* and feeding rate of *Coccinella septempunctata* on it in the Punjab. *International Journal of Pest Control*, 38: 26–27.

Saminathan, V.R., Mahadevan, N.R. & Muthukrishnan, N. (2003). Influence of prey density on the predatory potential and development of *Chrysoperla carnea*. *Indian Journal of Entomology*, 65(1): 1–6.

Sanjeev, R., Singh, N.N. & Monobrullah, M. (2007). Efficacy of management modules against *Lipaphis erysimi* on mustard. *Annals of Plant Protection Sciences*, 15: 251–253.

Sardana, H.R., Bambawale, O.M., Singh, D.K. & Kadu, L.N. (2006). Conservation of natural enemies through IPM in brinjal (*Solanum melongena* L.) fields. *Entomon*, 31: 83–88.

Satpute, P.P., Taley, Y.M., Nimbalkar, S.A. & Kadu, N.R. (1986). Natural enemies of citrus black fly, *Aleurocanthus woglumi* Ashby in Nagpur. *Bulletin of Entomology*, 27: 128–131.

Sattar, M. & Abro, G.H. (2009). Comparative effect of natural and artificial larval diets on biology of *Chrysoperla carnea* (Stephens) (Neuroptera: Chrysopidae). *Pakistan Journal of Zoology*, 41: 335–339.

Sattar, M., Fatima, B., Ahmed, N. & Abro, G.H. (2007). Development of larval artificial diet of *Chrysoperla carnea* (Stephens) (Neuroptera: Chrysopidae). *Pakistan Journal of Zoology*, 39: 103–107.

Sattar, M., Abro, G.H. & Syed T.S. (2011). Effect of different hosts on biology of *Chrysoperla carnea* (Stephens) in laboratory conditions. *Pakistan Journal of Zoology*, 43(6): 1049–1054.

Satyanarayana, J., Singh, K.M., Singh, R.N. & Singh, S. (1998). Influence of intercropping and insecticide, II. Natural enemy complex in rice bean *Vigna umbellata*. *Indian Journal of Entomology*, 60: 69–73.

Schilick, E.C., Toscano, L.C., Souza, G.D., Peres, A.J.A., Dias, P.M. & Maryama, W.I. (2015). Compatibility of *Metarhizium anisopliae* (Metschniko) Sorokin (Hypocreales: Clavicipitaceae) with *Chrysoperla externa* (Hagen) (Neuroptera: Chrysopidae). *EntomoBrasilis*, 8: 189–195.

Selvakumaran, S., Kallil, M. & Devasahayam, S. (1996). Natural enemies of *two* major species of scale insects infesting black pepper (Piper nigrum L.) in India. *Pest Management in Horticultural Ecosystem*, 2: 79–83.

Shah, V., Paul, B., Pandi, G.G.P. & Shankarganesh, K. (2013). Biology and predatory potential of green lacewing, *Chrysoperla sp.* (*carnea* group) on different aphid species. *Annals of Plant Protection Science*, 21(1): 9–12.

Shanmugam, P.S., Balagurunathan, R. & Sathiah, N. (2006). Biointensive integrated pest management in Bt cotton. *International Journal of Zoological Research*, 2: 116–122.

Sharanabasappa, H., Kulkarni, K.A., Mallapur, C.P. & Goud, K.B. (2008). Population dynamics of sugarcane woolly aphid, *Ceratovacuna lanigera* Zehntner and its natural enemies. *Annals of Plant Protection Sciences*, 16: 384–388.

Sharma, H.C., Arora, R. & Pampapathy, G. (2007). Influence of transgenic cottons with *Bacillus thuringiensis* cry1Ac gene on the natural enemies of *Helicoverpa armigera*. *BioControl*, 52: 469–489.

Silva, P.S., Albuquerque, G.S., Tauber, C.A. & Tauber, M.J. (2007). Life history of a widespread Neotropical predator, *Chrysopodes (Chrysopodes) lineafrons* (Neuroptera: Chrysopidae). *Biological Control*, 41: 33–41.

Singh, P.P. & Varma, G.C. (1989). Evaluation of the suitability of spent or dead *Corcyra cephalonica* (Pyralididae: Lepidoptera) moths as food for rearing *Chrysoperla carnea* (Chrysopidae: Neuroptera). *Biological Waste*, 28: 309–311.

Singh, S.P. & Narasimham, A.U. (1992). Indian chrysopidae. Technical Bulletin, Biological Control Centre, NCIPM, Faridabad, 34 pp.

Singh, J., Sohi, A.S., Dhaliwal, Z.S. & Mann, H.S. (1993). Comparative incidence of *Helicoverpa armigera* (Hub.) and other insect pests on okra and sunflower intercrops in cotton under Punjab conditions. *Journal of Insect Science*, 6: 137–138.

Singh, S.P., Bhummannavar, B.S., Jalai, S.K., Bakthavatsalam, N. & Pushpalatha, N.A. (1994a). Production and use of chrysopid predators. Technical Bulletin, Project Directorate of Biological Control, No 10: 14 p.

Singh, S.P., Bhummannavar, B.S., Jalai, S.K., Bakthavatsalam, N. & Pushpalatha, N.A. (1994b). Chrysopids and trichogrammatids: Strain selection and Utilization. Technical Bulletin, Project Directorate of Biological Control, No 9: 45 p.

Skirvin, D.J. & Fenlon, J.S. (2003). The effect of temperature on the functional response of *Phytoseiulus persimilis* (Acari: Phytoseiidae). *Experimental and Applied Acarology*, 31: 37–49.

Srivastava, R.C., Quereshi, Q.G. & Mathur, N.M. (1996). A new field application technology for *Chrysoperla carnea*. *Journal of Biological Control*, 10: 53–59.

Stange, L.A. (2004). A systematic catalog, bibliography, and classification of the world antlions (Insecta: Neuroptera: Myrmeleontidae). *Memoirs of the American Entomological Institute*, 74: 1–565.

Surulivelu, T. & Kumaraswami, T. (1989). Effect of skip row coverage of insecticides application on some sucking pests and their predators in cotton. *Journal of Biological Control*, 3: 77–79.

Suryawanshi, D.S., Bhede, B.V., More, D.G. & Zanwar, P.R. (2007). Management of insecticide resistance in cotton bollworms in Hingoli and Amravati districts of Maharashtra (India). *Resistant Pest Management Newsletter*, 16: 16–18.

Tanwar, R.K., Bambawale, O.M., Jeyakumar, P., Dhandapani, A., Vikas, K., Sharma, O.P. & Monga, D. (2007). Impact of IPM on natural enemies in irrigated cotton of North India. *Entomon*, 32: 25–32.

Tauber, M.J. & Tauber, C.A. (1983). Life history traits of *Chrysopa carnea* and *Chrysopa rufilabris* (Neuroptera: Chrysopidae): Influence of humidity. *Annals of the Entomological Society of America*, 76: 282–285.

Tauber, M.J. & Tauber, C.A. (2015). Phenological responses of *Pseudomallada* (Neuroptera: Chrysopidae): Comparative data from three Nearctic species and interspecific hybrids. *European Journal of Entomology*, 112: 49–62.

Tauber, M.J., Albuquerque, G.S. & Tauber, C.A. (1997). Storage of non diapausing *Chrysoperla externa* adults: Influence on survival and reproduction. *Biological Control*, 10: 69–72.

Tauber, M.J., Tauber, C.A. & Gardescu, S. (1993). Prolonged storage of *Chrysoperla carnea* (Neuroptera: Chrysopidae). *Enviromental Entomology*, 22: 843–848.

Tauber, M.J., Tauber, C.A., Daane, K.M. & Hagen, K.S. (2000). Commercialization of predators: Recent lessons *from* green lacewings (Neuroptera: Chrysopidae). *American Entomologist*, 46: 26–37.

Tauber, C.A., Tauber, M.J. & Albuquerque, G.S. (2003). Neuroptera (Lacewings, Antlions). In: *Encyclopedia of Insects* (Eds. V.H. Resh & R.T. Carde). New York: Academic Press, pp: 785–798.

Tauber, M.J., Tauber, C.A. & Hilton, T.W. (2006). Life history and reproductive behavior of the endemic Hawaiian *Anomalochrysa hepatica* (Neuroptera: Chrysopidae): A comparative approach. *European Journal of Entomology*, 103: 327–336.

Tauber, C.A., Tauber, M.J. & Albuquerque, G.S. (2009). Neuroptera (Lacewings, Antlions). In: *Encyclopedia of Insects* (Eds. V.H. Resh & R.T. Carde). New York: Academic Press, pp. 695–707.

Thakur, J.N., Pawar, A.D. & Rawat, U.S. (1988). Observations on the correlation between population density of apple wooly aphid and its natural enemies and their effectiveness in Kullu Valley (H.P). *Plant Protection Bulletin*, 40: 13–15.

Tjeder, B. (1960). Neuroptera from Newfoundland, Miquelon, and Labrador. *Opuscula Entomologica*, 25: 146–149.

Tóth, M., Szentkiraslyi, F., Vuts, J.D., Letardi, A., Tabilio, M.R., et al. (2009). Optimization of a phenylacetaldehyde-based attractant for common green lacewings (*Chrysoperla carnea* s.l.). *Journal of Chemical Ecology*, 35: 449–458.

van Lenteren, J.C. (2012). The state of commercial augmentative biological control: Plenty of natural enemies, but a frustrating lack of uptake. *BioControl*, 57: 1–20. doi: 10.1007/s10526-011-9395-1.

Varma, G.C. & Shenhmar, M. (1983). Some observations on the biology of *Chrysoperla carnea* (Stephens) (Chrysopidae: Neuroptera). *PAU Agricultural Research Journal*, 20: 222–223.

Vasquez, E.E., Schuster, D.J. & Hunter, W.B. (1998). Alternative method for encapsulation of artificial diet used in rearing *Ceraeochrysa cubana* (Hagen) larvae (Neuroptera: Chrysopidae). *Journal of Entomological Science*, 33: 316–318.

Venkatesan, S., Balasubramanian, G., Babu, P.C.S. & Sivaram, M.R. (1997). Use of nuclear polyhedrosis virus and green lacewing, *Chrysoperla carnea* Stephens for *Helicoverpa armigera* (Huebner) management on sunflower. *Pest Management and Economic Zoology*, 5: 63–66.

Venkatesan, T., Poorani, J., Murthy, K.S., Jalali, S.K., Kumar, G.A., Lalitha, Y. & Rajeshwari, R. (2008). Occurrence of *Chrysoperla zastrowi arabica* Henry et al. (Neuroptera: Chrysopidae), a cryptic song species of *Chrysoperla (carnea* group), in India. *Journal of Biological Control*, 22: 143–147.

Venkatesan, T., Jalali, S.K., Murhty, K.S., Rabindra, R.J. & Lalitha, Y. (2009). Occurrence of insecticide resistance in field populations of *Chrysoperla zastrowi arabica* (Neuroptera; Chrysopidae) in India. *Indian Journal of Agricultural Sciences*, 79: 910–912.

Vennila, S. (1998). Relationship between sucking pests *Amrasca biguttula biguttula*, *Aphis gossyii*, and their predators (*Cheilomenes sexmaculata*, *Chrysoperla carnea*) on cotton cultivars. *Journal of Entomological Research*, 22: 349–353.

Vennila, S., Biradar, V.K. & Panchbhai, P.R. (2007). Coccinellids and chrysopids as native predators of sucking pests in relation to rainfed cotton production system. *Journal of Biological Control,* 21(Special): 65–71.

Vidya, M., Ramegowda, G.K., Patil, R.K. & Lingappa, S. (2008a). Studies on ovipositional preference of brown lacewing, *Micromus igorotus* Banks (Neuroptera: Hemerobiidae). *Journal of Entomological Research*, 32: 135–137.

Vidya, M., Patil, R.K., Ramegowda, G.K. & Lingappa, S. (2008b). Record of brown lacewing, *Micromus australis* Hagen (Neuroptera: Hemerobiidae) from cotton and sorghum ecosystem. *Journal of Biological Control*, 22: 189–190.

Volkovich, T.A. (1998). Environmental control of seasonal cycles in green lacewings (Neuroptera, Chrysopidae) from the forest-steppe zone of Russia. *Acta Zoologica Fennica*, 209: 263–275.

Volkovich, T.A. (2006). Seasonal development of the lacewing *Chrysopa dorsalis* Burmeister (Neuroptera, Chrysopidae) in the forest-steppe zone of Russia. *Entomological Review*, 86: 741–750.

Volkovich, T.A. & Arapov, V.V. (1996). Features of the photoperiodic response of the lacewing *Chrysopa perla* L. (Neuroptera, chrysopidae) at constant and variable temperatures. *Entomological Reviews*, 75: 1–16.

Volkovich, T.A. & Blumental, N.A. (1997). Photo-thermo periodic responses in some species of lacewings (Chrysopidae): Their role in diapause induction. *European Journal of Entomology*, 94: 435–444.

Wadhwa, S. & Gill, R.S. (2007). Effect of Bt-cotton on biodiversity of natural enemies. *Journal of Biological Control*, 21: 9–16.

Wang, R. & Nordlund, D.A. (1994). Use of *Chrysoperla* spp. (Neuroptera: Chrysopidae) in augmentative release programmes for control of arthropods pests. *Biocontrol News and Information*, 15: 51N–57N.

Wells, P.M., Baverstock, J., Clark, S.J., Giggins, F.M., Roy, H.E. & Pell, J.K. (2017). Determining the effects of life stage, shared prey density and host plant on intraguild predation of a native lacewing (*Chrysoperla carnea*) by an invasive coccinellid (*Harmonia axyridis*). *Biocontrol*, 62: 373–384.

Winterton, S. & Freitas, S. (2006). Molecular phylogeny of the green lacewings (Neuroptera: Chrysopidae). *Australian Journal of Entomology*, 45: 235–243.

Winterton, S.L. & Makarkin, V.N. (2010). Phylogeny of moth lacewings and giant lacewings (Neuroptera: Ithonidae, Polystoechotidae) using DNA sequence data, morphology, and fossils. *Annals of the Entomological Society of America*, 103: 511–522.

Winterton, S.L., Hardy, N.B., & Wiegmann, B.M. (2010). On wings of lace: Phylogeny and Bayesian divergence time estimates of Neuropterida (Insecta) based on morphological and molecular data. *Systematic Entomology*, 35: 349–378.

Xu, Y.Y., Mu, J.Y., Hu, C. & Wang, H.G. (2004). Photoperiodic control of adult diapause in *Chrysoperla sinica* (Tjeder) (Neuroptera: Chrysopidae) - I. Critical photoperiod and sensitive stages of adult diapause induction. *Entomologia Sinica*, 11: 191–198.

Ye, J., Li, J., Li, Z. & Han, S. (2017). Rearing of *Mallada basalis* (Neuroptera: Chrysopidae) on modified artificial diets. *PLoS One*, 12(9): e0185223. doi: 10.1371/journal.pone.0185223.

Zade, S.M., Satpute, R.S., Bhagat, S.M. & Baheti, H.S. (2007). Predatory efficacy of *Mallada boninensis* (Okamoto) against *Aleurocanthus woglumi* Ashby on citrus. *International Journal of Agricultural Sciences*, 3: 164–166.

Zaki, F.N. & Gesraha, M.A. (2001). Production of the green lacewing *Chrysoperla caranea* (Steph.) (Neuropt., Chrysopidae) reared on semi-artificial diet based on the algae, *Chlorella vulgaris*. *Journal of Applied Entomology*, 125: 97–98.

Zarei, M., Madadi, H., Zamani, A.A. & Nedved, O. (2020). Intraguild Predation *between Chrysoperla carnea* (Neuroptera: Chrysopidae) and *Hippodamia variegate* (Coleoptera: Coccinellidae) at various extraguild prey densities and arena complexities. *Insects*, 11: 288. doi: 10.3390/insects11050288.

Zhang, Q.H., Chauhan, K.R., Erbe, E.F., Vellore, A.R. & Aldrich, J.R. (2004). Semiochemistry of the goldeneyed lacewing *Chrysopa oculata* (Neuroptera: Chrysopidae): Attraction of males to a male produced pheromone. *Journal of Chemical Ecology*, 30: 1849–1870.

Zhang, Q.H, Schneidmiller, R.G., Hoover, D., Young, K., Welshons, D., et al. (2006a). Male-produced pheromone of the green lacewing, *Chrysopa nigricornis* (Neuroptera: Chrysopidae). *Journal of Chemical Ecology*, 32: 2163–2176.

Zhang, Q.H., Sheng, M., Chen, G., Aldrich, J.R. & Chauhan, K.R. (2006b) Iridodial: A powerful attractant for the green lacewing, *Chrysopa septempunctata* (Neuroptera: Chrysopidae). *Naturwissenschaften*, 93: 461–465.

Zvereva, E.L. & Rank, N.E. (2003). Host plant effects on parasitoid attack on the leaf beetle *Chrysomela lapponica*. *Oecologia*, 135: 258–267.

7 Ground Beetles (Coleoptera: Carabidae)

Kolla Sreedevi and G. Mahendiran
ICAR-National Bureau of Agricultural Insect Resources
P. Sree Chandana
ICAR-Indian Institute of Horticultural Research

CONTENTS

LEARNING OBJECTIVES

1. Ground beetles' systematic position and their economic importance
2. Diagnostics features of Carabidae and important subfamilies of ground beetles
3. Important genera of ground beetles
4. Conservation of ground beetle or carabid species

DOI: 10.1201/9781003370864-7

7.1 INTRODUCTION

Ground beetles are one of the largest groups of beneficial insects belonging to the family Carabidae of the super family Caraboidea and the order Coleoptera with more than 40,000 species known worldwide (Zhang, 2011) and around 2,000 species in India (Saha et al., 1992).The adult beetles are very active, fast running and much diverse in their shape and size. The beetles have tough elytra that appear soldiered and possess grooves or row of punctures along the elytral length. The ground beetles are poor fliers and occasionally fly to migrate from one place to the other. These are mostly annuals having one generation per year but few are biennials and a few need 2–3 years to complete their generation. The ground beetles possess long legs and powerful mandibles, which make them voracious predators and important biocontrol agents in crop ecosystems (Snyder, 2019). Most of the beetles are nocturnal, active only during nights, and emerge from their hiding places in search of prey as soon as the light fails. A few beetles are diurnal and prey on insects and other arthropods during day time. The ground beetles' activities are at peak during April to October. The size of the beetles ranges from 0.7 to 66 mm (https://bugguide.net/node/view/186), and more generally it is between 2 and 35 mm. They are solitary in nature and congregate only during the mating period.

The adult beetles are important generalist predators feeding on insects, snails and slugs. The immature stages, larvae of these ground beetles, are also important predators and devour their prey with their pincher-like mandibles. The larvae remain below ground and sometimes in the dead leaves and in the surface soil. Most of these ground beetles are beneficial insects where the adults and larvae feed upon insect eggs, caterpillars, wireworms, maggots, grubs, ants, aphids and other soil-dwelling organisms including snails and slugs. Most of the time, the larvae and adults have the same diet; however, during certain times, the diet of larva may change depending upon the habitat it lives and is limited to the underground search range. The adult beetles consume up to their body weight daily.

7.2 HISTORY

The ground beetles (Carabidae) belong to the suborder Adephaga of Coleoptera and Adephagan insects are tracked back to 250 million years ago towards the end of the Permian era. Ground beetles are believed to have evolved in the latter Triassic period and diversified during the Jurassic era. The most extremely diverse and advanced lineage, the subfamily Harpalinae of Carabidae (which covers nearly half of the described species of ground beetles) expanded in the Cretaceous period. The phylogeny of the ground beetles has been worked out and it is understood that it consists of more basal lineages and the subfamily Harpalinae included several formerly considered independent families (Ōsawa et al., 2004).

7.3 SPECIES DIVERSITY

The ground beetles are one of the most species-rich group of soil-dwelling arthropods in agricultural and horticultural ecosystems. In 2017, it was known that the Carabidae comprises 40,000 valid species under 1,500 genera (Bouchard et al., 2017), classified under 86 tribes (Erwin, 1985). According to Anichtchenko (2007–2021), Carabidae comprises 40,227 species under 2,118 genera represented by 34 subfamilies, 92 tribes and 150 subtribes worldwide. In India, about 2,000 carabid species have been reported (Saha et al., 1992). They are the better-known taxa of soil fauna and cosmopolitan in distribution (Plate 7.1).

7.4 DIAGNOSTIC FEATURES OF CARABIDAE

The beetles can be distinguished from other groups of insects or arthropods by their hardened outer wings called elytra and well developed chewing type of mouth parts, which are characteristic features of the order Coleoptera.

PLATE 7.1 Diversity of ground beetles.

The ground beetles belong to the suborder Adephaga of Coleoptera. Earlier these were also known as Carnivora and Caraboidea.

In general, most adult beetles are black or dull brown in colour though few are iridescent with blue or green colours.

7.4.1 General Characters of the Family Carabidae Latreille, 1802

1. Antennae generally filiform, 11 segmented
2. Clypeus normal and not extending laterally before bases of antennae
3. Metasternum with a transverse suture before posterior coxae
4. The transverse suture extending across metasternum and the metasternum produced between the posterior coxae as an angular process
5. Visible abdominal ventrites to be six or seven, the first three connate (appear together) though the sutures are apparent
6. Hind wings with an oblongum
7. A groove in fore tibiae bearing a comb of hairs typically used for antennal cleaning (except in flanged bombardier beetles), Posterior coxae normal without any expansion, Tarsi 5-jointed
8. Larvae campodeiform

The false ground beetles or small ground beetles, which look similar to ground beetles, belong to the family Trachypachidae Thomson 1857. The closest living relatives of the ground beetles are false ground beetles and these beetles can be distinguished from ground beetles by their large coxae of hind legs and metaepisternum reaching the mesocoxal cavities. There are only six known living species of Trachypachidae.

Carabidae comprises two major subfamilies, Carabinae and Harpalinae, which are speciose. The two subfamilies, Carabinae and Harpalinae may be distinguished by the mesocoxal cavities, which are not entirely closed by the sterna in the former and fully closed by the sterna in the latter. Also by the mesepimera, which reaches the coxa in the former and the mesepimera do not reach the coxa in the latter (Andrews, 1929). The species under Harpalinae are found to be numerous in India than Carabinae.

Another subfamily of Carabidae, Brachininae, which occurs worldwide and India can be distinguished by its pubescent body, narrow head and prothorax, wide truncate elytra and apically protruding abdomen.

The distinguishing keys for the different tribes and genera have been provided by several researchers, the base one being the work by Andrews (1929).

7.5 BIOLOGY OF CARABIDS

Ground beetles are found in almost all habitats but mostly on moist stones, rocks, log of wood, moss, etc. where they take shelter during the day time. Certain species of the genus *Colpodes* live in the foliage (Andrews, 1929). They are most diverse and common in forests, but can be found on high mountains, in deserts and even on the seashores. The species of the genera *Bembidion* and *Tachys* are found to be dwelling along the banks of streams and lakes or in moist localities (Andrews, 1929) while the species of genus *Omophron* inhabit semiaquatic areas.

The ground beetles are holometabolous and hence have egg, larva, pupa and adult stages. The adult female beetle lays around 100–600 small, white and oval-shaped eggs in the soil or on leaves generally near a food source for larva to thrive upon. The female beetles feeding on diverse food types like insect eggs, caterpillars, nymphs and adults of aphids, etc. tend to lay more number and bigger size of eggs (Lövei and Sunderland, 1996). Parental care and egg guarding have been recorded in certain species. In about 7–10 days, the larvae hatch out from the eggs, which look like centipedes with black or brown segmented bodies. These larvae can be distinguished by their six thoracic legs and pincher-like mandibles on their heads. The larva passes through three instars and lives in debris or burrows in the soil. Pupation takes place in the soil or in the leaf debris. The duration of the life cycle varies with the species. The adults' life span varies from 1 to 4 years, the larger species and those that overwinter as larvae tend to have the longest life span (Lövei and Sunderland, 1996). The majority of species are reported to overwinter as adults in the soil or in sheltered sites.

7.5.1 Defence in Ground Beetles

The adult beetles possess a pair of pygidial glands in the lower back of the abdomen, which produce aversive and deadly secretions that deter the predators. The ground beetles of subfamily Brachininae of Carabidae are commonly known as "Bombardier beetles" (Fedorenko, 2013) because of their secretions as a defence mechanism. In species of bombardier beetles, these secretions are mixed up with certain volatile compounds and ejected with force as an explosive to injure and drive away its predators. In the case of *Anthia,* the adult beetles spray a jet of formic acid, which eventually causes blindness in the predator, if not treated timely. A few species have also been recorded to detect chemical cues from springtails, molluscs and aphids (Lövei and Sunderland, 1996). The species of *Calosoma,* the largest beetle of the group, produces a foul-smelling spray from their abdominal glands to deter the predators. The genus *Omphra* discharges fluid, which volatilizes on contact with the air and the emission is visible like smoke and also audible (Andrews, 1929).

7.5.2 Feeding Habits of Ground Beetles

As discussed under economic importance, the ground beetles are beneficial in many ways. Certain species feed on the organisms that come near them, a few hunt by sight and a few others climb the trees in search of prey and devour the prey. Based on the food preferences, the ground beetles can be grouped into carnivores, granivores and omnivores. Most of the carabids are generalist feeders feeding on whatever is available to them and only a few are specialized feeders. The generalist feeders are the most beneficial organisms in the agricultural and horticultural ecosystems as they consume the insect pest species and also weed seeds.

7.5.2.1 Carnivores

These species feed primarily on insects, their caterpillars, snails and slugs. *Calosoma* species, along with *Carabus nemoralis,* is being utilized in gypsy moth control. Certain species like *Scaphinotus* and *Spaeroderus* have specialized mouthparts to dig snails from their shells and feed on them. They have been used effectively to control slugs in greenhouses (Kromp, 1999), e.g., *Calosoma*, *Carabus*, *Bembidion* spp.

7.5.2.2 Granivores

These are generally called as seed predators that feed on seeds of weed species and are of great importance on North American farms as both larvae and adult beetles consume seeds, e.g., *Amara* and *Harpaluss* pp.

7.5.2.3 Omnivores

These are the most diverse feeders consuming insects, snails, slugs, weed seeds, etc. These beetle species can be few in number and act as both carnivores and granivores, e.g., *Pterostichus* and *Poecilus* spp.

7.6 ECONOMIC IMPORTANCE

The carabids or ground beetles are economically important for their role as predators of agricultural pests. Both larvae and adults are reported to be carnivorous (Saha et al., 1992), thus the ground beetles form an important component of biological control in the pest management alleviating the problems associated with chemical pesticide usage, especially hazards and residues. The use of biocontrol agents reduces the cost of plant protection and enhances the cost–benefit ratio. Clark et al. (1994) reported that the generalist predators collectively including carabid beetles can save the crop loss by up to 40%.Utilizing different species of insect predators at different times of the year will strengthen biological control (Tooley and Brust, 2002).

These beetles associated with almost all kinds of terrestrial habitats like field crops, horticultural crops, forest, etc. play an important role in providing ecosystem services. In this group, majority of them are generalist/polyphagous predators of arthropod pests in agro ecosystems (Kromp, 1999) and very few are reported as pests (herbivores) and some as granivores (Hurka and Jarosik, 2001). As granivores, few species of carabids feed on weed seeds in their adult stage and thus contribute to weed suppression. It has been suggested that both herbivory and granivory traits are largely underestimated in ground beetles (Tooley and Brust, 2002). A more number of carabid species in North America have been identified as weed seed predators, which were reported to consume a wide range of agriculturally important weed species seeds including common ragweed, common lambsquarters and giant foxtail (Lundgren, 2005). Certain species of ground beetles are found to synchronize their breeding period with the ripening of weed seeds as in the case of *Harpalus pensylvanicus,* which synchronizes its activity with the maturity of foxtail seeds (Tooley and Brust, 2002) and has been observed to feed on seeds that have been buried up to about 10 mm deep in the soil. Thus, the ground beetles become important in regulating the weed population (Liebman and Gallandt, 1997) apart from being efficient biocontrol agents. Carabids are also a very important bioindicator for ecological, environmental and climate changes as well as habitat qualities. Certain species of ground beetles exhibit habitat specificity and hence those species are used as bioindicators to assess land-use changes among different ecosystems.

7.7 IMPORTANT GENERA OF GROUND BEETLES

7.7.1 *Anthia* Weber, 1801

The beetles under this genus are fairly large, active, fast-moving, with prominent and sharp mandibles. Both nocturnal and diurnal species exist in this genus and diurnal predators are found more in semi-arid habitats. The genus comprises 50+ species worldwide. These prey on all types of insect eggs, aphids, caterpillars, etc. E.g., *Anthia sexguttata* –a six-spotted ground beetle commonly found in all habitats (Figure 7.1).

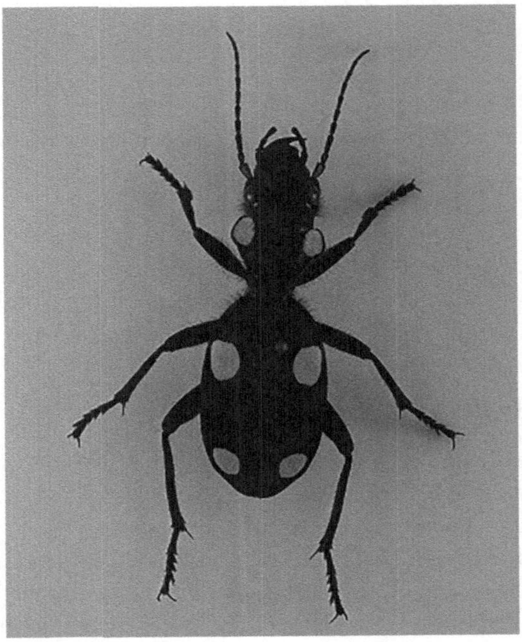

FIGURE 7.1 *Anthia sexguttata.*

7.7.2 *Amara* Bonelli, 1810

The species under this genus are medium sized ranging from 5 to 11 mm, mostly measuring 6–9 mm in size. These beetles are oval or pill shaped (Lindroth, 1961) and many species have a metallic sheen with coppery reflection. The ground beetles of this genus are both diurnal and nocturnal and predominantly weed seed feeders, a few feeding on grasshopper eggs (Blatchley, 1910). *Amara* spp. has been associated with general seed grain suppression (Bohan et al., 2011). *Amara* is found to be more abundant on farms during early summer. The genus comprises 600+ species under 47 subgenera worldwide (Bousquet, 2012). E.g., *Amara bifrons* (Gyllenhal, 1810).

7.7.3 *Bembidion* Latreille, 1802

It is the largest genus of the Carabidae comprising several species (Lindroth, 1985). The ground beetles of this genus are small sized measuring <8.5 mm and are swift movers (Lindroth, 1961). The species of this genus are good fliers and include both diurnal and nocturnal species (Luff, 1978). These form an excellent group of beneficial insects as they can chase and hunt and escape owing to their flight movement. Since these are small in size, they are more likely to be eaten by soil-dwelling predators like large ground beetles, spiders, etc. (Prasad and Snyder, 2004, 2006). The species of this genus are excellent predators of eggs and pupae of insect pests besides caterpillars, especially of borers and armyworm. E.g., *Bembidion quadrimaculatum* (Linnaeus, 1761): small-sized beetle measuring 3 mm, Elytra black to mid brown with four distinctive yellow spots and a constriction between the thorax and abdomen.

7.7.4 *Carabus* Linnaeus, 1758

This is the largest genus of subfamily Carabinae of Carabidae comprising more than 1,000 known species under 91 subgenera worldwide (Deuve et al., 2012). They are among the largest of ground beetles, ranging in size from 10 to 50 mm (Lindroth, 1961). The beetles have dark colouration with some species exhibiting iridescent and brighter coloration. Elytra possess distinctive deep punctuation. Owing to their large size, they prey upon slugs and late instar caterpillars (Hatteland et al., 2010, 2011) and are in an advantageous position than the smaller ground beetles in predation (Chaudhary et al., 2016). The species of this genus are nocturnal and get easily confused with another large genus, *Calosoma*. E.g., *Carabus coriaceus* Linnaeus, 1758.

7.7.5 *Calosoma* Weber, 1801

These are the large-sized ground beetles like *Carabus* but easily recognizable by their wider thorax, which is wider than the head and almost the size of abdomen. The species in the genus *Calosoma* are commonly called as caterpillar hunters. The genus is diverse and usually dark in colour, and for more information on this genus, book by Bruschi (2013) titled *Calosoma of the world: Coleoptera, Carabidae* may be referred. The adults and larvae are very active generalist predators. In 1905, *Calosoma sycophanta* was imported from Europe to New England to bring about the control of the gypsy moth. It has been reported that the larvae and adults of *C. sycophanta* consumed several hundred caterpillars. E.g., *Calosoma maderae* (Fabricius, 1775).

7.7.6 *Harpalus* Latreille, 1802

The species of this genus are medium-to large-sized beetles varying from 5.8 to 25.5 mm in size. The genus comprises more than 500 species worldwide. They are uniformly dark in coloration, occasionally metallic pale green and brown, with stout bodies and short legs (Lindroth, 1961). They resemble genus *Pterostichus* due to their coloration, but can easily be distinguished based on their body size and robustness. Species of this genus are found in almost all habitats but they mostly occur in open, dry and sandy areas and are nocturnal in habit (Luff, 1978). The adults of this genus are mainly seed feeders primarily on myriad weed seeds deposited in the soil; however, certain other species predate on insect pests both in their adult and larval stages (El-Danasoury et al., 2017);e.g., *Harpalus rufipes* De Geer (1774).

7.7.7 *Pterostichus* Bonelli, 1810

The species of this genus are mediumto large-sized ground beetles ranging from 12 to 28 mm in size. The genus comprises more than 1,000 species worldwide. Most of the species are dark coloured, black and a few may have metallic sheen . Most of the species are nocturnal while a few are diurnal (Luff, 1978). The species of this genus are potential biological control agents of slugs and other pests (Hatteland et al., 2010). Certain species are capable of climbing plants to lower heights and feed on foliar pests like aphids directly (Snyder and Ives, 2001, 2003). The large-sized species of this genus prey on smaller ground beetles thus ensuring biocontrol (Prasad and Snyder, 2006). Very few species of the genus are phytophagous and are reported to feed on strawberry. The species emerge in early summer and remain active through the autumn before overwintering. E.g., *Pterostichus melanarius* (Illiger, 1798).

For more details, please visit www.thewcg.org.uk/Carabidae/0118G.htm.

7.7.8 Genera Commonly Found in India

The subfamilies, Harpalinae and Carabinae, are speciose followed by Brachininae. The subfamily Brachininae comprises 14 genera under 2 tribes (Crepidogastrini and Brachinini) worldwide (Lorenz, 2016). Of these two tribes, Crepidogastrini is represented by a single genus, *Crepidogaster*, and Brachinini is represented by 4 genera and around 65 species in India.

The common genera found in India are *Brachinus*, *Bradybaenus*, *Chlaenius*, *Crepidogaster*, *Drypta*, *Eudema*, *Microcosmus*, *Ophionea*, *Pheropsophus*, *Scarites*, etc. (Andrews, 1929; Kushwaha et al., 2015); Fletcher (1919) in Andrewes (1929), Madhukar and Archana (2018), Chanu and Swaminathan (2017), Mili et al. (2018) and Akhil and Thomas (2018) (Plate 7.2).

7.8 CONSERVATION OF BENEFICIAL GROUND BEETLES

Conserving ground beetles through habitat manipulations and cultural practices can enhance the natural regulation of arthropod pest and weed populations, reducing the need for chemical controls.

These important beneficial insects can be conserved by adopting simple farm management practices like avoiding frequent tillage, not using broad-spectrum soil insecticides or fumigants and avoiding disruptive gardening practices. Their survival can be enhanced by providing refuges like growing tall grasses (Hajek and Eilenberg, 2018) and overwintering sites such as hedgerows or mulches that increase the abundance and diversity of ground beetles.

The mulches and hedge rows serve as beetle banks that not only enable movement of ground beetles to nearby farm areas but also nurture alternative prey for the ground beetles in times of low pest prey densities (MacLeod et al., 2004). This may be as simple as straw mulch, which provides humid, sheltered hiding places for nocturnal predators. Mulches can be with plants like red clover and white clover that are grown in between crop rows to provide an ideal microclimate and attract huge numbers of ground beetles (Blubaugh et al., 2016).These beetle banks have been shown to enhance biological pest control.

Another good practice can be facilitating suitable habitat and/or non-crop refuge habitats that nurture diverse food sources for the ground beetles to thrive well. These habitats can either be field edges or areas within fields that are undisturbed and accommodate ground beetles. Leaving strips of sown mixed grass seeds in between also enables a diverse flora that encourages a wide range of

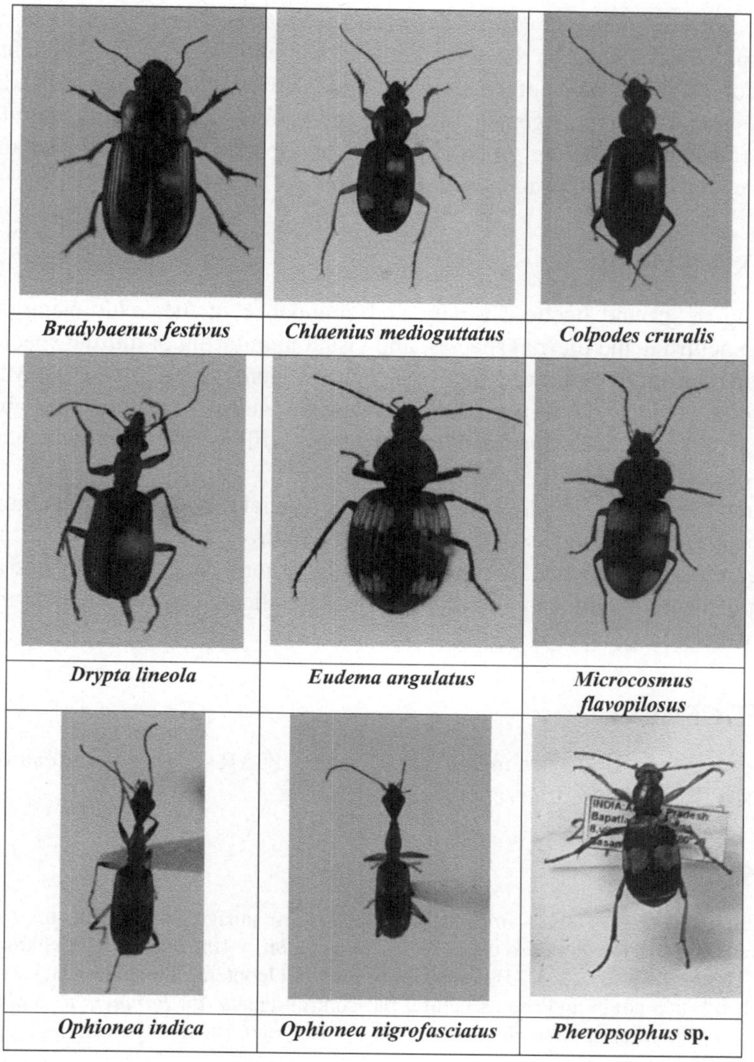

PLATE 7.2 Common genera of ground beetles found in India.

ground beetle species. Care should be taken with respect to tillage practices, avoidance of pesticide applications and burning in the fields. Minimum tillage operations have been found effective in building up of ground beetle communities (Shearin et al., 2014).

Habitat management with compost being used around the plants will aid in an increase of carabid populations compared to bare soil, which again provides a mulch effect and its benefits (Renkema et al., 2010). The organic farming that nurtures good biodiversity than conventionally managed fields also benefits ground beetles (Tuck et al., 2014; Caro et al., 2016; Adhikari and Menalled, 2018). Establishment and management of suitable and beneficial habitats with more natural flora encourage the ground beetles abundance and diversity.

7.9 CONCLUSION

Ground beetles belonging to Carabidae of the order Coleoptera are important beneficial insects that bring about biocontrol of crop pests and form an important component of Integrated Pest Management. Most of the species are generalist predators preying upon a wide range of crop pests including the recently introduced fall armyworm and are farmer friends in the fields. Conservation of these predatory beetles reduces the pesticide load and enhances the cost–benefit ratio besides several other ecological and environmental advantages. Conservation of ground beetles depends on their responsiveness to farm management practices and habitat manipulations. Regular monitoring through pitfall traps and promoting ground beetle population build-up in cropped and forest areas will aid in the regulation of insect and weed population.

POINTS TO REMEMBER

1. Monitoring of ground beetle diversity and abundance in the fields using pitfall traps (for both nocturnal and diurnal species) and visual monitoring of diurnal species.
2. Utilization of ground beetles as effective biocontrol agents in Integrated Pest Management.
3. Identification of bioindicator species of ground beetles and assessment of the habitat quality.
4. Avoidance of destructive farm practices like intensive tillage, clean cultivation, indiscriminate use of pesticides, etc.
5. The ground beetles that are predators of other beneficial insects need to be monitored and their populations regulated as they can negate biocontrol.
6. Effective habitat management for encouraging the ground beetle diversity and abundance.
7. Growing of beetle banks and adopting farming practices like mulches and mixed floral strips in the field.

ACKNOWLEDGEMENT

The authors express their sincere thanks to the Director, ICAR –National Bureau of Agricultural Insect Resources, Bengaluru for the support extended.

REFERENCES

Adhikari, S. & Menalled, F. D. (2018). Impacts of dryland farm management systems on weeds and ground beetles (Carabidae) in the Northern Great Plains. *Sustainability* 10:2146. doi: 10.3390/su10072146.

Akhil, S. V. & Thomas, S. K. (2018). Bombardier beetles (Coleoptera: Carabidae: Brachininae) of India: Notes on habit, taxonomy and use as natural bio-control agents. In: *Frontiers in Biological Research, Proceedings of the National Seminar*, Department of Zoology, University of Calicut, pp. 10–25. ISBN: 978-93-5321-240-7.

Andrews, H. E. (1929). The Fauna of British India including Ceylon & Burma (Coleoptera, Carabidae). *Carabinae* 1:1–383.

Anichtchenko, A. (2007–2021). Carabidae of the World. Available at: http://carabidae.org (accessed 5 March 2021).

Blatchley, W. S. (1910). *An Illustrated Descriptive Catalogue of the Coleoptera or Beetles (Exclusive of the Rhynchophora) Known to Occur in Indiana*. Nature Pub. Co: Indianapolis.

Blubaugh, C. K., Hagler, J. R., Machtley, S. A. & Kaplan, I. (2016). Cover crops increase foraging activity of omnivorous predators in seed patches and facilitate weed biological control. *Agriculture, Ecosystems & Environment* 231:264–270. doi: 10.1016/j.agee.2016.06.045.

Bohan, D. A., Boursault, A., Brooks, D. R. & Petit, S. (2011). National-scale regulation of the weed seedbank by carabid predators. *Journal of Applied Ecology* 48:888–898. doi: 10.1111/j.1365-2664.2011.02008.x.

Bouchard, P., Smith, A. B. T., Douglas, H., Gimmel, M. L., Brunke, A. J. & Kanda, K. (2017). Biodiversity of coleopteran. In: Foottit, R. G. & Adler, P. H. (Eds.), *Insect Biodiversity: Science and Society*, Volume I, Second Edition. John Wiley & Sons: Hoboken, NJ, 337–417 pp.

Bousquet, Y. (2012). Catalogue of Geadephaga (Coleoptera, Adephaga) of America, north of Mexico. *ZooKeys* 245:1–1722. doi: 10.3897/zookeys.245.3416.

Bruschi, S. (2013). *Calosoma of the World (Coleoptera, Carabidae)*. Natura Edizione Scienttifiche: Bologna.

Caro, G., Marrec, R., Gauffre, B., Roncoroni, M., Augiron, S. & Bretagnolle, V. (2016). Multi-scale effects of agri-environment schemes on carabid beetles in intensive farmland. *Agriculture, Ecosystems & Environment* 229:48–56. doi: 10.1016/j.agee.2016.05.00.

Chanu, N. Y. & Swaminathan, R. (2017): Notes on species of the genus Chlaenius Bonelli, 1810 (Coleoptera: Carabidae, Chlaeniini) from four agro-climatic zones of Rajasthan, with description of two new species. *Zootaxa* 4237(3):545–566. doi: 10.11646/zootaxa.4237.3.5.

Chaudhary, D. D., Kumar, B. & Mishra, G. (2016). Food resource exploitation in ladybirds: Consequences of prey species and size. *Current Science* 110:1343–1349. https://www.jstor.org/stable/24908024?seq=1.

Clark, M. S., Luna, J. M., Stone, N. D. & Youngman, R. R. (1994). Generalist predator consumption of armyworm (Lepidoptera: Noctuidae) and effect of predator removal on damage in no-till corn. *Environmental Entomology* 23:617–622.

Deuve T., Cruaud, A., Genson, G., & Rasplus, J. Y. (2012). Molecular systematics and evolutionary history of the genus Carabus (Col. Carabidae). *Molecular Phylogenetics and Evolution* 65(1):259–275. doi: 10.1016/j.ympev.2012.06.015.

Divya, M. & Thomas, S. K. (2020). Checklist of Indian Pterostichinae Bonelli, 1810 (Carabidae). *Oriental Insects*. doi: 10.1080/00305316.2020.1786476.

El-Danasoury, H., Cerecedo, C., Córdoba, M. & Iglesias-Piñeiro, J. (2017). Predation by the carabid beetle *Harpalus rufipes* on the pest slug *Deroceras reticulatum* in the laboratory. *Annals of Applied Biology* 170:251–262. doi: 10.1111/aab.12337.

Erwin, T. L. (1985). The taxon pulse: A general pattern of lineage radiation and extinction among carabid beetles. *Taxonomy, Phylogeny, and Zoogeography of Beetles and Ants* 30:437–488.

Fedorenko, D. N. (2013). New species of bombardier beetles of the genera *Brachinus* and *Pheropsophus* (Coleoptera: Carabidae: Brachininae) from Vietnam. *Zoosystematica Rossica* 22(2):271–284.

Fletcher, T. B. (1919). Second hundred notes. *Agricultural Research Institute Pusa Bulletin* 89:31–35.

Hajek, A. E. & Eilenberg, J. (2018). *Natural Enemies: An Introduction to Biological Control*. Cambridge University Press: Cambridge.

Hatteland, B. A., Grutle, K., Mong, C. E., Skartveit, J., Symondson, W. O. C. & Solhøy, T. (2010). Predation by beetles (Carabidae, Staphylinidae) on eggs and juveniles of the Iberian slug *Arion lusitanicus* in the laboratory. *Bulletin of Entomological Research* 100:559–567. doi: 10.1017/S0007485309990629.

Hatteland, B. A., Symondson, W. O. C., King, R. A., Skage, M., Schander, C. & Solhøy, T. (2011). Molecular analysis of predation by carabid beetles (Carabidae) on the invasive Iberian slug *Arion lusitanicus*. *Bulletin of Entomological Research* 101:675–686. doi: 10.1017/S0007485311000034.

Hurka, K. & Jarosik, V. (2001). Development, breeding type and diet of members of the *Amara communis* species aggregate (Coleoptera: Carabidae). *Acta Societatis Zoologicae Bohemicae* 65:17–23.

Kromp, B. (1999). Carabid beetles in sustainable agriculture: A review on pest control efficacy, cultivation impacts and enhancement. *Agriculture, Ecosystems & Environment* 74:187–228.

Kushwaha, R. K., Hegde, V. D. & Chandra, K. (2015). Ground Beetles (Coleoptera: Carabidae) from Chhattisgarh, India. *Zoo's Print* XXX:20–23.

Liebman, M. & Gallandt, E. R. 1997. Many little hammers: Ecological approaches for management of crop-weed interactions. In: Jackson L. E. (Ed) *Ecology in Agriculture*. Academic Press: San Diego.

Lindroth, C. H. (1961). The ground beetles (Carabidae, excl. Cicindelinae) of Canada and Alaska, Part 2–6. *Opuscula Entomologica Supplementa*, 1–1192. Entomoligiska Sallskapet.

Lindroth, C. H. (1985). *The Carabidae (Coleoptera) of Fennoscandia and Denmark.* Brill - Scandinavian Science Press: Leiden, Copenhagen, pp. 129–199. ISBN: 90-04-07727-8.

Lorenz, W. (2005). *Systematic list of extant ground beetles of the world (Insecta Coleoptera "Geadephaga": Trachypachidae and Carabidae incl. Paussinae, Cicindelinae, Rhysodinae)*, 2nd Edition. W. Lorenz: Tutzing, Germany, 530 pp.

Lorenz, W. (2016). CarabCat: Global database of ground beetles (version Sep 2013). In: Roskov, Y., et al. (Eds), *Species 2000 and ITIS Catalogue of Life*. Naturalis: Leiden, The Netherlands. ISSN 2405–8858. Available from: http://www.catalogueoflife.org/col.

Lövei, G. L. & Sunderland, K. D. (1996). Ecology and behavior of ground beetles (Coleoptera: Carabidae). *Annual Review of Entomology* 41:231–256. doi: 10.1146/annurev.en.41.010196.001311.

Luff, M. L. (1978). Diel activity patterns of some field Carabidae. *Ecological Entomology* 3:53–62. doi: 10.1111/j.1365-2311.1978.tb00902.x.

Lundgren, J. G. (2005). Ground beetles as weed control agents: Effects of farm management on granivory. *American Entomology* 51:224–226.

MacLeod, A., Wratten, S. D., Sotherton, N. W. & Thomas, M. B. (2004). 'Beetle banks' as refuges for beneficial arthropods in farmland: Long-term changes in predator communities and habitat. *Agricultural and Forest Entomology* 6:147–154. doi: 10.1111/j.1461-9563.2004.00215.x.

Madhukar, F. J. & Archana, H. B. (2018). Occurrence of carabid beetles (Coleoptera: Carabidae in rice fields during Kharif season. *International Journal of Researches in Biosciences, Agriculture and Technology* Special issue, 6–10.

Mili, P., Devee, A. & Saikia, D. K. (2018). Diversity of carabids in *Gerbera* and Gladiolus crops of Jorhat, Assam. *Journal of Entomology and Zoology Studies* 6(2):116–124.

Ōsawa, S., Su, Z.-H. & Inmura, Y. (2004). *Molecular Phylogeny and Evolution of Carabid Ground Beetles*. Springer: Berlin, Germany. ISBN 4-431-00487-4.

Prasad, R. P. & Snyder, W. E. (2004). Predator interference limits fly egg biological control by a guild of ground-active beetles. *Biological Control* 31:428–437. doi: 10.1016/j.biocontrol.2004.07.005.

Prasad, R. P. & Snyder, W. E. (2006). Polyphagy complicates conservation biological control that targets generalist predators. *Journal of Applied Ecology* 43:343–352. doi: 10.1111/j.1365-2664.2006.01129.x.

Renkema, J. M., Walde, S. J., Lynch, D. H., Cutler, G. C. & MacKenzie, K. (2010). Ground beetles (Carabidae) are affected by mulch in organic highbush blueberries. *Acta Horticulturae* 933:447–453. doi: 10.17660/ActaHortic.2012.933.58.

Saha, S. K., Mukherjee, A. K. & Sengupta, T. (1992). Carabidae (Coleoptera: Insecta) of Calcutta. *Records of the Zoological Survey of India, Occasional Paper* 144:1–63.

Samal, P. & Mishra, B. C. (1978). *Casnoidea indica* (Thunb.) carabid ground beetle predating on brown plant hopper, *Nilaparvata lugens* (Stal.) of rice. *Current Science* 47:688–689.

Shearin, A. F., Reberg-Horton, S. C. & Gallandt, E. R. (2014). Direct effects of tillage on the activity density of ground beetle (Coleoptera: Carabidae) weed seed predators. *Environmental Entomology* 36:1140–1146. doi: 10.1603/0046-225X(2007)36[1140:DEOTOT]2.0.CO;2.

Singh, S., Sharma, A. K., Saxena, A. K., Panday, A. K. & Kakade, S. H. (2018). Taxonomic analysis of phototactic beneficial insects as biocontrol agents (Predators and parasites) collected in light trap in rice ecosystem at Jabalpur. *Journal of Entomology and Zoology Studies* 6(3):850–853.

Snyder, W. E. (2019). Give predators a complement: Conserving natural enemy biodiversity to improve biocontrol. *Biological Control* 135:73–82. doi: 10.1016/j.biocontrol.2019.04.017.

Snyder, W. E., & Ives, A. R. (2001). Generalist predators disrupt biological control by a specialist parasitoid. *Ecology* 82:705–716. doi: 10.1890/0012-9658(2001)082[0705:GPDBCB]2.0.CO;2.

Snyder, W. E., & Ives, A. R. (2003). Interactions between specialist and generalist natural enemies: Parasitoids, predators, and pea aphid biocontrol. *Ecology* 84:91–107. doi: 10.1890/0012-9658(2003)084[0091:IBSAGN]2.0.CO;2.

Thomas, S. R. (2001). Assessing the value of beetle banks for enhancing farmland biodiversity. Doctoral thesis. University of Southampton, Southampton, England. https://eprints.soton.ac.uk/426725/.

Thomas, S., Goulson, D. & Holland, J. (2000). Spatial and temporal distributions of predatory Carabidae in a winter wheat field. *Aspects of Applied Biology* 62:55–60.

Tooley, J. & Brust, G. E. (2002). Weed seed predation by Carabid beetles. In: Holland, J.M. (Ed.), *The Agroecology of Carabid Beetles*. Intercept Ltd: Andover, pp. 215–230.

Tuck, S. L., Winqvist, C., Mota, F., Ahnström, J., Turnbull, L. A. & Bengtsson, J. (2014). Landuse intensity and the effects of organic farming on biodiversity: A hierarchical metaanalysis. *Journal of Applied Ecology* 51:746–755. doi: 10.1111/1365-2664.12219.

Zhang, Z.-Q. (2011). Animal biodiversity: An outline of higher-level classification and survey of taxonomic richness. *Zootaxa* 3148:237 p.

8 Ladybird Beetles (Coleoptera: Coccinellidae)

Bhupendra Kumar
Banaras Hindu University

Omkar
University of Lucknow

CONTENTS

DOI: 10.1201/9781003370864-8

LEARNING OBJECTIVES

1. To learn about potential biocontrol agent ladybird beetles and their prey.
2. To understand prey classification and prey searching behaviour of ladybirds.
3. To know the factors affecting prey preferences, like the prey quality; prey quantity; morphology, mobility and defence of prey; learning ability, memory, morphological characters and previous feeding experience of predators; various environmental parameters; and the genetic factors.
4. To explore such factors so that suitable species of ladybirds may be selected for mass multiplication in laboratories against the pests of agricultural importance.

8.1 INTRODUCTION

Ladybird beetles, commonly known as coccinellid beetles, ladybeetles, ladybugs or ladybirds, are an important group of predatory insects. These beetles are the most successful biocontrol agents of aphids and other soft-bodied insect pests (Hodek and Honek, 1996; Michaud, 2012; Rosagro et al., 2019). They belong to family Coccinellidae and superfamily Cucujoidea of the suborder Polyphaga and order Coleoptera. The family includes ~6,000 described species of ladybirds in some 360 genera and 42 tribes (Hodek et al., 2012). The Indian ladybird fauna is also rich and varied but is poorly studied. From the Indian subcontinent about 550 species of ladybirds are known from 90 genera, 16 tribes and 2 subfamilies (Poorani, 2019). These beetles evolved in lower Permian period, about 280 million years ago. The 'Lady' part of their name was initially given to the cosmopolitan ladybird species, *Coccinella septempunctata*, and came from the Virgin Mary for the resemblance of the scarlet colour of elytra to the cloak in Renaissance era paintings of Virgin Mary; seven spots are supposed to reflect the seven joys and sorrows of Virgin Mary's life, and it is believed that they bring good luck and bountiful harvests (Majerus, 1994). Ladybirds are the most fascinating, most recognized and highly loved insects across the world, and have always been culturally associated with good luck charms.

While majority of the ladybirds are considered as potential biological control agents of agricultural pests, some members of subfamilies Epilachninae and Coccinellinae of family Coccinellidae are phytophagous and harmful to crop plants (Hodek et al., 2012). Ladybirds are holometabolous insects. They have a 'complete metamorphosis' and pass through the following stages: egg, larva, pupa and adult. Their egg stage lasts for 15%–20% of the total pre-imaginal developmental time, larva 55%–65% and pupa 20%–25% (Honek and Kocourek, 1990; Dixon, 2000). The eggs are elongate, oval or elliptic and vary in colour from transparent (like *Scymnus louisianae*; Brown et al., 2003), light grey (*Stethorus*), yellowish (*Halyzia*), bright yellow (most species), dark orange (Chilocorini), to greenish (Klausnitzer, 1969). Eggs are either laid singly in most coccidophagous groups, or in clusters (=batch, clutch) in Coccinellinae (aphidophagous) and Epilachninae (phytophagous) groups (Klausnitzer and Klausnitzer, 1986). The surface of the coccinellid egg (chorion, egg shell) is usually smooth, and contains defensive or signalling alkanes (Hemptinne et al., 2000). A few days or hours before hatching, the eggs become greyish. The first-instar larvae possess special structures on head and prothorax called egg teeth that help in hatching. Hatched larvae rest for some time on the empty egg shells and may eat the shells (Rieder et al., 2008).

Besides the viable hatching eggs, ladybirds also produce non-hatching eggs, which are either infertile or non-viable. Such eggs are considered as trophic eggs that serve as the first food for the newly born sibling larvae (Hodek et al., 2012). In well-fed larvae, the first instar takes about 24% of the total developmental time, the second 17%, the third 19% and the fourth 40% (Honek, 1996). The fourth instar eventually stops feeding and attaches to a substrate, forming the prepupal stage. The duration of larval development is species specific and depends upon the ambient temperature, and the quality and quantity of food. The pupae of ladybird beetles are of the type *pupa adectica obtecta*, which means that all appendages (antennae, limbs, and wing pads) are glued to the body by exuvial fluid. The pupa is attached to the substrate by the tip of the abdomen. Male and female pupae cannot be distinguished by their external morphology, but female pupae are larger and heavier than the male pupae (Ueno, 2003). During pre-adult development, mortality is highest in the first instar, possibly higher than in eggs, is least in prepupae and increases somewhat during the pupal stage (Omkar and Pervez, 2004).

The adult ladybirds have variable sizes depending upon the type of the species. They have round or short oval to distinctly elongate body outline. The dorsal surface is weakly to very strongly convex, glabrous and shiny, or with sparse to dense pubescence. The dorsum has fine to coarse punctures, but never has distinct striations and sculptures. The inter-spaces between the punctures are smooth, mat-like or with microsculpture, and can only be observed under a microscope. The length of the smallest ladybird ranges from 0.7 to 1.0 mm, while that of the largest ladybird ranges from 25 to 28 mm. Majority of the species have bright red, orange or yellow elytra with distinctive spots, stripes or other markings. The structural features that distinguish ladybirds from other insect species are: (1) presence of five pairs of abdominal spiracles, (2) absence of tentorial bridge, (3) anterior tentorial branches are separate, (4) frontoclypeal suture is absent, (5) apical segment of maxillary palps are not needle-like, (6) galea and lacinia are separated, (7) mandible has reduced mola, (8) procoxal cavities open posteriorly, (9) middle coxal cavities open outwardly, (10) meta-epimeron are parallel-sided, (11) femoral lines are present on the abdominal sternite 2, (12) tarsal formula is 4-4-4 or 3-3-3, tarsal segment two, usually strongly dilated below, and (13) male genitalia with tubular curved sipho, distally embraced by the tegmen. Larvae are equipped with setae and setose processes, their antennae have 1–3 segments and the frontoclypeal suture is absent (excluding some Epilachnini). Pupae are of the type *pupa adectica obtecta* – all appendages are attached to the body by exuvial fluid. The pupa is attached to the substrate by the tip of the abdomen (Hodek et al., 2012).

The predaceous ladybirds are highly voracious and exhibit rapid functional and numerical responses (Bayoumy, 2011; Bayoumy and Michaud, 2012; Pervez et al., 2018). They have a wide prey range that includes sternorrhyncha hemiptera, thysanoptera, phytophagous mites, young instars of holometabolous insects and even fungi (Evans, 2009; Hodek et al., 2012). Based on their prey type, the ladybirds are categorized as aphidophagous, coccidophagous, acarophagous or mycophagous beetles. Although ladybirds are more specialized to feed upon the aphids (Pettersson et al., 2005; Evans, 2008), the short life span and patchy distribution of aphids make their availability quite uncertain in the agricultural fields (Hodek et al., 2012). Under such circumstances, these predaceous beetles use non-prey food as the integral components of their diet for their survival. The non-prey food includes the nectar, honeydew, pollen, fungal spores and other vegetation (Omkar, 2006; Lundgren, 2009; Sutherland and Parrella, 2009). Frequent consumption of inorganic materials by ladybirds have also been reported previously (Triltsch, 1999), but the roles of inorganic materials in nutritional ecology of the ladybirds are entirely unknown (Figures 8.1 and 8.2).

Ladybirds not only feed upon the prey and non-prey species, sometimes they even exhibit cannibalism and intraguild predation (IGP) (Toosi et al., 2019), especially when the prey is scarce or its nutritional quality decreases. Both cannibalism and IGP are alternative strategies to sustain and promote development and survival in ladybirds (Snyder et al., 2000; Kumar et al., 2014b). Cannibalism enables the larvae and adults to optimally utilize the food resources within a patch of prey, while IGP promotes the dominant species to monopolize the food resources and decreases the abundance of competitors (Hodek et al., 2012; Ovchinnikov et al., 2019). Both the mechanisms determine the efficiency with which resources at a specific feeding site are utilized. Consequently, the aphid prey is less attacked in agricultural fields, and the ladybird species exhibit the predatory interactions (Phoofolo and Obrycki, 1998). The preference for non-prey species by ladybirds may provide an initial advantage in fulfilling their nutritional requirements, to complete their development, to get them their first meal or to eliminate their potential competitors. While the non-prey range of ladybirds is broad and easy to be encountered, such preys are substantially less suitable for the ladybirds. In this chapter, we have made attempts to emphasize on the life attributes and the feeding potential of ladybirds so that they may be mass multiplied for augmentative biological control programmes against insect pests.

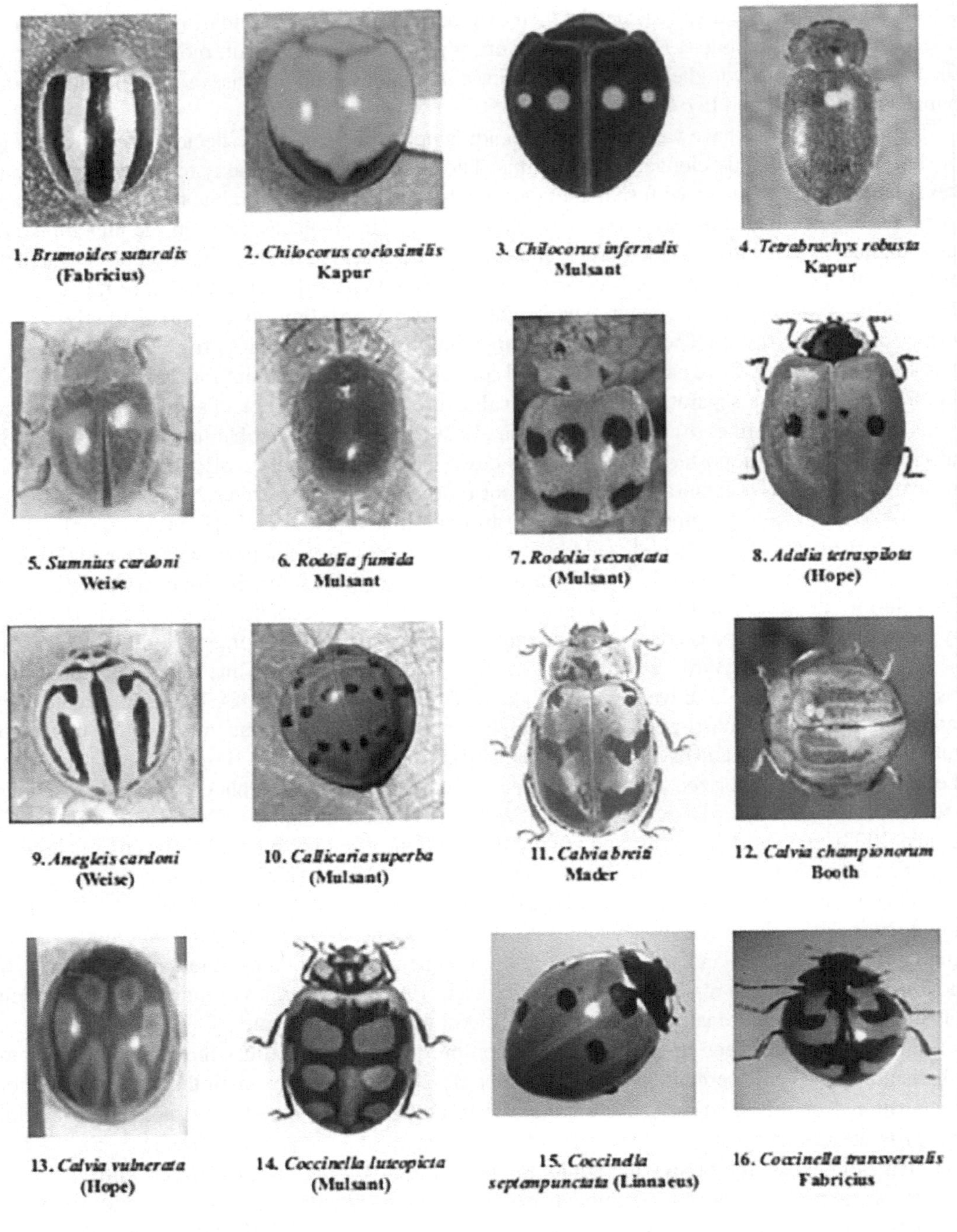

FIGURE 8.1 Ladybird beetles of the Indian subcontinent (contd…).

8.2 PREY CLASSIFICATION

The prey range of aphidophagous ladybirds is categorized as essential, alternative (accepted but suboptimum) or rejected (Hodek and Honek, 1996). This categorization of prey is based on the growth, development, survival and reproduction of the ladybirds when fed on different aphid species. Ladybirds feeding on the essential prey have better survival as larvae and are more fecund

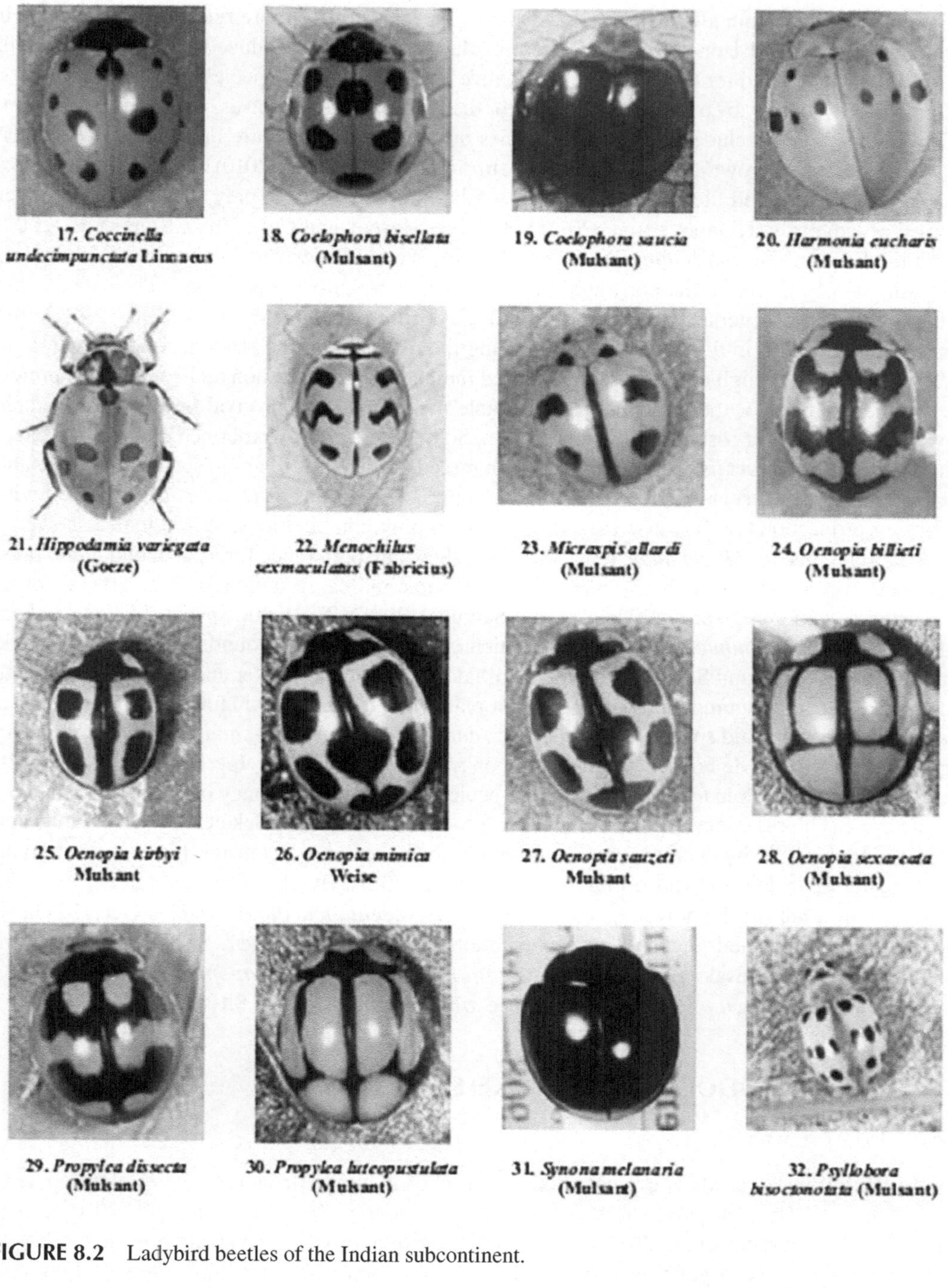

FIGURE 8.2 Ladybird beetles of the Indian subcontinent.

as adults. Essential prey is further categorized as optimal, adequate and marginal ones (Michaud, 2005). This categorization is based on the differences in their nutritive values and the ways by which the nutrients are assimilated and utilized after prey consumption (Phoofolo et al., 2007; Castro-Guedes et al., 2016; Gupta and Kumar, 2017). Ladybirds feeding on alternative prey can only survive but cannot grow and reproduce (Mills, 1981; Hodek and Honek, 1996). Rejected prey species are unpalatable for the ladybirds owing to their intensive or aposematic colouration, or due to the presence of certain allelochemicals. Such prey species are therefore rejected even after their encounter (Hodek and Honek, 1996). Later on, Malcolm (1992) introduced the terms included, peripheral and excluded for the essential, alternative and rejected prey species.

The selection of prey by ovipositing females for their offspring is termed as 'nursery prey' (Dixon, 2000). Nursery prey includes those aphid species on which the larvae are more likely to develop the best in terms of growth, development and survival (Wilder et al., 2010). Certain studies have even suggested a 'mixed diet' composed of 'essential' and 'alternative' prey species for the better survival of ladybirds (Kianpour et al., 2011; Schuldiner-Harpaz and Coll, 2017). Rana et al. (2002) reared the ladybird, *Adalia bipunctata* (L.) on its more suitable (*Acyrthosiphon pisum* Harris) and least suitable (*Aphis fabae* Scopoli) aphid prey for six consecutive generations. They found that with furtherance of generations, the mortality of ladybird on *A. fabae* reduced from 53.8% in the first generation to 0% in the sixth one. On shifting the offspring of the sixth generation reared on *A. fabae* to *A. pisum*, high mortality was recorded indicating specialization and conditioned preference of the ladybird for the previously less suitable prey. This poor survival on the most suitable prey was identified as a cost of prey specialization. Soares et al. (2004) explained dietary self-selection in ladybird, *H. axyridis*. When provided with mixed populations of aphid species, *A. fabae* and *Myzus persicae* (Sulzer) at different densities, the prey feeding rates of *H. axyridis* increased with increasing proportions of *M. persicae* in the provided prey combinations. While males showed a constant preference for *M. persicae*, females did not show any preference for a particular aphid prey.

Certain prey species, which are harmful to ladybirds and cause their mortality after feeding, are termed as toxic prey species (Nedved and Salvucci, 2008). The aphid species, *Aphis sambuci* L. feeds on elder (=*Sambucus nigra* Elder), which contains toxic compounds like cyanoglycoside sambunigrin (Nedved and Salvucci, 2008) and alkaloid sambucine (Auster and Schafer, 1956). The aphid sequesters these compounds from the host plants and is highly toxic for the ladybird species as a prey. The toxic aphid species either cause gradual poisoning (e.g. *A. sambuci*) or acute toxicity (e.g. *Aphis nerii* Boyer de Fonscolombe feeding on *Nerium oleander* L.) (Iperti, 1966) in ladybirds. In addition, ladybirds even feed on certain prey species that are not toxic but worsen their life-history parameters. Such prey species are considered as 'problematic prey' (Hodek et al., 2012). Adults and larvae of ladybird, *A. bipunctata* were occasionally found feeding on tetranychid mites but could not complete their development on the prey (Robinson, 1951). By changing the diet between the larvae and the adults of the polyphagous ladybird, *Coleomegilla maculata lengi* De Geer, Michaud and Jyoti (2008) revealed 'dietary complementation across the life stages'. Similar findings were also observed by Hamasaki and Matsui (2006) on ladybird, *Propylea japonica* (Thunberg) when fed on *Ephestia kuehniella* Zeller eggs combined with pea aphids (Table 8.1).

8.3 FACTORS MODULATING PREY PREFERENCES

8.3.1 Prey Quality

Prey suitability in ladybirds is evaluated on the basis of feeding attributes, growth, development, reproduction and life-table parameters (Stamp and Meyerhoefer, 2004; Berkvens et al., 2010; Kumar et al., 2013; Papachristos et al., 2015). The comprehensive approach to evaluate prey suitability of ladybirds on the basis of preference (prey biomass consumed and oviposition site selection) and performance (development, survival, reproduction, mortality and fertility life tables and individual fitness levels of predator) was established by Omkar and Mishra (2005). High-quality and the most suitable prey provides better growth and development, and induces lower mortality in ladybirds (Omkar and Mishra, 2005; Omkar et al., 2009a, 2011; Bista and Omkar, 2013; Kumar et al., 2013; Patel et al., 2017; Canovai et al., 2019). Studies on food assimilation and energy allocation have revealed higher conversion efficiencies and growth rates in ladybirds when fed on high-quality prey than on low-quality prey (Mishra et al., 2011, 2012; Kumar et al., 2014b). The most suitable prey species contain high levels of nutrients (Atwal and Sethi, 1963; Kumar et al., 2013), while the least suitable prey species sequester chemicals from the host plants that are rich in certain alkaloids and

TABLE 8.1
Some Common Prey Species of Ladybird Species (Omkar and Mishra, 2013)

Prey Species	Host Plant
	A. Aphids
Acyrthosiphon kondoi Shinji	*Medicago sativa* (Alfalfa), *Trifolium* spp. (Clover)
Acyrthosiphon pisum (Harris)	*Pisum sativum* (Pea), *Vicia faba* (Bakla), *Vigna unguiculata* (Cowpea or Lobia), *Dolichos lablab* (Broad Bean or Sem), *Lens esculentum* (Lentil or Masur), *Trigonella foenum-graecum* (Fenugreek or Methi) and all leguminous plants
Aphis citricola van der Goot	*Spiraea* spp.
Aphis craccivora Koch	Bakla, Pigeon pea or Arhar, Broad Bean or Sem, Cowpea or Lobia, Lentil or Masur and other leguminous plants
Aphis gossypii Glover	*Abelmoschus esculentus* (Lady's finger or Bhindi), *Citrullus vulgaris* (Round gourd or Tinda), *Coccinia cordifolia* (Little gourd or Kundru), *Coccinia grandis* (Ivy gourd), *Cucumis melo* (Musk melon or Kharbuja), *Cucumis sativus* (Cucumber or Kheera), *Cucurbita moschata* (Pumpkin or Kaddu), *Gamolepis* spp. (Bush daisy), *Gossypium* spp. (Cotton), *Holoptelea integrifolia* (Indian elm or Chilbil), *Lagenaria vulgaris* (Bottle gourd or Lauki), *Phragmites communis* (reed), *Solanum melongena* (Brinjal or Baigan)
Aphis pomi de Geer	Most plants of Rosaceae (roses)
Aphis rumicis Linnaeus	*Phaseolus limensis* (Lima beans)
Aphis sambuci Linnaeus	*Sambucus* spp.
Aphis solanella Theobald	Ivy gourd
Aphis spiraecola Patch	*Citrus* spp. (Orange or Santara, Lemon, Sweet lime, Grapefruit, etc.)
Brevicoryne brassicae (L.)	*Brassica juncea* (Black mustard), *Brassica campestris* (Mustard), *Brassica oleracea* (Cabbage or Patta gobhi), *Brassica rapa* (Turnip or Shalgam) and other crucifers
Ceratovacuna lanigera Zehnter	*Saccharum officinarum* (Sugarcane or Ganna)
Cryptosiphum artemisiae Buckton	*Artemisia* spp. (Mugworts or sage brush)
Diuraphis noxia (Mordvilko)	*Triticum aestivum* (Wheat)
Eriosoma lanigerum (Hausmann)	Apple or Seb, Peach or aadoo, Plum or Aalu Bukhara, etc.
Hyadaphis coriandri (Das)	*Coriandrum sativum* (Coriander)
Hyalopterus pruni (Geoffroy)	Almond or Badaam, Peach or Aadoo, Plum or Aalu bukhara, etc. *Beta vulgaris* (Beetroot or chukandar), Reed
Hyperomyzus lactucae (L.)	*Sonchus oleraceus* (Sow thistle)
Hysteroneura setariae (Thomas)	Almond or Badaam, Peach or Aadoo, Plum or Aalu bukhara, etc. Beet, Reed
Lipaphis erysimi (Kaltenbach)	Black mustard, Yellow mustard, Radish
Lipaphis pseudobrassicae (Kaltenbach)	Arhar, Radish, Cabbage or Patta gobhi
Macrosiphum euphorbiae	*Solanum tuberosum* (Potato or Aalu)
Macrosiphum rosae (L.)	Beetroot or chukandar, Rose or Gulab
Macrosiphum rosaeformis Das	Roses
Macrosiphum miscanthi (Takahashi)	Most plants of Graminae, e.g., cereals, wheat, barley, etc.
Myzus nicotianae	Tobacco
Myzus ornatus Laing	Peach or Aadhu, Plum or Aalu bukhara, Cherry, Almond or Badaam, etc.
Myzus persicae (Sulzer)	Brinjal or Baigan, *Solanum nigrum* (Makoi), *Nigella sativa* (Black cumin or Kalonji), *Raphanus sativum* (Radish or Mooli), *Chenopodium album* (Lamb's quarters or Bathua)
Pseudaulacaspis pentagona Targioni	Tozzetti- Potato or Aalu, *Morus alba* (Mulberry or Shehtoot)
Rhopalosiphum maidis (Fitch)	Barley or Jowar, *Zea mays* (Maize), *Pennisetum typhoides* (Spiked millet or Bajra)
Rhopalosiphum nymphaeae (L.)	*Euryale ferox* (Makhana). Peach or Aadhu, Plum or Aalu bukhara, Cherry, Almond or Badaam, etc.

(Continued)

TABLE 8.1 (*Continued*)
Some Common Prey Species of Ladybird Species (Omkar and Mishra, 2013)

Prey Species	Host Plant
Rhopalosiphum padi (Linnaeus)	Rice
Schizaphis graminum (Rondani)	Most plants of Gramineae
Sitobion avenae (Fabricius)	Most plants of Graminae
Subovatomyzus leucosceptri Basu	*Schima wallichii* (Needlewood, Chilauni)
Toxoptera citricida (Kirkaldy)	Citrus fruits
Toxoptera aurantii (Boyer de Fonscolmbe)	*Theobroma cacao* (Cocoa), Clover, Alfalfa, Coffee, Orange or Santara, Lemon, Lime and other citrus fruits
Toxoptera citricidus (Kirkaldy)	Orange or Santara, Lemon, Lime and other citrus fruits
Uroleucon compositae Theobald	*Tagetes erecta* (Genda), Guldawadi, and all compositae plants
	Scale insects (Coccidae)
Aspidiotus destructor (Signoret)	*Artocarpus altilis* (Jackfruit or Kathal), *Carica papaya* (Papaya or Papita), *Cocos nucifera* (Coconut or Nariyal), *Elaeis guineensis* (Palm), *Persea americana* (Avocado or Makhanphal), *Psidium guajava* (Guava or Amrood), *Zingiber officinale* (Ginger or Adrakh), Sugarcane or Ganna, Mango or Aam, Banana or Kela, Orange or Santara, Apple or Seb.
Ferrisia virgata (Cockerell)	*Ipomoea batatas* (Sweet potato), *Manihot esculenta* (Cassava), *Coffea* sp. (Coffee), Citrus fruits
Pseudococcus spp.	Coffee
Quadraspidiotus perniciosus (Comstock)	*Pyrus communis* (Pear or Nashpati), Apple or Seb, Willows, Sweet cherry, Almond or Badaam, Peach or Aadoo, Plum or Aalu bukhara, Grape or Angoor, Rose or Gulab
	Mealybugs (Pseudococcidae)
Drosicha stebbingi Green	Cassava, Fig, Mango or Aam, Orange or Santara, Lemon, Lime and other citrus fruits, Shireesh, Jacalin, Sal, Palas, Shisham, Mulberry or Shehtoot, Indian beech, Kejari, Tamarind or Imli, Arjun, India mahogany, Chinee apple, Babul
Ferrisia virgata (Cockerell)	Custard apple or Sharifa, Plum or Aalu bukhara, Cherry, Almond or Badaam, Peach or Aadoo, Orange or Santara, Lime, Grapefruit, *Vitis vinifera* (Grape or Angoor), etc.
	Psyllids
Diaphorina citri Kuwayama	Citrus fruits
Heteropsylla cubana Crawford	*Leucaena* sp.
Trioza viridula Zett.	*Daucus carota* (Carrot)
	Whiteflies
Aleurocanthus spp.	*Acacia senegal* (Gum senegal tree), *Acacia tortilis* (Umbrella thorn), *Prosopis juliflora* (Bayahonda, vilayati kikar). Coconut or Nariyal, Coffee, Orange, Lime, Lemon, etc.
Aleurodicus dispersus	*Pterocarpus indicus* (red sandal), *Terminalia catappa* (Indian almond), Guava, Papaya or Papita, etc.
Aleurolobus barodensis	Sugarcane or Ganna
	Leafhoppers
Empoascanara indica	Plum or Aalu bukhara, Peach or Aadoo, Almond or Badaam, etc.
Idioscopus clypealis	Mango or Aam
Pyrilla perpusilla	Sugarcane
	Adelgids
Adelges piceae	*Abies balsamea* and *Abies fraseri*
Adelges tsugae Annand	Forest and ornamental hemlock trees
Adelgis spp. – Spruce or Douglas fir, etc.	

(*Continued*)

TABLE 8.1 (*Continued*)
Some Common Prey Species of Ladybird Species (Omkar and Mishra, 2013)

Prey Species	Host Plant
	Coleoptera
Hypera variabilis Herbst	Alfalfa
Meligethes aeneus Fabricius	Rapeseed
	Lepidoptera
Helicoverpa armigera	Legumes and cotton
Papilio demoleus (Linnaeus)	Citrus fruits
Phthorimaea operculella (Zeller)	Potato or Aalu, Tobacco
Plutella maculipennis (Curtis)	Cabbage or Patta gobhi
Polychrosis botrana Schiff	Grape or Angoor
Pyrausta nubilalis Hubner	Corn
Spodoptera litura (Fabricius)	Cabbage or Patta gobhi, Turnip or Shalgam
	Collembola
Sminthurus viridis Linnaeus	Clover
	Mites
Eutetranychus orientalis (Klein)	Orange or Santara, Lemon or Neembu, etc.
Tetranychus bioculatus	Marigold, Rose or Gulab, Cosmos
Tetranychus ludeni	Sugarcane or Ganna, Cotton
Tetranychus neocaledonicus Koch	Apple, Mango or Aam, Orange or Santara, Grape or Angoor, Potato or Aalu, Tomato or Tamatar, Bottle gourd or Lauki, Wheat, Rice, etc.
Tetranychus turkestani atlanticus McGregor	Apple or Seb, Peach or Aadoo, Plum or Aalu Bukhara, etc.
Tetranychus urticae Koch	Vegetables, food crops

are toxic for many ladybirds (Hodek, 1956; Okamoto, 1966). Consequently, ladybirds avoid feeding on such prey species (Kumar et al., 2013).

The ladybird species that are efficient in utilizing prey biomass to their own biomass maximally have high reproductive rates. They are also efficient in converting prey biomass maximally to the number of offspring produced (Omkar et al., 2005a). High fecundity, maximum egg viability and more clutch number and clutch size of *C. septempunctata* and *C. transversalis* on *A. pisum* and *A. craccivora*, respectively, substantiate that these aphid species contain essential nutrients, resulting in their higher palatability by the respective ladybirds. Perhaps, it has taken several generations for *C. septempunctata* to get adapted to metabolize the *A. pisum* toxins, and for *C. transversalis* to acquire the ability to detoxify the toxins of *A. craccivora* (Bista and Omkar, 2013).

Prey quality and/or its suitability to individual ladybirds, *viz.*, *A. bipunctata* (Ferrer et al., 2008), *Anegleis cardoni* (Weise) (Omkar et al., 2009a, 2011), *Brumoides suturalis* (Fabricius) (Bista et al., 2012; Lima et al., 2017), *Chilocorus bipustulatus* (L.) (Uygun and Elekcioglu, 1998), *C. septempunctata* (Omkar and Srivastava, 2001, 2003a, b; Srivastava and Omkar, 2004; Kianpour et al., 2011; Bista and Omkar, 2013; Kumar et al., 2013, 2015; Arshad et al., 2017), *C. transversalis* (Omkar and James, 2004a; Gupta et al., 2006; Bista and Omkar, 2013; Kumar et al., 2013, 2015), *H. axyridis* (Osawa, 1989; Castro-Guedes et al., 2016), *Menochilus sexmaculatus* (F.) (Omkar and Bind, 2004; Patel et al., 2017), *Hyperaspis notata* Mulsant (Dreyer et al., 1997), *Micraspis discolor* (Fab.) (Omkar and Pervez, 2001; Omkar, 2006), *P. dissecta* (Mulsant) (Pervez and Omkar, 2004a; Omkar and Mishra, 2005), *P. japonica* (Zhang et al., 2007), *Propylea quatuordecimpunctata* (L.) (Keshavarz et al., 2015), *Rodolia iceryae* Janson (Kairo and Murphy, 1995), *Sphaerobothria hoffmanni* Karsch (Kawauchi, 1985), or within conspecific/heterospecific groups (Omkar et al., 2005b; Kumar et al., 2015) are well studied and their prey suitability in terms of developmental attributes, life-table parameters and reproductive performances are well documented. Many studies have also evaluated prey quality in terms of mortality parameters (Iverson and Hardling, 2007; Nielsen et al., 2008; Haye et al., 2010; Borges et al., 2011; Kianpour et al., 2011; Ugine, 2012). The findings of all such studies indicate that growth, development, reproduction and longevity are highest on the most suitable prey and lowest on the least suitable prey.

Predatory performances of ladybirds are well studied in terms of prey consumption rates (Omkar and Srivastava, 2001, 2003a, b; Finlayson et al., 2010; Kumar et al., 2013; Patel et al., 2017), consumption indices, conversion efficiencies and/or growth rates (Lundgren and Weber, 2010; Kumar et al., 2013; Patel et al., 2017; Jalali et al., 2019). The findings reveal that all prey species are not equally suitable for different ladybird species (Omkar et al., 2011; Zhang et al., 2012; Bista and Omkar, 2013) and they perform the best on the highly palatable prey that diminishes the population pressure on the larvae, resulting in their fast pre-imaginal development (Omkar and James, 2004a; Ferrer et al. 2008; Maurice and Kumar, 2012; Bista and Omkar, 2013). Within aphidophagous guilds formed by the larvae and adult ladybirds exploiting the same prey resource (Yasuda and Shinya, 1997; Dixon, 2000), abundance of highly palatable prey reduces the incidences of cannibalism and IGP amongst ladybirds (Agarwala and Dixon, 1992; Snyder et al., 2000; Cottrell, 2005; Kumar et al., 2014b).

On evaluating the performance of ladybird *A. cardoni* on three aphid species (i.e. *Aphis gossypii* Glover, *A. craccivora* and *Lipaphis erysimi* Kaltenbach), Omkar et al. (2009a) found maximum reproductive parameters and highest intrinsic rate of increase on *A. gossypii,* and lowest on *L. erysimi.* Omkar et al. (2011) calculated life-table parameters of *A. cardoni* on ten aphid species and revealed that the overall mortality of *A. cardoni* larvae was lowest on *Uroleucon compositae* (Theobald) and highest on *Ceratovacuna silvestrii* (Takahashi). Bista et al. (2012) investigated the development and survival rates of ladybird, *B. suturalis* on two aphid species, *A. pisum* and *A. gossypii.* They found that larvae developed faster on *A. pisum* but had higher survival rates on *A. gossypii.* Studies on fertility life table of *Hippodamia variegata* (Goeze) on three aphid species, viz., *A. gossypii*, *A. fabae* and *M. rosae*, revealed that the net reproduction rate (R_0) of the ladybird beetle was highest on *A. fabae* and lowest on *Macrosiphum rosae* (L.) (Golizadeh and Jafari-Behi, 2012). Similarly, findings on life-history parameters of *C. transversalis* and *C. septempunctata* indicated pea aphid, *A. pisum*, to be the most suitable prey for *C. septempunctata* and bean aphid, *A. craccivora* for *C. transversalis* (Kalushkov and Hodek, 2004; Bista and Omkar, 2013) (Figures 8.3 and 8.4).

Keshavarz et al. (2015) assessed feeding preferences of ladybird *P. quatuordecimpunctata* collected from two different localities on *A. fabae* and *A. gossypii.* They found that *P. quatuordecimpunctata* from two different localities preferred *A. gossypii* over *A. fabae.* Kumar et al. (2013, 2015) reported higher predation attributes of *C. transversalis* larvae on *A. craccivora* and *C. septempunctata* larvae on *L. erysimi*, but when the larvae of both the species were kept together in heterospecific combinations, the larvae exhibited additive effects on *L. erysimi* but antagonistic effects on *A. craccivora.* However, Patel et al. (2017) found higher feeding attributes of *M. sexmaculatus* larvae on *A. craccivora* than on *L. erysimi* under prey fluctuating conditions. Hosseini et al. (2019) manipulated the quality of host plant (i.e. fertilized with different nitrogen levels), and aphid species (i.e. *Aphis gossypii*) collected from different host plants caused alteration of life-history parameters of the ladybird *Hippodamia variegata* when fed on the aphid prey. Thus, life time aphid consumption and intrinsic rate of increase in *H. variegata* were lowered when host plants were treated with fertilizers containing high nitrogen content than those treated with medium or low nitrogen content. Previous studies have further suggested that prey species may differ in terms of their physical structure, size, chemical defence, behaviour, ability to escape capture and ability to derive nutrients from the host plants. Such features either individually or in combination affect the selection of particular prey by the ladybird species in natural and artificial agroecosystems.

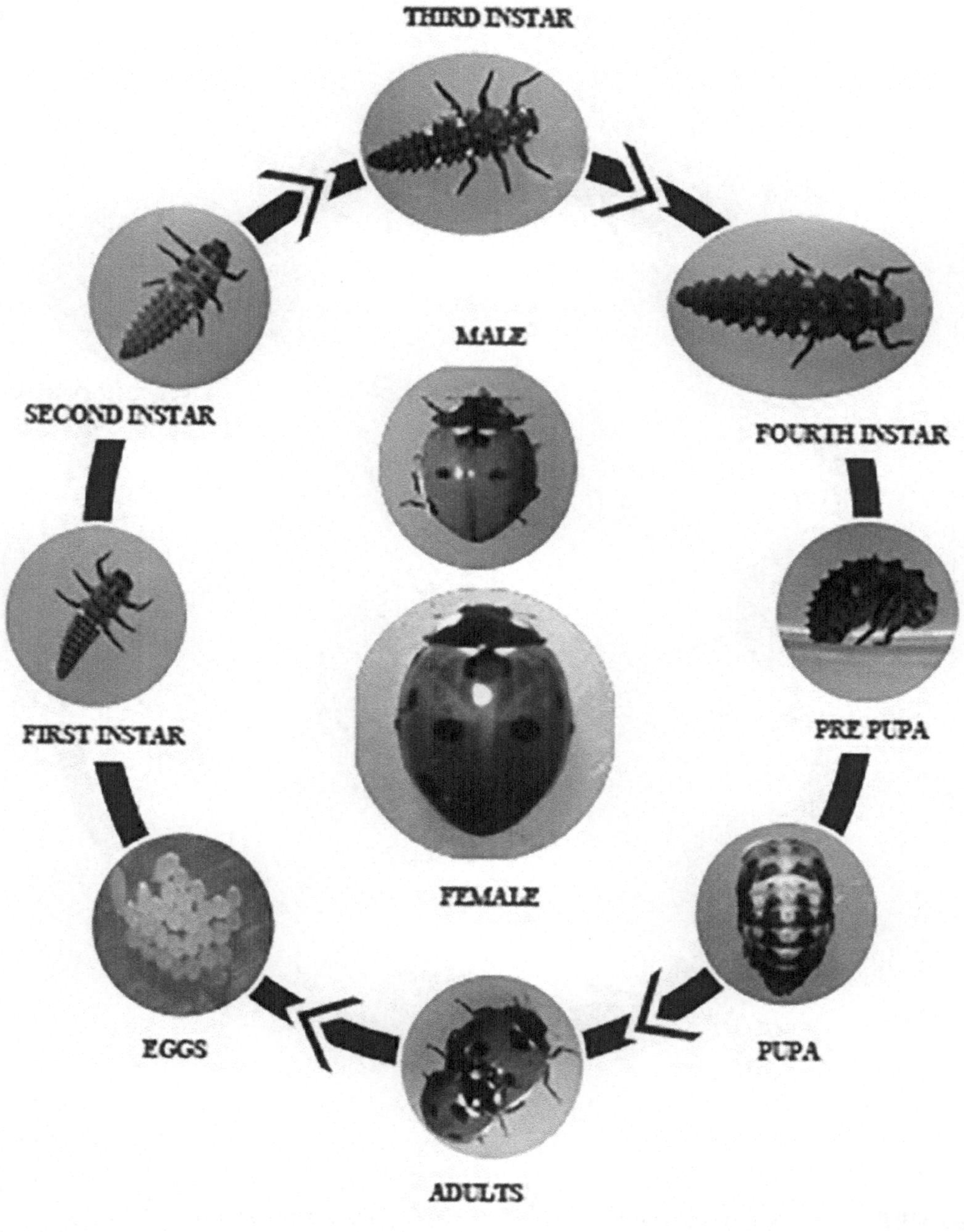

FIGURE 8.3 Life cycle of a cosmopolitan ladybird beetle, *Coccinella septempunctata*.

8.3.2 Prey Quantity

Prey–predator relationships involving aphid prey species and ladybird predators are well investigated. The availability of aphids in agroecosystems fluctuates with space and time, and the deprivation of aphids severely affects the life parameters of ladybirds (Omkar and Pervez, 2003; Phoofolo et al., 2008; Santos-Cividanes et al., 2011; Chaudhary et al., 2015; Patel et al., 2017). Consequently, predatory responses of ladybirds differ under variable quantities of aphid prey. Such variations in predation attributes of ladybirds govern complex interactions (e.g. intra- and

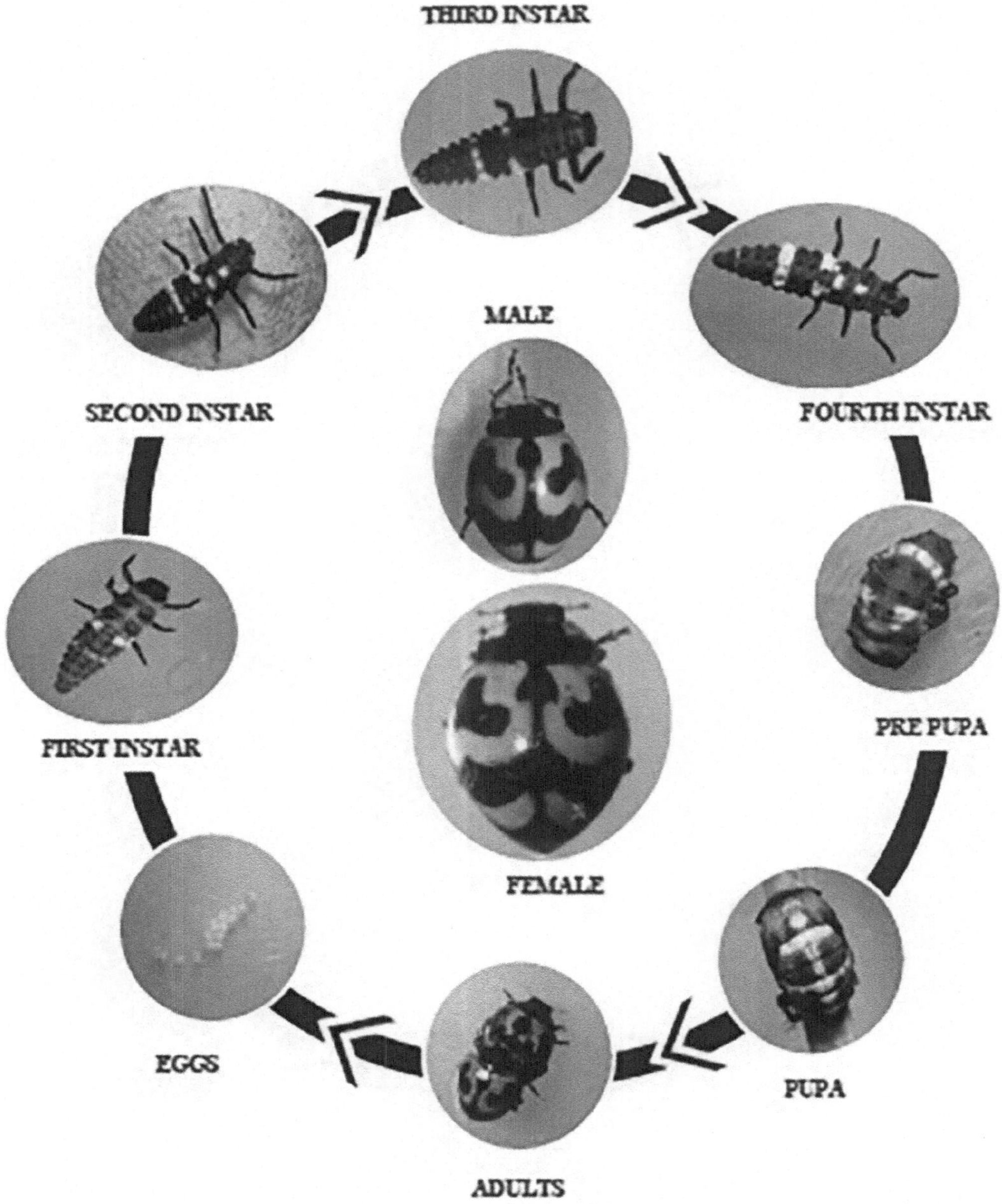

FIGURE 8.4 Life cycle of an indigenous ladybird beetle, *Coccinella transversalis*.

inter-specific competitions) between them when preying on a common prey resource. Therefore, investigating the nature and intensity of such complex interactions between the ladybird predator and the aphid prey in relation to prey scarcity or abundance is of great relevance as they may affect the structures of predatory guilds.

8.3.2.1 Functional and Numerical Responses

Most studies on prey quantity have assessed the functional responses of ladybirds under different prey densities to understand the interactions between the predaceous ladybirds and their prey species (Omkar and Kumar, 2013; Kumar et al., 2014a; Chaudhary et al., 2015; Patel et al., 2017;

Jalali and Ziaaddini, 2017; Wu et al., 2018; Crookes et al., 2019). Solomon (1949, 1964) has described a predator's functional response as the rate of prey consumption by the predator with respect to the available prey density. However, Holling (1959a, b, 1965) proposed three basic types of functional responses, i.e., as the prey density increases the number of prey consumed by a predator may either increase linearly (type I), in a decelerating fashion (type II) or in a sigmoidal fashion (type III). Later studies reported five types of predatory response curves, *viz.*, Type I, which is linear, Type II, which is curvilinear, Type III, which is sigmoidal (Holling, 1959a), Type IV, which is dome shaped (Luck, 1985; Bressendorff and Toft, 2011) and Type V, which is negative exponential (Sabelis, 1992). In general, Holling's Type II functional response curve is reported in ladybirds (e.g. Omkar and Pervez, 2011; Gupta et al., 2012; Omkar and Kumar, 2013; Chaudhary et al., 2015; Latifian, 2017; Qin et al., 2017; Rahimi et al., 2017), characterized by a curvilinear shape, where a predator is assumed to feed on prey without any initial delay in learning ways to avoid difficulties (Sarmento et al., 2007).

The functional response of ladybirds is not stage specific and both the larvae and adults show similar pattern of functional response curves (Omkar and Srivastava, 2001, 2002; Omkar and Kumar, 2013), characterized by an increased predation with increasing prey density, reaching an asymptote at higher prey densities and thereafter remaining unaffected due to satiation (Mills, 1982; Jeschke et al., 2002, Jalali et al., 2010). Some studies have reported a modified Type II functional response, i.e., consisting of a decelerating (type II) functional response when provided with low to optimal supply of prey, followed by an accelerating (type III) functional response when provided with an optimal to abundant supply of prey (Kumar et al., 2014a; Chaudhary et al., 2015; Patel et al., 2017). However, few studies have also found Type III response (i.e. sigmoidal curve) in ladybirds (e.g. Sarmento et al., 2007; Abdollahi et al., 2010; Latifian, 2017), where a predator is assumed to learn to evade the difficulties associated with prey capture (Schenk and Bacher, 2002).

In recent years, the functional responses in ladybirds have been viewed broadly as the rate of prey consumption by an individual predator (Hodek et al., 2012). The functional response can accordingly be modelled, as it varies not only with the prey density but also with the other interacting factors that may influence the foraging behaviour of predator, like the predator density (Mills and Lacan, 2004; Schenk et al., 2005), density of alternate prey (e.g. Tschanz et al., 2007), density of predator's natural enemies (e.g. Krivan and Sirot, 2004), foraging substrate and the abiotic factors (Jalali et al., 2010; Khan and Khan, 2010). The functional response alone may not provide sufficient information about the population-level effects of natural enemies; and to predict the influence of prey density on oviposition, numerical responses of ladybird species need to be studied. Several studies have evaluated the functional and numerical responses of ladybirds either individually (Sarmento et al., 2007; Abdollahi et al., 2010; Bayoumy, 2011; Bayoumy and Michaud, 2012; Omkar and Kumar, 2013; Patel et al., 2017; Jalali and Ziaaddini, 2017; Rahimi et al., 2017) or within guilds (Snyder and Ives, 2003; Costamagna et al., 2007; Snyder, 2009; Omkar and Pervez, 2011; Kumar et al., 2014a; Chaudhary et al., 2015).

Solomon (1949, 1964) defined the numerical response as the change in predator density (through immigration or aggregation and reproduction) in response to prey density. This may further be simplified as the rate of increase in oviposition of predators with increase in prey density. Numerical response is an important tool in determining the nature of interactions between predator and the prey species where the latter affects the reproduction of the former (Vieira et al., 2012). This may occur via predators' aggregation in prey hotspots or an increase in predator's reproduction and survival. The mechanism of numerical recruitment driven by immigration of predators is known as *aggregative numerical response*, whereas numerical recruitment driven by increased reproduction of predator is considered as *reproductive numerical response* (Vieira et al., 2012). Factors like attack strategies of predators and the quality of prey significantly affect the numerical responses of predators (Hemptinne et al., 1992; Palmer and Sheppard, 2002). Both cyclicity and variability in prey dynamics are also known to strengthen the numerical response of predators. While in nature, predators respond both functionally and numerically on their prey populations, the functional response is the key factor that determines the numerical response of predators (Keith et al., 1977). Therefore, assessing both functional and numerical responses of predators towards changing densities of prey populations may help the agriculturalists to evaluate the suitability of predators as successful biological control agents.

8.3.2.2 Prey Quantity and Life Attributes

The quantity of prey intake, its digestibility and utilization significantly affect the growth, development, body biomass and survivability of ladybird predators (Rath, 2010). Under prey-scarce conditions, both cannibalism and IGP are the important mechanisms that influence the intra- and inter-specific relationships of ladybirds (Kumar et al., 2014b; Rocca et al., 2017; Yang et al., 2017). Atlihan and Guldal (2009) found increased prey consumption by the ladybird, *Scymnus subvillosus* (Goeze) with increasing prey density. Similar results were also obtained by Britto et al. (2009) in ladybird, *Stethorus tridens;* Gordon, and Reis et al. (2007) in predaceous mite, *Amblyseius herbicolus* (Chant) when fed on tomato red spider mite (*Tetranychus evansi*) and false spider mite (*Brevipalpus phoenicis*), respectively. Increased prey consumption by ladybirds with increasing prey density has also been reported earlier by Omkar and Srivastava (2001, 2003a, b), Omkar and James (2003), Jalali and Ziaaddini (2017) and Rahimi et al. (2017).

The larvae of *H. axyridis* compensated for prey shortage by limiting their growth rates than those fed on optimal or abundant prey (Dmitriew and Rowe, 2007). Similarly, adults of *H. axyridis* reduced their efficiencies in converting prey biomass to egg biomass when supplied with limited prey (Agarwala et al., 2008). In contrast, Schuder et al. (2004) reported that larvae of *H. axyridis* compensated for low prey availability by reducing their development rates, but enhanced their prey exploitation efficiencies up to 100% and prey biomass conversion efficiencies over 40%. Similar increased rates of efficiency of conversion of ingested (ECI) material at low levels of prey consumption were found in aphidophagous mirids and chrysopids (Zheng et al., 1993a, b). It therefore appears that the prey biomass conversion rate or ECI varies with prey availability so as to overcome the adverse effects of prey scarcity (Smith, 1965; Waldbauer, 1968; Bista and Omkar, 2014).

Previous studies have also revealed that reduced prey consumption slows down the rate of development and enhances mortality in ladybirds (Phoofolo et al., 2008; Bista and Omkar, 2014). Phoofolo et al. (2007) reared larvae of *Hippodamia convergens* Guérin-Méneville on *ad libitum* versus limited quantity of aphid species, *Schizaphis graminum* Rondani and *Rhopalosiphum padi* (L.) collected from winter wheat. The larvae grew with the same rate on *ad libitum* diets of the two aphid species, but the development was relatively slower on limited diets of *R. padi* and *S. graminum*. Dmitriew and Rowe (2007) investigated the compensatory abilities of *H. axyridis* larvae in response to temporary prey shortage during development. They reported that temporary prey shortage resulted in lower lifetime fitness in terms of prolonged larval period and greater larval mortality.

Prey deprivation severely affects the growth, development and survival of ladybirds (Omkar and Pervez, 2003; Schuder et al., 2004). Under prey-scarce conditions, ladybirds utilize their entire energy to maximize survival and reproduction (Legaspi and Legaspi, 1998). Thus, prey scarcity can affect the fitness of adult ladybirds by decreasing their developmental rates and reducing the final body size of their offspring (Agarwala et al., 2001; Seagraves, 2009). While evaluating the developmental duration and survival rate of ladybird, *S. subvillosus*, on different aphid densities, Atlihan and Guldal (2009) found that increased prey density reduced both developmental durations and mortality factors in *S. subvillosus*. Maurice and Kumar (2011) also reported that durations of development in *C. septempunctata* and *C. transversalis* larvae were short and the larvae pupated earlier when prey was present in abundant quantity.

While studying the performance of larvae and adults of *H. axyridis* on abundant and limited prey supply, Agarwala et al. (2008) reported that on limited prey the larvae developed slower and surviving adults were smaller than those on abundant prey. While studying intra- and inter-specific interactions on larval survival of two aphidophagous ladybirds, *C. septempunctata brucki* and *H. axyridis* at two prey densities, Sato et al. (2009) concluded that when reared together, the larval density of both the species decreased mainly due to cannibalism. But when prey was abundant, the survival of both the species improved. However, when prey was scarce, both cannibalism and IGP became the crucial factors in deciding intra- and inter-specific relationships between the two species (Rocca et al., 2017). Thus, when prey is abundant, the intensity of both intra- and inter-specific competitions is low and does not limit the predation rates even for functionally similar predators (Casula et al., 2006). But when prey is scarce, the competition becomes much stronger (Yang et al., 2017; Ducatti et al., 2017), particularly among the functionally similar predators (Casula et al., 2006), and in such a case, the functionally diverse predators consume more biomass of prey (Werling et al., 2012).

Individuals accumulating abundant prey biomass from a resource-rich environment are benefitted by transferring their resources to their offspring, thereby enhancing their progeny fitness (Qvarnstrom and Price, 2001; Bonduriansky and Head, 2007). However, under low prey biomass consumption, females invest less energy in egg production (Evans, 2003). Studies conducted on *Cingula trifasciata* and *Cingula californica* revealed that females produced eggs in constant ratio to the amount of food provided to them (Ives, 1981). The investment of energy in egg production by females was limited in the absence of aphid prey, which further was enhanced upon locating the patches of high aphid density (Evans, 2000). Females therefore use the resources from limited prey consumption to support the modest egg production. Engelmann (1970) reported that the number of ovarioles per ovary was reduced when the supply of food during larval development was limited. During the time of food scarcity, many ladybird species utilize their energy reserves to maintain their survival. Consequently, to carry out their normal metabolic processes, they utilize their fat reserves. As a result, they lower their fecundity and use the scant energy available for keeping themselves alive (Ricklefs, 1990).

Both the clutch size and oviposition rates of ladybirds are highly influenced by the prey quantity available to females at the time of oviposition (Dixon and Guo, 1993; Dixon, 2000; Ware et al., 2008). The availability of optimal prey during immature development is highly essential even when adult diet is optimal. However, the development of offspring is influenced by both the nutrient reserves in eggs and the food provided to them during their development (Bertram and Strathmann, 1998). Dixon and Guo (1993) have suggested that the larval diet may not only affect the body size of larvae, but may also affect the rate of ovariole maturation during their adulthood. Agarwala et al. (2009) reported that females of *Harmonia dimidiata* (F.) matured earlier and produced more eggs at high prey (i.e. *Cervaphis quercus* Takahashi) density and *vice versa* at low prey density. While assessing the performance of larvae and adults of *H. axyridis* on abundant and scarce prey availability, Agarwala et al. (2008) found that on scarce prey the larvae grew slower and surviving adults were smaller with prolonged pre-reproductive periods, reduced fecundity, and increased longevity than those reared on abundant prey. While the quantity of prey affect growth, development, survival and reproduction of predators (e.g. Zheng et al., 1993a, b; Dmitriew and Rowe, 2011; Bista and Omkar, 2014) studies assessing predators responses to prey availability are highly needed for understanding their ecology, foraging efficiency and rate of prey consumption. As the performance of both larvae and adults are constrained by the quantity of prey (Lee and Kang, 2004; Seagraves, 2009; Bista and Omkar, 2014), the prey quality as a factor is very important in understanding the potential of ladybird predators for their use in biocontrol programmes against insect pests (Agarwala and Bhowmik, 2011).

8.3.3 Prey Mobility and Prey Defence

The mobility of prey is another factor that influences the prey selection by ladybird predators. Several studies have found that prey mobility increases the prey searching and prey handling time, thereby reducing the net energetic gain by predators (e.g. Allan et al., 1987; Nordlund and Morrison, 1990). Rosenheim and Corbett (2003) have demonstrated that for a sit-and-wait predator, the high prey mobility may increase the encounter rate and attack probability. The selection of prey by ladybird predators is highly affected by the prey activity (Provost et al., 2006); and ladybirds generally attack the most vulnerable prey than the active prey. However, they select preferentially the active prey in the presence of immobilized prey. Provost et al. (2006) have further reported that in the presence of active prey, the ladybird *H. axyridis* was unable to capture the mirid prey, *Hyaliodes vitripennis* (Say), because the prey was highly mobile and was using defensive mechanisms to escape the predatory attacks.

Several studies have demonstrated that mobile prey generally increases the foraging cost, reduces the probability of capture and results in a lower net energetic gain for the predator (e.g. Allan et al., 1987; Nordlund and Morrison, 1990). The defensive mechanisms displayed by prey species may also influence the prey selection by the predator. Roger et al. (2000) observed reduced prey capture rates with effective defensive responses by the ladybird, *Coleomegilla maculata* (DeGeer). In contrast, Soares et al. (2004) observed that distribution and defensive strategies of prey species, *M. persicae* and *A. fabae*, did not provide any escape advantage from the ladybird, *H. axyridis aulica*. Kumar et al. (2014b) also reported that the immobility and defensive surface semiochemicals of conspecific and heterospecific eggs, rich in alkanes and alkaloids (Hemptinne and Dixon, 2000; Hemptinne et al., 2000; Omkar et al., 2004), reduced their consumption and utilization by the larvae of *C. septempunctata* and *C. transversalis*.

Studies have further found that the least suitable prey sequesters chemicals from its host plant, which contains certain alkaloids that are toxic for many ladybird species (Hodek, 1956; Okamoto, 1966). The pungent smell of mustard aphid may be due to the presence of allyl isothiocyanates, and is responsible for its reduced consumption as a prey by the ladybird (Joshi et al., 1999; Noble et al., 2002; Omkar et al., 2009a). There are possibilities that the allelochemicals found in mustard aphid may not be detoxified by many ladybird predators causing its lower palatability (Kumar et al., 2013).

Studies have reported that the bean aphid, *A. craccivora*, reared on bean plants (i.e. *D. lablab* Linnaeus; Fabaceae) is highly suitable and palatable for *C. septempunctata* (Omkar and Srivastava, 2003a, b), *C. transversalis* (Bista and Omkar, 2013; Kumar et al., 2013), *M. sexmaculatus* (Omkar and Bind, 2004; Kumar et al., 2013; Patel et al., 2017) and *P. dissecta* (Omkar and Mishra, 2005; Kumar et al., 2013). However, when reared on host plants, such as *Robinia pseudoacacia* L. (Fabaceae) (Obatake and Suzuki, 1985), *Astragalus sinicus* L. (Fabaceae), *Vicia hirsuta* (L.) (Fabaceae), *Capsella bursa-pastoris* (L.) (Brassicaceae), *Cardamine flexuosa* With. (Brassicaceae), *Rorippa palustris* (L.) (Brassicaceae) and *Hemistepta carthamoides* Kuntze (Asteraceae) (Hukusima and Kamei, 1970), the aphid species is reported to be highly toxic for ladybirds due to the presence of amines, canavanine and ethanolamine found in the aphid species (Obatake and Suzuki, 1985).

8.3.4 Morphology, Age and Parasitization Status of Prey

Prey size also affects the prey selection by ladybird predators. Previous studies have found a relationship between the prey size and the nutrient gain, i.e., larger prey provides higher energetic gain to predators than the smaller prey (Roger et al., 2000; Sloggett, 2008a, b; Skelhorn and Rowe, 2006; Skelhorn et al., 2016). Houck (1991) found that the ladybird, *Stethorus punctum* (LeConte), selected its prey according to its quality and the nutritional needs. While a larger prey has higher nutrients than the smaller prey, the attack and capture of larger prey may be more expensive and risky (Allan et al., 1987; Roger et al., 2000). Harmon et al. (1998) measured the feeding behaviour of *C. septempunctata*, *H. axyridis*, *H. convergens* and *C. maculata* on red- and green-coloured morphs of pea aphid on red, green and white backgrounds. They reported that *C. septempunctata* consumed more of the aphid morph that contrasted with the background colour, whereas *H. axyridis* consumed significantly more number of red morph aphids regardless of the background colour. The remaining two species showed no significant differences in their consumption rates on the two aphid morphs. This variation in the use of visual cues by the predators suggests that different species of predators exhibit different foraging behaviour when searching for the same prey.

Prey age also influences the feeding behaviour of ladybird predators. Zarghami et al. (2014) examined the feeding behaviour of coccidophagous ladybird, *Nephus arcuatus* Kapur, on two mealybugs, *Nipaecoccus viridis* (Newstead) and *Planococcus citri* Risso. They reported that *N. arcuatus* consumed significantly more number of first-instar nymphs than the eggs, and more eggs than the adult females of both the prey species. There are reports indicating that ladybirds discriminate their prey based on the parasitization status of the prey (Al-Zyoud and Sengonca, 2004). Discrimination could be based either on the physical changes in the prey after parasitization or on the presence of external or internal markings left by the female parasitoid during or after oviposition (van Baaren et al., 1995).

The larvae of ladybird, *C. maculata*, can discriminate between the eggs of *Trichoplusia ni* (Hubner) (Lepidoptera: Noctuidae) of different qualities. Eggs that are parasitized by the parasitoid, *Trichogramma turkestanica* (=*evanescens*) Meyer (Hymenoptera: Trichogrammatidae), are less accepted by *C. maculata* larvae as the immature development of parasitoid reaches towards completion (Roger et al., 2001; Al-Zyoud and Sengonca, 2004). Although *C. maculata* larvae prefer young eggs as their prey, the pre-imaginal development prolongs, the larval food intake increases and the survival rate decreases when *C. maculata* larvae are reared on old parasitized eggs rather than on young unparasitized eggs (Roger et al., 2001; Al-Zyoud and Sengonca, 2004). Thus, parameters like size, age and parasitization status of the pest species need to be assessed so as to successfully implement the biocontrol strategies for insect pest management.

8.3.5 Stage, Size, Sex and Age of Predators

In an agroecosystem, ladybirds partition their prey on the basis of their own size and/or the size of their prey. When kept together, ladybirds of small size capture smaller prey while those of large size capture the larger prey (Sloggett, 2008a, b; Costa, 2009; Patel et al., 2018). Optimal foraging theory also suggests that large predators prefer larger prey to maximize their energy intake (Stephens and Krebs, 1986). Besides the body size, food resource partitioning in ladybirds also depends upon the prey density. Small ladybirds have less energy requirements; they therefore rely upon low densities of smaller prey. However, they require high densities of larger prey that contain sufficient smaller instars, which can effectively fulfil their required energy needs (Dixon, 2007). The low densities of smaller prey cannot fulfil the high energy requirements of large ladybirds, and the large ladybirds therefore require either: (1) the low density of larger prey, or (2) the high density of smaller prey so that they can easily capture even the biggest prey instars to fulfil their high energy needs (Dixon, 2007; Sloggett, 2008a, b).

Chaudhary et al. (2015) found that when given choices between the prey instars, the small and large females of *M. sexmaculatus* consumed higher numbers of smaller and larger prey instars, respectively. However, small females exhibited modified Type II functional response on the smaller prey instar and Type II functional response on the larger prey instar. In contrast, the large females exhibited only Type II functional responses when provided with either the smaller or larger prey instar. While in small females, oviposition was positively correlated with the number of both smaller and larger prey instars consumed, the relationship was stronger for the smaller prey instar. However, in large females, the oviposition was positively correlated with only the number of larger prey instars consumed. It therefore appears that the prey-abundant conditions are efficiently exploited by the large ladybirds; however, the small ladybirds are more effective in exploiting the prey under scarce prey conditions (Dixon, 2007; Sloggett, 2008a, b). Investigations have further found that prey colonies are initially attacked by the small and then the large ladybird species (Dixon, 2007; Sloggett, 2008a, b), indicating the former to be more effective during exploitative competition than the latter (Omkar et al., 2014). Therefore, the abundance or scarcity of prey governs the coexistence of ladybirds within agroecosystems.

Prey preferences in ladybirds are also affected by their stage, sex and age. Previous findings have found that in ladybirds the fourth instars have higher prey consumption rates than the early instars. Contrary to the fourth instar, the early instars have higher prey conversion efficiencies and growth rates (Omkar et al., 2014; Kumar et al., 2014b). Higher prey consumption by the fourth instars and adult females over the males have also been reported earlier (Omkar and Srivastava, 2001; Kumar et al., 2013). Moreover, not all prey species are equally preferred by both larvae and adults. While evaluating the preference of *A. gossypii* in *C. transversalis* and *M. sexmaculatus*, Veeravel and Baskaran (1995) found that only the larvae showed preference for the prey and not the adult stage. Soares et al. (2004) observed that prey preference of *H. axyridis aulica* differed between genders.

No feeding preference was observed for the females in the presence of both *M. persicae* and *A. fabae*; however, males showed preferences for *M. persicae*. While assessing the feeding habits of ladybird, *N. arcuatus*, on the two mealybugs, *N. viridis* and *P. citri*, Zarghami et al. (2014) found that the prey preference of females did not change with the developmental stage of the mealybug. However, males reared on either *N. viridis* or *P. citri* showed significant preferences for the first-instar nymphs of *P. citri* over those of *N. viridis*.

In addition, a decline in prey consumption and the rate of food assimilation was reported in ladybirds with an increase in their age (Veeravel and Baskaran, 1995; Dixon and Agarwala, 2002). Mishra et al. (2012) evaluated stage- and age-specific predatory patterns and nutritional ecology of the four co-occurring ladybirds, viz., *C. septempunctata*, *C. transversalis*, *M. sexmaculatus* and *P. dissecta* on pea aphid. Their results found that small ladybirds (*M. sexmaculatus* and *P. dissecta*) and early larval instars had lower consumption rates but had higher food conversion efficiencies and growth rates than the large ladybirds (*C. septempunctata* and *C. transversalis*) and late instars. Moreover, the prey consumption rates of the larvae of the four ladybird species increased with increasing age until pupation. The adult females of the four ladybirds had higher consumption rates at the early age than at the later age, which further decreased with increasing age.

Bista and Omkar (2014) observed similar findings while assessing the predation potential and reproductive attributes of *C. septempunctata* and *C. transversalis* under low, medium and high prey densities. Their findings also revealed that the females of both the ladybirds exhibited triangular fecundity function with age, and the highest reproductive performance was recorded during the middle age on the three prey densities; being highest on the high prey density. On the three prey densities, females also exhibited higher body biomass conversion efficiency during pre- and post-oviposition periods, and higher egg biomass conversion efficiency during the oviposition period; being highest on the high prey density. Such findings therefore suggest that the middle-aged ladybirds reared on high prey density may suppress the pest populations better than the early and late aged ladybirds reared on low or medium prey density.

8.3.6 Learning Ability and Memory of Ladybirds

Ladybirds respond to diverse external cues while in search of food, shelter, oviposition sites or potential mates (Dukas, 2008). Most of these cues are innate, but some may be learned by the predator during its lifetime (Remen, 2004; Giurfa, 2013, 2015). If the learnt information is integrated into the prey searching process, the rate of encountering the prey species by predators may get increased (Papaj, 1993). Optimal foraging theory suggests that the predator has to maximize its encounter rate while searching for the most suitable and most profitable prey, and avoid searching for poor-quality prey (Stephens and Krebs, 1986). For generalist ladybirds that feed on numerous prey species living in different microhabitats, the recognition of specific cues to identify the most suitable prey is unlikely (Papaj, 1993; Hirvonen et al., 1999). Under such conditions, learning may be adaptive because it allows the ladybirds to acquire the neuronal representation of spatial environmental configuration, sensory information or associations between the perceived stimuli and the environmental state (Dukas, 2008), and to adapt to a variety of situations that may be encountered during the prey search (Lewis et al., 1990).

Predaceous ladybirds use olfactory and visual cues to detect resource-rich habitats and, with experience, they learn to associate these cues with the most profitable habitats containing the suitable prey (Hodek et al., 2012; Rondoni et al., 2017). Learning is therefore likely to be an important capability to increase the overall foraging efficiency of ladybirds. Studies have shown that the pre-imaginal and adult learning is widespread among phytophagous insects (Dukas and Bernays, 2000; Xue et al., 2007), parasitoids (Stireman, 2002; van Baaren et al., 2005; Dauphin et al., 2009) and social insects (Farina et al., 2005). However, the role of learning in prey selection by ladybirds has received very little attention (Boivin et al., 2010; Hodek et al., 2012). Studies have reported an improvement in searching behaviour (Ettifouri and Ferran, 1993), capture efficiency (Guillette et al., 2009), and prey recognition in adult predators (Guershon and Gerling, 2006) due to previous learning experiences (Dejean et al., 2003). Papaj and Prokopy (1989) have proposed that learning occurs when an individual's behaviour: (1) changes in a repeatable way as a consequence of experience, (2) changes gradually with continued experience and (3) vanishes in the absence of continued experience of the same type, or as a consequence of a novel experience or trauma (Boivin et al., 2010). Thus, the evolution of learning depends on the predictability of the environment within or between generations (Boivin et al., 2004), but learning is more likely to occur when the environment changes between generations but remains constant within a generation (Stephens, 1991).

Boivin et al. (2010) reported that hunger and experience with other food types may have little effect on the rate of learning, but learning may change the initial level of rejection of low-quality prey. The initial level of rejection of low-quality eggs was higher in *Coleomegilla maculata lengi* Timberlake (Coccinellidae) larvae previously fed on aphids than in starved larvae. However, opposite was reported when larvae were previously fed on low-quality eggs (Vet et al., 1990; van Baaren et al., 2005). Although, on the basis of their previous experiences, larvae can select their prey more effectively in mixed patches of prey, learning may further improve their overall prey searching efficiency (Punzo and Garman, 1989; Papaj and Vet, 1990; Johnson, 1991; Boivin et al., 2010).

8.3.7 Environmental Factors

The environment consists of both abiotic and biotic factors. Among the various abiotic factors, temperature and photoperiod are the crucial ones that directly affect the population of insect pests and their predators, viz., the ladybird beetles (Omkar et al., 2009b; Dell et al., 2013; Kumar and Omkar, 2015; Omkar and Kumar, 2016; Morales-Ramos and Rojas, 2017).

8.3.7.1 Temperature

A broad temperature range may affect the geographic range of insect populations by either causing their mortality or limiting the range of host plants or the prey species. Ladybirds and their prey species are poikilothermic in nature. The temperature significantly affects the various life processes, viz., growth, development, survival, longevity, fecundity and fertility (Omkar and James, 2004b; Pervez and Omkar, 2004b; Omkar et al., 2009a; Ramalho et al., 2009; Regniere et al., 2012; Majumder and Agarwala, 2013; Jarosík et al., 2014; Omkar and Kumar, 2016; Morales-Ramos and Rojas, 2017). Studies have revealed that increased environmental temperature increases the metabolic rates of ladybirds and finally their rates of prey intake (Ba M'Hamed and Chemseddine, 2001). The fitness of ladybirds decreases as temperature increases above optimal (Denis et al., 2011). Consequently, ladybirds avoid low temperature by either migrating to optimal temperature sites or undergoing diapause. Temperature also sets the limits of performance of ladybirds in terms of low and high thermal thresholds, and these thresholds are the major determinants of their various life processes.

The rates of development (Perdikis and Lykouressis, 2002) and food consumption (Anderson et al., 2001) in ladybirds generally increase with increasing temperature (Kumar and Omkar, 2015; Omkar and Kumar, 2016). The metabolism rates also vary with changing temperature (Sentis et al., 2012). Investigations on ladybirds have found that their predation rate, which is a modified Holling's type II functional response, is temperature dependent, and changing temperature influences the predator's time to search, capture and handle the prey (Gillooly et al., 2002; Rahimi et al., 2017). Low temperature slows down the rate of development, and strong fluctuations in daily temperature do not allow ladybirds to meet their nutritional needs, leading to their higher mortality. Studies have shown that the impact of temperature on the rate of prey consumption differs amongst species. Moreover, a temperature regime suitable for one ladybird species may be less or unsuitable for the other species (Allawi, 2006).

While studying the temperature effects on functional response of two-spotted ladybird, *A. bipunctata* on green peach aphid, *M. persicae*, Jalali et al. (2010) reported that the ladybird exhibited type II functional response on temperature ranging from 19°C to 27°C. Although the number of prey captured by the predator increased with increase in temperature, the prey handling time significantly decreased as temperature increased from 19°C to 27°C.

Studies have further shown that the ladybirds kept at high temperature complete their development earlier than the ones kept at low temperature (Omkar and Kumar, 2016). However, the proportion of total time required to complete the development per stage remains the same, a phenomenon known as 'development rate isomorphy' (Jarosík et al., 2014). In such a case, the lower thermal threshold, i.e., temperature below which development ceases, is uniform amongst the developmental stages within a population and within a species (Dixon et al., 1997). While assessing predation and developmental attributes of two ladybird species, *C. transversalis* and *C. septempunctata* on pea aphid, Omkar and Kumar (2016) found the attributes to be highest at 25°C than at 15°C, 20°C, 30°C and 35°C. Non-linear regression models further extrapolated ~11°C and ~10.5°C as lower development thresholds and ~41°C and ~43°C as upper development thresholds for *C. transversalis* and *C. septempunctata*, respectively. The study further revealed that a ubiquitous ladybird is less susceptible to climatic changes than an indigenous ladybird and the former is widely distributed to different geographical areas, thereby behaving as a good biocontrol agent than the latter.

High temperature accelerates development in ladybirds and causes early emergence of larvae from the egg stage (Omkar and Pervez, 2004; Damos and Savopoulou-Soultani, 2012). Temperature variations affect the daily activity periods, causing either the ladybird predator or the prey species to be more or less active. Consequently, the predator search time is affected by the temperature (Logan et al., 2006). The energy partitioning model also explains the effect of temperature and age in shaping the fecundity function in ladybirds (Kindlmann et al., 2001). Temperature-dependent age-specific fecundity function of ladybirds is triangular in shape with initial increase in daily oviposition rate, reaching a peak and thereafter declining with further increase in age (Pervez and Omkar, 2004b). Gradual increase in temperature increases this peak of oviposition rate towards the younger side of ageing ladybirds. At optimum temperature (i.e. 27°C–30°C), the oviposition peak is attained at the youngest age (Pervez and Omkar, 2004b). Thereafter, this peak gets delayed and shortened with further increase in temperature.

Increasing temperature also accelerates the progeny development in ladybirds and reduces their developmental durations and mortality. This has earlier been reported in *A. bipunctata* larvae by Schuder et al. (2004) and *Stethorus tridens* Gordon by Fiaboe et al. (2007). Developmental durations of immature stages of ladybird, *S. subvillosus* reared on plum aphid, *Hyalopterus pruni* (L.) decreased with increasing temperature (Atlihan and Chi, 2008). Similarly, the total developmental duration of *C.s bipustulatus* reared on oleander scale, *Aspidiotus nerii* Bouché was the lowest at 32.5°C and highest at 15°C (Eliopoulos *et al.*, 2010). Survival rate of larvae of *P. quatuordecimpunctata* was lowest (14%) at 32.5°C and the highest (96%) at 25°C. In addition, temperature displayed a profound effect on female longevity, which decreased from 123.06 days at 17°C to 33.53 days at 30°C (Papanikolaou et al., 2013). Both, longevity and feeding rates of ladybird, *M. discolor* were lower at 23.5°C when compared with those reared at 30.71°C (Hong et al., 2013). Khan and Spooner-Hart (2017) studied the influence of temperature on the development of the ladybird, *Stethorus vagans* Blackburn, fed on *T. urticae* at seven constant (10°C, 12°C, 15°C, 20°C, 25°C, 30°C and 35°C) and fluctuating (12.7°C and –32.1°C) temperatures and found a strong positive correlation between the rate of development and temperature. Thus, the pest biocontrol strategy may be at the threshold levels if prey–predator relationships are established at the optimum temperature range.

8.3.7.2 Photoperiod and Light Intensity

Light is one of the major abiotic factors that determine the timing of a ladybird's activity synchronized with its environment (Luker et al., 2002; Hao et al., 2013). The duration of light in a day–night cycle (24 h) is termed as photoperiod. Light is described in terms of its properties of periodicity (seasonal day length changes), quality (i.e. spectral profile) and quantity (i.e. light intensity) (Downing and Litvak, 2002). Insects predict the future seasonal environmental conditions through the alteration in day length. They perceive alteration in photoperiod very efficiently, even when the differences from day to day are relatively small at some times of the year and at lower latitudes. Chunju et al. (2012) reported that photoperiod, as an ambient environmental factor, was important for the stimulation of alate aphids in *S. graminum* population. Based on the photoperiodic assumptions given by Danks (1987), insects are classified into the following categories: (1) Long photoperiod/daylength inducing development, (2) short photoperiod/daylength inducing development, and (3) intermediate photoperiod/day length (close to the critical day length) inducing development. Most insect species have accelerated development under short days in comparison to long days (Lopatina et al., 2007).

Danks (1991) found that long photoperiodic condition was accelerating growth in young larvae but decelerating the growth in older ones. When reared under two different photoperiodic conditions, *viz.*, 14L:10D and 24L:0D, larvae of ladybird, *P. dissecta* had faster development with higher survival rates at 14L:10D photoperiod (Mishra and Omkar, 2005). In ladybirds, *Coelophora saucia* (Mulsant) (Omkar and Pathak, 2006) and *H. axyridis* (Berkvens et al., 2008), the egg hatching rate was not consistently different under long and short photoperiods. But Omkar and Pathak (2006) found maximal mortality in *C. saucia* under short-day conditions followed by continuous day and long-day conditions. Omkar and Singh (2007) found high incidence of pupation during late photophase and early scotophase in *M. sexmaculatus*. Under short-day conditions, the pre-imaginal development was faster than under long-day conditions in *H. axyridis* (Reznik and Vaghina, 2011). However, Omkar and Kumar (2016) found high prey consumption rates at long-day lengths than at short-day lengths in ladybirds, *C. septempunctata* and *C. transversalis*. Such a behaviour in ladybirds further reveals that long photoperiodic conditions facilitate foraging and higher prey exploitation than the short photoperiodic conditions (Kumar and Omkar, 2015; Omkar and Kumar, 2016; Singh et al., 2016a,b).

Light intensity also affects the bioattributes of ladybird beetles. Since ladybirds are diurnal insects and depend on visual cues, the presence of light is essential to complete activities, like predation, mating, moulting and pupation (Omkar and Pathak, 2006; Omkar et al., 2006; Nalepa, 2013; Wang et al., 2014; Singh et al., 2016a,b). The white light is the more suitable light component essential for biological activities in ladybirds than the red and blue components of the visible spectrum (Omkar and Pathak, 2006). Thus, light influences many aspects of an insect's life. Investigations of the complete photoperiodic and/or light intensity responses of insect pests and their predators are necessary to predict their phenology and potential geographic distribution (Koch and Galvan, 2008).

8.3.8 Genetic Factors

The local populations of polyphagous and generalist ladybirds that display variations in the use of host species undergo natural selection to improve their ability to use the host (Via, 1990; Ueno et al., 2001). Within a population, the variation in the ability to use different hosts may reflect the past selective forces. When assessing the genetic differences in performance of herbivorous ladybird beetle, *Epilachna vigintioctomaculata* Motschulsky, on both the normal and the novel host, Ueno et al. (2001) found higher heritability in performance on the novel host than the normal host. The results suggested that natural selection might improve the ability of the ladybird population to use a particular host plant that had reduced genetic variation within the population in the ability to use the host plant.

Thus, estimation of genetic variations within and among the populations of polyphagous or generalist ladybirds is essential to understand the different adaptations in host/prey species (Via, 1990). Understanding host/prey species adaptation requires the explanation of the presence and nature of genetic variation and covariation in host/prey use within these populations of polyphagous ladybirds. Since selection changes gene frequencies, the genetic variation changes per generation of selection. Natural selection reduces the genetic variation through fixation and elimination of alleles in the population, the continual selection results in an evolutionary plateau, as confirmed in studies with small artificial populations (Hill, 1982). However, the effect of selection on genetic variation for natural populations have yet to be established where the effective population size is often large (Stearns, 1992). In a large population, it takes a long time for selection to eliminate the genetic variations; mutation may regenerate certain variations for polygenic characters.

8.4 PREY SEARCHING BEHAVIOUR OF LADYBIRDS

The classification of environmental levels apparent to foragers, as suggested by Hassell and Southwood (1978), in terms of prey, patch and habitat has provided a framework for discussing the foraging behaviour of ladybirds. Adult beetles locate the prey patch and their larvae feed on the prey in these patches. Movement within the aggregate of prey is referred to as local or *intensive search* (Bond, 1980) or area restricted search (Curio, 1976), while those between aggregates within a patch is known as ranging or *extensive search.* Furthermore, the movement between the patches is known as *dispersal,* and that from patches to hibernation sites is referred to as *migration* (Ferran and Dixon, 1993; Hodek and Evans, 2012).

8.4.1 Pattern of Foraging

Predators initially locate the prey by extensive search and thereafter switch over to intensive search once the feeding initiates (Jander, 1975). If no further prey is captured, the predators gradually change their prey search mode from intensive to extensive search. Extensive search is characterized by a relatively linear and fast movement, whereas the intensive search is marked by sinuous slow tracks interrupted by numerous stops. The length of time spent in intensive search is called the 'giving up time'. During a stop, the larvae cast their body upwards and sidewards, thereby increasing the area within which the prey may be located (Banks, 1957; Chandler, 1969). Intensive search is considered as adaptive when the predators hunt the clumped prey as it increases the chances of finding the prey (Dixon, 1959; Ferran and Dixon, 1993).

8.4.2 Prey Recognition

While undergoing extensive and intensive searches, ladybirds require both the environmental and internal informations (Papaj and Prokopy, 1989). These informations orientate the ladybirds towards their prey. They are classified into two main groups (Ferran and Dixon, 1993):

1. **External sensory information**: It is typically the visual, chemical or tactile information. This information may be directional (visual stimuli, a trail, veins and edge of leaves) or nondirectional (pheromone).
2. **Internally stored information**: It is transmitted from proprioreceptor in or near the locomotory organs and is stored when the individuals move. It may be acquired through learning or heredity.

8.4.3 Extensive Search

Previous studies have identified that ladybirds do not perceive their prey before making any contact (Banks, 1957; Kaddou, 1960; Kehat, 1968). The efficiency of this random search is enhanced by both the predator and the prey, being negatively geotactic and positively phototactic.

8.4.3.1 Visual Cues

Studies have suggested that adult ladybirds orientate visually towards their prey (Nakamuta, 1984, 1985). Both adults and larvae perceive the colour contrast and orientate towards the objects with sharpest colour contrast relative to the background. They prefer the objects with a complex shape (Said et al., 1985; Harmon et al., 1998; Wang et al., 2015). Nakamuta (1985) has suggested that ladybirds can visually discriminate the prey from the non-prey object by its shape and size. Obata (1986) suggests that *H. axyridis* utilizes the visual cues in locating its prey. Still evidences are lacking that can indicate that larvae use visual cues when searching for the prey.

8.4.3.2 Olfactory Cues

Prey species secrete alarm pheromones in response to attack by predators that causes the other prey to disperse (Nault et al., 1973). Synthetic (E)-farnesene, which is present in aphid alarm pheromone, behaves as the principle olfactory cue for ladybirds (Nakamuta, 1991; Francis et al., 2004; Hatano et al., 2008; Singh et al., 2016a,b). Aphid alarm pheromones have been reported to attract *A. bipunctata* (Francis et al., 2004), *C. septempunctata* (Al Abassi et al., 2000; Ninkovic et al., 2001), *C. maculata* (Zhu et al., 1999), *Cryptolaemus montrouzieri* (Meidari and Copland, 1992), *H. axyridis* (Verheggen et al., 2007), *H. convergens* (Acar et al., 2001) and *M. sexmaculatus* (Singh et al., 2016a,b). Aphid odour greatly influences the oviposition site selection strategy in ladybirds (Evans and Dixon, 1986; Sarmento et al., 2007). The ovipositing ladybirds therefore assess the aphid patch quality for the age of the aphid colony (Hemptinne et al., 2000), aphid density (Oliver et al., 2006) and its suitability as a prey (Fréchette et al., 2006), prior to oviposition.

Wang et al. (2015) evaluated the role of olfactory cues using foraging-naıve and foraging-experienced *P. japonica* adults and cotton plants with or without infestation by the cotton aphid. Their findings revealed that experienced beetles were more responsive to olfactory cues than the naıve beetles. Singh et al. (2018) found that ladybird, *M. sexmaculatus*, was able to perceive the presence of prey in its environment and modify its oviposition accordingly. The findings therefore suggest that besides the actual prey consumed, the perception of prey may also influence oviposition in ladybirds. In addition, the host plants may also provide olfactory and visual cues that are recognized by predators in the field. These host plant–based cues, which are altogether absent in laboratory experiments, may affect the predator's choice for a particular prey (Ninkovic et al., 2001; Zhu and Park, 2005; Rondoni et al., 2017).

8.4.3.3 Gustatory Cues

Larvae of *C. septempunctata* mark the host plant chemically while searching for the prey (Marks, 1977; Kumar and Mishra, 2014). This enables subsequent short-term recognition and avoidance of areas that have previously been searched by the larvae. The semiochemical tracks that are left by the larvae during foraging may fulfil the following functions: (1) means of avoiding cannibalism and intraspecific competitions (Magro et al., 2010), (2) reduce overexploitation of food resources that are limited in time and space (Nufio and Papaj, 2001) and (3) signals for other species to prevent IGP (Meisner et al., 2011). Such markings, however, do not affect the prey searching ability of other larvae.

8.4.4 Intensive Search

The switching over to intensive search mode by the ladybirds after ingesting the prey indicates the switch from extensive to intensive search results from the ingestion of prey (Nakamuta, 1984). Consequently, the intensive search is also called as '*success motivated search*' (Vinson, 1977). As most ladybirds are polyphagous in nature, chemical substances common to many aphid species might stimulate the changes in movements of ladybirds (Hodek, 1973). Dixon (1958) noted that as soon as *Adalia decempunctata* L. penetrated the body wall of aphid, *H. pruni*, the prey was rejected. In contrast, *A. bipunctata* could not distinguish between the toxic prey, *Megoura viciae* and the non-toxic prey, *A. pisum* (Blackman, 1967).

It appears that the chemicals excreted by aphids might induce intensive search in ladybirds. Carter and Dixon (1984) reported that the presence of aphid honeydew increased the time *C. septempunctata* spent searching for the prey on the host plant. In addition, the major internal factor that affects intensive search in ladybirds is the hunger (Carter and Dixon, 1982). Starvation causes the ladybirds to adopt an intensive search mode before feeding, and to search more intensively with less prey specificity. But, remaining without food for too long a period leads to gradual decrease in activity and finally the death of the ladybird (Nakamuta, 1987).

8.4.5 Learning

If the ladybird species profits from experience, learning can improve its foraging efficiency by responding appropriately to the changing environment (Drost et al., 1988). *Conditioning* is the most common type of learning and is previously reported in order Coleoptera (Papaj and Prokopy, 1989). Conditioning may bring about changes in the feeding preferences of ladybirds. Each ladybird species prefers a particular habitat (Iperti, 1966), like *C. septempunctata* is found mainly on herbaceous plants, whereas *A. bipunctata* inhabits the trees. In their respective habitats, these ladybirds eat all the aphid species they encounter. However, the foraging behaviour of larvae varies according to the food they had previously eaten (Ettifouri and Ferran, 1993). Studies have found that after feeding the same food, the larvae changed their search from extensive to intensive mode (Hattingh and Samways, 1995), but when they were fed a different food, the larvae either continued extensive search mode or adopted a less sinuous path than in the typical intensive search mode. While these changes are temporary, they may however alter the predatory responses (Ferran and Dixon, 1993).

8.4.6 Factors Affecting Foraging Behaviour in Ladybirds

The foraging behaviour in ladybirds is affected by both the abiotic and biotic factors, and also by the physiological state of the predators.

8.4.6.1 Abiotic Factors

Environmental factors, like temperature, humidity and light intensity, significantly affect the prey searching behaviour of ladybird predators by modifying their mobility rates. Ladybirds warm up by absorbing the solar radiations (Jankowsky, 1973). Rapid movements under high temperature may be adaptive because predators need more and more prey to overcome the depletion of their energy and water reserves. Desiccation often stimulates the searching for prey by ladybirds because aphid prey is an important source of water. Climatic conditions may also affect the distribution of predators on host plants through phototropic and geotropic responses. Curio (1976) suggested that daily pattern in the search for prey is more or less dependent on the opportunities to feed rather than rigidly linked to circadian activity. Thus, foraging behaviour of ladybirds follows a rhythmical pattern controlled by a combination of endogenous factors and exogenous climatic components (Omkar and Pathak, 2006).

8.4.6.2 Biotic Factors

8.4.6.2.1 Host Plant

Ladybird species having limited sensory abilities depend on the physical and chemical characteristics of the host plant to orient towards their prey (Carter et al., 1984; Jalali and Ziaaddini, 2017). Plant density affects both the ladybird abundance and the prey species composition. In cereal crops, the thermophilic ladybird, *C. septempunctata*, is more abundant at low-density strands and *P. quatuordecimpunctata* prevails in dense strands (Honek, 1983). The rate at which a predator encounters a prey can be affected by the presence of hairs, pubescence, density of trichomes and waxes on the host plant (e.g. Kumar et al., 1999; Timms et al., 2008; Karami and Shishehbor, 2012). Banks (1957) found that the larvae of ladybird, *P. quatuordecimpunctata* moved much slowly on the hairy leaves of potato than on the glabrous leaves of bean. Ladybirds generally concentrate their search along the veins of leaves where aphid density remains the highest (Dixon, 1959). On leaves lacking the prominent veins, ladybirds search the edges of leaves and the petiole (Carter et al., 1984).

Even the physiological state of the host plant may affect the foraging behaviour of ladybirds. Kesten (1969) found that adults of ladybird, *Anatis ocellata* L., assessed the pine needles with their mandibles and searched the new needles only when they had to reduce the amount of time in searching areas where the prey was absent. Plant structures may also protect the aphids from their predators. The third and fourth larval instars were unable to enter the space between the stem and the ear of wheat where the abundance of aphids was maximal.

8.4.6.2.2 Prey Species

The quality of prey significantly affects the duration and speed of prey search, its capture rate and the movement of ladybird from intensive to extensive search. Majority of the studies on prey quality have correlated its nutritional value with the predation rates of the ladybird species (Omkar, 2006; Kumar et al., 2013; Patel et al., 2017). As the adoption of an intensive search mode follows the ingestion of suitable prey (Nakamuta, 1984), the locomotory pattern of predators may be linked to the prey quality (Houck, 1986; Ettifouri and Ferran, 1993). The response to prey quality possibly depends on the food specificity of ladybirds (Pervez and Omkar, 2004a; Omkar et al., 2009a, 2011). Intensive search by ladybirds is also related to the size of prey consumed. *Coccinella septempunctata brucki* showed significantly longer periods of intensive search when fed with a smaller thereafter a larger aphid than the *vice versa* (Nakamuta, 1985). The rate of prey consumption usually increases along with prey density until ladybirds get satiated. This prey consumption rate may further help the ladybirds to:

1. Learn to find prey more readily at some critical prey encounter rate and then concentrate feeding on it.
2. Spend less time in non-foraging activities at high prey encounter rates.
3. Emigrate from prey patches more readily at low prey densities.

Distance between patches and distribution of prey within patches may also affect the foraging behaviour and searching success of ladybirds. Murakami and Tsubaki (1984) assessed the searching efficiency of *C. septempunctata brucki* when attacking the prey populations with different distributions and densities. They found that searching efficiency was correlated with prey distribution. At low prey density, the ladybird was highly efficient in exploiting the prey when the prey were distributed uniformly. However, at high prey density, the ladybird was mostly efficient when prey was highly aggregated.

Many aphid species respond visually to walking ladybirds (Klingauf, 1967), to the tactile stimuli, the plant vibrations induced by predators (Brodsky and Barlow, 1986) and the olfactory stimuli like the alarm pheromones emitted by the captured aphids (Ferran and Deconchat, 1992). The most active form of avoidance behaviour shown by aphid prey is walking away, kicking and dropping. Dropping from plants is a fairly common response but is not without risks (Dixon, 1958), as aphids frequently fail to find another plant (Roitberg et al., 1979) and may be captured by ladybirds walking on the ground (Ferran et al., 1991). Morphology of prey, like thickness of cuticle, length of appendages and body size, may affect the preference of ladybirds for the particular prey (Nedved and Salvucci, 2008). Being toxic, unpalatable and defended chemically may also help the prey to escape from the ladybird. The siphuncular wax is the most obvious chemical defence of many aphids against the ladybird larvae (Pasteels, 2007).

8.4.6.2.3 Predators

All species show certain specific patterns of movement during feeding, and the area they traverse per unit time depends on their size or the age. Newly hatched larvae remain on the egg cluster for some time after hatching and feed on unhatched eggs for their survival. The first-instar larvae are poor at capturing the prey, and this stage is a critical period in a ladybird's development (Dixon, 1958, 1959). The crowding of conspecific larvae promotes social feeding for growth, development and survival (Omkar and Pathak, 2009). However, above optimum level, crowding promotes intraspecific competition among the larvae, negatively affecting their growth and development (Omkar and Afaq, 2009; Bista et al., 2012).

While exploiting the aphid colonies both larvae and adults of ladybirds release semiochemicals as they walk (Kumar and Mishra, 2014). These secretions contain non-volatile hydrocarbons and act as signals for ladybirds to identify the previously searched areas (Magro et al., 2007; Ferrero et al., 2011). Chemical analyses show that semiochemical tracks of different ladybird species vary greatly (Hemptinne et al., 2001; Michaud and Jyoti, 2007). Females refrain from laying eggs and larvae desist from feeding

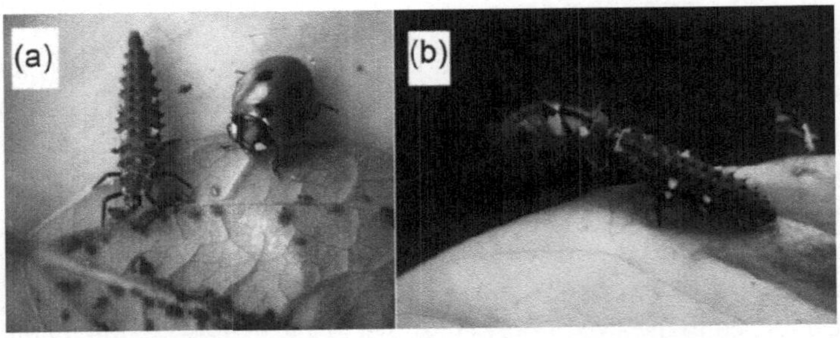

FIGURE 8.5 (a) Group feeding by larva and adult of *Coccinella septempunctata* on a patch of aphid prey; (b) intraspecific interaction between larva and adult of *Coccinella septempunctata* for the common aphid prey.

on the aphid colonies where they perceive the presence of semiochemical tracks left by conspecific or heterospecific ladybirds (see Kumar and Mishra, 2014).

8.4.6.2.4 Intraspecific Competition

Ladybirds aggregate on those plants where the prey abundance is high (Huffaker et al., 1971). When satiated, the intensive search mode and inactivity of ladybirds result in their aggregation in areas of high prey abundance (Dixon, 1959; Kareiva and Odell, 1987). As the density of ladybirds rises, the overall total number of prey consumed increases; however, the rate of prey consumption per individual decreases (Bista et al., 2012). Consequently, there is rapid prey depletion and the incidence of mutual interference between ladybirds, causing the ladybirds to scatter (Eveleigh and Chant, 1982). Such mutual interferences cause the ladybirds to fall off a plant, switch their prey searching mode from intensive to extensive, and ultimately ladybirds stop feeding (Ferran and Dixon, 1993) (Figure 8.5).

8.5 CONCLUSIONS

Analysis of food relationships in ladybirds may be helpful in forecasting their activities in agricultural fields and improving their impact in relation to conservative and augmentative biocontrol strategies for insect pests. While the prey of ladybirds is classified as essential, alternative, rejected and toxic, the essential prey is preferred by ladybirds to perform their best in terms of egg maturation, oviposition and development. Till date, numerous factors have been explored that affect prey preference in ladybirds, viz., prey species, prey abundance, prey mobility and its defence, morphological characters and previous feeding experiences of predators, size of predator and prey, and environmental and genetic factors. In addition, numerous cues guide the feeding behaviour of ladybirds, like the volatiles of the host plant and the prey species, aphid alarm pheromones and the oviposition-deterrent semiochemicals of conspecific or heterospecific ladybirds. Since the food resources of phytophagous and coccidophagous ladybirds are rather stable, specialized adaptive characteristics have evolved in aphidophagous species under the pressure exerted by inherently ephemeral occurrence of their prey. This constraint has led to the evolution of adaptive life-history traits, like fast development, high adult mobility and high speed of larval movement (see Hodek et al., 2012). Future studies in ladybirds may be designed targeting such evolutionary aspects. Studies related to ladybird orientation to volatiles have not been done beyond laboratory experiments, and field trials are still needed to validate the laboratory findings. Moreover, suitability of any prey is generally established after assessing the life-history parameters of ladybirds, such as their successful completion of larval development and high reproductive performance. While earlier researches have worked with single prey, some recent studies have revealed the importance of mixed food. Modern serological and molecular methods have also been developed to detect the types of prey consumed by the ladybird predators (Rondoni *et al.*, 2015). Future studies on prey–predator relationships in Coccinellidae may also be designed to develop and/or validate such sophisticated methods.

8.6 POINTS TO REMEMBER

i. Ladybird beetles are predaceous and polyphagous insects. Their prey range is classified as essential, alternative or rejected.
ii. Ladybirds feeding on essential prey survive better as larvae and are more fecund as adults.
iii. Those feeding on alternative prey can only survive, but they neither grow nor reproduce. Rejected prey species are unpalatable and are rejected even after their encounter.
iv. Ladybirds accept some prey species, which are non-toxic, but worsen their life-history parameters. Such prey species are considered as 'problematic prey'.
v. Factors that affect prey preferences of ladybirds, include: (1) prey quality, (2) prey quantity, (3) morphology, mobility and defence of prey, (4) learning ability, memory, morphological characters and previous feeding experience of predator, (5) environmental parameters (temperature and photoperiod) and (6) genetic factors.
vi. While ladybirds locate their prey by extensive search, they switch over to intensive search once feeding of prey is initiated.
vii. During both the extensive and intensive searches, ladybirds sense environmental and inner cues that orientate them towards the prey.
viii. In addition, the abiotic factors, like temperature, humidity and light intensity; and the biotic factors, like the prey species and co-existing predators affect the searching behaviour of ladybirds by modifying their rate of movement.
ix. The factors that affect prey–predator relationships in ladybirds need to be fully explored so that suitable species of ladybirds may be selected for mass multiplication in laboratories against the pests of agricultural importance.

REFERENCES

Abdollahi, A.G., Afshari, A., Baniameri, V., Dadpour, H., Asadeh, G. & Yazdanian, M. (2010). Functional response of fourth larval instars and female adults of *Cryptolaemus montrouzieri* Mulsant (Col.: Coccinellidae) to citrus mealybug, *Planococcus citri* (Risso) (Hom.: Pseudococcidae) in laboratory conditions. *Proceedings of 19th Iranian Plant Protection Congress*, Tehran, pp. 1–69.

Acar, E.B., Medina, J.C., Lee, M.L., & Booth, G.M. (2001). Olfactory behavior of convergent lady beetles (Coleoptera: Coccinellidae) to alarm pheromone of green peach aphid (Hemiptera: Aphididae). *The Canadian Entomologist*, 133(3), 389–397.

Agarwala, B.K. & Dixon, A.F.G. (1992). Laboratory study of cannibalism and interspecific predation in ladybirds. *Ecological Entomology*, 17: 303–309.

Agarwala, B.K., & Bhowmik, P.J. (2011). Effect of resource gradient on age and size at maturity and their influence on early-life fecundity in the predatory Asian lady beetle. *Harmonia axyridis*. *Entomologia Experimentalis et Applicata*, 141(2): 97–102.

Agarwala, B.K., Bardhanroy, P., Yasuda, H. & Takizawa, T. (2001). Prey consumption and oviposition in aphidophagous predator *Menochilus sexmaculatus* (Coleoptera: Coccinellidae) in relation to prey density and adult size. *Environmental Entomology*, 30: 1182–1187.

Agarwala, B.K., Yasuda, H. & Sato, S. (2008). Life history response of a predatory ladybird, *Harmonia axyridis* (Pallas) (Coleoptera: Coccinellidae) to food stress. *Applied Entomology and Zoology*, 43 (2): 183–189.

Agarwala, B.K., Singh, T.K., Lokeshwari, R.K. & Sharmila, M. (2009). Functional response and reproductive attributes of the aphidophagous ladybird beetle, *Harmonia dimidiata* (F.) in oak trees of sericultural importance. *Journal of Asia Pacific Entomology*, 12: 179–182.

Al Abassi, S., Birkett, M.A., Pettersson, J., Pickett, J.A., Wadhams, L.J. & Woodcock, C.M. (2000). Response of the *seven*-spot ladybird to an aphid alarm pheromone and an alarm pheromone inhibitor is mediated by paired olfactory cells. *Journal of Chemical Ecology*, 26: 1765–1771.

Allan, J.D., Flecker, A.S. & McClintock, N.L. (1987). Prey preference of stoneflies: Sedentary vs mobile prey. *Oikos*, 49: 323–331.

Allawi, T.F. (2006). Biological and ecological studies on *Scymnus syriacus* and *Scymnus levaillanti* (coleoptera: coccinellidae). *European Journal of Entomology*, 103: 501–503.

Al-Zyoud, F.A. & Sengonca, C. (2004). Prey consumption preferences of *Serangium parcesetosum* Sicard (Col., Coccinellidae) for different prey stages, species and parasitized prey. *Journal of Pest Science*, 77: 197–204.

Anderson, D.R., Link, W.A., Johnson, D.H. & Burnham, K.P. (2001). Suggestions for presenting the results of data analyses. *Journal of Wildlife Management*, 65: 373–378.

Arshad, M., Khan, H.A.A., Hafeez, F., Sherazi, R. & Iqbal, N. (2017). Predatory potential of *Coccinella septempunctata* L. against Four Aphid Species. *Pakistan Journal of Zoology*, 49(2): 623–623.

Atlihan, R. & Chi, H. (2008). Temperature-dependent development and demography of *Scymnus subvillosus* (Coleoptera: Coccinellidae) reared on *Hyalopterus pruni* (Homoptera: Aphididae). *Journal of Economic Entomology*, 101: 325–333.

Atlihan, R. & Guldal, H. (2009). Prey density-dependent feeding activity and life history of *Scymnus subvillosus*. *Phytoparasitica*, 37: 35–41.

Atwal, A.S. & Sethi S.L. (1963). Biochemical basis for the food preference of a predator beetle. *Current Science*, 32: 511–512.

Auster, F. & Schafer, J. (1956). Arzneipflanzen (7. Lieferung), Veb Georg Thieme, Leipzig.

Ba M'Hamed, T. & Chemseddine, M. (2001). Assessment of temperature effects on the development and fecundity of *Pullus mediterraneus* (Col., Coccinellidae) and consumption of *Saissetia oleae* eggs (Hom., Coccoida). *Journal of Applied Entomology*, 125: 527–531.

Banks, C.J. (1957). The behaviour of individual coccinellid larvae on plants. *British Journal of Animal Behaviour*, 5: 12–24.

Bayoumy, M.H. (2011). Functional response of the aphelinid parasitoid, *Aphytis diaspidis*: Effect of host scale species, *Diaspidiotus perniciosus* and *Hemiberlesia lataniae*. *Acta Phytopathologica Entomologica Hungarica*, 46(1): 101–113.

Bayoumy, M.H. & Michaud, J.P. (2012). Parasitism interacts with mutual interference to limit foraging efficiency in larvae of *Nephus includens* (Coleoptera: Coccinellidae). *Biological control*, 62(2): 120–126.

Berkvens, N., Bonte, J., Berkevens, D., Deforce, K., Tirry, L. & De Clercq, P. (2008). Pollen as an alternative food for *Harmonia axyridis*. In: Roy, H.E. & Wajnberg, E. (Eds.), *From Biological Control to Invasion: The Ladybird Harmonia Axyridis as a Model Species*. Springer Science & Business Media: Berlin/Heidelberg, Germany, pp. 201–210.

Berkvens, N., Bale, J.S., Berkvens, D., Tirry, L. & De Clercq, P. (2010). Cold tolerance of the harlequin ladybird in Europe. *Journal of Insect Physiology*, 56: 438–444.

Bertram, D.F. & Strathmann, R.R. (1998). Effects of maternal and larval nutrition on growth and form of planktotrophic larvae. *Ecology*, 79: 315–327.

Bista, M. & Omkar (2013). Effects of body size and prey quality on the reproductive attributes of two aphidophagous Coccinellidae (Coleoptera) species. *The Canadian Entomologist*, 145(5): 566–576.

Bista, M. & Omkar (2014). Consumption, developmental and reproductive attributes of two con-generic ladybird predators under variable prey supply. *Biological Control*, 74: 36–44.

Bista, M., Mishra, G. & Omkar (2012). Influence of crowding and diet on the development and survival of the ladybird *Brumoides suturalis* (Coleoptera: Coccinellidae) reared on two aphid species. *International Journal of Tropical Insect Science*, 32(1): 64–68.

Blackman, R.L. (1967). Selection of aphid prey by *Adalia bipunctata* L. and *Coccinella septempunctata* L. *Annals of Applied Biology*, 59: 331–338.

Boivin, G., Fauvergue, X. & Wajnberg, E. (2004). Optimal patch residence time in egg parasitoids: innate versus learned estimate of patch quality. *Oecologia*, 138: 640–647.

Boivin, G., Roger, C., Coderre, D. & Wajnberg, E. (2010). Learning affects prey selection in larvae of a generalist coccinellid predator. *Entomologia Experimentalis et Applicata*, 135: 48–55.

Bond, A.B. (1980). Optimal foraging in a uniform habitat: The search mechanism of the green lacewing. *Animal Behavior*, 28: 10–19.

Bonduriansky, R.A. & Head, M. (2007). Maternal and paternal condition effects on offspring phenotype in *Telostylinus angusticollis* (Diptera: Neriidae). *Journal of Evolutionary Biology*, 20: 2379–2388.

Borges, I., Soares, A.O., Magro, A. & Hemptinne, J.L. (2011). Prey availability in time & space is a driving force in life history evolution of predatory insects. *Evolutionary Ecology*, 25: 1307–1319.

Bressendorff, B.B. & Toft, S. (2011). Dome-shaped functional response induced by nutrient imbalance of the prey. *Biological Letters*, 7(4):517–520.

Britto, E.P.J, Gondim, M.G.C., Torres, J.B., Fiaboe, K.K.M., Moraes, G.J. & Knapp, M. (2009). Predation and reproductive output of the ladybird beetle *Stethorus tridens* preying on tomato red spider mite *Tetranychus evansi*. *BioControl*, 54: 363–368.

Brodsky, L.M. & Barlow, C.A. (1986). Escape responses of the pea aphid, *Acyrthosiphon pisum* (Harris) (Homoptera:Aphididae): Influence of predator type and temperature. *Canadian Journal of Zoology*, 64: 937–939.

Brown, G.C., Sharkey, M.J. & Johnson, D.W. (2003). Bionomics of *Scymnus (Pullus) louisianae* J. Chapin (Coleoptera: Coccinellidae) as a predator of the soybean aphid, *Aphis glycines* Matsumura (Homoptera: Aphididae). *Journal of Economic Entomology*, 96: 21–24.

Canovai, R., Benelli, G., Ceragioli, T., Lucchi, A., & Canale, A. (2019). Prey selection behaviour in the multicoloured Asian ladybird, *Harmonia axyridis* (Coleoptera: Coccinellidae). *Applied Entomology and Zoology*, 54: 213–222.

Carter, M.C. & Dixon, A.F.G. (1982). Habitat quality and the foraging behaviour of coccinellid larvae. *Journal of Animal Ecology*, 51:865–878.

Carter, M.C. & Dixon, A.F.G. (1984). Honeydew: an arrestant stimulus for coccinellids. *Ecological Entomology*, 9: 350–383.

Carter, M.C., Sutherland, D. & Dixon, (1984). Plant structure and the searching efficiency of coccinellid larvae. *Oecologia*, 63: 394–397.

Castro-Guedes, C.F., de Almeida, L.M., Penteado, S.D.R.C., & Moura, M.O. (2016). Effect of different diets on biology, reproductive variables and life and fertility tables of *Harmonia axyridis* (Pallas) (Coleoptera, Coccinellidae). *Revista Brasileira de Entomologia*, 60(3): 260–266.

Casula, P., Wilby, A. & Thomas, M.B. (2006). Understanding biodiversity effects on prey in multi-enemy systems. *Ecological Letters*, 9: 995–1004.

Chandler, A.E.F. (1969). Locomotory behaviour of first instar larvae of aphidophagous Syrphidae (Dipt.) after contact with aphids. *Animal Behavior*, 17: 676–678.

Chaudhary, D.D., Kumar, B., Mishra, G. & Omkar (2015). Resource partitioning in a ladybird, *Menochilus sexmaculatus*: function of body size and prey density. *Bulletin of Entomological Research*, 105: 121–128.

Chunju, A., Xiaodong, F., Wenfeng, C. & Zhangwu, Z. (2012). The integrative effects of population density, photoperiod, temperature, and host plant on the induction of alate aphids in *Schizaphis graminum*. *Archives of Insect Biochemistry and Physiology*, 79(4–5): 198–206.

Costa, G.C. (2009). Predator size, prey size, and dietary niche breadth relationships in marine predators. *Ecology*, 90(7): 2014–2019.

Costamagna, A.C., Landis, D.A. & Difonzo, C.D. (2007). Suppression of soybean aphid by generalist predators results in a trophic cascade in soybeans. *Ecological Applications*, 17(2): 441–451.

Cottrell, T.E. (2005). Predation and cannibalism of lady beetle eggs by adult lady beetles. *Biological Control*, 34: 159–164.

Crookes, S., DeRoy, E.M., Dick, J.T. & MacIsaac, H.J. (2019). Comparative functional responses of introduced and native ladybird beetles track ecological impact through predation and competition. *Biological Invasions*, 21: 519–529.

Curio, E. (1976). *The Ethology of Predation*. Springer-Verlag: Berlin, 249 pp.

Damos, P. & Savopoulou-Soultani, M. (2012). Temperature driven models for insect development and vital thermal requirements. *Psyche*, 2012: 1–13.

Danks, H.V. (1987). *Insect Dormancy: An Ecological Perspective*. Biological Survey of Canada (Terrestrial Arthropods): Ottawa, 439 pp.

Danks, H.V. (1991). Winter habitats and ecological adaptations for winter survival. In: Lee, R.E., Jr & Denlinger, D.L. (Eds.), *Insects at Low Temperatures*. Chapman and Hall: New York, pp. 231–259.

Dauphin, G., Coquillard, P., Colazza, S., Peri, E. & Wajnberg, E. (2009). Host kairomone learning and foraging success in an egg parasitoid: A simulation model. *Ecological Entomology*, 34: 193–203.

Dejean, A., Gibernau, M., Lauga, J. & Orivel, J. (2003). Coccinellid learning during capture of alternative prey. *Journal of Insect Behavior*, 16: 859–864.

Dell, A.I., Pawar, S. & Savage, V.M. (2013). Temperature dependence of trophic interactions are driven by asymmetry of species responses and foraging strategy. *Journal of Animal Ecology*, 83(1): 70–84.

Denis, D., Pierre, J.S., van Baaren, J. & van Alphen, J.J.M. (2011). How temperature and habitat quality affect parasitoid lifetime reproductive success-A simulation study. *Ecological Modelling*, 222: 1604–1613.

Dixon, A.F.G. (1958) The escape response shown by certain aphids to the presence of the coccinellid *Adalia bipunctata* L. *Transactions of the Royal Entomological Society London*, 110: 319–334.

Dixon, A.F.G. (1959). An experimental study of the searching behaviour of the predatory coccinellid beetle *Adalia decempunctata* (L.). *Journal of Animal Ecology*, 28: 259–281.

Dixon, A.F.G. (2000). *Insect Predator-Prey Dynamics, Ladybird Beetles and Biological Control*. Cambridge University Press: Cambridge, 257 pp.

Dixon, A.F.G. (2007). Body size and resource partitioning in ladybirds. *Population Ecology*, 49: 45–50.

Dixon, A.F.G. & Agarwala, B.K. (2002). Triangular fecundity function and ageing in ladybird beetles. *Ecological Entomology*, 27: 433–440.

Dixon, A.F.G. & Guo, Y. (1993). Egg and cluster size in ladybird beetles (Coleoptera: Coccinellidae): The direct and indirect effects of aphid abundance. *European Journal of Entomology*, 90: 457–463.

Dixon, A.F.G., Hemptinne, J.L. & Kindlmann, P. (1997). Effectiveness of ladybirds as biological control agents: Patterns and processes. *Entomophaga*, 42(1–2): 71–83.

Dmitriew, C. & Rowe, L. (2007). Effects of early resource limitation and compensatory growth on lifetime fitness in the ladybird beetles (*Harmonia axyridis*). *Journal of Evolutionary Biology*, 20: 1298–1310.

Dmitriew, C. & Rowe, L. (2011). The effects of larval nutrition on reproductive performance in a food-limited adult environment. *PLoS One*, 6(3): e17399. doi: 10.137 1/journal.pone.0017399.

Downing, G. & Litvak, M.K. (2002). Effects of light intensity, spectral composition and photoperiod on development and hatching of haddock (*Melanogrammus aeglefinus*) embryos. *Aquaculture*, 213(1–4): 265–278.

Dreyer, B.S., Neuenschwander, P., Bouyjou, B., Baumgärtner, J. & Dorn, S. (1997). The influence of temperature on the life table of *Hyperaspis notata*. *Entomologia Experimentalis et Applicata*, 84: 85–92.

Drost, Y.C., Lewis, W.J. & Tumlinson, J.H. (1988). Beneficial arthropod behaviour mediated by airborne semiochemicals. V. Influence of rearing method, host plant and adult experience on host searching behaviour of *Microlitis croceipes* (Cresson) a larval parasitoid of *Heliothis sp*. *Journal of Chemical Ecology*, 14: 1607–1616.

Ducatti, R.D.B., Ugine, T.A. & Losey, J. (2017). Interactions of the Asian Lady Beetle, *Harmonia axyridis* (Coleoptera: Coccinellidae), and the North American Native Lady Beetle, *Coccinella novemnotata* (Coleoptera: Coccinellidae): Prospects for Recovery Post-Decline. *Environmental Entomology*, 46(1): 21–29.

Dukas, R. (2008). Evolutionary biology of insect learning. *Annual Review of Entomology*, 53: 145–160.

Dukas, R. & Bernays, E.A. (2000). Learning improves growth rate in grasshoppers. *Proceedings of the National Academy of Sciences of the USA*, 97: 2637–2640.

Eliopoulos, P.A., Kontodimas, D.C. & Stathas, G.J. (2010). Temperature-dependent development of *Chilocorus bipustulatus* (Coleoptera: Coccinellidae). *Environmental Entomology*, 39: 1352–1358.

Engelmann, F. (1970). *The Physiology of Insect Reproduction*. Pergamon: Oxford, 307 pp.

Ettifouri, M. & Ferran, A. (1993). Influence of larval rearing diet on the intensive searching behaviour of *Harmonia axyridis* larvae. *Entomophaga*, 38: 51–59.

Evans, E.W. (2000). Egg production in response to combined alternative foods by the predator *Coccinella transversalis*. *Entomologia Experimentalis et Applicata*, 94: 141–147.

Evans, E.W. (2003). Searching and reproductive behaviour of female aphidophagous ladybirds (Coleoptera: Coccinellidae): A review. *European Journal of Entomology*, 100: 1–10.

Evans, E.W. (2008). Multitrophic interactions among plants, aphids, alternate prey and shared natural enemies: A review. *European Journal of Entomology*, 105: 369–380.

Evans, E.W. (2009). Lady beetles as predators of insects other than Hemiptera. *Biological Control*, 51: 255–267.

Evans, E.W. & Dixon, A.F.G. (1986). Cues for oviposition by ladybird beetles (Coccinellidae): Response to aphids. *Journal of Animal Ecology*, 55: 1027–1034.

Eveleigh, E.S. & Chant, D.A. (1982). Experimental studies on acarina predator-prey interactions: The effect of predator density on prey consumption, predator searching efficiency and the functional response to prey density (Acarina, Phytoseiidae). *Canadian Journal of Zoology*, 60: 611–629.

Farina, W.M., Gruter, C. & Diaz, P.C. (2005). Social learning of floral odours inside the honeybee hive. *Proceedings of the Royal Society of London Series B*, 272: 1923–1928.

Ferran, A. & Deconchat, M. (1992). Exploration of wheat leaves by *Coccinella septempunctata* L. (Col., Coccinellidae). *Journal of Insect Behavior*, 5: 147–159.

Ferran, A. & Dixon, A.F.G. (1993). Foraging behaviour of ladybird larvae (Coleoptera: Coccinellidae). *European Journal of Entomology*, 90: 383–402.

Ferran, A., Iperti, G., Lapchin, L. & Rabasse, J.M. (1991). La localization, le comportement et les relations "proie-predateur" chez *Coccinella septempunctata* L. dans un champ de ble. *Entomophaga*, 36: 213–225.

Ferrer, A., Dixon, A.F.G. & Hemptinne, J.L. (2008). Prey preference of ladybird larvae and its impact on larval mortality, some life-history traits of adults and female fitness. *Bulletin of Insectology*, 61: 5–10.

Ferrero, D., Lemon, J., Fluegge, D., Pashkovski, S., Korzan, W., Datta, S., Spehr, M., Fendt, M. & Liberles, S. (2011). Detection and avoidance of a carnivore odor by prey. *Proceedings of the National Academy of Sciences USA*, 108: 11235–11240.

Fiaboe, K.K.M., Gondim, M.G.C. Jr., de Moraes, G.J., Ogol, C.K.P.O. & Knapp, M. (2007). Bionomics of the acarophagous ladybird beetle *Stethorus tridens* fed *Tetranychus evansi*. *Journal of Applied Entomology*, 131: 355–361.

Finlayson, C.J., Alyokhin, A.V., Gross, S. & Porter, E.W. (2010). Differential consumption of four aphid species by four lady beetle species. *Insect Science*, 10(31): 1–10.

Francis, F., Lognay, G. & Haubruge, E. (2004). Olfactory responses to aphid and host plant volatile releases: (E)-β-farnesene an effective kairomone for the predator *Adalia bipunctata*. *Journal of Chemical Ecology*, 30: 741–755.

Fréchette, B., Dixon, A.F.G., Alauzet, C., Boughenou, N. & Hemptinne, J.L. (2006). Should aphidophagous ladybirds be reluctant to lay eggs in the presence of unsuitable prey? *Entomologia Experimentalis et Applicata*, 118: 121–127.

Gillooly, J.F., Charnov, E.L., West, G.B., Savage, V.M. & Brown, J.H. (2002). Effects of size and temperature on developmental time. *Nature*, 417: 70–73.

Giurfa, M. (2013). Cognition with few neurons: Higher-order learning in insects. *Trends in Neurosciences*, 36: 285–294.

Giurfa, M. (2015). Learning and cognition in insects. *WIRES Cognitive Science*, 6(4): 383–395.

Golizadeh, A. & Jafari-Behi, V. (2012). Biological traits and life table parameters of variegated lady beetle, *Hippodamia variegata* (Coleoptera: Coccinellidae) on three aphid species. *Applied Entomology and Zoology*, 47: 199–205.

Guershon, M. & Gerling, D. (2006). Effects of plant and prey characteristics on the predatory behavior of *Delphastus catalinae*. *Entomologia Experimentalis et Applicata*, 121: 15–21.

Guillette, L.M., Hollis, K.L. & Markarian, A. (2009). Learning in a sedentary insect predator: Antlions (Neuroptera: Myrmeleontidae) anticipate a long wait. *Behavioural Processes*, 80: 224–232.

Gupta, G. & Kumar, N.R. (2017). Growth and development of ladybird beetle *Coccinella septempunctata* L. (Coleoptera: Coccinellidae), on plant and animal based protein diets. *Journal of Asia-Pacific Entomology*, 20(3): 959–963.

Gupta, A.K., Srivastava, S., Mishra, G., Singh, K. & Omkar (2006). Survival, development and life table of two con-generic ladybirds in aphid guilds. *Insect Science*, 13(2): 119–126.

Gupta, R.K., Pervez, A., Guroo, M.A. & Srivastava, K. (2012). Stage specific functional response of an aphidophagous ladybird, *Coccinella septempunctata* (Coleoptera: Coccinellidae), to two aphid species. *International Journal of Tropical Insect Science*, 32(3): 136–141.

Hamasaki, K. & Matsui, M. (2006). Development and reproduction of an aphidophagous coccinellid, *Propylea japonica* (Thunberg) (Coleoptera: Coccinellidae), reared on an alternative diet, *Ephestia kuehniella* Zeller (Lepidoptera: Pyralidae) eggs. *Applied Entomologie and Zoologie*, 41: 233–237.

Hao, Z., AghaKouchak, A. & Phillips, T.J. (2013). Changes in concurrent monthly precipitation and temperature extremes. *Environmental Research Letters*, 8(3): 034014.

Harmon, J.P., Losey, J.E. & Ives, A.R. (1998). The role of vision and color in the close proximity foraging behavior of four coccinellid species. *Oecologia*, 115: 287–292.

Hassell, M.P. & Southwood, T.R.E. (1978). Foraging strategies of insects. *Annual Review of Ecology System*, 9: 75–98.

Hatano, E., Kunert, G., Michaud, J.P. & Weisser, W.W. (2008). Chemical cues mediating aphid location by natural enemies. *European Journal of Entomology*, 105: 797–806.

Hattingh, V. & Samways, M.J. (1995). Visual and olfactory location of biotopes, prey patches, and individual prey by the ladybeetle *Chilocorus nigritus*. *Entomologia Experimentalis et Applicata*, 75: 87–98.

Haye, T., Mason, P.G., Dosdall, L.M. & Kuhlmann, U. (2010). Mortality factors affecting the cabbage seedpod weevil, *Ceutorhynchus obstrictus* (Marsham), in its area of origin: A life table analysis. *Biological Control*, 54(3): 331–341.

Hemptinne, J.L. & Dixon, A.F.G. (2000). Defence, oviposition and sex: Semiochemical parsimony in two species of ladybird beetles (Coleoptera: Coccinellidae)? A short review. *European Journal Entomology*, 97: 443–447.

Hemptinne, J.L., Dixon, A.F.G. & Coffin, J. (1992). Attack strategy of ladybird beetles (Coccinellidae): Factors shaping their numerical response. *Oecologia*, 90: 238–245.

Hemptinne, J.L., Gaudin, M., Dixon, A.F.G. & Lognay, J. (2000). Social feeding in ladybird beetles: Adaptive significance and mechanism. *Chemoecology*, 10: 149–152.

Hemptinne, J.L., Lognay, G., Doumbia, M. & Dixon, A.F.G. (2001). Chemical nature and persistence of the oviposition deterring pheromone in the tracks of the two spot ladybird, *Adalia bipunctata* (Coleoptera: Coccinellidae). *Chemoecology*, 11: 43–47.

Hill, W.G. (1982). Predictions of response to artificial selection from new mutations. *Genetics Research*, 40(3): 255–278.

Hirvonen, H., Ranta, E., Rita, H. & Peuhkuri, N. (1999). Significance of memory properties in prey choice decisions. *Ecological Modelling*, 15: 177–189.

Hodek, I. (1956). The influence of *Aphis sambuci* L. as prey of the ladybird beetle *Coccinella septempunctata* L. *Vestnik Zoologii*, 20: 62–74.

Hodek, I. (1973). *Biology of Coccinellidae.* Academia, Prague and Junk: The Hague, 260 pp.
Hodek, I. & Evans, E.W. (2012). Food relationships. In: Hodek, I., van Emden, H.F., Honěk, A. (Eds), *Ecology and Behaviour of the Ladybird Beetles (Coccinellidae).* Wiley-Blackwell: Chichester, pp. 141–274.
Hodek, I. & Honek, A. (1996). *Ecology of Coccinellidae.* Kluwer Academic Publishers: Dordrecht, Boston, London, 464 pp.
Hodek, I., Van Emden, H.F. & Honek, A. (2012). *Ecology and Behavior of the Ladybird Beetles (Coccinellidae).* Blackwell Publishing Limited: Hoboken, NJ, 600 pp.
Holling, C.S. (1959a). Some characteristics of simple types of predation and parasitism. *The Canadian Entomologist*, 91(7): 385–398.
Holling, C.S. (1959b). The components of predation as revealed by a study of small-mammal predation of the European Pine Sawfly. *The Canadian Entomologist*, 91(5): 293–320.
Holling, C.S. (1965). The functional response of predators to prey density and its role in mimicry and population regulation. *Memoirs of the Entomological Society of Canada*, 45: 3–60.
Honek, A. (1983). Factors affecting the distribution of larvae of aphid predators (Col., Coccinellidae and Dipt., Syrphidae) in cereal stands. *Zeitschrift fur Angewandte Entomologie*, 95: 336–345.
Honek, A. (1996). Geographical variation in thermal requirements for insect development. *European Journal of Entomology*, 93: 303–312.
Honek, A. & Kocourek, F. (1990). Temperature and development time in insects: A general relationship between thermal constants. *Zoologische Jahrbucher Abteilung fur Systematik und Okologie der Tiere*, 117: 401–439.
Hong, B.M., Binh, T.T.T. & Hang, V.T.T. (2013). Effect of temperature on the life cycle and predatory capacity of ladybird beetle *Micraspis discolor* Fabricius (Coleoptera: Coccinellidae). *Journal of Biology*, 35(1): 37–42.
Hosseini, A., Hosseini, M., Michaud, J.P., Awal, M.M., & Ghadamyari, M. (2019). Life history responses of *Hippodamia variegata* (Coleoptera: Coccinellidae) to changes in the nutritional content of its prey, *Aphis gossypii* (Hemiptera: Aphididae), mediated by nitrogen fertilization. *Biological Control*, 130: 27–33.
Houck, M.A. (1986). Prey preference in *Stethorus punctum* (Col., Coccinellidae). *Environmental Entomology*, 15: 967–970.
Houck, M.A. (1991). Time and resource partitioning in *Stethorus punctum* (Coleoptera: Coccinellidae). *Environmental Entomology*, 20: 494–497.
Huffaker, C.B., Messenger, P.S. & Debach, P. (1971). The natural enemy component in natural control and the theory of biological control. In: Huffaker, C.B. (Ed.) *Biological Control.* Plenum Press: New York, pp. 16–67.
Hukusima, S. & Kamei, M. (1970). Effects of various species of aphids as food on development, fecundity and longevity of *Harmonia axyridis* Pallas (Coleoptera: Coccinellidae). *Research Bulletin of the Faculty of Agriculture, Gifu University*, 29: 53–66.
Iperti, G. (1966). Comportement naturel des coccinelles aphidiphages du sud-est de la France. Leur type de specificite, leur action predatrice sur *Aphis fabae* Scop. *Entomophaga,* 11: 203–210.
Iverson, T. & Harding, S. (2007). Life table parameters affecting the population development of the woolly beech aphid. *Phyllaphis fagi. Entomologia Experimentalis et Applicata,* 123(2): 109–117.
Ives, P.M. (1981). Feeding and egg production of two species of coccinellids in the laboratory. *The Canadian Entomologist*, 113: 999–1005.
Jalali, M.A. & Ziaaddini, M. (2017). Effects of host plant morphological features on the functional response of *Adalia bipunctata* (Coleoptera: Coccinellidae) to *Myzus persicae* (Hemiptera: Aphididae). *International Journal of Pest Management*, 63(4): 309–315.
Jalali, M.A., Tirry, L. & De Clercq, P. (2010). Effect of temperature on the functional response of *Adalia bipunctata* to *Myzus persicae. BioControl*, 55: 261–269.
Jalali, M.A., Reitz, S., Mehrnejad, M.R., Ranjbar, F. & Ziaaddini, M. (2019). Food utilization, development, and reproductive performance of *Coccinella septempunctata* (Coleoptera: Coccinellidae) feeding on an aphid or psylla prey species. *Journal of Economic Entomology*, 112: 571–576.
Jander, R.J. (1975). Ecological aspects of spatial orientations. *Annual Review of Ecological System*, 6: 171–188.
Jankowsky, H.D. (1973). Body temperature and energy budget. In: Precht, H. et al. (Eds.), *Temperature and Life*. Springer: Berlin, Heidelberg, New York, pp. 87–101.
Jarosík, V., Honek, A. & Dixon, A.F.G. (2002). Developmental rate isomorphy in insects and mites. *The American Naturalist*, 160: 497–510.
Jarosík, V., Kumar, G., Omkar & Dixon, A.F.G. (2014). Are thermal constants constant? A test using two species of ladybird. *Journal of Thermal Biology*, 40: 1–8.
Jeschke, J.M., Kopp, M. & Tollrian, R. (2002). Predator functional responses: Discriminating between handling and digesting prey. *Ecological Monograph*, 72: 95–112.

Johnson, R.A. (1991). Learning, memory, and foraging efficiency in two species of desert seed-harvester ants. *Ecology*, 72: 1408–1419.

Joshi, S., Ballal, C.R. & Rao, N.S. (1999). Biotic potential of three coccinellid predators on six different aphid hosts. *Journal of Entomological Research*, 23: 1–7.

Kaddou, I.K. (1960). The feeding behaviour of *Hippodamia quinquesignata* (Kirby) larvae. *The University of California Publications in Entomology*, 16: 181–182.

Kairo, M.T.K. & Murphy, S.T. (1995). The life history of *Rodolia iceryae* Janson (Coleoptera: Coccinellidae) and the potential for use in innoculative releases against *Icerya pattersoni* Newstead (Homoptera: Margarodidae) on coffee. *Journal of Applied Entomology,* 119: 487–491.

Kalushkov, P. & Hodek, I. (2004). The effect of thirteen species of aphids on some life history parameters of the ladybird *Coccinella septempunctata*. *Biocontrol*, 49: 21–32.

Karami, J.T. & Shishehbor, P. (2012). Host plant effects on the functional response of *Stethorus gilvifrons* to strawberry spider mites. *Biocontrol Science and Technology*, 22(1): 101–110.

Kareiva, P. & Odell, G. (1987). Swarms of predators exhibit "preytaxis" if individual predators use area restricted search. *The American Naturalist*, 130: 233–270.

Kawauchi, S.E. (1985). Comparative studies on the fecundity of three aphidophagous Coccinellids (Coleoptera: Coccinellidae). *Japanese Journal of Applied Entomology and Zoology*, 29(33): 203–209.

Kehat, M. (1968). The feeding behaviour of *Pharoscymnus numidicus* (Coccinellidae) predator of the date palm scale *Parlatoria blanchardi*. *Entomologia Experimentalis et Applicata*, 11: 30–42.

Keith, L.B., Todd, A.W., Brand, C.J., Adamcik, R.S. & Rusch, D.H. (1977). An analysis of predation during a cyclic fluctuation of snowshoe hares. *Proceedings of the International Congress of Game Biologists*, 13: 151–175.

Keshavarz, M., Seiedy, M. & Allahyari, H. (2015). Preference of *two* populations of *Propylea quatuordecimpunctata* (Coleoptera: Coccinellidae) for *Aphis fabae* and *Aphis gossypii* (Homoptera: Aphididae). *European Journal of Entomology*, 112(3): 560–563.

Kesten, U. (1969). Zur Morphologie and Biologie of *Anatis ocellata* L. (Col., Coccinellidae). *Zeitschrift fur Angewandte Entomologie*, 63: 412–445.

Khan, M.R. & Khan, M.R. (2010). The relationship between temperature and the functional response of *Coccinella septempunctata* (L.) (Coleoptera: Coccinellidae). *Pakistan Journal of Zoology*, 42(4): 461–466.

Khan, I. & Spooner-Hart, R. (2017). Temperature-dependent development of immature stages of predatory ladybird beetle Stethorus vagans (Coleoptera: Coccinellidae) at constant and fluctuating temperatures. *Acta Zoologica Academiae Scientiarum Hungaricae*, 63(1): 83–96.

Kianpour, R., Fathipour, Y., Kamali, K. & Omkar (2011). Effects of mixed prey on the development and demographic attributes of a generalist predator, *Coccinella septempunctata* (Coleoptera: Coccinellidae). *Biocontrol Science and Technology*, 21(4): 435–447.

Kindlmann, P., Dixon, A.F.G. & Dostálková, I. (2001). Role of ageing & temperature in shaping reaction norms and fecundity functions in insects. *Journal of Evolutionary Biology,* 14: 835–840.

Klausnitzer, B. (1969). Zur Unterscheidung der Eier mitteleuropaischer Coccinellidae. *Acta Entomologica Bohemoslovaca*, 66: 146–149.

Klausnitzer, B. & Klausnitzer, H. (1986). *Marienkafer (Coccinellidae)*. A. Ziemsen Verlag, Wittenberg: Lutherstadt, 104 pp.

Klingauf, F. (1967). Abwehr and Meidereaktionen von Blattlausen (Aphididae) bei Bedrohung durch Rauber und Parasiten. *Zeitschrift fur Angewandte Entomologie*, 60: 269–317.

Koch, R.L. & Galvan, T.L. (2008). Bad side of a good beetle: The North American experience with *Harmonia axyridis*. *Biocontrol*, 53: 23–35.

Krivan, V. & Sirot, E. (2004). Do short-term behavioural responses of consumers in tri-trophic food chains persist at the population time-scale? *Evolutionary Ecology Research*, *6*: 1063–1081.

Kumar, B. & Mishra, G. (2014). Larval and female footprints as feeding deterrent cues for immature stages of two congeneric ladybird predators (Coleoptera: Coccinellidae). *Bulletin of Entomological Research*, 104(5): 652–660.

Kumar, B. & Omkar (2015). Temperature and photoperiod influence prey consumption and utilization by two sympatric *Coccinella* species (Coleoptera: Coccinellidae) in conspecific and heterospecific combinations. *Acta Entomologica Sinica*, 58(3): 297–307.

Kumar, A., Kumar, N., Siddiqui, A. & Tripathi, C.P.M. (1999). Prey–predator relationship *between Lipaphis erysimi* Kalt.(Hom., Aphididae) and *Coccinella septempunctata* L (Col., Coccinellidae). II. Effect of host plants on the functional response of the predator. *Journal of Applied Entomology*, 123: 591–596.

Kumar, B., Pandey, G., Mishra, G. & Omkar (2013). Predatory performance of *four* aphidophagous ladybirds: A measure of prey suitability. *International Journal of Tropical Insect Science*, 33(2): 120–126.

Kumar, B., Mishra, G. & Omkar (2014a). Functional response and predatory interactions in conspecific and heterospecific combinations of two congeneric species (Coleoptera: Coccinellidae). *European Journal of Entomology*, 111(2): 257–265.

Kumar, B., Bista, M., Mishra, G. & Omkar (2014b). Stage specific consumption and utilization of aphids, conspecific and heterospecific eggs by two species of *Coccinella* (Coleoptera: Coccinellidae). *European Journal of Entomology*, 111(3): 363.

Kumar, B., Mishra, G. & Omkar (2015). Prey species modify interactions within two predator conspecific and heterospecific combinations: A case study using two sympatric *Coccinella* species (Coleoptera: Coccinellidae). *Journal of Asia-Pacific Entomology*, 18(2): 109–116.

Latifian, M. (2017). Foraging and functional response of the predator, *Stethorus gilvifrons* Mulsant. (Coleoptera: Coccinellidae), fed on the Date Palm Spider Mite, *Oligonychusa frasiaticus* McGregor (Acari: Tetranychidae). *Egyptian Journal of Biological Pest Control*, 27(1): 93.

Lee, J.H. & Kang, T.J. (2004). Functional response of *Harmonia axyridis* (Pallas) (Coleoptera: Coccinellidae) to *Aphis gossypii* Glover (Homoptera: Aphididae) in the laboratory. *Biological Control*, 31: 306–310.

Legaspi, J.C. & Legaspi, B.C. (1998). Life history trade-offs in insects, with emphasis on *Podisus maculiventris* (Heteroptera: Pentatomidae). In: Coll, M. & Ruberson, J.R. (Eds.), *Predatory Heteroptera: Their Ecology and Use in Biological Control*. Entomological Society of America, Thomas Say Publications in Entomology: Lanham, MD, pp. 71–87.

Lewis, W.J., Vet, L.E.M., Tumlinson, J.H., van Lenteren, J.C. & Papaj, D.R. (1990). Variations in parasitoid foraging behavior: Essential element of a sound biological control theory. *Environmental Entomology*, 19: 1183–1193.

Lima, M.S., Melo, J.W.S. & Barros, R. (2017). Alternative food sources for the ladybird *Brumoides foudrasii* (Mulsant) (Coleoptera: Coccinellidae). *Brazilian Journal of Biology*. doi: 10.1590/1519-6984.02816.

Logan, J.D., Wolesensky, W. & Joern, A. (2006). Temperature-dependent phenology and predation in arthropod systems. *Ecological Modelling*, 196: 471–482.

Lopatina, E.B., Balashov, S.V. & Kipyatkov, V.E. (2007). First demonstration of the influence of photoperiod on the thermal requirements for development in insects and in particular the linden-bug, *Pyrrhocoris apterus* (Heteroptera, Pyrrhocoridae). *European Journal of Entomology*, 104: 23–31.

Luck, R.F. (1985). Principles of arthropod predation. In: Huffaker, C.B. & Rabb, R.L. (Eds.), *Ecological Entomology*. Wiley Interscience: New York, pp. 497–530.

Luker, L.A., Hatle, J.D. & Juliano, S.A. (2002). Reproductive responses to photoperiod by a South Florida population of the grasshopper *Romalea microptera* (Orthoptera: Romaleidae). *Environmental Entomology*, 31: 702–707.

Lundgren, J.G. (2009). Nutritional aspects of non-prey foods in the life histories of predaceous Coccinellidae. *Biological Control*, 51: 294–305.

Lundgren, J.G. & Weber, D.C. (2010). Changes in digestive rate of a predatory beetle over its larval stage: Implications for dietary breadth. *Journal of Insect Physiology*, 56: 431–437.

Magro, A., Tene, J.N., Bastin, N., Dixon, A.F.G. & Hemptinne, J.L. (2007). Assessment of patch quality by ladybirds: Relative response to conspecific and heterospecific larval tracks a consequence of habitat similarity? *Chemoecology*, 17: 37–45.

Magro, A., Ducamp, C., Ramon-Portugal, F., Lecompte, E., Crouau-Roy, B., Dixon, A.F.G. & Hemptinne, J.L. (2010). Oviposition deterring info chemicals in ladybirds: The role of phylogeny. *Evolutionary Ecology*, 24: 251–271.

Majerus, M.E.N. (1994). *Ladybirds*. Harper Collins: London, 367 pp.

Majumder, J. & Agarwala, B.K. (2013). Biology and population dynamics of giant ladybird predator, *Anisolemnia dilatata* (F.) (Coleoptera: Coccinellidae): A specialized predator of wooly aphids of bamboo plants in northeast India. *World Journal of Zoology*, 8(1): 55–61.

Malcolm, C.B. (1992). Prey defence and predator foraging. In: Crawley, M.J. (Ed.), *Natural Enemies: The Population Biology of Predators, Parasites and Diseases*. Blackwell Scientific Publication: Oxford, UK, pp. 458–475.

Marks, R.J. (1977). Laboratory studies of plant searching behaviour by *Coccinella septempunctata* L. larvae. *Bulletin of Entomological Research*, 67: 235–341.

Maurice, N. & Kumar, A. (2011). Effect of quantity and consumption of food on body weight and development of two species of ladybird beetles. *Annals of Plant Protection Sciences*, 19(1): 59–62.

Maurice, N.G. & Kumar, A. (2012). Development of two species of ladybird beetles (Coleoptera: Coccinellidae) on aphids as well as conspecific eggs. *International Journal of Advance Pharmaceutical and Biological Sciences*, 2(4): 272–279.

Meidari, M. & Copland, M.J.W. (1992). Host finding by *Cryptolaemus montrouzieri* (Col., Coccinellidae) a predator of mealybugs (Hom., Pseudococcidae). *Entomophaga*, 37: 621–625.

Meisner, M.H., Harmon, J.P. & Anthony, R.I. (2011). Response of coccinellid larvae to conspecific and heterospecific larval tracks: A mechanism that reduces cannibalism and intraguild predation. *Environmental Entomology*, 40: 103–110.

Michaud, J.P. (2005). On the assessment of prey suitability in aphidophagous Coccinellidae. *European Journal of Entomology*, 102: 385–390.

Michaud, J.P. (2012). Coccinellids in biological control. In: Hodek, I., van Emden, H.F., & Honek, A. (Eds.), *Ecology and Behaviour of the Ladybird Beetles (Coccinellidae)*. John Wiley: Chichester, UK, pp. 488–519.

Michaud, J.P. & Jyoti, J.L. (2007). Repellency of conspecific and heterospecific larval residues to *Hippodamia convergens* (Coleoptera: Coccinellidae) ovipositing on sorghum plants. *European Journal of Entomology*, 104: 399–405.

Michaud, J.P. & Jyoti, J.L. (2008). Dietary complementation across life stages in the polyphagous lady beetle *Coleomegilla maculata*. *Entomologia Experimentalis et Applicata*, 126: 40–45.

Mills, N.J. (1981). Essential and alternative foods for some British Coccinellidae (Coleoptera). *Entomological Gazette*, 32: 197–202.

Mills, N.J. (1982). Satiation and the functional response: A test of a new model. *Ecological Entomology*, 7(3): 305–315.

Mills, N.J. & Lacan, I. (2004). Ratio dependence in the functional response of insect parasitoids: Evidence from *Trichogramma minutum* foraging for eggs in small host patches. *Ecological Entomology*, 29: 208–216.

Mishra, G. & Omkar (2005). Influence of components of light on the life attributes of an aphidophagous ladybird, *Propylea dissecta* (Coleoptera: Coccinellidae). *International Journal of Tropical Insect Science*, 25(1): 32–38.

Mishra, G., Kumar, B., Shahid, M. & Singh, D. (2011). Evaluation of four co-occurring ladybirds for use as biocontrol agents of the pea aphid, *Acyrthosiphon pisum* (Homoptera: Aphididae). *Biocontrol Science and Technology*, 21: 991–997.

Mishra, G., Omkar, Kumar, B. & Pandey, G. (2012). Stage and age specific predation in four aphidophagous ladybird beetles. *Biocontrol Science and Technology*, 22: 463–476.

Morales-Ramos, J.A. & Rojas, M.G. (2017). Temperature-dependent biological and demographic parameters of *Coleomegilla maculata* (Coleoptera: Coccinellidae). *Journal of Insect Science*, 17(2): 55; 1–9.

Murakami, Y. & Tsubaki, Y. (1984). Searching efficiency of the ladybeetle *Coccinella septempunctata* larvae in a uniform and patchy environment. *Journal of Ethology*, 2: 1–6.

Nakamuta, K. (1984). Visual orientation of the ladybeetle *Coccinella septempunctata* L. (Col., Coccinellidae) towards its prey. *Applied Entomology and Zoology*, 19: 82–86.

Nakamuta, K. (1985). Mechanism of the switchover from extensive to area concentrated search behaviour of the ladybeetle *Coccinella septempunctata bruckii*. *Journal of Insect Physiology*, 31: 849–856.

Nakamuta, K. (1987). Diel rhythmicity of prey search activity and its predominance over starvation in the ladybeetle *Coccinella septempunctata bruckii*. *Physiological Entomology*, 12: 91–98.

Nakamuta, K. (1991). Aphid alarm pheromone component (E)-Beta-farnesene and local search by a predatory lady beetle *Coccinella septempunctata bruckii* Mulsant (Col., Coccinellidae). *Applied Entomology and Zoology*, 26: 1–17.

Nalepa, C.A. (2013). Coccinellidae captured in blacklight traps: Seasonal and diel pattern of the dominant species *Harmonia axyridis* (Coleoptera: Coccinellidae). *European Journal of Entomology*, 110(4): 593–597.

Nault, R., Edwards, L.J. & Styer, W.E. (1973). Aphid alarm pheromones: Secretion and reception. *Environmental Entomology*, 2: 101–105.

Nedved, O. & Salvucci, S. (2008). Ladybird *Coccinella septempunctata* (Coleoptera: Coccinellidae) prefers toxic preyin laboratory choice experiment. *European Journal of Entomology*, 105: 431–436.

Nielsen, A.L., Hamilton, G.C. & Matadha, D. (2008). Developmental rate estimation and life table analysis for *Halyomorpha halys* (Hemiptera: Pentatomidae). *Environmental Entomology*, 37(2): 348–355.

Ninkovic, V., AlAbassi, S. & Pettersson, J. (2001). The influence of aphid-induced plant volatiles on ladybird beetle searching behavior. *Biological Control*, 21: 191–195.

Noble, R.R.P., Harvey, S.G. & Sams, C.E. (2002). Toxicity of Indian mustard and allyl isothiocyanate to masked chafer beetle larvae. *Plant Health Progress*. doi: 10.1094/PHP-2002-0610-01-RS.

Nordlund, D.A. & Morrison, R.K. (1990). Handling time, prey preference, and functional response for *Chrysoperla rufilabris* in the laboratory. *Entomologia Experimentalis et Applicata*, 57(3): 237–242.
Nufio, C.R. & Papaj, D.R. (2001). Host marking behavior in phytophagous insects and parasitoids. *Entomologia Experimentalis et Applicata*, 99: 273–293.
Obata, S. (1986). Mechanisms of prey finding in the aphidophagous ladybird beetle *Harmonia axyridis* (Col., Coccinellidae). *Entomophaga*, 31: 303–311.
Obatake, H. & Suzuki, H. (1985). On the isolation and identification of canavanine and ethanolamine contained in the young leaves of *black* locus, *Robinia pseudoacacia*, lethal for the lady beetle, *Harmonia axyridis*. *Technical Bulletin of Faculty Agricultural Kagawa University*, 36: 107–115.
Okamoto, H. (1966). Three problems of prey specificity of aphidophagous coccinellids. In: Hodek, I. (Ed.), *Ecology of Aphidophagous Insects*. Academia, Prague and W. Junk: The Hague, pp. 45–46.
Oliver, T.H., Timms, J.E.L., Taylor, A. & Leather, S.R. (2006). Oviposition responses to patch quality in the larch ladybird *Aphidecta obliterata* (Coleoptera: Coccinellidae): Effects of aphid density, and con-and heterospecific tracks. *Bulletin of Entomological Research*, 96: 25–34.
Omkar (2006). Suitability of different foods for a generalist ladybird, *Micraspis discolor* (Coleoptera: Coccinellidae). *International Journal Tropical Insect Science*, 26(1): 35–40.
Omkar & Afaq, U. (2009). Intraspecific competition in the *Parthenium* beetle *Zygogramma bicolorata* (Coleoptera: Chrysomelidae): Effect of larval crowding on life history traits. *International Journal of Tropical Insect Science*, 29(1): 40–47.
Omkar & Bind, R.B. (2004). Prey quality dependent growth, development and reproduction of a biocontrol agent, *Cheilomenes sexmaculata* (Fabricius) (Coleoptera: Coccinellidae). *Biocontrol Science and Technology*, 14: 665–673.
Omkar & James, B.E. (2003). Searching and feeding efficiency of *Coccinella transversalis* Fabricius on aphid, *Aphis gossypii* Glover. *Journal of Biological Control*, 17(2): 107–112.
Omkar & James, B.E. (2004a). Influence of prey species on immature survival, development, predation and reproduction of *Coccinella transversalis* Fabricius (Col., Coccinellidae). *Journal of Applied Entomology*, 28: 150–157.
Omkar & James, B.E. (2004b). Influence of temperature on the survival, development of immature stages and reproduction of a ladybeetle, *Coccinella transversalis* Fabricius. *Entomon*, 29(1): 13–23.
Omkar & Kumar, G. (2013). Responses of an aphidophagous ladybird beetle, *Anegleis cardoni*, to varying densities of *Aphis gossypii*. *Journal of Insect Science*, 13(24): 1–12.
Omkar & Kumar, B. (2016). Effects of temperature and photoperiod on predation attributes and development of two *Coccinella* species (Coleoptera: Coccinellidae). *Acta Entomologica Sinica*, 59(1): 64–76.
Omkar & Mishra, G. (2005). Preference–performance of a generalist predatory ladybird: A laboratory study. *Biological Control*, 34: 187–195.
Omkar & Mishra, G. (2013). *Predaceous Ladybirds of Uttar Pradesh*. Centre of Excellence in Biocontrol of Insect Pests, Department of Zoology, University of Lucknow: Lucknow, 48 pp.
Omkar & Pathak, S. (2006). Effects of different photoperiods and wavelengths of light on life-history traits of an aphidophagous ladybird, *Coelophora saucia* (Mulsant). *Journal of Applied Entomology*, 130(1): 45–50.
Omkar & Pathak, S. (2009). Crowding affects the life attributes of an aphidophagous ladybird beetle, *Propylea dissecta*. *Bulletin of Insectology*, 62(1): 35–40.
Omkar & Pervez, A. (2001). Prey preference of a ladybeetle, *Micraspis discolor* (Fabricius). *Entomon*, 26(2): 195–197.
Omkar & Pervez, A. (2003). Influence of prey deprivation on biological attributes of pale morphs of the ladybeetle, *Propylea dissecta* (Mulsant). *Insect Science and Application*, 23(2): 143–148.
Omkar & Pervez, A. (2004). Temperature-dependent development and immature survival of an aphidophagous ladybeetle, *Propylea dissecta* (Mulsant). *Journal of Applied Entomology*, 128(7): 510–514.
Omkar & Pervez, A. (2011). Functional response of two aphidophagous ladybirds searching in tandem. *Biocontrol Science and Technology*, 21: 101–111.
Omkar & Singh, K. (2007). Rhythmicity in life events of an aphidophagous ladybird beetle, *Cheilomenes sexmaculata*. *Journal of Applied Entomology*, 131(2): 85–89.
Omkar & Srivastava, S. (2001). Comparative predatory potential of a ladybird beetle, *Coccinella septempunctata* Linn. on six prey species. *Biological Memoirs*, 27(2): 59–63.
Omkar & Srivastava, S. (2002). Functional response of the ladybeetle, *Coccinella septempunctata* Linnaeus on *Uroleucon compositae* (Theobald) and *Aphis nerii* Boyer de Fonscolombe. Prof. S.B. Singh Commomarium, Zoological Society of India, Lucknow, pp. 115–128.

Omkar & Srivastava, S. (2003a). Comparative prey consumption and searching efficiency of ladybeetles, *Coccinella septempunctata* Linnaeus and *Coccinella transversalis* Fabricius for different aphid species. *Journal of Biological Control*, 17(1): 35–41.

Omkar & Srivastava, S. (2003b). Influence of six aphid prey species on development and reproduction of a ladybird beetle, *Coccinella septempunctata. BioControl*, 48: 379–393.

Omkar, Pervez, A. & Gupta, A.K. (2004). Role of surface chemicals in egg cannibalism and intraguild predation by neonates of two aphidophagous ladybirds, *Propylea dissecta* and *Coccinella transversalis. Journal of Applied Entomology*, 128: 691–695.

Omkar, Pervez, A., Mishra, G., Srivastava, S., Singh, S.K. & Gupta, A.K. (2005a). Intrinsic advantages of a ladybird, *Cheilomenes sexmaculata* over the relatively bigger two co-occurring *Coccinella* species. *Insect Science*, 12(3): 179–184.

Omkar, Gupta, A.K. & Pervez, A. (2005b). Attack, escape and predation rates of larvae of two aphidophagous ladybirds during conspecific and heterospecific interactions. *Biocontrol Science and Technology*, 16(3): 295–305.

Omkar, Pervez, A. & Gupta, A.K. (2006). Why do neonates of aphidophagous ladybird beetles preferentially consume conspecific eggs in presence of aphids? *Biocontrol Science and Technology*, 16(3): 233–243.

Omkar, Kumar, G. & Sahu, J. (2009a). Performance of a predatory ladybird beetle, *Anegleis cardoni* (Weise) (Coleoptera: Coccinellidae) on three aphid species. *European Journal of Entomology*, 106: 565–572.

Omkar, Rastogi, S. & Pandey, P. (2009b). Effect of temperature on reproductive attributes of the Mexican beetle, *Zygogramma bicolorata* Pallister (Coleoptera: Chrysomelidae). *International Journal of Tropical Insect Science*, 29(1): 48–52.

Omkar, Kumar, G. & Sahu, J. (2011). Monotypic prey-mediated development, survival and life table attributes of a ladybird beetle *Anegleis cardoni* (Coleoptera: Coccinellidae) on different aphid species. *International Journal of Tropical Insect Science*, 31(3): 162–173.

Omkar, Mishra, G., Kumar, B., Singh, N. & Pandey, G. (2014). Risks associated with tandem release of large and small ladybirds in heterospecific aphidophagous guilds. *The Canadian Entomologist*, 146(1): 52–66.

Osawa, N. (1989). Sibling and non-sibling cannibalism by larvae of a lady beetle *Harmonia axyridis* Pallas (Coleoptera, Coccinellidae) in the field. *Research in Population Ecology*, 31: 153–160.

Ovchinnikov, A.N., Belyakova, N.A., Ovchinnikova, A.A., & Reznik, S.Y. (2019). Factors determining larval cannibalistic behavior in invasive and native populations of the multicolored Asian ladybird, *Harmonia axyridis*. *Entomologia Generalis*, 38(3): 243–254.

Palmer, D.J. & Sheppard, J.L. (2002). Mass rearing *Pseudoscymnus tsugae* at the New Jersey Department of Agriculture: challenges and lessons. In *Proceedings of the Hemlock Woolly Adelgid in the Eastern United States Symposium*. New Jersey Agricultural Experiment Station and Rutgers University, 214–220 pp.

Papachristos, D.P., Katsarou, I., Michaelakis, A. & Papanikolaou, N.E. (2015). Influence of different species of aphid prey on the immature survival and development of four species of aphidophagous coccinellids (Coleoptera: Coccinellidae). *European Journal of Entomology*, 112(3): 440.

Papaj, D.R. (1993). Automatic behaviour and the evolution of instinct: Lessons from learning in parasitoids. In: Papaj, D.R. & Lewis, A.C. (Eds.), *Insect Learning Ecological and Evolutionary Perspectives*. Chapman and Hall: London, 243–272 pp.

Papaj, D.R. & Prokopy, R.J. (1989). Learning in phytophagous insects. *Annual Review of Entomology*, 34: 315–350.

Papaj, D.R. & Vet, L.E.M. (1990). Odor learning and foraging success in the parasitoid, *Leptopilina heterotoma*. *Journal of Chemical Ecology*, 16: 3137–3150.

Papanikolaou, N.E., Milonas, P.G., Kontodimas, D.C., Demiris, N. & Matsinos, Y.G. (2013). Temperature-dependent development, survival, longevity, and fecundity of *Propylea quatuordecimpunctata* (Coleoptera: Coccinellidae). *Annals of the Entomological Society of America*, 106(2): 228–234.

Pasteels, J.M. (2007). Chemical defence, offence and alliance in ants–aphids–ladybirds relationships. *Population Ecology*, 49: 5–14.

Patel, P., Kumar, B & Kumar, D. (2018). Does body size of coccinellid predators (coleopteran: Coccinellidae) influence their preference to eat well-defended prey? *Acta Entomologica Sinica*, 61: 704–711.

Patel, P., Kumar, B. & Kumar, D. (2017). Prey species based resource fluctuation affects predation attributes of a ladybird beetle, *Menochilus sexmaculatus* (Fabricius). *Journal of Applied Bioscience*, 43: 57–62.

Perdikis, D.C. & Lykouressis, D.P. (2002). Thermal requirements for development of the polyphagous predator *Macrolophus pygmaeus* (Hemiptera: Miridae). *Environmental Entomology*, 31: 661–667.

Pervez, A. & Omkar (2004a). Prey-dependent life attributes of an aphidophagous ladybird beetle, *Propylea dissecta* (Coleoptera: Coccinellidae). *Biocontrol Science and Technology*, 14(4): 385–396.

Pervez, A. & Omkar (2004b). Temperature dependent life attributes of an aphidophagous ladybird beetle, *Propylea dissecta* (Mulsant). *Biocontrol Science and Technology,* 14(6): 587–594.

Pervez, A., Singh, P.P. & Bozdoğan, H. (2018). Ecological perspective of the diversity of functional responses. *European Journal of Environmental Sciences*, 8: 97–101.

Pettersson, J., Ninkovic, V., Glinwood, R., Birkett, M.A. & Pickett, J.A. (2005). Foraging in a complex environment–semiochemicals support searching behaviour of the seven spot ladybird. *European Journal of Entomology*, 102: 365–370.

Phoofolo, M.W. & Obrycki, J.J. (1998). Potential for intraguild predation and competition among predatory Coccinellidae and Chrysopidae. *Entomologia Experimentalis et Applicata*, 89: 47–55.

Phoofolo, M.W., Giles, K.L. & Elliott, N.C. (2007). Quantitative evaluation of suitability of the greenbug, *Schizaphis graminum*, and the bird cherry-oat aphid, *Rhopalosiphum padi*, as prey for *Hippodamia convergens* (Coleoptera: Coccinellidae). *Biological Control*, 41: 25–32.

Phoofolo, M.W., Giles, K.L. & Elliott, N.C. (2008). Larval life history responses to food deprivation in three species of predatory lady beetles (Coleoptera: Coccinellidae). *Environmental Entomology*, 37: 315–322.

Poorani, J. (2019). *Coccinellidae of the Indian subcontinent. In Indian Insects*. CRC Press: Boca Raton, FL, 223–246 pp.

Provost, C., Lucas, E., Coderre, D. & Chouinard, G. (2006). Prey selection by the lady beetle *Harmonia axyridis*: The influence of prey mobility and prey species. *Journal of Insect Behavior*, 19: 265–277.

Punzo, F. & Garman, B. (1989). Effects of encounter experience on the hunting behavior of the spider wasp, *Pepsis formosa* (Say) (Hymenoptera: Pompilidae). *Southwest Naturalist*, 34: 513–518.

Qin, Z.Q., Wei, J.J., Song, X.P., Luo, Y.W., Liu, L. & Deng, Z.Y. (2017). Efficacy of the Ladybird Beetle Cryptolaemus montrouzieri Mulsant for Control of *Saccharicoccus sacchari* (Cockerell). *Sugar Tech*, 19: 599–603.

Qvarnstrom, A. & Price, T.D. (2001). Maternal effects, paternal effects and sexual selection. *Trends in Ecology and Evolution,* 16: 95–100.

Rahimi, R., Mahdian, K. & Noghabi, S.S. (2017). Effect of temperature on functional response of coccinellid (*Coccinula elegantula*) on Aphis gossypii. *Journal of Science and Technology of Greenhouse Culture*, 7(28): 125–133.

Ramalho, F.S., Wanderley, P.A., Malaquias, J.B., Rodrigues, K.E.V., Souza, J.V.S. & Zanuncio, J.C. (2009). Temperature-dependent development rates of *Bracon vulgaris*, a parasitoid of boll weevil. *Phytoparasitica*, 37: 17–25.

Rana, J.S., Dixon, A.F.G. & Jarosik, V. (2002). Costs and benefits of prey specialization in a generalist insect predator. *Journal of Animal Ecology*, 71: 15–22.

Rath, S.S. (2010). Food utilization efficiency in *Antheraea mylitta* fed on *Terminalia arjuna* leaves. *Academic Journal of Entomology*, 3(1): 23–28.

Regniere, J., Powell, J., Bentz, B. & Nealis, V. (2012). Effects of temperature on development, survival and reproduction of insects: Experimental design, data analysis and modeling. *Journal of Insect Physiology*, 58: 634–647.

Reis, P.R., Teodoro, A.V., Neto, M.P. & Da Silva, E.A. (2007). Life history of *Amblyseius herbicolus* (Chant) (Acari: Phytoseiidae) on coffee plants. *Neotropical Entomology*, 36: 282–287.

Remen, C. (2004). Associated learning of colour and odour in the seven-spotted ladybird *Coccinella septempunctata* (L.): An olfactometer experiment. Inst. f. Entomol. Sveriges lantbruksuniv, 1–33 pp.

Reznik, S.Y.A. & Vaghina, N.P. (2011). Photoperiodic control of development and reproduction in *Harmonia axyridis* (Coleoptera: Coccinellidae). *European Journal of Entomology*, 108: 385–390.

Ricklefs, R.E. (1990). *Ecology*. W.H. Freeman & Co.: New York.

Rieder, J.P., Newbold, T.A.S., Sato, S., Yasuda, H. & Evans, E.W. (2008). Intraguild predation and variation in egg defence between sympatric and allopatric populations of two species of ladybird beetles. *Ecological Entomology*, 33: 53–58.

Robinson, A.G. (1951). Annotated list of predators of Tetranychid mites in Manitoba. *Report on Entomological Society of Ontario*, 82: 33–37.

Rocca, M., Rizzo, E., Greco, N. & Sánchez, N. (2017). Intraezon interspecific interactions between aphidophagous ladybirds: The role of prey in predator coexistence. *Entomologia Experimentalis et Applicata*, 162(3): 284–292.

Roger, C., Coderre, D. & Boivin, G. (2000). Differential prey utilization by the generalist predator *Coleomegilla maculata lengi* according to prey size and species. *Entomologia Experimentalis et Applicata*, 94: 3–13.

Roger, C., Coderre, D., Vigneault, C. & Boivin, G. (2001). Prey discrimination by a generalist coccinellid predator: Effect of prey age or parasitism? *Ecological Entomology*, 26: 163–172.

Roitberg, B.D., Myers, J.H. & Frazer, B.D. (1979). The influence of predators on the movement of apterous pea aphids between plants. *Journal of Animal Ecology*, 48: 111–122.

Rondoni, G., Athey, K.J., Harwood, J.D., Conti, E., Ricci, C. & Obrycki, J.J. (2015). Development and application of molecular gut-content analysis to detect aphid and coccinellid predation by *Harmonia axyridis* (Coleoptera: Coccinellidae) in Italy. *Insect Science*, 22: 719–730.

Rondoni, G., Ielo, F., Ricci, C. & Conti, E. (2017). Behavioural and physiological responses to prey-related cues reflect higher competitiveness of invasive vs. native ladybirds. *Scientific Reports*, 7: 3716.

Rosagro, R.M., Borges, I., Vieira, V., Solé, G.P., & Soares, A.O. (2019). Evaluation of *Scymnus nubilus* (Coleoptera: Coccinellidae) as a biological control agent against *Aphis spiraecola* and *Cinara juniperi* (Hemiptera: Aphididae). *Pest Management Science*. doi: 10.1002/ps.5585.

Rosenheim, J.A. & Corbett, A. (2003). Omnivory and the indeterminacy of predator function: Can a knowledge of foraging behavior help? *Ecology*, 84: 2538–2548.

Sabelis, M.W. (1992). Predatory arthropods. In: Crawley, M.J. (Ed.), *Natural Enemies: The population Biology of Predators, Parasites and Diseases*. Wiley: Oxford, UK, pp. 225–264.

Said, K.K., Ali Shah, M. & Baloch, U.K. (1985). Optical orientation in predatory coccinellids. *Pakistan Journal of Agricultural Research*, 6: 40–44.

Santos-Cividanes, T.M., Dos Anjos, A.C.R., Cividanes, F.J. & Dias, P.C. (2011). Effects of food deprivation on the development of *Coleomegilla maculata* (De Geer) (Coleoptera: Coccinellidae). *Neotropical Entomology*, 40: 112–116.

Sarmento, R.A., Pallini, A., Venzon, M., Souza, O.F.F.D., Molina-Rugama, A.J. & Oliveira, C.L.D. (2007). Functional response of the predator *Eriopis connexa* (Coleoptera: Coccinellidae) to different prey types. *Brazilian Archives of Biology and Technology*, 50: 121–126.

Sato, S., Shinya, K., Yasuda, H., Kindlmann, P. & Dixon, A.F.G. (2009). Effects of intra and interspecific interactions on the survival of two predatory ladybirds (Coleoptera: Coccinellidae) in relation to prey abundance. *Applied Entomology and Zoology*, 44: 215–221.

Schenk, D. & Bacher, S. (2002). Functional response of a generalist insect predator to one of its prey species in the field. *Journal of Animal Ecology*, 71: 524–531.

Schenk, D., Bersier, L.F. & Bacher, S. (2005). An experimental test of the nature of predation: Neither prey-nor ratio dependent. *Journal of Animal Ecology,* 74: 86–91.

Schuder, I., Hommes, M. & Larink, O. (2004). The influence of temperature and food supply on the development of *Adalia bipunctata* (Coleoptera: Coccinellidae). *European Journal of Entomology*, 101: 379–384.

Schuldiner-Harpaz, T. & Coll, M. (2017). Effect of diet history on prey and pollen food choice by two lady beetle species. *Journal of Insect Behavior*, 30(4), 432–438.

Seagraves, M.P. (2009). Lady beetle oviposition behavior in response to the trophic environment. *Biological Control*, 51: 313–322.

Sentis, A., Hemptinne, J.L. & Brodeur, J. (2012). Using functional response modeling to investigate the effect of temperature on predator feeding rate and energetic efficiency. *Oecologia*, 169: 1117–1125.

Singh, N., Mishra, G. & Omkar (2016a). Effect of photoperiod on slow and fast developing individuals in aphidophagous ladybirds, *Menochilus sexmaculatus* and *Propylea dissecta* (Coleoptera: Coccinellidae). *Insect Science*, 23(1): 117–133.

Singh, S., Mishra, G. & Omkar (2016b). Perceived prey quantity modulates oviposition in the ladybird *Menochilus sexmaculatus*. *Journal of Ethology*, 34(1): 59–64.

Singh, S., Mishra, G. & Omkar (2018). Plasticity in reproductive output and development in response to thermal variation in ladybird beetle, *Menochilus sexmaculatus*. *Journal of Thermal Biology*, 71: 180–188.

Skelhorn, J. & Rowe, C. (2006). Predator avoidance learning of prey with secreted or stored defences and the evolution of insect defences. *Animal Behaviour*, 72(4): 827–834.

Skelhorn, J., Halpin, C.G. & Rowe, C. (2016). Learning about aposematic prey. *Behavioral Ecology*, 27(4): 955–964.

Sloggett, J.J. (2008a). Habitat and dietary specificity in aphidophagous ladybirds (Coleoptera: Coccinellidae): Explaining specialization. *Proceedings of the Netherlands Entomological Society Meeting*, 9: 95–113.

Sloggett, J.J. (2008b). Weighty matters: Body size, diet and specialization in aphidophagous ladybird beetles (Coleoptera: Coccinellidae). *European Journal of Entomology*, 105: 381–389.

Smith, B.C. (1965). Effect of food on the longevity, fecundity and development of adult coccinellids. *The Canadian Entomologist*, 97: 760–768.

Snyder, W.E. (2009). Coccinellids in diverse communities: Which niche fits? *Biological Control,* 51(2): 323–335.

Snyder, W.E. & Ives, A.R. (2003). Interactions between specialist and generalist natural enemies: Parasitoids, predators, and pea aphid biocontrol. *Ecology*, 84(1): 91–107.

Snyder, W.E., Joseph, S.B., Preziosi, R.F. & Moore, A.J. (2000). Nutritional benefits of cannibalism for the lady beetle *Harmonia axyridis* (Coleoptera: Coccinellidae) when prey quality is poor. *Environmental Entomology*, 29: 1173–1179.

Soares, A.O., Coderre, D. & Schanderl, H. (2004). Dietary self-selection behaviour by the adults of the aphidophagous ladybeetle *Harmonia axyridis* (Coleoptera: Coccinellidae). *Journal of Animal Ecology*, 73(3): 478–486.

Solomon, M.E. (1949). The natural control of animal populations. *Journal of Animal Ecology*, 18: 1–35.

Solomon, M.E. (1964). Analysis and processes involved in the natural control of insects. In: Cragg, J.B. (Ed), *Advances in Ecological Research*. Academic Press: London and New York, pp. 1–54.

Srivastava, S. & Omkar (2004). Fertility & mortality life–tables of an aphidophagous ladybird beetle, *Coccinella septempunctata* Linnaeus. *Entomon*, 29(2): 101–110.

Stamp, N.E. & Meyerhoefer, B. (2004). Effects of prey quality on social wasps when given a choice of prey. *Entomologia Experimentalis et Applicata*, 110: 45–51.

Stearns, S.C. (1992). *The Evolution of Life Histories*. Oxford University Press: Oxford.

Stephens, D.W. (1991). Change, regularity and value in the evolution of animal learning. *Behavioral Ecology*, 2: 77–89.

Stireman, J.O. (2002). Learning in the generalist tachinid parasitoid *Exorista mella* Walker (Diptera: Tachinidae). *Journal of Insect Behavior*, 15: 689–707.

Stephens, D.W. & Krebs, J.R. (1986). *Foraging Theory*. Princeton University Press: Princeton, NJ, 247 pp.

Sutherland, A.M. & Parrella, M.P. (2009). Mycophagy in Coccinellidae: Review and synthesis. *Biological Control*, 51: 284–293.

Timms, J.E., Oliver, T.H., Straw, N.A. & Leather, S.R. (2008). The effect of host plant on the coccinellid functional response: Is the conifer specialist *Aphidecta obliterata* (L.) (Coleoptera: Coccinellidae) better adapted to spruce than the generalist *Adalia bipunctata* (L.) (Coleoptera: Coccinellidae)? *Biological Control*, 47: 273–281.

Toosi, M., Rasekh, A. & Osawa, N. (2019). Effects of intraguild predation on the life history traits and progeny of the ladybird beetle *Hippodamia variegata*. *Bulletin of Insectology*, 72: 161–168.

Triltsch, H. (1999). Food remains in the guts of *Coccinella septempunctata* (Coleoptera: Coccinellidae) adults and larvae. *European Journal of Entomology*, 96: 355–364.

Tschanz, B., Bersier, L.F. & Bacher, S. (2007). Functional responses: A question of alternative prey and predator density. *Ecology*, 88: 1300–1308.

Ueno, H. (2003). Genetic variation in larval period and pupal mass in an aphidophagous ladybird beetle *(Harmonia axyridis)* reared in different environments. *Entomologia Experimentalis et Applicata*, 106: 211–218.

Ueno, H., Hasegawa, Y., Fujiyama, N. & Katakura, H. (2001). Comparison of genetic variation in growth performance on normal and novel host plants in a local population of a herbivorous ladybird beetle. *Epilachna Vigintioctomaculata Heredity*, 87: 1–7.

Ugine, T.A. (2012). Developmental times and age-specific life tables for *Lygus lineolaris* (Heteroptera: Miridae), reared at multiple constant temperatures. *Environmental Entomology*, 41(1): 1–10.

Uygun, N. & Elekcioglu, N.Z. (1998). Effect of three diaspididae prey species on development and fecundity of the ladybeetle *Chilocorus bipustulatus* in the laboratory. *BioControl*, 43: 153–162.

van Baaren, J., Boivin, G. & Outreman, Y. (2005). Patch exploitation strategy by an egg parasitoid in constant or variable environment. *Ecological Entomology*, 30: 502–509.

van Baaren, J., Nenon, J.P. & Boivin, G. (1995). Comparison of oviposition behavior of a solitary and a gregarious parasitoid (Hymenoptera: Mymaridae). *Journal of Insect Behavior*, 8: 671–686.

Veeravel, R. & Baskaran, P. (1995). Effect of prey and predator age on the feeding preference and rate of predation by two predators *Coccinella transversalis* Fab. And *Cheilomenes sexmaculatus* (Coleoptera: Coccinellidae). *Journal of Biological Control*, 9: 26–29.

Verheggen, F.J., Fagel, Q., Heuskin, S., Lognay, G., Francis, F. & Haubruge, E. (2007). Electrophysiological and behavioral responses of the multicolored Asian lady beetle, *Harmonia axyridis* Pallas, to sesquiterpene semiochemicals. *Journal of Chemical Ecology*, 33(11): 2148–2155.

Vet, L.E.M., Lewis, W.J., Papaj, D.R. & van Lenteren, J.C. (1990). A variable- response model for parasitoid foraging behaviour. *Journal of Insect Behavior*, 3: 471–490.

Via, S. (1990). Ecological genetics and host adaptation in herbivorous insects: The experimental study of evolution in natural and agricultural systems. *Annual Review of Entomology*, 35: 421–446.

Vieira, L.C., Salom, S.M. & Kok, L.T. (2012). Functional and numerical response of *Laricobius spp*. Predators (Coleoptera: Derodontidae) on hemlock woolly adelgid, *Adelges tsugae* (Hemiptera: Adelgidae). *Biological Control*, 61: 47–54.

Vinson, S.B. (1977). Behavioural chemicals in the augmentation of natural enemies. In: Ridgeway, R.L. & Vinson, S.B. (Eds.), *Biological Control by Augmentation of Natural Enemies*. Plenum: New York, pp. 237–279.

Waldbauer, G.P. (1968). The consumption and utilization of food by insect. *Advances in Insect Physiology*, 5: 229–288.

Wang, S., Wang, K., Michaud, J.P., Zhang, F. & Tan, X.L. (2014). Reproductive performance of *Propylea japonica* (Coleoptera: Coccinellidae) under various light intensities, wavelengths and photoperiods. *European Journal of Entomology*, 111(3): 341–347.

Wang, P., Su, J., Ouyang, F. & Ge, F. (2015). Orientation behavior of *Propylea japonica* toward visual and olfactory cues from its prey-host plant combination. *Entomologia Experimentalis et Applicata*, 155: 162–166.

Ware, R.L., Yguel, B. & Majerus, M.E.N. (2008). Effects of larval diet on female reproductive output of the European coccinellid *Adalia bipunctata* and the invasive species *Harmonia axyridis* (Coleoptera: Coccinellidae). *European Journal of Entomology*, 105: 437–443.

Werling, B.P., Lowenstein, D.M., Straub, C.S. & Gratton, C. (2012). Multi–predator effects produced by functionally distinct species vary with prey density. *Journal of Insect Science*, 12: 1–7.

Wilder, S.M., Mayntz, D., Toft, S., Rypstra, A.L., Pilati, A. & Vanni, M.J. (2010). Intraspecific variation in the quality of prey: A comparison of nutrient presence in prey and nutrient extraction by spiders. *Oikos*, 119: 350–358.

Wu, P., Zhang, J., Haseeb, M., Yan, S., Kanga, L., & Zhang, R. (2018). Functional responses and intraspecific competition in the ladybird *Harmonia axyridis* (Coleoptera: Coccinellidae) provided with *Melanaphis sacchari* (Homoptera: Aphididae) as prey. *European Journal of Entomology*, 115: 232–241.

Xue, H.J., Egas, M. & Yang, X.K. (2007). Development of a positive preference-performance relationship in an oligophagous beetle: Adaptive learning? *Entomologia Experimentalis et Applicata*, 125: 119–124.

Yang, F., Wang, Q., Wang, D., Xu, B., Xu, J., Lu, Y. & Harwood, J.D. (2017). Intraguild predation among three common coccinellids (Coleoptera: Coccinellidae) in China: Detection using DNA-based gut-content analysis. *Environmental Entomology*, 46(1): 1–10.

Yasuda, H. & Shinya, K. (1997). Cannibalism and interspecific predation in two predatory ladybirds in relation to prey abundance in the field. *Entomophaga*, 42: 153–163.

Zarghami, S., Kocheili, F., Mossadegh, M.S., Allahyari, H. & Rasekh, A. (2014). Effect of temperature on population growth and life table parameters of *Nephus arcuatus* (Coleoptera: Coccinellidae). *European Journal of Entomology*, 111, 199–206.

Zhang, S.Z., Zhang, F. & Hua, B.Z. (2007). Suitability of various prey types for the development of *Propylea japonica* (Coleoptera: Coccinellidae). *European Journal of Entomology*, 104: 149–152.

Zhang, S.Z., Li, J.J., Shan, H.W., Zhang, F. & Liu, T.X. (2012). Influence of five aphid species on development and reproduction of *Propylaea japonica* (Coleoptera: Coccinellidae). *Biological Control*, 62: 135–139.

Zheng, Y., Hagen, K.S., Daane, K.M. & Mittler, T.E. (1993a). Influence of larval dietary supply on the food consumption, food utilization efficiency, growth and development of the lacewing *Chrysoperla carnea*. *Entomologia Experimentalis et Applicata*, 67: 1–7.

Zheng, Y., Daane, K.M., Hagen, K.S. & Mittler, T.E. (1993b). Influence of larval food consumption on the fecundity of the lacewing *Chrysoperla carnea*. *Entomologia Experimentalis et Applicata*, 67: 9–14.

Zhu, J. & Park, K.-C. (2005). Methyl salicylate, a soybean aphid-induced plant volatile attractive to the predator *Coccinella septempunctata*. *Journal of Chemical Ecology*, 31: 1733–1746.

Zhu, J., Cosse, A.A., Obrycki, J.J., Boo, K.S. & Baker, T.C. (1999). Olfactory reactions of the twelve-spotted beetle, *Coleomegilla maculata* and the green lacewing, *Chrysoperla carnea* to semiochemicals released from their prey and host plant: Electroantennogram and behavioral responses. *Journal of Chemical Ecology*, 25: 1163–1177.

9 Syrphid Predators (Diptera: Brachycera)

Sunil Joshi, K. J. David, and K. Sachin
ICAR-National Bureau of Agricultural Insect Resources

CONTENTS

LEARNING OBJECTIVES

a. To study the diversity of Syrphidae and its feeding habits
b. To standardize mass production protocol of Syrphids, which can be used as biocontrol agents
c. To study the field efficacy of potential biocontrol agents of Syrphidae

9.1 INTRODUCTION

Syrphidae, commonly called hoverflies, are cosmopolitan in distribution with maximum diversity in Nearctic, Neotropical and Palaearctic regions. They play a key role in the ecosystem as larvae of syrphids have diverse feeding habits, *viz.* saprophytic, predatory and phytophagous, depending up on the taxa. Adult flies of several species are pollinators of cultivated crops and wild plants. They do mimic bees and wasps; hence, form an excellent model for studying mimicry. Syrphidae can be distinguished from closely related families by the presence of a spurious vein in the wing that bisects r-m cross-vein. They are presently classified in four subfamilies, namely, Syrphinae, Eristalinae, Pipizinae and Microdontinae (GBIF Secretariat, 2018) with 6,674 described species. Keys to aphidophagous syrphids have been published by Coe (1953), Dixon (1960), Chandeler (1968a) and Goeldlin de Tiefenau (1974). Stubbs and Falk (1983) published an introductory guide to U.K. fauna. Indian Syrphidae was reviewed by Brunetti (1923). Knutson et al. (1975) provided

DOI: 10.1201/9781003370864-9

correct nomenclature of Syrphidae of the Oriental region. Ghorpade (1973) conducted faunistic studies on hoverflies of south India; listed 542 species of host insects of 47 species of syrphids from India and neighbouring countries (Ghorpade, 1981). Thompson and Ghorpade (1992) published a paper on Oriental *Paragus*, wherein notes on Indian Paragini were provided. Later, Ghorpade (1994) provided diagnostic keys to genera and species of Syrphinae from the Indian subcontinent. A checklist of hoverflies occurring in Eastern Himalayas has also been published (Mitra et al., 2008). Checklist to Syrphidae of the Indian subcontinent, northwest India and peninsular India was provided by Ghorpade (2014, 2015, 2019), respectively. Studies on species occurring in a particular zoogeographic area in a state have also been listed (Baskaran et al., 2008). Several faunal studies have been conducted recently from several countries, like Turkey (Sarbyk, 2008; Bayrak and Hayat, 2008), East Azerbaijan province and Iran (Ehteshamnia et al., 2010; Khaghaninia, 2010). Genera-wise and crop-wise taxonomic revisions of Syrphidae are also available (Sorokina, 2009; Morales and Marinoni, 2009; Rossi et al., 2007; Sheridan et al., 2008). Mitra et al. (2015) listed 357species of Syrphidae from India in 14 tribes and three subfamilies.

9.2 DIVERSITY OF INSECTS AND PREY

The superfamily Syrphoidea belongs to the series Aschiza under suborder Cyclorrhapha of the order Diptera. Syrphidae is one of the most diverse families in Brachycera with more than 6,674 species in 493 genera in four subfamilies, namely, Syrphinae, Eristalinae, Pipizinae and Microdontinae (GBIF Secretariat, 2018). They are widely distributed with most of the species from Nearctic, Neotropical and Palaearctic regions. Syrphidae can be considered as dual service providers in nature, as they do comprise species that help in crop pollination and also act as predators forming an integral part of natural biological control. Adults of most of the species are flower visitors and do assist in the pollination of crops, whereas larvae are often predaceous (Larson et al., 2001; Rader et al., 2016). Larvae of species belonging to Syrphinae, Eristalinae and Pipizinae are major predators of Aphididae and other soft-bodied insects, like thrips (Skevington and Dang, 2002). 493 species in 107 genera have been recorded from the Indian subcontinent (Ghorpade, 2019, 2014). Major species of Syrphidae predators on aphid species are listed in Table 9.1.

TABLE 9.1
Major Species of Syrphidae as Predators in India and Its Prey Species

Sl. No.	Syrphidae	Prey Species	References
1	*Ischiodon scutellaris* (Figure 9.1d, g, h, i and j)	*Brevicoryne brassicae* *Lipaphis erysimi* *Toxoptera aurantii* *Aphis craccivora* *R. maidis* *Lipaphis erysimi*	Sharma and Bhalla (1988), Roy and Basu (1977), Radhakrishnan and Muraleedharan (1993), Joshi et al. (1999a), Singh and Mishra (1988), and Devjani and Singh (2006)
2	*Asiobaccha nubilipennis*	*Toxoptera aurantii*	Radhakrishnan and Muraleedharan (1993)
3	*Betasyrphus serarius*	*Toxoptera aurantii*	Radhakrishnan and Muraleedharan (1993)
4	*Betasyrphus fletcheri*	*Aphis craccivora*	Joshi et al. (1999b)
5	*Betasyrphus linga*	*Aphis craccivora*	Joshi et al. (1999b)
6	*Dideopsis aegrota* (Figure 9.1a and k)	*Aphis craccivora*	Joshi et al. (1999b), Radhakrishnan and Muraleedharan (1993)
7	*Paragus tibialis*	*Toxoptera aurantii*	Radhakrishnan and Muraleedharan (1993)
8	*Paragus serratus* (Figure 9.1e and p)	*Aphis craccivora*	Joshi et al. (1999a)
9	*Paragus yerburiensis* (Figure 9.1f)	*Aphis craccivora*	Joshi et al. (1999b)

(Continued)

TABLE 9.1 (*Continued*)
Major Species of Syrphidae as Predators in India and Its Prey Species

Sl. No.	Syrphidae	Prey Species	References
10	*Scaeva pyrastri*	*Brevicoryne brassicae*	Sharma and Bhalla (1988)
11	*Eupeodes confrater* (Figure 9.1b, l. m and n)	*Brevicoryne brassicae* *Lipaphis erysimi* *Myzus rosae*	Sharma and Bhalla (1988), Roy and Basu (1977), and Verma and Sharma (2006)
12	*Episyrphus balteatus* (Figure 9.1c and o)	*Brevicoryne brassicae* *Lipaphis erysimi* *Toxoptera aurantii* *Myzus persicae*	Sharma and Bhalla (1988), Roy and Basu (1977), Radhakrishnan and Muraleedharan (1993), and Devjani and Singh (2006)
13	*Eupeodes corollae*	*Brevicoryne brassicae*	Verma and Makhmoor (1987)
14	*Allograpta javana*	*Aphis craccivora*	Patro and Behera (2002)

9.3 BASIC BIOLOGY

Like any other holometabolous insect, syrphids do have four life stages; egg, larva, puparium and adult. In the case of aphidophagous syrphids, the entire life cycle is completed within 20–30 days. Gravid females prefer to lay eggs on plants with aphid infestations. Eggs are often laid in groups and usually hatch in 2–6 days. The eggs are laid singly inside or near the aphid colony, or in groups side by side on the leaf surface and stem. The eggs are white and elongate oval in shape with sculpturing on chorion. Larvae are often fusiform except in Eristalinae, and head and mouthparts are not conspicuous and in most species they are retractile. The larval instars are difficult to differentiate as moulted skins are very thin and transparent, and remain adherent to the surface on which they are laid. Moulting of syrphid larva is peculiar. Exuviae are not obtained in the early winter season. Exuviae of trachea and bucco-pharyngeal armature are also shed. A small longitudinal slit can be seen at the anterior end of moulted skin through which the larva has emerged. Prior to each moulting, the larva passes the faeces and empties the gut contents. The larva remains inactive before moulting. It occasionally feeds at that time and secretes a colourless moulting fluid. Larvae have three instars, which turn into puparium in 7–16 days. Puparium is often tear shaped, flat ventrally with dorsal and lateral portions inflated. As soon as they emerge, syrphid larvae feed on aphids.; pupariation often last for 9–14 days. After pupation, adult syrphids shift from a predatory diet to feeding on nectar and pollen from flowers. Adult flies may live for a week to 2 weeks (Joshi and Ballal, 2013). The syrphid adults are distinguished from other dipterans by the presence of a false or spurious vein in the wing, crossing r-m between R_{4+5} and M_{1+2}. The R_5 cell is closed. Proboscis is short, face is narrow and grooves are not present below the antennae (Borror et al., 1976). The adults of syrphids are characterized by hovering behaviour near the pollen and prey source, and hence commonly known as hoverflies or flower flies. The adults have bright coloured patches on black abdomen. In many cases, they resemble wasps and honeybees. The adults are active during daytime and are found visiting flowers to feed on pollen and nectar.

9.4 ECOLOGY AND BEHAVIOUR

Syrphidae has two key roles in the ecosystem: primarily as a biocontrol agent of soft-bodied insects, like aphids, and as pollinators of various crops.

9.4.1 As a Potential Biocontrol Agents

One-third of Syrphidae are aphidophagous; adults of most of these species have been adapted to aphid-finding, such as identifying prey pheromones, which makes aphidophagous syrphids potential biocontrol agents (Dib et al., 2010; Mizuno et al., 1997). Syrphid larvae have unique adaptation to ensure their feeding; highly sticky secretions from the larval mouth facilitate them to hold on to the aphid prey firmly. First instar of *E. balteatus* and *E. confrater* were able to prey only the small-sized aphids, later instars (second and third instars) efficiently lifted the prey irrespective of its size (Kumar et al., 1996). In contrary to these observations, Joshi et al. (1999a) found no marked preference for specific stage of aphid by particular instar of *I. scutellaris* larvae. Even freshly hatched larvae fed on gravid aphid female much larger than their own size.

After lifting the prey, the syrphid larvae make a hollow cup by retracting their anterior thoracic segments for resting the prey in it. The larva then pierces mouth hooks of bucco-pharyngeal armature into the body of the prey and slowly evacuates the whole of the inner body contents. After sucking the body fluids of the prey, the larva discards it when shrivelled and dry (Kumar et al., 1996; Joshi et al., 1999a). Prey consumption time varies with species; larvae of *I. scutellaris* spent least time (642.2 seconds), followed by *E. confrater* (702.8 seconds) and *E. balteatus* (1122.1 seconds).

Feeding efficiency of syrphids shows variations with respect to the species and the prey. Larvae of *E. corollae* consumed on an average 346 *Capitophorus eleagni* (Del Guercio) in 8.6 days of larval period (Tawfik et al., 1974), while *E. balteatus* fed on 416 *Aphis pomi* (De Geer) during its development (Wnuk, 1977). *Scaeva pyrastri* (Linnaeus) ate 550 *B. brassicae* (Wnuk and Fuchs, 1977). The larvae of *Platycheirus clypeatus* (Meigen) and *Melanostoma mellinum* (Linnaeus) (Tribe Melastomini) consume fewer aphids, i.e., 135 and 150, respectively (Bankowska et al., 1978). Studies on feeding potential and biology by Joshi et al. (1999a) of *P. serratus* and *I. scutellaris* on six species of aphids revealed that *I. scutellaris* is superior to *P. serratus* in feeding potential, fecundity and longevity, thereby proving to be a potential biocontrol agent. *A. craccivora* was the most preferred and *Uroleucon compositae* (Theobald) was the least preferred prey in the study.

Feeding potential and development of Syrphidae are influenced by prey density and other abiotic factors, like temperature and relative humidity. Developmental period of *I. scutellaris* decreased with increase in the population density of *Rhopalosiphum maidis* (Fitch) (Singh and Mishra, 1988). Studies on functional response of *I. scutellaris* on *L. erysimi* had shown a positive linear relationship between the prey density and the number of prey. However, a negative relationship was observed between the aphid density and percentage of consumption (Devi and Singh, 2000). In *E. corollae*, voracity of the syrphid larvae increased with their growth, and the third instar was the most voracious consumer (80%–90%) of the total food consumed (Benestad, 1970). In the case of *E. balteatus,* first, second and third instars consumed 3%, 13% and 84%, respectively, of the total kill. Joshi et al. (1999b) observed a gradual increase in aphid consumption and the last instar larva was the most voracious, consuming 46.99%–62.43% of total aphids by *B. linga*, *B. fletcheri*, *D. aegrota*, *I. scutellaris*, *P. serratus* and *P. yerburiensis* on *A. craccivora*. Adams (1984) found that the daily kill is influenced only by ambient temperature and relative humidity and is not affected by prey density. Several researchers recorded an indirect effect of prey availability on development. During instances of shortage of food, more energy is allocated to body maintenance and larval development is longer and pupae are smaller (Ruzicka and Cairo, 1976; Barlow, 1979; Cornelius and Barlow, 1980).

Biocontrol potential of natural enemies has been influenced by the fertility schedules of the species selected for biocontrol. Makhmoor and Verma (1989), Sharma et al. (1994) and Sharma and Bhalla (1995) studied the intrinsic rate of natural increase of three species, namely, *Eupeodes confrater*, *Eupeodes frequens* and *Eupeodes corollae*, respectively, which indicated that these predators have the potential to rapidly increase their population size with a strong possibility of bringing about an effective check on aphid population. *I. scutellaris* had higher net reproductive rate in winter than in summer when population indices for two seasons (winter and summer) were compared (PDBC, 1999). Similar studies on *P. serratus* and *P. yerburiensis* revealed that these species could perpetuate at a faster rate when environmental conditions are favourable (Joshi et al., 2000). This attribute makes species of *Paragus* as additional predators along with *I. scutellaris*, which has the ability to produce a large population but at a relatively lower pace (Joshi et al., 2001).

9.4.2 As Pollinators

Apart from being predators, adults of Syrphidae play a key role as crop pollinators. Both males and females of aphidophagous and non-aphidophagous syrphids feed on pollen and nectar (Proctor and Yeo, 1973). They are often classified as generalist flower visitors. Species belonging to *Eupeodes*, *Sphaerophoria*, *Syrphus*, *Eristalis* and *Episyrphus* visited at least 40 different species of plants in not <15 families (Dunn et al., 2020). Bees often carry more pollen than syrphids owing to their morphological adaptations, such as hairiness and size. Despite morphological limitations, syrphids can carry a high number of pollen grains as wild bees and honey bees. In a study conducted on *Brassica rapa* var *chinensis*, it was observed that *Eristalis tenax* deposited the same amount of pollen on stigma as honey bees did (Rader et al., 2009). In a study conducted by Rader et al. (2011) to compare the pollen transport by bees and flies, they could find that syrphids could carry viable pollen up to 400 m against 300 m by bees. Even though syrphids are considered general predators, their attraction to flowers varies upon the flower colour and accessibility of nectar. Usually, aphidophagous syrphids have shorter proboscis, hence they prefer to feed on flowers with short corolla. For example, *Episyrphus balteatus* and *E. corollae* were abundant in fields with high cover of flowers with open nectar and shorter corollas. Gilbert (1981) while studying foraging ecology of syrphids correlated the length of the tongue to the depth of the corolla and suggested that, this is the factor that determines the selection of flower species visited by syrphids. He further discussed diel activity of different syrphid species in relation to their size, temperature and light intensity (Gilbert, 1985). Presence of flowering plants in the vicinity of aphid-infested crop has been suggested as a means of increasing the effectiveness of syrphids as predators. However, Chandler (1968b) did not find any increase in the number of eggs in plots of Brussels sprouts with additional flowers placed in them. On the other hand, the presence of nectar (adult food) may be utilized to retain adult syrphids in areas with low prey density. Hagen et al. (1972) used sprays of sucrose and yeast to increase the number of adult syrphids in cotton (Figure 9.1).

9.5 LIFE HISTORY

Most of the species are multivoltine with several generations per year. Adults are often associated with flowers and aid in pollination. Though larvae of Syrphinae are generally carnivorous and are predatory on other soft-bodied insects, they do exhibit a wide array of feeding habits, like saprophagy, coprophagy, mycetophagy, phytophagy and specialized inquilines in social insect nests. Larvae of Eristalinae are aquatic and are often termed as rat tail maggots owing to their elongate posterior extension (Mitra et al., 2015). Studies on the life history of syrphids have been extensively undertaken on aphidophagous species. Adults of syrphids are active flers; Aubert and Tiefenau (1981) studied flight activity in Swiss Alps by marking adults and by actual observation. They suggested that hoverflies are strong flyers and can travel long distances (up to 50 km) in migratory flights. Mating in syrphids occur both during flight as well as at rest. Before mating, the male and female of *I. scutellaris* showed courtship behaviour of hovering mouth to mouth in the air (Joshi et al., 1999a). Mating duration varies in different species. Verma and Makhmoor (1987) observed repeated matings in *E. balteatus*, each lasting for 1–2 seconds, whereas in *E. corollae*, it lasted for 1–6 hours. In *B. serarius* and *P. tibialis*, copulation lasts for 10–13 minutes and in *A. nubilipennis* it lasts for 30 minutes. Repeated mating, each lasting for 50–65 minutes, was observed in *I. scutellaris*, whereas in *P. serratus* and *P. yerburiensis*, it lasted for 15–20 minutes and in *D. aegrota*, mating lasted for 20–30 minutes (Joshi et al., 1999b). Pre-oviposition period varies between 2 and 5 days in different species. Mated females can be easily recognized by inflated abdomen bearing eggs, which is whitened from underside. The gravid female hovers around aphid-infested plants and walks slowly near the aphid colony. While searching, the female moves the antennae rapidly and extends its abdomen just before oviposition. The eggs are generally laid in the aphid colony, but as the females get older, the eggs are deposited on the surfaces of the rearing cage. These eggs fail to

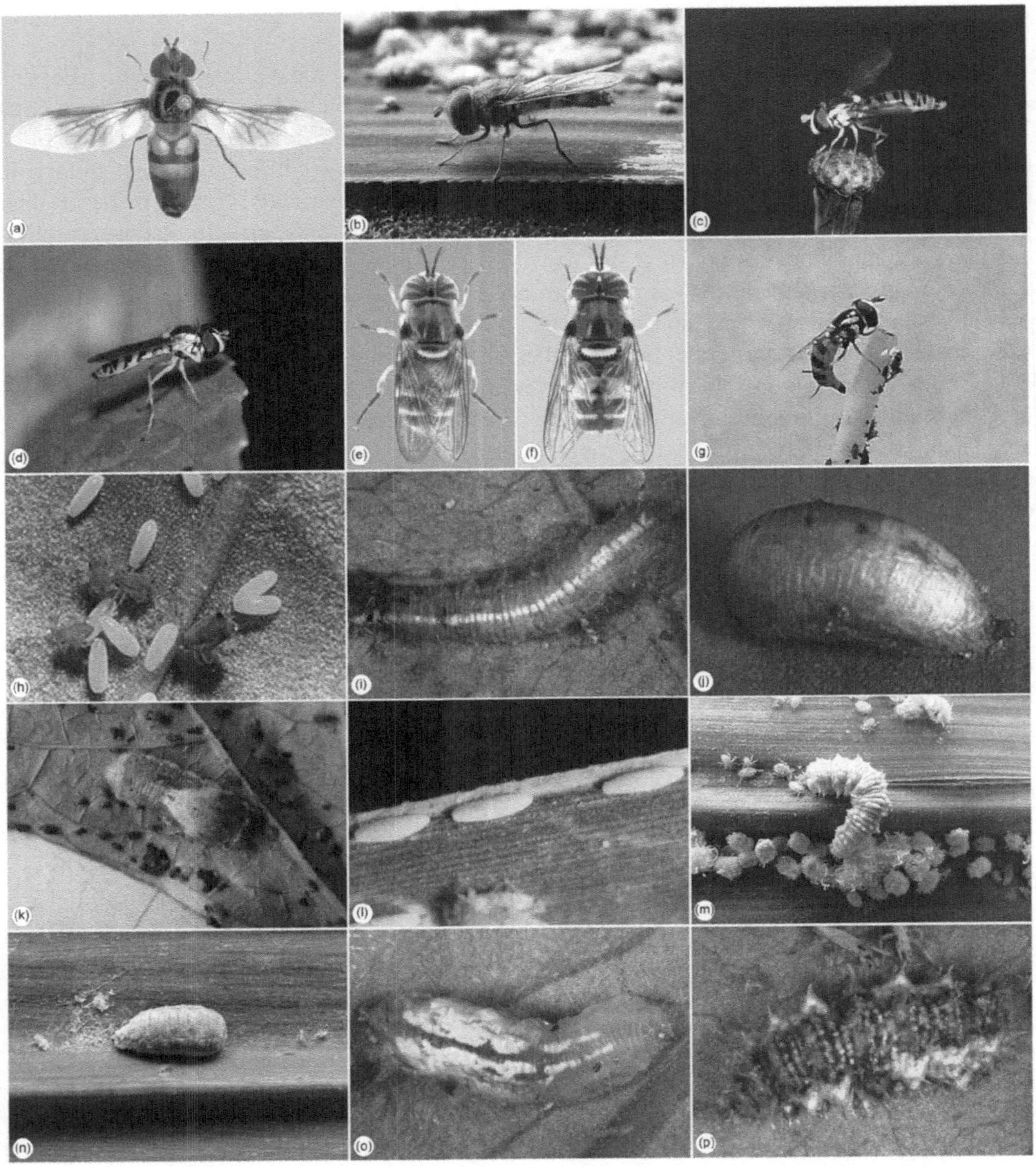

FIGURE 9.1 (a) *Dideopsis aegrota* adult; (b) *Eupeodes confrater* adult; (c) *Episyrphus balteatus* adult; (d) *Episyrphus balteatus* adult; (e) *Paragus serratus* adult; (f) *Paragus yerburiensis* adult; (g) Female of *I. scutellaris* ovipositing on seedling of cowpea infested with *Aphis craccivora*; (h) Egg of *I. scutellaris*; (i) Larva of *I. scutellaris*; (j) Pupa of *I. scutellaris*; (k) Larva of *D. aegrota*; (l) Eggs of *E. confrater*; (m) Larva of *E. confrater*; (n) Pupa of *E. confrater*; (o) Larva of *E. balteatus*; and (p) Larva of *Paragus serratus.*

hatch as they shrink due to lack of moisture (unpublished observation). The oviposition period varies considerably among different species. It lasts for about 49.7 days in *E. balteatus* and 16.9 days in *E. confarater* (Verma and Makhmoor, 1987). Joshi et al. (1999b) recorded an oviposition period of 3–4, 4–5, 5–6 and 22–23 days in *P. yerburiensis, D. aegrota, P. serratus* and *I. scutellaris*, respectively. Joshi et al. (1998) studied life cycle parameters of *I. scutellaris* and found egg, larval, pupal period and female and male adult longevity to be 3–4, 9–12, 9–12, 35–42 and 30–35 days, respectively (Figure 9.2).

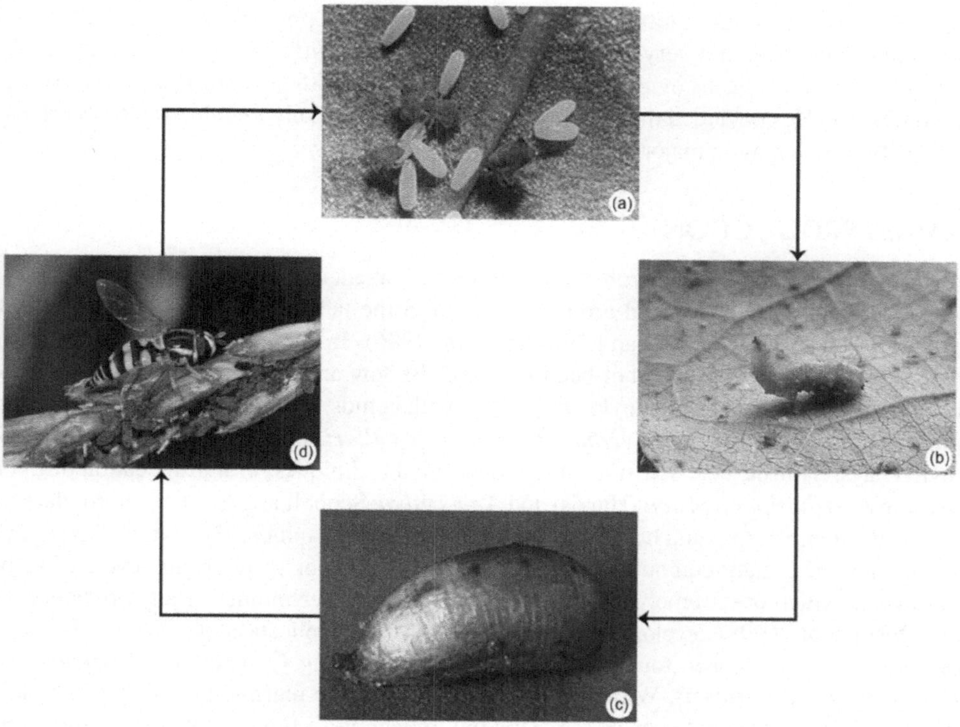

FIGURE 9.2 Life cycle of *Ischiodon scutellaris* (Fabricius): (a) eggs; (b) larva; (c) pupa; (d) adult.

Olfactory stimuli originating from aphid honeydew and cornicle secretion, and visual and mechanical cues play an important role in determining oviposition site. Eggs of Syrphini, such as *E. luniger* (Meigen), *I scutellaris*, *E. corollae*, *E. balteatus*, and Paragini, such as *P. serratus*, *P. yerburiensis* and *P. tibialis*, are laid close to or among the aphid colonies. But, members of Melanostomini are known to deposit their eggs in groups away from aphid colonies or on adjacent uninfested plants (Chandler, 1968b, c). In addition to this, light, shade and position of aphid colony also influence the choice of oviposition site. Sanders (1980, 1981) found that *E. corollae* preferred dark to lighter areas when given a choice and also the females exhibited preference for oviposition beside colonies on vertical rather than horizontal surface. Larval feeding history is generally thought to affect the reproductive output of the female; however, in *E. corollae*, Scott and Barlow (1984) found that when more aphids were offered, larvae consumed more and produced heavier pupae. Fecundity of adults reared from these pupae was extremely variable (range, 0–1,488 eggs per female) and depended mostly on adult longevity and not on larval feeding history and weight of pupae. It has been found that the hoverfly females avoid ovipositing in aphid colonies in which conspecific larvae or their tracks are already present, suggesting that this behaviour constitutes a strategy that enables females to optimize their oviposition site and reduce competition suffered by their offspring (Almohamad et al., 2010a). It has also been found that *E. balteatus* females lay fewer eggs at sites where there are heterospecific larval tracks of *Harmonia axyridis* (Almohamad et al., 2010b).

Smith (1969, 1976) investigated the effects of weeds in the crop on oviposition of syrphids, *viz.*, *Melanostoma* spp., *Platycheirus* sp. and *E. balteatus* in plots of Brussels sprouts. Only *Melanostoma* laid more eggs in weedy plots as compared to bare plots, whereas *E. balteatus* laid eggs preferentially in the weedy plots only when aphid density was low. On the other hand, *Platycheirus* sp. preferred sprouts grown in bare soil. Horn (1981) found higher population of syrphids in plots with weeds, despite high densities of *M. persicae* in weed-free plots. Habitat preference was studied by

Pollard (1971) using heavily infested Brussels sprouts grown in pots. *E. balteatus* laid more eggs on plants in crop habitat and very few in woodland, whereas *Platycheirus scutains* (Meigen) and *Leucozona lucorum* Linnaeus oviposited in hedge row habitat with fewer eggs at an increasing distance from hedge. He concluded that habitat outside a crop is unlikely to influence the syrphids. It is aphid density, which plays a major role in oviposition of syrphids.

9.6 MASS PRODUCTION

Although species belonging to Syrphidae are important agents in natural control of aphids, they have never been mass cultured and are rarely reared in the laboratory. *E. corollae* represents an exception (Barlow, 1979; Barlow and Whittingham, 1986). In these publications also, complete techniques for mass rearing have not been revealed. Barlow and Whittingham (1986) mentioned about collection of syrphid eggs in nylon bags filled with aphids. However, in our laboratory studies on multiplication of *P. serratus*, *P. yerburiensis* and *I. scutellaris* positive results were not achieved with nylon bags. Samsoe-Petersen et al. (1989) developed techniques for multiplication of *E. corollae* by using *Acyrthosiphon pisum* (Harris) and *Aphis fabae* Scopoli as prey. They used glasshouses with controlled temperature and humidity conditions for rearing aphids. This article, however, does not provide any information about handling of different stages of syrphids and cycle of activities to be followed, which is extremely important in multiplication techniques of any predatory insect. In India, Joshi et al. (1998) developed techniques for mass multiplication of syrphids. These methods of rearing have also been found to be successful for rearing *I. scutellaris*, *Paragus serratus* (Fabricius) and *P. yerburiensis*. With suitable modifications, the rearing units may prove satisfactory for rearing other syrphid species too. This multiplication technique has been summarized in the form of schematic presentation in Figure 9.3.

Two species of syrphids, *viz.*, *Episyrphus balteatus* (de Geer) and *Eupeodes bucculatus* (Rondani), have been reared on artificial diet consisting of drone honey bee powder (DP) (Iwai et al., 2007).

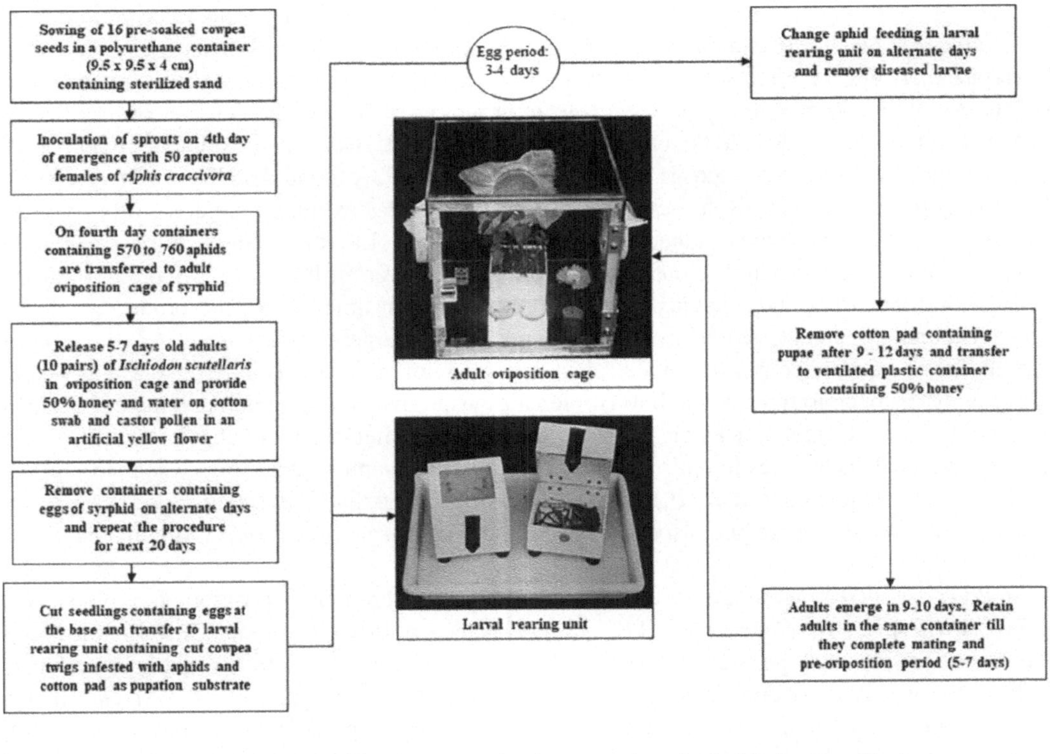

FIGURE 9.3 Schematic presentation of mass production procedure for *Ischiodon scutellaris*.

Later, it was found that addition of linoleic acid and oleic acid could improve adult emergence rate and adult body size (Iwai et al., 2009). Artificial diet for adults has been found to substantially increase the oviposition by *P. serratus*. This diet includes addition of multivitamins, clomiphane citrate and tocopheryl acetate to 50% honey (Baskaran et al., 2009a). The same group of workers have also studied the influence of pollen grain on fecundity of *P. serratus* (Baskaran et al., 2009b).

9.7 EFFECTIVENESS OF SYRPHIDS AS BIOCONTROL AGENTS IN THE FIELD

Chambers (1988) defined effectiveness as the degree to which aphid increase is impeded by predators; the extent to which increasing aphid number is slowed down, held steady or reversed by action of predators (syrphids), and whether this keeps the aphid population density below the economic injury level. He outlined two methods, which are generally used for predator evaluation: qualitative and quantitative methods. Examination of aphid colonies for the presence of predators, predator exclusion or inclusion and observational methods with graphical analysis are qualitative methods, whereas manipulation of initial prey and predator ratios and the observation of outcome or estimate of abundance of predator are the methods employed under quantitative analysis.

9.7.1 Predator Exclusion Method

Pollard (1969) removed syrphid larvae from Brussels sprouts by hand and examined daily aphid population. *E. balteatus* and *Sphaerophoria scripta* (Linnaeus) were found to be the predators bringing down aphid population. Way and Banks (1968) used sleeve cages to exclude predators of *A. fabae* and found syrphids to be a major factor in reducing aphid population. Similar techniques to exclude predators of *M. persicae* were adopted by Tamaki (1973). They found both syrphid and coccinellids to reduce aphid production by 95%. Similar observations were recorded by Chambers et al. (1983) in winter wheat. There are other methods of exclusion of predators by using selective pesticides. Carbaryl was used by Smith (1981) to kill syrphid larvae in wheat plots. Death of the larvae resulted in higher aphid numbers in sprayed plots.

9.7.2 Observational Studies

These studies tried to correlate the presence of syrphid larvae with a decline in aphid population. Such experiments cannot be concluded with confidence, as there are several factors that contribute to population decline. These factors are emigration of aphid alates, change in host plant physiology, impact of adverse weather conditions on aphid multiplication and effect of other natural enemies. However, such studies are still considered for the assessment of predators as these experiments are relatively easy to conduct. Horej (1982) recorded the presence of three species in sugar beet on *A. fabae* and related aphid decline with it. According to Chambers et al. (1986), synchronization of larval population of syrphids with *S. avenae* population results in decline of the latter. In India, Kotwal et al. (1984), and Sharma and Bhalla (1991) found the association of different syrphid species with cabbage aphid in mid-hill regions of Himachal Pradesh. They found peak population of syrphids coinciding with peak aphid population. Radhakrishnan and Muraleedharan (1993) stated that the population density of the various syrphid predators synchronizes with the abundance of *Toxoptera aurantii* (Boyer de Fonscolombe) in the tea fields at Valparai.

9.7.3 Predator: Prey Ratios

Such types of studies are conducted on a small scale. Wnuk (1977) found that *A. pomi* population can be totally taken care of if *E. balteatus* and its prey population are maintained at 1:50 at the initial infestation level and 1:200 at the later part of infestation. Chambers et al. (1986) found that increasing predator:prey ratios are required to control *A. gossypii* on cucumber. In India, Kotwal et al. (1984)

found that syrphids could contain *B. brassicae* population at a predator:prey ratio of 1:129, but failed to do so when the ratio was 1:1,163. Verma and Makhmoor (1987) studied the relative abundance of nine species of syrphids preying on *B. brassicae* in cauliflower. They found that high activity of predators always coincided with peak aphid population but failed to check aphid population late in season. In spite of very high larval counts, predator: prey ratio never reached a desirable level. Predator: prey ratios recorded in the field by Kumar et al. (1988) indicated that the level of *B. brassicae* influenced the number of syrphid species active on the mustard crop. Egg as well as larval population positively correlated with the number of aphid colonies. Dispersion of egg and larval population was non-aggregated at very low and very high aphid densities, and aggregated at intermediate densities. Ratios of predator and prey population were desirable at the beginning but were not so late in the crop season. While studying relative abundance of six species of syrphids on *A. craccivora* infesting cowpea, Joshi et al. (1999b) found a prey-dependant syrphid predator growth, often failing to reach ideal predator:prey ratio of 1:100. This emphasizes the need for aphid management by way of augmentation of syrphid predators.

9.7.4 Prediction Model

Tamaki (1974) proposed a model for the prediction of performance of predators in the fields. It was used to assess the reductive power of complex of predators. This model was based on relative feeding capacities and efficacy estimated in the field. In addition to this model, Raworth (1984) suggested a model in which he employed rates of kill change with size of larvae of the predator and used an estimate of larval voracity expressed as biomass of aphids consumed. However, this model could not account for the impact of the syrphid on *B. brassicae*. Another simulation model by Rabbinge et al. (1979) used published data on life histories and feeding potential of syrphids. But, this model could not provide accurate data for field kills in relation to population density. According to Raworth (1984), there is a need for studies that estimate quantitative effectiveness of hoverflies, which use functional and numerical responses of the predators.

9.8 CONCLUSIONS

Of the known syrphid diversity, only a few species have been studied in detail, which limits the inclusion of syrphids in Integrated Pest Management (IPM) programmes. Prey consumption rates, habitat requirements, flower preferences and pollination potential of the vast majority of syrphid species needs to be thoroughly investigated. Since most of the experiments on syrphids were laboratory based, findings have not been translated to the crop environment fully. Results of laboratory experiments may not always be indicative of what will be successful in the field (Nelson et al., 2012) and they often fall short of considering other biotic and abiotic factors that may occur in the field (Eckberg et al., 2014). Field-based and glasshouse trials are necessary to determine the efficacy of syrphids as biocontrol agents and crop pollinators. Even though syrphids are not the most efficient pollinators, alluring them to the field can lead to an increase in overall pollinator diversity. Hence, encouragement of syrphids along with more traditional pollinators, such as honeybees, is likely to foster outcomes for food production systems. Inoculative release of syrphids is another option for use in crop environments. Unfortunately, rearing predators that require both prey and floral resources across different life stages can be challenging. Several aspects of syrphid biology including nutritional requirements need to be investigated thoroughly for mass-rearing programmes to be successfully integrated into IPM strategies. Extensive research in the field of syrphid ecology, like searching behaviour, factors affecting their foraging and the influence of aphid semiochemicals on their effectiveness needs to be undertaken. It is essential to formulate a methodology for determining the effect of temperature and other factors on their efficiency and on the factors that limit their functional and numerical responses. Development of techniques for their effective transport and release are necessary. Release rates, stage to be released and the stage, which can be stored at low temperature, needs to be identified. More conservation studies on the effect of weeds, source of nectar and pollen and inter-cropping on the effectiveness of syrphid are desirable.

POINTS TO REMEMBER

a. Syrphidae are dual service providers in the ecosystem as crop pollinators and predators
b. Species of subfamily Syrphinae are predominantly aphidophagous and can be potential biocontrol agents of aphids
c. As syrphids are more abundant during winter, they should be targeted against aphids infesting winter Cole crops, like cabbage and cauliflower, and oilseeds like mustard
d. Mass production and procedures for shipment of syrphids needs to be standardized to make it an effective biocontrol agent.

ACKNOWLEDGEMENTS

The authors are thankful to Dr Chandish R. Ballal, Director, ICAR – National Bureau of Agricultural Insect Resources, Bengaluru, India for providing facilities and to Dr J. Poorani, Principal Scientist, ICAR – National Research Centre for Banana, Tiruchirapalli, Tamil Nadu, India for allowing us to use photographs of syrphids captured by her (photos (i), (o) & (p) of Figure 9.1 and photo (d) of Figure 9.2).

REFERENCES

Adams, T.H.L. (1984). The effectiveness of aphid-specific predators in preventing outbreaks of cereal aphids. Ph.D. Thesis, University of East Anglia, Norwich, Great Britain, 129 p.

Almohamad, R., Verheggen, F.J., Francis, F. & Haubruge, E. (2010a). Assessment of oviposition site quality by aphidophagous hoverflies: Reaction to conspecific larvae. *Animal Behaviour*, 79(3): 589–594.

Almohamad, R., Verheggen, F.J., Francis, F. & Haubruge, E. (2010b). Intraguild interactions between the predatory hoverfly *Episyrphus balteatus* (Diptera: Syrphidae) and the Asian ladybird, Harmonia axyridis (Coleoptera: Coccinellidae): Effect of larval tracks. *European Journal of Entomology*, 107(1): 41–45.

Aubert, J. & Tiefenau, P.G. (1981). Observations surles migrations de Syrphides (Dipt.) dans les Alpes de la Suisse occidentale. *Mitteilungen Schweizerischen Entomol Gesellschaft*, 54: 377–388.

Bankowska, R., Mikolajczyk, W., Palmowska, J. & Trojan, P. (1978). Aphid-aphidophage community in alfalfa cultures (*Medicago sativa* L.) in Poland. Part 3. Abundance regulation of *Acyrthosiphon pisum* (Harr.) in a chain of oligophagous predators. *Annales Zoologici (Warsaw)*, 34: 39–77.

Barlow, C.A. (1979). Energy utilization by larvae of the flower fly, *Syrphus corollae* (Dipt. Syrphidae). *Canadian Entomologist*, 111: 897–904.

Barlow, C.A. & Whittingham, J.A. (1986). Feeding economy of larvae of a flower fly, *Metasyrphus corollae* (Dipt.: Syrphidae): Partial consumption of prey. *Entomophaga*, 31: 49–57.

Baskaran, R.K.M., Sasikumar, S., Rajavel, D.S. & Suresh K. (2008). Species diversity of aphidophagous syrphids in Southern Districts of Tamil Nadu. *Insect Environment*, 14(3): 120–122.

Baskaran, R.K.M., Sasikumar, S., Rajavel, D.S. & Suresh, K. (2009a). Influence of semi-synthetic diet on fecundity of *Paragus serratus*. *Annals of Plant Protection Sciences*, 17(1): 235–236.

Baskaran, R.K.M., Sasikumar, S., Rajavel, D.S. & Suresh, K. (2009b). Influence of pollen grains on fecundity of aphidophagous syrphid, *Paragus serratus* (Fab.) (Diptera: Syrphidae). *Hexapoda*, 16(1): 66–67.

Bayrak, N. & Hayat, R. (2008). Faunistic studies on the species of Syrphidae (Diptera) in Kayseri province. *Bitki Koruma Bulteni*, 48(4): 35–49.

Benestad, E. (1970). Food consumption at various temperature conditions in larvae of *Syrphus corollae* (Fabr.) (Dipt.: Syrphidae). *Norsk Entomologisk Tidsskrift*, 17: 87–91.

Borror, D., Delong, D.M. & Triplehorn, C.A. (1976). *An Introduction to the Study of Insects*. Rinechart & Co: New York.

Brunetti, E. (1923). Pipunculidae, syrphidae, conopidae, oestridae. Xii+424 figs, 7 pls. In: *The Fauna of British India Including Ceylon and Burma Diptera Volume-3*. Taylor & Francis): London, p. 85.

Chambers, R.J. (1988). Syrphidae. In: *Aphids, World Crop Pests Volume-B* (Eds. Minks, A.K. & Harrewijn, P.). Elsevier Science: B.V. Amsterdam, 2B, 310 pp.

Chambers, R.J., Sunderland, K.D., Wyatt, I.J., & Vickerman, G.P. (1983). The effects of predator exclusion and caging on cereal aphids in winter wheat. *Journal of Applied Ecology*, 20: 209–224.

Chambers, R.J., Sunderland, K.D., Stacey, D.L., & Wyatt, I.J. (1986). Control of cereal aphids in winter wheat by natural enemies: Aphid-specific predators, parasitoids and pathogenic fungi. *Annals of Applied Biology*, 108: 219–231.

Chandler, A.E.F. (1968a). A preliminary key to the eggs of some of the common aphidophagous Syrphidae (Diptera) occurring in Britain. *Transactions of Royal Entomological Society of London*, 120: 199–217.

Chandler, A.E.F. (1968b). Some factors influencing the site and occurrence of oviposition by aphidophagous Syrphidae (Diptera). *Annals of Applied Biology*, 61: 435–446.

Chandler, A.E.F. (1968c). Some host plant factors affecting oviposition by aphidophagous Syrphidae (Diptera). *Annals of Applied Biology*, 61: 415–423.

Coe, R.L. (1953). Diptera: Syrphidae. In: *Handbooks for the Identification of British Insects*, Volume-10(1). Royal Entomology Society: London, p. 98.

Cornelius, M. & Barlow, C.A. (1980). Effect of aphid consumption by larvae on development and re productive efficiency of a flower fly, *Syrphus corollae* (Diptera: Syrphidae). *Canadian Entomologist*, 112: 989–992.

Devi, C.L. & Singh, T.K. (2000). The functional response of *Ischiodon scutellaris* (Fabricius) (Diptera: Syrphidae), a predator of mustard aphid, *Lipaphis erysimi* (Kalt.) (Hemiptera: Aphididae). *Journal of Aphidology,* 14: 73–76.

Devjani, P. & Singh, T.K. (2006). Larval development and prey consumption of *five* aphidophagous predators in relation to *two* aphid species. *Advances in Indian Entomology: Productivity and Health a Silver Jubilee*, II(3): 121–125.

Dib, H., Simon, S., Sauphanor, B. & Capowiez, Y. (2010). The role of natural enemies on the population dynamics of the rosy apple aphid, *Dysaphisplantaginea passerini* (Hemiptera: Aphididae) in organic apple orchards in south-eastern France. *Biological Control*, 55: 97–109.

Dixon, T.J. (1960). Key to and descriptions of the third instar larvae of some species of Syrphidae (Diptera) occurring in Britain. *Transactions of Royal Entomological Society of London*, 112: 345–379.

Dunn, L., Manuel, L., Chris, R.R. & Tanya, L. (2020). Dual ecosystem services of syrphid flies (Diptera: Syrphidae): Pollinators and biological control agents. *Pest Management Science*, 76: 1973–1979.

Eckberg, J.O., Peterson, J.A., Borsh, C.P., Kaser, J.M., Johnson, G.A., Luhman, J.C. (2014). Field abundance and performance of hoverflies (Diptera: Syrphidae) on soybean aphid. *Annals of Entomological Society of America*, 108: 26–34.

Ehteshamnia, N., Khaghaninia, S. & Pourabad, R.F. (2010). Some hoverflies of subfamily Syrphinae of Qurigol fauna in East Azerbaijan province. *Iran (Diptera: Syrphidae). Munis Entomology & Zoology Journal*, 5(2): 499–505.

GBIF Secretariat (2018). GBIF backbone taxonomy checklist dataset. doi: 10.15468/39omei accessed via GBIF.org on 2021-01-02.

Ghorpade, K.D. (1973). A faunistic study of the hover flies (Diptera: Syrphidae) of Bangalore, southern India. M.Sc, University of Agricultural Sciences, Bangalore, India, x+208 p.

Ghorpade, K.D. (1981). Insect prey of Syrphidae (Diptera) *from* India and neighboring countries: A review and bibliography. *Tropical Pest Management*, 27: 62–82.

Ghorpade, K.D. (1994). Diagnostic keys to new and known genera of species of Indian subcontinent Syrphinae (Diptera: Syrphidae). *Colemania*, 3: 1–15.

Ghorpade, K.D. (2014). An updated check-list of the Hover-flies (Diptera-Syrphidae) recorded in the Indian sub-continent. *Colemania*, 44: 1–30.

Ghorpade, K.D. (2015). Hover flies (Diptera: Syrphidae) documented *from* the northwest frontier of the Indian subcontinent: A circumstantial history and inclusive bibliography. *Colemania*, 50: 1–151.

Ghorpade, K.D. (2019). Hover-Flies (Diptera: Syrphidae) Recorded *from* "Dravidia," or Central and Peninsular India and Sri Lanka, an Annotated Checklist and Bibliography. In: *Indian Insects Diversity and Science, a Festschrift for Professor C. A. Viraktamath's 75th Birthday* (Eds. Ramani, et al.). CRC Press, Taylor & Francis Group: Boca Raton, FL, London, New York, pp. 325–388.

Gilbert, F.S. (1981). Foraging ecology of hoverflies: Morphology of the mouthparts in relation to feeding on nectar and pollen in some common urban species. *Ecological Entomology*, 6: 245–262.

Gilbert, F.S. (1985). Diurnal activity patterns in hover flies (Diptera, Syrphidae). *Ecological Entomology*, 10: 385–392.

Goeldlin de Tiefenau, P. (1974). Contribution a l'etude systematique et'ecologique des Syrphidae (Dipt.) de la Suisse occidentale. *Mitteilungen der Schweizerischen Entomologischen Gesellschaft*, 47: 151–152.

Hagen, K.S., Sawall, E.F. & Tassan, R.L. (1972). The use of food sprays to increase effectiveness of entomophagous insects. In: *Proceedings of Tall Timbers Conference on Ecological Animal Control by Habitat Management*, Tallahassee, FL, pp. 59–81.

Horej, M. (1982). Natural control of *Aphis fabae* Scop. population by Syrphidae (Diptera) on sugar beet. *Polskie Pismo Entomologiczne*, 52: 287–294.

Horn, D.J. (1981). Effect of weedy backgrounds on colonisations of collards by *green* peach aphid, *Myzus persicae* and its major predators. *Environmental Entomology*, 10: 285–289.

Iwai, H., Niijima, K. & Matsuka, M. (2007). An artificial diet for aphidophagous syrphids, *Episyrphus balteatus* (de Geer) and *Eupeodes bucculatus* (Rondani) (Diptera: Syrphidae) using drone honeybee brood powder. *Applied Entomology & Zoology*, 42(2): 167–172.

Iwai, H., Niijima, K. & Matsuka, M. (2009). Improvement of artificial diet for aphidophagous syrphids, *Episyrphus balteatus* (de Geer) and *Eupeodes bucculatus* (Rondani) (Diptera: Syrphidae) additional effects of fatty acids and preservatives. *Applied Entomology & Zoology*, 44(3): 439–446.

Joshi, S. & Ballal, C.R. (2013). Syrphid predators for biological control of aphids. *Journal of Biological Control*, 27(3): 151–170.

Joshi, S., Ballal, C.R. & Rao, N.S. (1998). An efficient and simple mass culturing technique for *Ischiodon scutellaris* (Fabricius), an aphidophagous syrphid. *Indian Journal of Plant Protection*, 26: 56–61.

Joshi, S., Ballal, C.R. & Rao, N.S. (1999a). Evaluation of potential of syrphid predators *Ischiodon scutellaris* (Fabricius) and *Paragus serratus* (Fabricius) (Diptera: Syrphidae). *Journal of Aphidology*, 13: 9–16.

Joshi, S., Ballal, C.R. & Rao, N.S. (1999b). Species complex, population density and dominance structure of aphidophagous syrphids in cowpea ecosystem. *Entomon*, 24: 203–213.

Joshi, S., Ballal, C.R. & Rao, N.S. (2000). The age specific life-table of *Paragus serratus* (Fabricius) and *Paragus yerburiensis* Stuckenberg (Diptera: Syrphidae), predators of *Aphis craccivora* Koch (Homoptera: Aphididae). *Journal of Aphidology*, 14: 67–72.

Joshi, S., Ballal, C.R. & Rao, N.S. (2001). Intrinsic rate of natural increase of *Ischiodon scutellaris* (Fabricius) (Syrphidae), a predator of Aphis craccivora Koch. *Journal of Insect Science*, 14(1/2): 15–18.

Khaghaninia, S. (2010). Faunistic study on flower flies of Zunuz region in East Azerbaijan province-Iran (Diptera; Syrphidae). *Munis Entomology & Zoology Journal*, 5(2): 586–593.

Knutson, L.V., Thompson, F.C. & Vockeroth, J.R. (1975). Family syrphidae. In: *A Catalog of the Diptera of the Oriental Region Volume-2* (Eds. Delfinado, M.D. & Hardy, D.E.). University Press of Hawaii: Honolulu, Hawaii, pp. 307–374.

Kotwal, D.R., Bhalla, O.P. & Verma, A.K. (1984). Natural enemies of the cabbage aphid, *Brevicoryne brassicae* (Linn.) in the mid-hill regions of Himachal Pradesh. *Indian Journal of Agricultural Sciences*, 54: 1011–1012.

Kumar, A., Kapoor, V.C. & Mahal, M.S. (1988). Population buildup and dispersion of immature stages of aphidophagous syrphids (Syrphidae: Diptera) on Raya, *Brassica juncea* Coss. *Journal of Insect Science*, 1: 39–84.

Kumar, A., Kapoor, V.C. & Mahal, M.S. (1996). Feeding behaviour and efficacy of *three* aphidophagous syrphids. *Journal of Insect Science*, 9: 15–18.

Larson, B., Kevan, P. and Inouye, D.W. (2001). Flies and flowers: Taxonomic diversity of anthophiles and pollinators. *Canadian Entomologist*, 133: 439–465.

Makhmoor, H.D. & Verma, A.K. (1989). The intrinsic rate of natural increase of *Metasyrphus confrater* (Wied.) (Diptera: Syrphidae) a predator of the cabbage aphid (Homoptera: Aphididae). *Proceedings of Indian National Academy of Sciences*, 55: 79–84.

Mitra, B., Mukherjee, M. & Banerjee, D. (2008). A check list of hoverflies (Diptera: Syrphidae) of Eastern Himalayas. *Records of Zoological Survey of India*, 284: 1–47.

Mitra, B., Sankarsan, R., Imtiaz, I. & Maitree G. (2015). A review of the hover flies (Syrphidae: Diptera) *from* India. *International Journal of Fauna and Biological Studies*, 2(3): 61–73.

Mizuno, M., Itioka, T., Tatematsu, Y. & Ito, Y. (1997). Food utilization of aphidophagous hoverfly larvae (Diptera: Syrphidae, Chamaemyiidae) on herbaceous plants in an urban habitat. *Ecological Research*, 12: 239–248.

Morales, M.N. & Marinoni, L. (2009). Cladistic analysis and taxonomic revision of the *scutellaris* group of *Palpada Macquart* (Diptera: Syrphidae). *Invertebrate Systematics*, 23(4): 301–347.

Nelson, E.H., Hogg, B.N., Mills, N.J. & Daane, K.M. (2012) Syrphid flies suppress lettuce aphids. *Biological Control*, 57: 819–826.

Patro, B. & Behera, M.K. (2002). Biology and feeding potential of *Sphaerophoria javana* Weid. (Diptera: Syrphidae) on the bean aphid, *Aphis craccivora* Koch. *Journal of Biological Control*, 16: 165–167.

PDBC (1999). Annual Report, 1998–1999. Project Directorate of Biological Control, ICAR, Bangalore, India, 234 p.

Pollard, E. (1969). The effects of removal of arthropod predators on an infestation of *Brevicoryne brassicae* (Hemiptera, Aphididae) on Brussels sprouts. *Entomologia Experimentalis et Applicata*, 12: 118–124.

Pollard, E. (1971). Hedges, VI. Habitat diversity and crop pests. A study of *Brevicoryne brassicae* and its syrphid predators. *Journal of Applied Ecology*, 8: 751–780.

Proctor, M. & Yeo, P. (1973). *The Pollination of Flowers*. New Naturalist Series, No. 54. Collins: London, p. 418.

Rabbinge, R., Ankersmit, G.W. & Pak, G.A. (1979). Epidemiology and simulation of population development of *Sitobion avenae* in winter wheat. *Netherlands Journal of Plant Pathology*, 85: 197–220.

Rader, R., Howlett, B.G., Cunningham, S.A., Westcott, D.A., Newstrom- Lloyd, L.E. & Walker, M.K. (2009). Alternative pollinator taxa are equally efficient but not as effective as the honeybee in a mass flowering crop. *Journal of Applied Ecology*, 46: 1080–1087.

Rader, R., Edwards, W., Westcott, D.A., Cunningham, S.A. & Howlett, B.G. (2011). Pollen transport differs among bees and flies in a human-modified landscape. *Diversity and Distributions*, 17: 519–529.

Rader, R., Bartomeus, I., Garibaldi, L.A., Garratt, M.P., Howlett, B.G. & Winfree, R. (2016). Non-bee insects are important contributors to global crop pollination. *Proceedings of National Academy of Sciences*, 113: 146–151.

Radhakrishnan, B. & Muraleedharan, N. (1993). Bio-ecology of *six* species of syrphid predators of the tea aphid, *Toxoptera aurantii* (Boyer de Fonscolombe) in southern India. *Entomon*, 18: 175–180.

Raworth, D.A. (1984). Population dynamics of the cabbage aphid, *Brevicoryne brassicae* (Homoptera: Aphididae) at Vancouver, British Columbia, V. A simulation model. *Canadian Entomologist*, 116: 895–911.

Rossi, J., Gamba, U., Pinna, M., Spagnolo, S., Visentin, C. & Alma, A. (2007). Survey of syrphid fauna (Diptera: Syrphidae) in organic apple orchards in North Western Italy. *Bolletino-di-Agric Biol*, 3: 23–34.

Roy, P. & Basu, S.K. (1977). Bionomics of aphidophagous syrphid flies. *Indian Journal of Entomology*, 39: 165–174.

Ruzicka, Z. & Cairo, V.G. (1976). The effect of larval starvation on the development of *Metasyrphus corollae* (Diptera). *Vestnik Ceskoslovenske Spolecnosti Zoolozicke*, 40: 206–213.

Samsoe-Petersen, L., Bigler, F., Bogenschutz, H., Brun, J., Hassan, S.A., Helyer, N.L., Kuhner, C., Mansour, E., Naton, E., Oomen, P.A., Overmeer, W.P.J., Polgar, L., Rieckmann, W., & Staubli, A. (1989). Laboratory rearing techniques for 16 beneficial arthropd species and their prey/hosts. *Zeits Pflanz Pflanz*, 96: 289–316.

Sanders, W. (1980). Das Eiablageverhalten der Schwebfliege *Syrphus corollae Fabr*. in Abhangigkeit von der raumlichen Lage der Blattlauskolonie. *Zeitschrift Ang Zool*, 67: 35–46.

Sanders, W. (1981). Der Einfluss von weissen und schwarzen Flachen auf das Verhalten der Schwebfliege *Syrphus corollae* Fabr. *Zeitschrift Ang Zool*, 68: 307–314.

Sarbyk, S. (2008) Contributions to the Syrphidae fauna of Turkey (Diptera: Syrphidae). *Entomological News*, 119(5): 501–508.

Scott, S.M. & Barlow, C.A., (1984). Effect of prey availability during development on the reproductive output of *Metasyrphus corollae* (Diptera: Syrphidae). *Environmental Entomology*, 13: 669–674.

Sharma, K.C. & Bhalla, O.P. (1988). Biology of *six* syrphid predators of cabbage aphid (*Brevicoryne brassicae*) on cauliflower (*Brassica oleracea var. botrytis*). *Indian Journal of Agricultural Science*, 58: 652–654.

Sharma, K.C. & Bhalla, O.P. (1991). Predatory potential of syrphid species on different aphids of cruciferous crops in the mid hill regions of Himachal Pradesh. *Indian Journal of Plant Protection*, 19: 73–75.

Sharma, K.C. & Bhalla, O.P. (1995). Life table on *Eupeodes corollae* (Fabricius) (Diptera: Syrphidae), a predator of the cabbage aphid, Brevicoryne brassicae (Linnaeus) (Homoptera: Aphididae). *Journal of Biological Control*, 9: 78–81.

Sharma, K.C., Bhalla, O.P. & Chauhan, U. (1994). Life-table and intrinsic rate of natural increase of *Eupeodes frequens* Matsumura: A predator of *Brevicoryne brassicae* (Aphididae: Homoptera) infesting cauliflower. *Journal of Biological Control*, 8: 56–58.

Sheridan, H., Finn, J.A. & O'Donovan, G. (2008). Diversity of Syrphidae (Diptera) on *two* Irish grassland farms. *Tearmann*, 6: 69–82.

Singh, R. & Mishra, S. (1988). Development of a syrphid fly, *Ischiodon scutellaris* (Fabricius) on *Rhopalosiphum maidis* (Fitch). *Journal of Aphidology*, 2: 28–34.

Skevington, J.H. & Dang, P.T. (2002). Exploring the diversity of flies (Diptera). *Biodiversity*, 3(4): 3–27.

Smith, J.G. (1969). Some effects of crop background on populations of aphids and their natural enemies on Brussels sprouts. *Annals of Applied Biology,* 63: 326–329.

Smith, J.G. (1976). Influence of crop background on natural enemies of aphids on Brussels sprouts. *Annals of Applied Biology*, 83: 15–29.

Smith, R.K. (1981). Studies on the ecology of cereal aphids and prospects for integrated control. Ph.D. Thesis, Imperial College, London.

Sorokina, V.S. (2009). Hover flies of the genus *Paragus Latr*. (Diptera, Syrphidae) of Russia and adjacent countries. *Entomologicheskoe Obozrenie*, 88(2): 466–487.

Stubbs AE, Falk SJ. (1983). *British Hover Flies; an Illustrated Identification Guide*. British Entomological and Natural History Society: London, p. 253.

Tamaki, G. (1973). Spring populations of the *green* peach aphid on peach trees and the role of natural enemies in their control. *Environmental Entomology*, 2: 186–191.

Tamaki, G. (1974). Life system analysis of the autumn population of *Myzus persicae* on peach trees. *Environmental Entomology*, 3: 221–226.

Tawfik, M.F.S., Azab, A.K. & Awadallah, K.T. (1974). Studies on the life histories and description of the immature forms of the Egyptian aphidophagous syrphids: *Syrphus corollae* Fabr. (Diptera: Syrphidae). *Bulletin de la Societe Entomologique d'Egypte* 58: 1–16.

Thompson, F.C. & Ghorpade K.D. (1992). A new coffee aphid predator, with notes on other Oriental species of *Paragus* (Diptera: Syrphidae). *Colemania*, 5: 1–24.

Verma, A.K. & Makhmoor, H.D. (1987). Relative abundance of syrphid predators of *Brevicoryne brassicae* (L.) in cauliflower seed crop ecosystem. *Journal of Biological Control,* 1: 98–103.

Verma, J.S. & Sharma, K.C. (2006). Biology and predatory potential of *Metasyrphus confrater* (Wiedemann) on aphid, *Macrosiphum rosae* L. infesting Rosa spp. *Journal of Entomological Research*, 30(1): 31–32.

Way, M.J. & Banks, C.J. (1968). Population studies on the active stages of the *black* bean aphid, *Aphis fabae* Scop., on its winter host *Euonymus europaeus* L. *Annals of Applied Biology*, 62: 177–197.

Wnuk, A. (1977). The natural enemy *Episyrphus balteatus* (Deg.) (Diptera, Syrphidae) limiting *Aphis pomi* Deg. (Hom. Aphididae). *Polski Pismo Entomologizne*, 47: 455–460.

Wnuk, A. & Fuchs, R. (1977). Observations on the effectiveness of the limitation of *Brevicoryne brassicae* (L.) by Syrphidae. *Polski Pismo Entomo ligizne*, 47: 147–156.

10 Predators of Mite Pests

Shima Yazdanpanah and Yaghoub Fathipour
Tarbiat Modares University

CONTENTS

LEARNING OBJECTIVES

i. Phytophagous mites are important agricultural pests that feed on a wide range of host plant species throughout the world.
ii. Chemical pesticides are the most common control measures against mite pests, which result in environmental pollution and elimination of many natural enemies.
iii. One of the safe methods to control mite pests is using biocontrol agents including predatory mites and insects.
iv. Most important predators of phytophagous mites belong to Acari, especially the family Phytoseiidae, as well as insect orders of Coleoptera, Thysanoptera, Diptera, and Hemiptera.
v. In biocontrol of phytophagous mites, cost-effective mass production of predators received priority; therefore, in this study, importance and mass production of some commercial predators are discussed.

10.1 INTRODUCTION

Among pests, the phytophagous mites are ubiquitous and economically important agricultural pests with a global distribution, which feed on a wide range of host plant species throughout the world. The use of chemicals is the most common method for controlling these mites, with the annual value of pesticides used against them constituting hundreds of million dollars. Extensive use of chemical pesticides for the management of phytophagus mites resulted in the development of resistance to almost all the available groups of acaricides, only after a few applications, because of their rapid developmental rate and high reproduction. Furthermore, widespread use of acaricides has eliminated many of the natural enemies of spider mites, resulting in a reduction in predation pressure on these mites. These problems primarily necessitate the development of alternative pest control strategies, such as integrated pest management (IPM) and biocontrol agents for the management of mite pests.

DOI: 10.1201/9781003370864-10

FIGURE 10.1 Some important families of the predators of phytophagous mites.

The integrated pest control is based on a strategy to control pest populations using all possible means (chemical, biological, biotechnical, cultural, etc.). The IPM strategy does not exclude the use of chemicals but recommends to use them as a last resort after a systematic monitoring of pest populations in order to minimize their use as minimum as possible and if the natural control factors will pose the need. However, in this case, the rapid development of populations, increase in the damage level, and small size of mites make the monitoring difficult. Biological, technical, and chemical methods must be balanced on environmental protection, income, and social needs (Vacante, 2010). Although biocontrol has actually solved many entomological problems, even today, the use of chemicals seems to be necessary. However, the exploitation of natural enemies as a tool in pest management strategies is essential for sustainability and food security. Biological control provides an environmentally safe, cost-effective, and energy-efficient means of pest control, either alone or as a component of an IPM program.

The natural enemies of mite pests include pathogens, viruses, bacteria, and fungi; predatory mites, such as Phytoseiidae, Stigmaeidae, Anystidae, Cheyletidae, Erythraeidae, Bdellidae, Tydeidae, and Cunaxidae; spiders; and various insects, such as Coleoptera (Coccinellidae and Staphylinidae), Neuroptera, Thysanoptera, Diptera (Cecidomyiidae), Dermaptera, and Hemiptera (Anthocoridae, Miridae, Nabidae, and Lygaeidae) (Vacante, 2010) (Figure 10.1). Of the above mentioned organisms, the most important predators for phytophagous mites belong to the family Phytoseiidae. They play a major role in keeping phytophagous mite populations at low densities and decreasing their impact (Al-Azzazy and Alhewairini, 2020). The phytoseiid mites have been intensely studied for nearly six decades because of their recognized potential as biocontrol agents of mites as well as of thrips and whiteflies. In most ecosystems, this family is first represented among all families of acarine predators, as the prime exploiter of the foliage habitat (McMurtry et al., 2013).

Among biocontrol agents, predators receive priority (Fathipour and Maleknia, 2016); therefore, in this chapter, we review important mite predators and their efficacy in control of mite pests. The future of biocontrol of mites and marketing of biocontrol agents is also discussed.

10.2 IMPORTANT MITE PEST FAMILIES AND SPECIES

Several mite families cause economic damage in agricultural ecosystems (Figure 10.2). The most important pest mites are spider mites of the family Tetranychidae [e.g. *Tetranychus urticae* Koch,

FIGURE 10.2 Important mite pest families.

Tetranychus cinnabarinus (Boisduval), *Tetranychus turkestani* Ugarov & Nikolski, *Tetranychus ludeni* Zacher, *Tetranychus evansi* Baker and Pritchard, *Panonychus ulmi* (Koch), *Panonychus citri* (McGregor), *Bryobia rubrioculus* (Scheuten), *Bryobia praetiosa* Koch, *Petrobia latens* (Müller), *Eutetranychus orientalis* (Klein), *Eotetranychus lewisi* (McGregor), *Eotetranychus yumensis* (McGregor), *Eotetranychus frosti* (McGregor), *Schizotetranychus oryzae* Rossi de Simons, *Oligonychus punicae* (Hirst), *Oligonychus afrasiaticus* (McGregor), and *Oligonychus sacchari* McGregor], false spider mites of the family Tenuipalpidae [e.g. *Brevipalpus californicus* (Banks), *Brevipalpus lewisi* McGregor, *Brevipalpus obovatus* Donnadieu, *Brevipalpus phoenicis* (Geijskes), *Tenuipalpus heveae* Baker, *Raoiella indica* Hirs, and *Cenopalpus irani* Dosse], thread-footed mites of the family Tarsonemidae [e.g. *Phytonemus pallidus* (Banks), *Polyphagotarsonemus latus* (Banks), *Steneotarsonemus spinki* Smiley, and *Steneotarsonemus furcatus* De Leon], and gall mites of the family Eriophyidae [e.g. *Aceria tristriata* (Nalepa), *Aceria guerreronis* Keifer, *Aceria sheldoni* (Ewing), *Aceria tosichella* Keifer, *Aceria tulipae* (Keifer, *Aceria ficus* (Cotte), *Aceria litchii* (Keifer), *Eriophyes pyri* (Pagenstecher), *Phyllocoptruta oleivora* (Ashmead), *Aculops lycopersici* (Massee), *Acalitus vaccinii* (Keifer), and *Calepitrimerus vitis* (Nalepa)]. Some mite pests with less economical importance belong to other families, such as peacock mites of the family Tuckerllidae [e.g. *Tuckerella pavoniformis* (Ewing), *Tuckerella knorri* Baker & Tuttle, and *Tuckerella japonica* Ehara], earth mites of the family Penthaleidae [e.g. *Halotydeus destructor* (Tucker), *Penthaleus major* (Dugès)], and leaf vagrant mites of the family Diptilomiopidae (= Rhyncaphytoptidae) [e.g. *Rhyncaphytoptus ficifoliae* Keifer].

10.3 BIOLOGICAL CONTROL OF MITE PESTS

Biological control is the use of an organism to reduce the population density of another organism. It is one of the most environmentally safe and most economical modes of pest management for growers. Biological control has been in use for about two millennia, and has become widely used in pest management since the end of the 19th century. The following types of biological control can be distinguished: natural, conservation, inoculative (=classical), and augmentative biological control. Natural biological control is the reduction of pest organisms by their natural enemies and has been occurring since the evolution of the first terrestrial ecosystems some 500 million years ago. Conservation biological control consists of human actions that protect and stimulate the performance of naturally occurring natural enemies. In inoculative biological control, natural enemies are collected in an exploration area and then released in new areas, where the pest was accidentally introduced. The aim is that the offspring of the released natural enemies build up populations, which are large enough for suppression of pest populations during many subsequent years. As it was the first type of biological control practiced widely, it is also called 'classical' biological control. Inoculative biological control is estimated to be used on 10% of land under cultivation and, over the last 120 years, 165 pest species have been brought under long-term control. In augmentative biological control, natural enemies are mass reared in biofactories for release in large numbers to obtain an immediate control of pests. The history of commercial mass production and sale of natural enemies span a period of roughly 120 years. Compared to natural and inoculative biological control, commercial augmentative biological control is applied on a very limited scale, i.e., only on 16 million ha, which is 0.4% of the cultivated land with crops on which this type of control could be used. In commercial augmentative biological control, natural enemies are mass reared in biofactories for release in large numbers to obtain an immediate control of pests (van Lenteren, 2012).

The biocontrol industry has made great advances in the last decades in the identification of natural enemies and development of commercial products, leading to more than 230 species of natural enemies available for augmentative biological control worldwide (Calvo et al., 2015). Natural enemies can be native or introduced. The use of natural enemies prevents environmental risks associated with chemical pesticides, while sustainably protecting the crop, given that the biological control agent does not cause any harmful nontarget effects (Onzo et al., 2012). Natural enemies can be separated in two groups: species where very large numbers need to be released per unit area in order to obtain sufficient control (acarids and nematodes), and species that are released at a low density on very large areas (coleopterans, heteropterans, and hymenopterans) in crops. This means that simply looking at the numbers produced does not give a good indication of the market value of a species (van Lenteren, 2012).

Within the arthropods, four taxonomic groups provided most natural enemies: first of all, Hymenoptera (52.2%, 120 species), next Acari (13.1%, 30 species), followed by Coleoptera (12.2%, 28 species) and Heteroptera (8.3%, 19 species). The first species used in augmentative biological control were a hymenopteran (*Metaphycus lounsburyi* (Howard)) and a coleopteran (*Chilocorus circumdatus* (Gyllenhal)), both in 1902. The most widely used natural enemies in inoculative weed and insect control (e.g. *Aphelinus mali* (Haldeman), *Aphytis lingnanensis* Compere, *Cotesia flavipes* Cameron, *Cryptolaemus montrouzieri* Mulsant, *Rodolia cardinalis* (Mulsant), and *Teleonemia scrupulosa* Stal) have been introduced in more than 20 countries/regions worldwide and resulted in permanent control of pests (van Lenteren, 2012). Economically important families of mite pests and their predators are listed in Table 10.1.

10.3.1 Mite Predators

10.3.1.1 Acari

Predatory mites are popular because they can easily be mass reared, can be released by mechanical means, may control several pest species, do not spread actively over large distances, and are relatively small. The last two characteristics help to prevent negative effects on nontarget species. Predatory mites are important natural enemies of small arthropods (mite and insect pests). As the second largest group of biological control agents after the hymenopterans, the family Phytoseiidae includes about 30 species commercially produced worldwide (Onzo et al., 2012; Yang et al., 2019).

TABLE 10.1
The Most Important Mite Pests and Their Predators

a. Family	Pest	Predator
b. Tetranychidae	*Tetranychus urticae*	*Phytoseiulus persimilis, Ph. macropilis, Amblyseius andersoni, A. tsugawai, A. chiapensis, Euseius nicholsi, Typhlodromus pyri, Neoseiulus californicus, N. womersleyi, N. longispinosus, N. cucumeris, Cydnoseius negevi, Iphiseius degenerans* (Phytoseiidae); *Agistemus buntex* (Stigmaeidae); *Stethorus punctillum, S. gilvifrons* (Coccinellidae); *Conwentzia psociformis* (Chrysopidae); *Orius vicinus, O. minutus, O. albidipennis, O. laevigatus* (Anthocoridae)
	Tetranychus evansi	*Phytoseiulus longipes* (Phytoseiidae)
	Tetranychus bastosi	*Stethorus tridens* (Coccinellidae)
	Tetranychus cinnabarinus	*Phytoseiulus persimilis* (Phytoseiidae)
	Eutetranychus orientalis	*Typhlodromus athiasae, Neoseiulus barkeri, Iphiseius degenerans* (Phytoseiidae)
	Eutetranychus lewisi	*Neoseiulus californicus* (Phytoseiidae)
	Eotetranychus frosti	*A. swirskii* (Phytoseiidae)
	Oligonychus punicae	*Amblyseius largoensis* (Phytoseiidae)
	Oligonychus afrasiaticus	*Cydnoseius negevi* (Phytoseiidae); *Spinibdella cronini* (Bdellidae)
	Oligonychus coffeae	*Euseius finlandicus, Typhlodromus psyllakisi* (Phytoseiidae); *Anystis baccarum* (Anystidae); *Tydeus caudatus* (Tydeidae)
	Panonychus citri	*Amblyseius largoensis, A. swirskii, Typhlodromalus peregrinus, Phytoseiulus persimilis, Galendromus occidentalis, Neoseiulus californicus* (Phytoseiidae); *Agistemus buntex* (Stigmaeidae)
	Panonychus ulmi	*Amblyseius swirskii, Typhlodromus caudiglans* (Phytoseiidae), *Zetzellia mali* (Stigmaeidae)
	Bryobia rubrioculus	*Zetzellia mali, Z. pourmirzai* (Stigmaeidae)
Tenuipalpidae	*Raoiella indica*	*Amblyseius caudatus, A. channabasavanna, A. largoensis, A. hainanensis* (Phytoseiidae); *Oligota* sp. (Staphylinidae); *Aleurodothrips fasciapennis* (Phlaeothripidae); *Ceraeochrysa claveri, Chrysopodes collaris* (Chrysopidae)
	Cenopalpus irani	*Amblyseius swirskii, Typhlodromus bagdasarjani* (Phytoseiidae)
	Tenuipalpus heveae	*Metaseiulus camelliae* (Phytoseiidae)
	Brevipalpus pulcher	*Amblyseius swirskii* (Phytoseiidae)
	Brevipalpus obovatus	*Amblyseius hainanensis* (Phytoseiidae)
	Brevipalpus Phoenicis	*Neoseiulus californicus* (Phytoseiidae), *Agistemus brasiliensis* (Stigmaeidae)
Penthaleidae	*Halotydeus destructor*	*Amblyseius victoriensis, A. messor, A. masiaka, A. dieteri, Typhlodromus occidentalis* (Phytoseiidae); *Balaustium murorum* (Erythraeide); *Walzia australica, Anystis wallacei, A. baccarum, Erythracarus* sp. (Anystidae); *Chrysopa* spp. (Chrysopidae); *Hemerobius* spp. (Hemerobiidae); *Harmonia* sp., *Diomus* sp. (Coccinellidae);
Tarsonemidae	*Polyphagotarsonemus latus*	*Euseius scutalis* (Phytoseiidae); *Stethorus* sp. (Coccinellidae); *Chrysoperla carnea* (Chrysopidae)
	Steneotarsonemus spinki	*Amblyseius taiwanicus, Galendromus alveolaris, G. longipilus, Neoseiulus paraibensis, N. baraki, N. paspalivorus, Proprioseiopsis asetus, Typhlodromus* sp. (Phytoseiidae); *Hypoaspis* sp. (Laelapidae); *Cheiroseius serratus, Ch. napalensis, Lasioseius parberlesei, Aceodromus asternalis, Asca pineta, Proctolaelaps bickleyi* (Ascidae)
	Phytonemus pallidus	*Amblyseius andersoni, A. swirskii, A. reductus, Anthoseius rhenanus, Neoseiulus aurescens, N. barkeri, N. californicus N. cucumeris, N. fallacis, Typhlodromus pyri* (Phytoseiidae)

(*Continued*)

TABLE 10.1 (*Continued*)
The Most Important Mite Pests and Their Predators

a. Family	Pest	Predator
	Tarsonemus bilobatus	*Lasioseius chaudhrii* (Ascidae)
	Aceria litchii	*Amblyseius compositus, Euseius concordis, Neoseiulus californicus, Iphiseiodes zuluagai, Phytoseius intermedius, Ph. plumifer, Ph. finitimus* (Phytoseiidae)
Eriophyidae	*Aceria dioscoridis*	*Lasioseius athiasae* (Ascidae); *Agistemus exsertus* (Stigmaeidae)
	Aceria ailanthae	*Lasioseius* sp. (Ascidae)
	Aceria ficus	*Pronematus ubiquitous* (Tydeidae); *Agistemus exsertus* (Stigmaeidae); *Amblyseius swirskii, Neoseiulus californicus, Phytoseius plumifer* (Phytoseiidae)
	Aceria lycopersici	*Agistemus exsertus* (Stigmaeidae); *Homeopronematus anconai* (Tydeidae); *Amblyseius victoriensis, A. swirskii, A. fallacis, A. andersoni, Typhlodromips montdorensis, Typhlodromus athiasae, Neoseiulus cucumeris, N. californicus, Euseius concordis* (Phytoseiidae)
	Aceria guerreronis	*Amblyseius largoensis, Neoseiulus baraki, N. paspalivorus* (Phytoseiidae), *Proctolaelaps bickleyi, P. bulbosus* (Ascidae)
	Aceria dioscoridis	*Agistemus exsertus* (Stigmaeidae)
	Aceria mangiferae	*Agistemus olive* (Stigmaeidae)
	Aculus fockeui	*Agistemus olivi* (Stigmaeidae)
	Colomerus vitis	*Agistemus exsertus* (Stigmaeidae)
	Aculops pelekassi	*Euseius sojaensis, Amblyseius eharai* (Phytoseiidae)
	Eriophyes lyeopemiei	*Amblyseius gossipi* (Phytoseiidae)
	Eriophyes pyri	*Typhlodromus pyri* (Phytoseiidae)
	Eriophyes olivi	*Agistemus exsertus* (Stigmaeidae)
	Eriophyes ficus	*Agistemus exsertus* (Stigmaeidae)
	Eriophyes dioscoridis	*Typhlodromus athiasae, Amblyseius cabonus* (Phytoseiidae)
	Phyllocoptruta oleivora	*Amblyseius citri, A. swirskii, A. victoriensis, Typhlodromalus peregrinus, Iphiseius degenerans, Typhlodromus athiasae, Euseius scutalis* (Phytoseiidae)
Rhyncaphytoptidae	*Rhyncaphytoptus ficifoliae*	*Pronematus ubiquitous* (Tydeidae); *Amblyseius swirskii, Phytoseius finitimus, Phytoseius plumifer, Neoseiulus californicus* (Phytoseiidae); *Agistemus exsertus* (Stigmaeidae)

10.3.1.1.1 Family Phytoseiidae

Adult females of phytoseiids are about 500 µm (0.5 mm) long, shiny, and pear-shaped. Males are smaller (about half the size of females), usually are oval, are often more active than females or nymphs, and can run rapidly. The life cycle consists of egg → larva → protonymph → deutonymph → adult males and females (Figure 10.3). Most phytoseiids have a female-biased sex ratio, commonly about 2.5 females to 1 male. Phytoseiids have a rapid life cycle, usually about a week at 25°C, which helps them to be effective natural enemies of pests and they usually have short generation times (Hoy, 2011).

Type I mites feed exclusively on web-producing spider mites, specializing on those in the genus *Tetranychus*. Type II mites have a broader host range than Type I mites, but still display a preference to prey on web-producing spider mites. Mites in Type III are categorized as generalist predators that feed on mites and other prey like insects (e.g. Thysanoptera), and often prefer or develop better on other prey than spider mites. Type IV mites are pollen-feeding generalist predators, most in the genus *Euseius* Wainstein (McMurtry and Croft, 1997; McMurtry et al., 2013).

Certain phytoseiids can be important natural enemies because they consume large numbers of prey, can maintain plant-feeding mites at low densities, can be mass reared for augmentative releases, have a high reproductive rate and rapid developmental rate comparable to their prey, and have a female-biased sex ratio equivalent to their prey, which allows them to respond

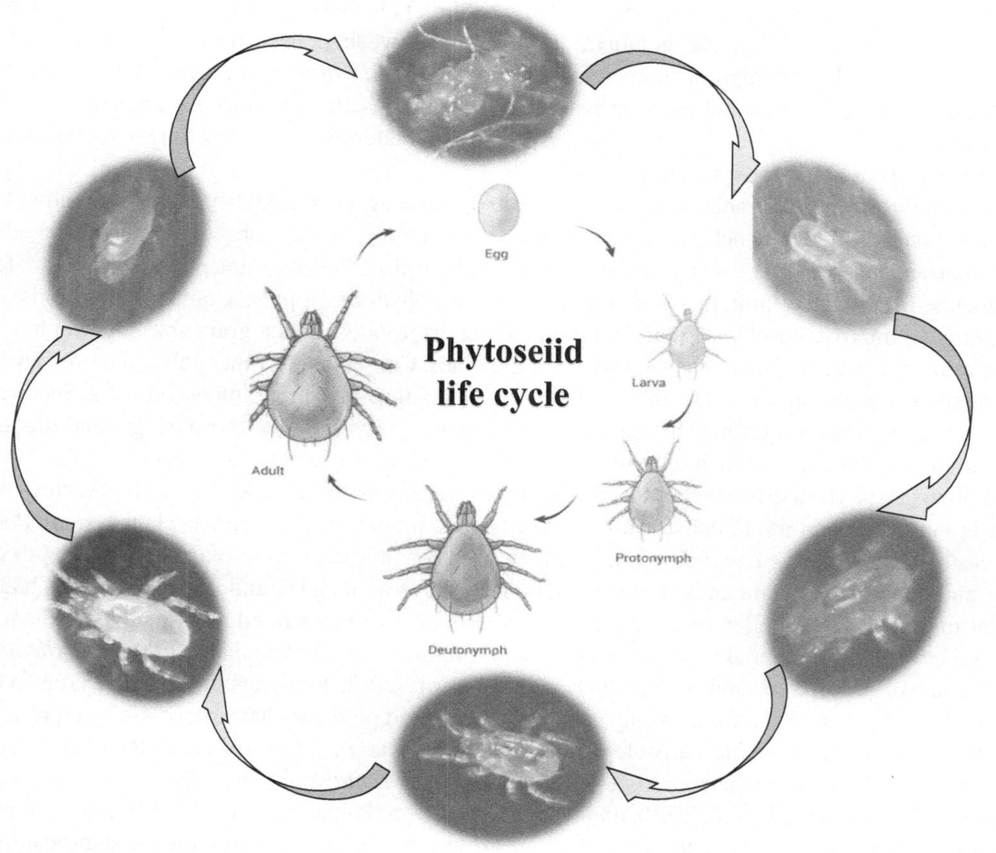

FIGURE 10.3 Life cycle of phytoseiid mites.

numerically to increased prey density (Hoy, 2011). The use of Phytoseiidae as biocontrol agents for insect pests started considerably later with the use of *Neoseiulus barkeri* Hughes and *Neoseiulus cucumeris* (Oudemans) (Acari: Phytoseiidae) for thrips control, which have been on the market since 1981 and 1985, respectively (Calvo et al., 2015). Some phytoseiids are found on crops in a few countries, but some premier species are used in integrated mite management (IMM) programs in multiple countries. These include *Neoseiulus barkeri*, *N. californicus*, *N. cucumeris*, *N. fallacis* (Garman), *Amblyseius degenerans* Berlese, *A. hibisci* (Chant), *A. potentillae* (Garman), *A. swirskii* Athias-Henriot, *Metaseiulus occidentalis* (Nesbitt), *Phytoseiulus persimilis* Athias-Henriot, and *Typhlodromus pyri* Scheuten (Hoy, 2011). Several phytoseiid mites are produced for mass release including *P. persimilis* that was developed as a product in 1968 and is now sold in over 20 countries. Over 20 species of phytoseiids have been developed as commercial products and four species appear on a list of the 25 most used products globally (i.e. *P. persimilis*, *A. swirskii*, *N. cucumeris*, and *N. californicus*) (McGregor et al., 2020).

At present, rearing phytoseiids on natural foods (prey or pollen) is expensive and time-consuming, and it requires substantial amount of greenhouse or other space if large-scale releases are to be made (millions of mites). Several artificial diets have been developed, but these are primarily used for the temporary feeding, shipping, and holding of phytoseiids that are obligatory predators. Effective and inexpensive artificial diets suitable for long-term rearing are needed for the phytoseiids that are obligatory predators. The potential of artificial diets have been evaluated on some phytoseiids, e.g., the effects of hemolymph of *Plusia gamma* L., multivitamin syrup, bovine serum albumin and bull sperm on *Typhlodromus bagdasarjani* Wainstein and Arutunjan

(Riahi et al., 2018), and on *N. californicus* (Khanamani et al., 2017a); the effects of honey, sucrose, tryptone, yeast extract, egg yolk, hemolymph from oak silkworm pupae (*Antheraea pernyi* (Guérin-Méneville)), and hemolymph of the black soldier fly, *Hermetia illucens* L. on *A. swirskii* (Nguyen et al., 2013) have been evaluated. However, numerous phytoseiid predators feed not only on their prey but also on nonprey food sources, particularly plant materials, including fungal spores, nectar, prey products (e.g. honeydew), and pollen.

An important and desirable characteristic of a biocontrol agent is its ability to use alternative food sources, such as pollen, which promotes its persistence in the field when prey is absent or scarce, and facilitates its mass rearing for augmentative biocontrol purpose. Pollen grains provide important food elements, for instance, proteins, free amino acids, carbohydrates, lipids, vitamins, flavonoids, and minerals to improve development and reproduction of arthropods. Pollen grain size does not have an important effect on its quality as food source for the mites. Consequently, other pollen features, such as exine thickness and structure and nutritional composition, may play more important roles. Successful pollen-feeding depends on matching mouthpart morphology with pollen morphology, and digestive metabolism in the mites, which requires further investigation (Ranabhat et al., 2014).

Pollen grains from different botanical families are nutritionally suitable to sustain predatory mite rearing in the laboratory. Pollen can also be available to predators in a 'banker' plant system. Pollen of castor bean (*Ricinus communis* L.) is largely used for predatory mite rearing, it is easy to collect and handle, the plant can be found abundantly in several habitats, and it can be used as banker plant in greenhouses. Pollen of cattail (*Typha* sp.) has been demonstrated as a suitable food source for several phytoseiid predators, such as *Euseius finlandicus* (Oudemans), *Anthoseius rhenanus* (Oudemans), *A. swirskii,* and *N. cucumeris* (Britto et al., 2012; Renkema et al., 2017; Gravandian et al., 2022; Hadadi et al., 2022). Many generalist phytoseiid predators have successfully reproduced on date palm pollen as an alternative food source, e.g., *A. swirskii, T. bagdasarjani* (Riahi et al., 2016, 2017a), *N. californicus* (Khanamani et al., 2017a), *Euseius scutalis, E. finlandicus, Proprioseiopsis asetus* (Chant) (Alatawi et al., 2018), and *N. cucumeris* (Yazdanpanah et al., 2022d). Populations of predatory mites, such as *A. swirskii, A. degenerans* and *N. cucumeris*, can increase significantly in the presence of supplemental foods, such as the pollen of cattail (*Typha latifolia* L.), caster bean or maize (Fadaei et al., 2018). The pollen grains of almond, date palm, castor bean, and sunflower are promising alternative food source for *N. cucumeris* (Yazdanpanah et al., 2021b).

Besides, astigmatid mites have been commonly used as an alternative prey for the commercial production of different mesostigmatids. Of the phytoseiids that are reared commercially on factitious prey, most are Type III mites. Because of specialization or preference for web-producing mites, Type I and II mites typically require spider mites for efficient rearing. Type I mites, like *P. persimilis*, can only be reared on *Tetranychus* species on plants. Because of their feeding preferences, development of grain mite rearing systems for Type II mites presents a substantial challenge even though these species will feed on prey other than *Tetranychus* species. Some progress has been made in developing factitious prey rearing for one Type II mite, *N. californicus* (McGregor et al., 2020). To date, a number of astigmatid mite species are used in the commercial mass production of several species of phytoseiid predators; these include *Carpoglyphus lactis* (L.), *Thyreophagus entomophagus* (Laboulbene), and *Tyrophagus putrescentiae* (Schrank) (Pirayeshfar et al., 2020).

Lasioseius fimetorum Karg, *A. swirskii*, and *N. cucumeris* have been developed and reproduced on *T. putrescentiae* as diet (Britto et al., 2012). However, astigmatid species causes health risks to production workers in the mass rearing of predatory mites, and sometimes causes allergies in farmers when released in protected crops, resulting in allergic contact dermatitis and asthma (Liu and Zhang, 2017). Nonnatural prey (pollen and artificial diets) are also advantageous over natural prey because they are not prone to respiratory problems created when the stored-product mites are used for mass production (Nemati and Riahi, 2019).

Good coverage and dispersal of predatory mites is important for biological control, especially in ornamentals, where the pest levels that can be tolerated are lower than in vegetables. Therefore, patches of prey can be sparsely distributed and there is often not enough prey to maintain a predator population. To achieve this, various release methods have been developed. Initially, the predatory mites were only available in bottles (Figure 10.4a) with carrier material, such as vermiculite, bran, or corn grits, and mites needed to be distributed by hand. One problem with the use of these dry carriers is that the predators can arrive partially desiccated and debilitated; however, the carrier does make it easier to shake the predators out evenly on the plants. Later, release systems based on mechanical blowers were developed, which distribute the mites more uniformly and reduce labor cost. Another release method is the 'slow release', using breeding sachets (Figure 10.4b). These sachets contain a carrier material, such as wheat bran, and a food source for the factitious prey mites. The grain mites, however, may cause damage to some plants, so caution should be exercised in such releases. Slow-release sachets should be hung on plants out of the direct sun, to prevent high temperatures from killing the mites inside. The predatory mites feed and multiply in the sachet and can leave through a small hole that is punched into the sachets and disperse into the crop for several weeks. Sachets are therefore a valuable option in systems where no alternative food in the form of pollen is available and preventive releases help to keep pest levels low. However, some species, such as *Euseius* species, could not be reared on any substrate except beeswax-coated paper, perhaps due to the need for these mites to imbibe liquids from leaves (Hoy, 2011; Calvo et al., 2015).

10.3.1.1.1.1 Amblyseius swirskii Athias-Henriot *Amblyseius swirskii* (Figure 10.5f) originates from the East Mediterranean coast and was described in 1962 from almond in Bet Dagan, Israel. In addition to Israel, it has been reported from other Middle Eastern countries; Southern Europe; West, Central, and East Africa; and North, Central, and South America. The predatory mite *A. swirskii* quickly became one of the most successful biocontrol agents in protected cultivation after its introduction into the market in 2005 (Calvo et al., 2015).

According to McMurtry et al. (2013), this species has a Type III lifestyle (subtype III-b), and it can develop and reproduce on a wide range of arthropod prey, such as *Frankliniella occidentalis* (Pergande), *Scirtothrips dorsalis* Hood, *Bemisia tabaci* (Gennadius), *Trialeurodes vaporariorum* (Westwood), *Polyphagotarsonemus latus, Diaphorina citri* Kuwayama, *Eotetranychus frosti*, and *Cenopalpus irani*. *Amblyseius swirskii* is a promising control agent for *P. oleivora* and *E. orientalis* on orange trees; the maximum daily consumption rate was recorded at 30°C and occurred during the oviposition period, when the females consumed an average of 169.40 specimens of *P. oleivora* and 24.58 of *E. orientalis* (Bazgir et al., 2020; Al-Azzazy and Alhewairini, 2020). The feeding capacity of *A. swirskii* was greater under complete darkness, but the fecundity was greater under a photoperiod of 16:8 hours (L:D). It can feed on *T. urticae*, but it is not able to control spider mites in the absence of

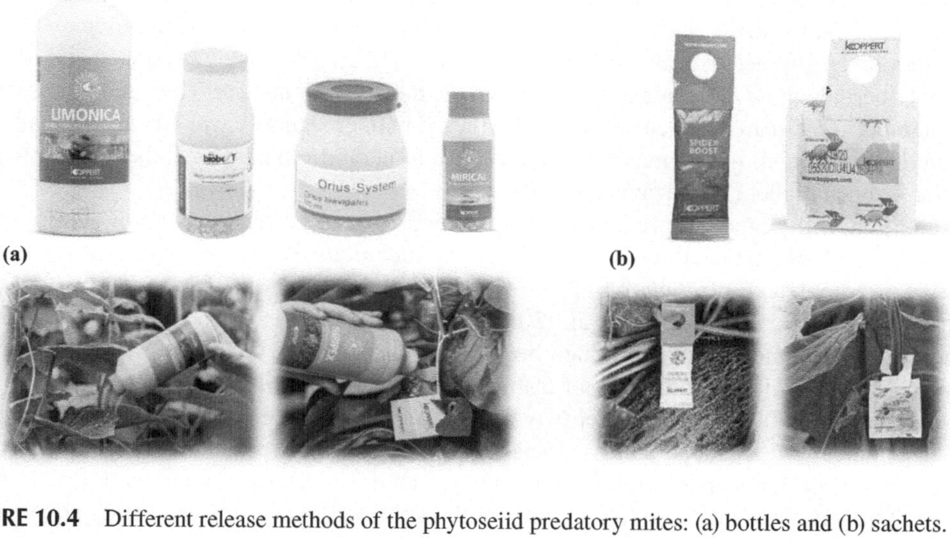

FIGURE 10.4 Different release methods of the phytoseiid predatory mites: (a) bottles and (b) sachets.

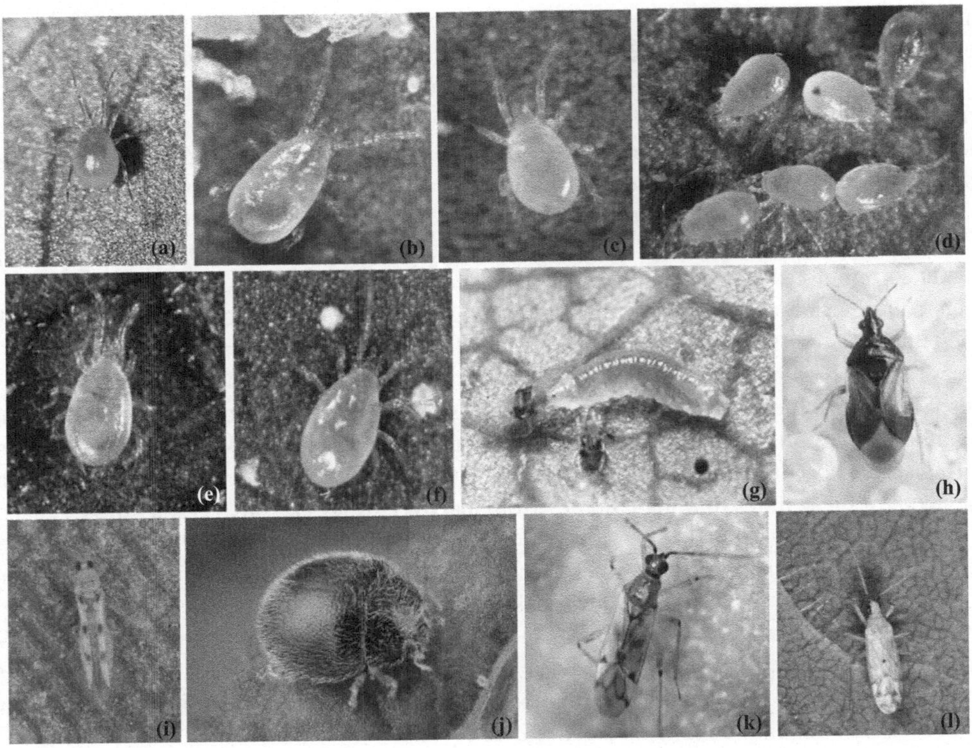

FIGURE 10.5 Some most important commercial predators; (a) *Phytoseiulus persimilis*, (b) *Neoseiulus cucumeris*, (c) *Amblydromalus limonicus*, (d) *Typhlodromus pyri*, (e) *Neoseiulus californicus*, (f) *Amblyseius swirskii* (Acari: Phytoseiidae), (g) *Feltiella acarisuga* (Diptera: Cecidomyiidae), (h) *Orius laevigatus* (Hemiptera: Anthocoridae), (i) *Scolothrips sexmaculatus* (Thysanoptera: Thripidae), (j) *Stethorus punctillum* (Coleoptera: Coccinellidae), (k) *Nesidiocoris tenuis*, and (l) *Macrolophus pygmaeus* (Hemiptera: Miridae).

other pests, mainly because the predatory mites cannot enter dense spider mite webbing and therefore can only feed on spider mites outside or near the edges of the webbing (Calvo et al., 2015).

Some studies show that although *A. swirskii* is able to reduce the spider mite, *T. urticae* population, the spider mite webbing has an adverse effect on the performance of the predator. Therefore, the presence of the predator's population at the time of the spider mite infestation is crucial to the success of biological control. The alternative food sources, such as pollen and greenhouse whitefly, *Trialeurodes vaporariorum*–produced honeydew play an important role in maintaining the predator population in the absence of pests. Moreover, the results indicate that using the pollen and presence of another pest along with the target pest can promote the predator density. This predator consumes lower numbers of *T. urticae* when concurrently offered with greenhouse whitefly and/or maize pollen. On the other hand, in the presence of alternative food or alternative prey, the fecundity of the predator is much higher (Mortazavi et al., 2019).

In addition, this generalist predatory mite can feed on many different nonprey foods, including eggs of the Mediterranean flour moth, *Ephestia kuehniella* Zeller, decapsulated dry cysts of the brine shrimp *Artemia franciscana* Kellogg, and mango powdery mildew, *Oidium mangiferae* Berthet (Calvo et al., 2015). Riahi et al. (2017b) found that *A. swirskii* individuals survived and reproduced on almond and maize pollen better than on factitious prey or even natural prey (*T. urticae*). An artificial diet consisting of 5% honey, 5% sucrose, 5% tryptone, 5% yeast extract, 10% egg yolk, and 70% distilled water (w/w) (formulated by Nguyen et al. (2013)) supplemented with 20% maize pollen is a suitable artificial diet for rearing this predator (Riahi et al., 2017c). Besides, cattail pollen, pepper pollen, and apple pollen are food sources of equal quality for *A. swirskii*. This mite was able to adapt to each of the six tested diets (one consisting solely of the eggs of *T. urticae* and the others being pollen of apricot, soybean, sesame, walnut, and date, each mixed with the eggs of *T. urticae*) (Fadaei et al., 2018). Feeding on some pollen grains, such as *Aesculus hippocastanum* L., *Crocus vernus* L., *Echinocereus* sp., and *Paulownia tomentosa* (Thunb), leads to the best performance of this mite (Goleva and Zebitz, 2013). Although it has been reported that the almond pollen as a good diet for longtime rearing of *A. swirskii* (Nemati and Riahi, 2019; Ansari-Shiri et al., 2022), a mass-rearing system based on the use of the factitious prey *Carpoglyphus lactis* and/or *Tyrophagus putrescentiae* would be easier (Onzo et al., 2012; Calvo et al., 2015). Oviposition rate of *A. swirskii* when fed with frozen larvae of *T. putrescentiae* was similar to the rate observed when they were fed with cattail pollen or living prey mites, but developmental time of *A. swirskii* was longer on this frozen diet than on a diet of living prey mites or pollen (Pirayeshfar et al., 2020).

A comperehensice picture of foraging strategies (Fathipour et al., 2017, 2021) of *A. swirskii*as well as demographic parameters (Ansari-Shiri et al., 2022; Hadadi et al., 2022; Rahimi et al., 2022) and intreguild predation experiments (Maleknia et al., 2016) show high efficacy of this predatory mite in different situations.

For *A. swirskii*, commercial slow-release sachets can contain the factitious prey, *C. lactis*, *Suidasia medanensis* (Oudemans), *Thyreophagus entomophagus*, and/or *Lepidoglyphus destructor* (Schrank) (Calvo et al., 2015). In a greenhouse study, the use of frozen prey mite stages as supplementary food on chrysanthemum plants allowed the population of *A. swirskii* to establish, but not increase; in contrast, provision of living prey mites and pollen increased *A. swirskii* populations on plants. Hence, living prey mites had the greatest potential as a supplementary food source for *A. swirskii* (Pirayeshfar et al., 2020).

10.3.1.1.1.2 Amblyseius andersoni (Chant) *Amblyseius andersoni* is reported to be one of the first predatory mites to migrate to fruit crops after insecticide treatments are reduced. It is primarily found in areas of grape and apple production but also in soft fruit, peaches, deciduous ornamentals, and conifers. The generalist predator *A. andersoni* (Type III lifestyle (subtype III-b)) is a cosmopolitan but commonly found mite in apple orchards in several countries, especially in Europe (Rahmani and Afshari, 2020).

Adult female *A. andersoni* lay single egg onto leaf hairs, which hatch after 2–3 days as larvae; they lay about 35 eggs during their lifetime. The total oviposition of *A. andersoni* is high at 25°C and low at 35°C, and the daily average oviposition increases as the temperature increases, but few eggs are laid at 17°C. The duration of the different stages is reduced when the temperature increases. Development from an egg to adult takes 8–11 days at 20°C–25°C, and it stops working at 35°C; adults live for about 3 weeks (Li et al., 2019).

Some authors showed that it can keep the population of *T. urticae*, *P. ulmi*, broad mites, cyclamen mites, boxwood bud mite (*Eriophyes canestrinii* (Nalepa)), and *Aculus schlechtendali* (Nalepa) under the economic injury levels in apple orchards. The ability of *A. andersoni* to control phytophagous mites depends on several factors, such as the climatic conditions in the area, the morphological characteristics of plant leaves, inter-specific competition and the impact of pesticides (Li et al., 2019).

This predatory mite feeds on alternative foods and reaches high population levels, besides feeding on pollen, thrips larvae, sugary excretions from pests, and fungi allowing the population to survive when mite pest populations decrease (Li et al., 2019). For best results, *A. andersoni* should be applied when mite pest numbers are low. The predatory mites will then be able to feed on small colonies of mites and prevent them from growing and causing major damage. They are available in loose shaker tubes and breeding sachets. The sachets can provide a release of mites over several weeks.

10.3.1.1.1.3 Amblyseius largoensis (Muma) This is a generalist species of Phytoseiidae predominantly found in coconut palm leaflets around the world. In several countries, *A. largoensis* has been reported as the most frequent and abundant predator associated with the red palm mite,

Raoiella indica, especially in the countries recently invaded by this pest, such as Caribbean islands, the United States of America, Cuba, and Brazil (Mendes et al., 2018). This species is the most common predator encountered on coconut mite, *Aceria guerreronis* on coconut palms, where it resides primarily on the leaves or exposed fruit surface and rarely under the bracts of coconuts. The distribution of *A. largoensis* indicates that it is a pantropical species, common in regions subjected to long rainy periods.

Amblyseius largoensis fails to complete development at or below 72% relative humidity (RH). Maximum population increase of the predator (about 26-fold in one generation) occurs at 91% RH (Mendes et al., 2017; Gómez-Moya et al., 2018). This mite is a biological control agent of *P. latus, Tetranychus gloveri* Banks eggs, and larvae of *O. punicae*, *P. citri*, and *Tetranychus* spp. (Carrillo et al., 2012). The development of immature stages of *A. largoensis* is faster and fecundity, survivorship, and intrinsic rate of natural increase are high when fed on red palm mite. These results suggest that despite earlier assessments, *A. largoensis* can play a role in controlling red palm mite (Carrillo et al., 2010).

Amblyseius largoensis is considered a generalist due to its ability to feed on multiple food sources (e.g. phytophagous mites, small insects, plant exudates, nectar, and pollen). Several studies have demonstrated that *A. largoensis* is able to feed, develop, and reproduce on a diet consisting of solely the red palm mite, showing a strong preference for red palm mite eggs over other stages of the pest (Mendes et al., 2017). Despite being so common and widespread, this predator has not been used for commercial purposes (Gómez-Moya et al., 2018).

10.3.1.1.1.4 Amblyseius tsugawai Ehara *Amblyseius tsugawai* was first described from specimens collected on apples in Japan in 1959. This species is known to be a generalist predator, feeding on a variety of foods, including spider mites, thrips, tea pollen, *Bemisia tabaci*, and eggs of *E. kuehniella*. Moreover, *A. tsugawai* can be reared on *C. lactis* (Yang et al., 2019).

10.3.1.1.1.5 Amblyseius herbicolus (Chant) *Amblyseius herbicolus* (lifestyle subtype III-c) is very similar to *A. largoensis* and widespread in tropical and Indo-Pacific countries. The intrinsic rate of increase of *A. herbicolus* is higher on the diet of broad mites, castor bean pollen or chili pepper than sunn hemp pollen. The predator shows high survival rates, short developmental time, and high reproductive success when feed on only *P. latus*. This performance confirms that *A. herbicolus* is a potential biological control agent of *P. latus* (Duarte et al., 2015).

10.3.1.1.1.6 Neoseiulus cucumeris (Oudemans) The first description of this predatory mite was based on specimens found in muskmelon infested with *Tetranychus* spider mites, collected in France (Beard, 1999). *Neoseiulus cucumeris* (Figure 10.5b) has a worldwide distribution because of its natural occurrence and commercial use in various parts of the world.

Neoseiulus cucumeris is a Type III lifestyle (subtype III-e) phytoseiid mite that feeds on mites, insects, and pollen grains (McMurtry et al., 2013). This predator is generally seen on the lower leaf surface or inside flowers. It was the first phytoseiid mite found to feed on thrips, but until 1980, its use in thrips management was little explored. It has been successfully used in greenhouses to control pests, such as western thrips, onion thrips, melon thrips, common blossom thrips, chili thrips, Asian citrus psyllid, cyclamen mites, and other plant damaging mites (Patenaude et al., 2020). When *N. cucumeris* is fed on *A. lycopersici* and *T. urticae,* the maximum reproduction was recorded at 35°C and 65% RH, while the minimum was at 25°C±1°C and 55%±5% RH, respectively (Al-Azzazy et al., 2018). When *N. cucumeris* is fed on the spider mite *Schizotetranychus nanjingensis* Ma & Yuan, a pest of the moso bamboo, the period from egg to first oviposition lasted for 7.7 days in the first generation and 7.8 days in the second generation, in which it was similar to that on its normal diet (*T. putrescentiae*) in the laboratory (7.8 days) at 27°C–28°C. Furthermore, the number of prey consumed by the predator with increasing prey density and the number of eggs produced were directly correlated with the number of prey consumed. *Neoseiulus cucumeris* females were unable to invade intact webnests of *S. nanjingensis*, but they were able to invade and liked to stay and lay eggs in broken nests with existing openings or holes (Zhang et al., 2000).

A comprehensive picture of foraging strategies of *N. cucumeris* on the two-spotted spider mite and western flower thrips has already been determined (Dalir et al. 2021a,b; Yazdanpanah et al., 2022b). The age-specific functional and numerical responses of *N. cucumeris* on eggs of *T. urticae* have been assessed, which indicated that the functional response of 28- and 40-day-old *N. cucumeris* was Type II, whereas the other ages displayed a Type III functional response (Dalir et al., 2021b). Furthermore, the generation-dependent functional and numerical responses of *N. cucumeris* on nymphs of *T. urticae* showed a stable predation up to 30 generations (Yazdanpanah et al., 2022b).

Neoseiulus cucumeris can be reared successfully on pollen (Yazdanpanah et al., 2021a), natural as well as facticious prey (Yazdanpanah et al., 2022a) for many generations, but it would be cost-effective on the latter one. The pollen grains of tulip, horse-chestnut, cattail, and birch are food sources for *N. cucumeris* (Delisle et al., 2015). In addition, almond pollen is a suitable diet for long-term rearing of this predator (Yazdanpanah et al., 2021a) and remain the predator's quality constant after switching to the natural prey (Yazdanpanah et al., 2023). However, results show that *N. cucumeris* feeds, develops and reproduces better when it was provided with plant and animal diets together, compared to pollen grains alone. When this mite is reared on maize and mungbean pollen along with *T. putrescentiae*, significantly shorter developmental time is obtained (Sarwar, 2016). Both species of *T. urticae* and *A. lycopersici*, particularly the eriophyid mite proved to be suitable prey for *N. cucumeris*, as a facultative predator (Al-Azzazy et al., 2018). The lifetable parameters of this predatory mite have been determined when provided with a liquid artificial diet (consisting of honey, sucrose, tryptone, yeast extract, and egg yolk) enriched with extract of dry decapsulated cysts of the brine shrimp, *Artemia franciscana*, and compared with those obtained on *Typha latifolia* L. pollen. Female developmental time is significantly shorter when it feeds on pollen compared with the artificial diet. The fecundity of *N. cucumeris* fed on pollen grains is significantly higher than that on the artificial diet (Nguyen et al., 2015).

Neoseiulus cucumeris is known to be sensitive to cold temperatures and its efficacy depends on temperature (Yazdanpanah et al., 2022c). The common recommendation in outdoor production is to release *N. cucumeris* when the risk of spring frost is over and/or after a minimum time of 3 weeks after an insecticide application (Patenaude et al., 2020). It has been reported that sachets can protect *N. cucumeris* from intraguild predation by *Dalotia coriaria* (Kraatz) (Staphylinidae), resulting in higher populations of mites. Therefore, sachets should be used in greenhouse biocontrol programs where *D. coriaria* is also released. Furthermore, breeder piles that provide 'quick-releases' or sachets that provide 'slow-releases' of mites should be considered when incorporating *N. cucumeris* into greenhouse thrips management programs (Pochubay et al., 2015).

10.3.1.1.1.7 Neoseiulus californicus (McGregor) *Neoseiulus californicus* (Figure 10.5e) was first described in 1954 from lemon in California. Its natural populations are found in Argentina, California, Chile, Florida, Japan, South Africa, Texas, parts of southern Europe, and all along the border of the Mediterranean Sea. It was first released under license as a biological control agent in the United Kingdom in 1991. A commercial strain of *N. californicus* (Spical@R) has been used in greenhouses and fields in Japan since 2003.

It completes a generation in 1–2 weeks depending on temperature. *Neoseiulus californicus* is tolerant of various temperatures and low humidity, but works best under warm to hot conditions. It can tolerate higher temperatures and lower humidity than *P. persimilis*. They prefer warm temperatures, but they can tolerate much colder temperatures for short periods of time. *Neoseiulus californicus* females can lay up to four eggs a day; however, two eggs per day is the average. Total developmental time can be as short as 4.0 days or as long as 12.0 days depending on the temperature. This mite develops more quickly at higher temperatures.

This predator can tolerate a wide range of RH (40%–80%) but prefers humidity at the upper end of this range. The survival curves and median survival time are significantly different between high and low RHs, and the mean survival time is 1.6–2.3 times longer at high RH than that at low RH, except in larvae, in which tolerances to high RH and/or low air temperature are not well developed. At high RH, the median survival time of the mated adult females is 63.1 days and is 1.4 times longer than that of unmated adult females. These results indicate that the suitable water content in the body was probably maintained by increasing the RH at low air temperature, which is considered to be a primary factor affecting the survival time of *N. californicus*. Moreover, the nutrient content in the body, such as eggs and male accessory gland proteins in the mated adult females, could have positive effects on prolonging their survival time. These findings help to construct an efficient long-term storage method for *N. californicus* (Ghazy et al., 2012).

Neoseiulus californicus has characteristics of both Type II specialist predatory mites and Type III generalist predatory mites. This predatory mite is one of the major phytoseiid predators of spider mites, and it can cut their dense webbing using its chelicerae (McMurtry et al., 2013). However, it will also successfully develop and reproduce when consuming other mite species including *Aculus schlechtendali*, *Oligonychus pratensis* (Banks), *Oligonychus perseae* Tuttle, *Oligonychus ilicis* (McGregor), *P. ulmi*, *Phytonemus pallidus*, *T. urticae*, *T. kanzawai*, *T. cinnabarinus*, *T. turkestani*, *T. evansi*, *T. ludeni*, *Polyphagotarsonemus latus*, and *E. lewisi*, and some insects, such as *Scolothrips longicornis* Priesner and *Thrips tabaci* Lindeman.

Neoseiulus californicus can survive longer under starvation conditions and can also live on a wide range of pollen types (Eini et al., 2022); they also will cannibalize their young ones to stay alive. After they run out of food and begin to starve, if a food source is then found, the females then fatten up and start to lay eggs again. Corn pollen, date pollen, and walnut pollen are suitable alternative foods for the mass rearing of this predator and between these pollen grains, date pollen is better than the others (Rezaie, 2017). In addition, based on Marafeli et al. (2014), castor bean pollen is a suitable food for this predator. Furthermore, Khanamani et al. (2017a, b) showed that *N. californicus* has high fecundity on the almond pollen, followed by the maize pollen and *T. urticae*, and the cohort reared on almond has the highest intrinsic rate of increase (*r*). High fecundity is obtained on pistachio pollen, followed by apricot pollen and *T. urticae*; besides, the population growth parameters are high on pistachio pollen+*T. urticae* diet (Soltaniyan et al., 2020). Eini et al. (2022) offered the pollen grains of 23 plant species to *N. californicus* as a food and concluded that feeding on pistachio and thorn apple pollen lead to the best performance of this predator and these pollens could be used to improve the experimental and commercial mass rearing programs of this predator or inter-planted to support the predator population in the field.

Although it has been reported that long-term rearing of *N. californicus* on a basic artificial diet (formulated by Nguyen et al. (2013)) enriched with *E. kuehniella* may have potential for use in the mass rearing of this predator (Song et al., 2019). Khanamani et al. (2017b) reported that rearing of *N. californicus* on this basic artificial diet enriched with 20% *E. kuehniella* eggs or *Artemia franciscana* cysts or maize pollen may not have potential for use in the mass rearing of this predator.

The recommended release ratio depends on pest species, pest density, and crop type. For example, in strawberry cropping system, a release ratio of one female *N. californicus* to ten two-spotted spider mites is recommended once mite pests are present. Slow-release sachet is the common release method for this mite, but best results are seen when *N. californicus* is allowed to build up before the spider mite populations are able to establish themselves (Akyazi et al., 2020).

10.3.1.1.1.8 Neoseiulus fallacis (Garman) *Neoseiulus fallacies* was first described in 1948 as *Iphidulus fallacies* Garman during a survey of mite species in Connecticut apple orchards. According to a classification of predatory mites based on feeding specialization developed by McMurtry et al. (2013), *N. fallacis* is a selective predator of the spider mite family (Type II). These mites are usually associated with web-producing spider mite species. They require lower numbers of prey than other phytoseiids to complete development but possess a high power of dispersal (McMurtry and Croft, 1997). In apple orchards, *N. fallacis* overwinter as adult females in ground cover and debris, and then migrate to trees in the spring after increasing the *P. ulmi* population.

The adaptability of *N. fallacis* to various habitats actually allows it to feed on a wide range of prey species. The predatory mite *N. fallacis* has been evaluated as a biocontrol agent of herbivorous mites on outdoor-grown ornamental landscape plants. *Neoseiulus fallacis* is most effective on shrubs and herbaceous perennials and less effective on conifers and shade trees; it is equally effective at controlling spider mites of *Tetranychus*, *Oligonychus* or *Schizotetranychus* species (Pratt et al., 2002). This mite has been shown to feed on a variety of food types including mites, insects, and pollen, but the grain mite of *T. putrescentiae* is not a suitable prey. *Lepidoglyphus destructor* has been mentioned as a suitable rearing prey for *N. fallacis* in a US patent application that includes data showing its oviposition in the presence of this grain mite species. *Neoseiulus fallacis* is currently reared in British Columbia (Canada) on *T. urticae* in a plant-based system. In addition, it can be reared on three astigmatid mites of *L. destructor*, *Aleuroglyphus ovatus* (Tropeau), and *Thyreophagus entomophagus*, although *N. fallacis* rejected *A. ovatus* and *T. entomophagus* as prey more often than *T. urticae*. For the three acarid mites tested, *L. destructor* is the most promising species for further development as a factitious prey for *N. fallacis* (McGregor et al., 2020).

10.3.1.1.1.9 Neoseiulus barkeri Hughes *Neoseiulus barkeri* is a predator, which feeds on a wide variety of mites, thrips, and citrus psyllids, and it was shown to be an effective predator of citrus red mite, *P. citri*. *Neoseiulus barkeri* has been commercialized in China and used as a biocontrol agent along with other species. For example, it was released along with *N. cucumeris* in controlling Asian citrus psyllid, *Diaphorina citri* (Liu et al., 2020). Although *N. barkeri* (lifestyle subtype III-e) has been shown to survive on the pollen of tulips and hazel, and subsequently reproduce successfully, they did not remain on plants when reared on cattail pollen. *Neoseiulus barkeri* did not reproduce at all when fed with castor bean and date pollen, while the reproduction was 1.9 eggs/female/day when fed on *T. urticae*. Type II functional response has been repoted for this predator at different temperatures and patch conditions (Jafari et al., 2012; Fathipour et al., 2021) and it seems that *N. barkeri* has inherent potential for the control of two-spotted spider mite at higher temperatures especially at temperatures between 30–35°C (Jafari et al., 2010).

10.3.1.1.1.10 Phytoseiulus persimilis Athias-Henriot This mite was accidentally introduced into Germany from Chile in 1958; it was subsequently shipped to other parts of the world, including California and Florida. The adult female may lay up to 60 eggs during her 50-day-long lifetime at 17°C–27°C. Generation time lasts from 7 to 17 days, depending on temperature and humidity. *Phytoseiulus persimilis* (Figure 10.5a) larvae or adults can be stored at 7.5°C for 4 weeks. They do not cannibalize their young ones or consume other food sources to stay alive; therefore, when *P. persimilis* starve down, they cannot recover their body size.

This mite is a Type I lifestyle (subtype I-a) phytoseiid mite that feeds exclusively on *Tetranychus* species, especially *T. urticae*, and its survival depends on the presence and quality of its prey (McMurtry et al., 2013). The predator–prey system of *P. persimilis* and *T. urticae* represents the most studied and the most known system among mites. *Phytoseiulus persimilis* used for biocontrol of *T. urticae* was one of the earliest biocontrol agents commercially available in Europe in 1968. This mite consumes up to seven adult spider mites or several dozen of their eggs in a day. Owing to the ability of *P. persimilis* to control spider mites, it is being mass produced and sold by many commercial biological control companies (Fathipour et al., 2018).

Population growth potential and foraging behavior of *P. persimilis* are affected by several factors. The lifetime-dependent functional response and predation rate of *P. persimilis* on eggs of *T. urticae* has been determined by Fathipour et al. (2018) and the results showed that on the 15th, 35th, and 40th days of the predator age, predation rate on *T. urticae* eggs increases, resulting in the Type III functional response, while at other ages, the functional response was Type II. The results of their study revealed that *P. persimilis*, especially at the middle and late ages, has a good predation potential on *T. urticae* eggs at higher prey densities. Alipour et al. (2019) evaluated the development rate of different life stages of *T. urticae* and its predator, *P. persimilis*, on the susceptible and resistant rose cultivars. Based on their results, resistance of the rose cultivar negatively affects the population growth parameters of *T. urticae* and its predator *P. persimilis*. Although sometimes the susceptible cultivar may cause the most favorable life-table parameters for *P. persimilis*, the predation rate on *T. urticae* would be frequent on the resistant cultivar (Fathipour et al., 2019).

Obligatory predators, such as *P. persimilis*, often are the most difficult to rear because they may not accept alternative foods, such as pollen or grain mites. Tritrophic interactions can provide important insights for improving the efficacy of pest control programs. Rearing phytoseiids that are obligatory predators of spider mites in the greenhouse is a tritrophic endeavor that involves rearing plants (free of mites, insects, and diseases), infesting them with a known number of prey mites (often *T. urticae*), allowing these mites to multiply (predators often prefer eggs or immatures), infesting the plants with the appropriate ratio of phytoseiids, harvesting the predators, and packaging and shipping them. The harvest of predators is a trade secret. The basic method, however, probably involves utilizing the behavior of phytoseiid females. Female phytoseiids tend to move up toward light if they are hungry. They may aggregate in specific sites where they can be harvested with a vacuum pump aspirator (Hoy et al., 1982).

Voracious foraging for spider mites enables *P. persimilis* to express rapid functional and numerical responses to changes in prey populations. *Phytoseiulus persimilis* swiftly overexploits the prey, and as it cannot reproduce on other diets, predator numbers then decline. They primarily are effective in situations when multiple releases can be made (inundative augmentation programs), such as greenhouse crops or vegetables, where it is important to get rapid control because the tolerance for damage is low. In many cases, the predator is used like a pesticide (hence being called a 'living pesticide' or 'biopesticide'). It is meant to provide a rapid solution to the pest problem, but may have only a short-term (or single-season) effect (Gerson et al., 2003).

10.3.1.1.1.11 Phytoseiulus longipes Evans The predatory mite, *Phytoseiulus longipes*, was described from Zimbabwe. Later, it was reported from other localities, e.g., Cape Province, South Africa, Rio Grande do Soul, Brazil, Los Andes, and Chile. The duration of the immature phase ranged from 3.1 to 15.4 days, at 30°C and 15°C, respectively, and the intrinsic rate of increase (*r*) of *P. longipes* ranged from 0.091 to 0.416 day^{-1}at 15°C and 30°C, respectively; the shortest duration and the highest reproduction were observed at 25°C and 30°C. *Phytoseiulus longipes* relatively tolerate the drought conditions; this drought-resistant species can perform better during the dry season when the RH drops to critical levels (Mutisya et al., 2012).

Phytoseiulus longipes is a Type I lifestyle (subtype I-a) phytoseiid mite, which has the potential to control *Tetranychus pacificus* McGregor, *T. evansi*, *T. urticae*, *T. marianae* McGregor, *O. punicae*, and *P. citri* (Mutisya et al., 2012). The predator has about the same oviposition rate on *T. evansi* and *T. urticae*. The provision of 10% sucrose solution and clover honey doubled the longevity of *P. longipes* females compared with no food, while it did not oviposite when fed on pollen of castor bean or cattail (Furtado et al., 2007).

10.3.1.1.1.12 Phytoseiulus macropilis (Banks) *Phytoseiulus macropilis* is categorized as a specialized predator of *Tetranychus* species (Type I, subtype I-a) (McMurtry et al., 2013). It has also been reported from several countries of Europe, Africa, and America. It was first noted as an important enemy of spider mites, and it is a specialist predator of *T. urticae*. However, *Mononychellus planki* (McGregor) (Tetranychidae), a phytophagous mite present mainly on common bean, soybean, and cottonweeds, is an alternative prey enhancing the survival of *P. macropilis* when *T. urticae* is absent in the environment (Majolo and Ferla, 2014).

The fecundity of females mated more than once was 1.7 times higher than in females mated only once, also oviposition period and adult longevity was longer. The adult longevity and life span of *P. macropilis* were similar when fed on mixed active stages, and eggs of *T. urticae*, but the total number of eggs deposited by females was significantly higher on eggs than on the mixed active stages of the mentioned prey. Life-table analysis showed that a diet of eggs of *T. urticae* had the longest generation time (11.07 days) and female longevity (33.54 days), as well as the highest intrinsic rate of natural increase ($r = 0.368$), and greatest total fecundity, which resulted in the highest net reproductive rate ($R_0 = 58.41$) value. The adult female consumes daily an average of 6.8 mixed active stages of *T. urticae* during its life cycle, while it increases to 13.6 during the oviposition period (Abdel Khalk and Momen, 2009).

10.3.1.1.1.13 *Amblydromalus limonicus* (Garman & McGregor) *Amblydromalu slimonicus* (Figure 10.5c) was first described in 1956 from citrus plants. It has been reported from cultivated and wild plants including trees, shrubs, and herbaceous plants in Afrotropic, Australasian, Neotropic, and Nearctic ecozones as well as in the Palearctic ecozone as an introduced species. It is known that in the early 1990s, the species was collected from California and New Zealand and brought to Europe (Netherlands and England) for mass rearing and used as a biological control agent of thrips and whiteflies. Since 2012, this mite has been commercially available in some European countries (Chora et al., 2016).

Amblydromalus limonicus is a generalist predator (subtype III-b) (McMurtry et al., 2013). It feeds and reproduces on whiteflies (*Bemisia tabaci* and *Trialeurodes vaporariorum*), thrips (*Frankliniella occidentalis, Thrips tabaci*, and *Retithrips syriacus* (Mayet)), eggs of the moth, *Prays citri* (Milliere), and spider mites (*O. punicae, P. citri*, and *E. orientalis*). In addition to various prey species, *A. limonicus* also feeds on pollen of different plants, extra floral nectar, and spores of fungi (Chora et al., 2016). Pollen significantly shortens the developmental time and increases the survival and oviposition of the predator when they are mixed with spider mites, although it negatively affects its predation rate. Nevertheless, pollen enhances the maintenance of juvenile predators on the leaf by substantially decreasing their dispersal rate in the mixed diet with spider mites (Samaras et al., 2019). Cattail pollen followed by olive pollen resulted in high intrinsic rate of increase, and are suitable for the development and reproduction of *A. limonicus* (Samaras et al., 2015).

This mite can feed on factitious food, such as eggs of *E. kuehniella*, cysts of *A. franciscana*, *C. lactis*, and eggs of Mediterranean fruit fly, *Ceratitis capitata* (Wiedemann). However, it has been reported that fecundity and oviposition rate are higher on *E. kuehniella* and *A. franciscana* than cattail pollen (Vangansbeke et al., 2014).

Amblydromalus limonicus individuals can be reared in enclosed cells with an ambient RH of 50% on the following diets: *Typha orientalis* pollen; *E. kuehniella* eggs, untreated, punctured or coated with bee honey solution (10%); and an artificial diet enriched with the pupal hemolymph of *Epiphyas postvittana* (Walker) (Tortricidae). Cattail pollen and *E. kuehniella* eggs (to a lesser extent) are suitable food sources for the augmentation of *A. limonicus* and can also be used as alternative diets for preventive releases in early crop seasons. The storage mite, *C. lactis* is conventionally used as the primary food source in the commercial production of *A. limonicus* (Liu and Zhang, 2017). The rearing media with rice husk is found to hold the highest populations of the predator, whereas the treatment with no substrate and wheat bran treatments have the lowest. The higher populations associated with the rice husk treatment are attributed to a decrease in encounter rates, the provision of additional shelter, and oviposition sites, and low mold infestation (Lee and Zhang, 2016).

10.3.1.1.1.14 *Galendromus (Metaseiulus) occidentalis* (Nesbitt) *Galendromus occidentalis* is an important and effective predator of mite pests in agricultural ecosystems around the world. This predator mite is distributed in Nearctic, Oriental, and Palearctic regions of the world. This species is well adapted to the arid environment of the major apple-growing regions, and has been incorporated into integrated control programs since the 1960s (Schmidt-Jeffris and Beers, 2015). *Galendromus occidentalis* reproduces very quickly, and there may be eight to ten generations per year in the field. Data on mortality of *G. occidentalis* females held at 7°C or 10°C at 75% or 95% RH were recorded. The presence of domatia was found to increase *G. occidentalis* reproduction at low humidity. However, variable photoperiods and egg consumption did not influence the fecundity of *G. occidentalis*. This mite does not consume pollen, and its oviposition may have been slowed by cool weather and limited use of pollen as food (Yanar and Marjorie, 2015; Schmidt-Jeffris and Beers, 2017).

Galendromus occidentalis is a specialist predator on spider mites of the genus *Tetranychus*; it is attracted to volatiles produced by mites while feeding and is capable of navigating the copious amounts of webbing that *Tetranychus* spp. produce. This places *G. occidentalis* in the 'Type II' specialist category, indicating that it is loosely specialized on tetranychids, but is commonly associated with web-producing *Tetranychus* spp. This predatory mite is capable of controlling the most abundant spider mite pest, such as *T. urticae*, *Tetranychus mcdanieli* McGregor, and *Tetranychus pacificus*. Although its preferred prey species are *Tetranychus* spp., it can consume *P. ulmi* and *Aculus schlechtendali* (Schmidt-Jeffris and Beers, 2015).

10.3.1.1.1.15 *Phytoseius finitimus* Ribaga *Phytoseius finitimus* is a generalist phytoseiid mite (Type III, subtype III-a) mainly recorded in the Mediterranean region on a variety of both cultivated and noncultivated plants, such as grapevine, hazelnut, citrus, elm, walnut, mulberry, peach, plum, pomegranate, cotton, fig, eggplant, and lantana (Pappas et al., 2013). *Phytoseius finitimus* can feed and reproduce on three major greenhouse pests, the two-spotted spider mite, the greenhouse whitefly, and the western flower thrips. It is a natural enemy of tetranychid and eriophyid mites, but it also feeds on pollen.

The attack rate (a) divided by the handling time (T_h) (a/T_h) value indicates that *P. finitimus* is the most effective against eriophyoids prey. The calculated maximum number of consumed prey per predator per day for *Aceria ficus* and *Rhyncaphytoptus ficifoliae* is 200 and 66.67 individuals, respectively (Pappas et al., 2013; Abou-Awad et al., 2018). Mobile stages of *A. ficus* and *R. ficifoliae* were offered as prey and the results showed that *A. ficus* ranked as the most suitable prey followed by the fig leaf mite. Females consume more spider mite eggs and larvae, as well as whitefly crawlers than thrips larvae, but they lay eggs when feeding on all prey mixed. When provided alone, cattail pollen sustained the development and reproduction of the predator. Mixing pollen with prey reduces prey consumption, although it increases the predator's egg production. Pollen grains of *Zea mays* (corn) are a better diet for the predator than castor bean in terms of higher population growth potential (Pappas et al., 2013; Abou-Awad et al., 2018).

10.3.1.1.1.16 *Euseius scutalis* (Athias-Henriot) *Euseius scutalis* was first described in Algeria as *Typhlodromus scutalis* in 1958. *Euseius scutalis* was reported to be a common phytoseiid mite in Middle East countries and North Africa on a variety of host plants. It is described (McMurtry et al., 2013) as a specialized pollen feeder/generalist predator (Type 1V predator) able to feed on a wide range of foods, but the reproductive potential is highest on pollen (Momen and Abdel-Khalek, 2008).

Previous studies have shown that *E. scutalis* is not only a predator of spider mites but also feeds on eggs and immature stages of whiteflies, thrips larvae, eriophyid mites, and scale insects (Momen and Abdel-Khalek, 2008). Biological aspects and life-table parameters of this predator consuming eggs of *Insulaspis pallidula* (Green) and *Phoenicococcus marlatti* Cockerell (Diaspididae, Phoenicococcidae) as well as pollen grains of date palm have been determined under laboratory conditions. The life-table parameters are the most favorable feeding on date palm pollen compared with eggs of *I. pallidula* and *P. marlatti* (Abou-Elella et al., 2013). Having a mixed diet of pollen, broad mite and whitefly is possibly better for *E. scutalis* population development, but it also might reduce broad mite predation due to satiation. Other food sources for the predator are castor bean pollen, eggs, and nymphs of *E. orientalis*, moving stages of *A. ficus* and *R. ficifoliae* and immature stages of *Icerya aegyptiaca* (Douglas). Development is faster and reproduction is higher when *E. scutalis* feeds on castor bean pollen, while the developmental time increases and reproduction decreases when it feeds on *R. ficifoliae*. The number of *A. ficus* consumed is significantly higher compared with *R. ficifoliae*. Life-table studies show that a diet of castor bean pollen provides the highest intrinsic rate of natural increase and greatest total fecundity, which results in a high net reproductive rate. The same results are acquired when the predator feeds on *A. ficus* and *R. ficifoliae*. Together, the eggs of *E. orientalis* are the least favorable source of food among all test foods (Momen and Abdel-Khalek, 2008).

10.3.1.1.1.17 *Euseius finlandicus* (Oudemans) *Euseius finlandicus* is one of the most important predators of phytophagous mites worldwide, especially in orchards. McMurtry et al. (2013) proposed four main types (Types I–IV), based primarily on differences in food utilization in which Type IV is a specialized pollen feeder/generalist predator represented by *Euseius* species only. *Euseius finlandicus* is widespread on deciduous fruit trees and one of the few *Euseius* species occurring in temperate climates on apple, peach, grape, and cherry (Abou-Elella et al., 2014).

Larvae of *E. finlandicus* have to feed to reach the protonymphal stage. *Euseius finlandicus* has a mean developmental time of 5.6 days when it feeds on San Jose scale, *Quadraspidiotus perniciosus* (Comstock) (Abdallah et al., 2001). Total egg production is highest (30.53 eggs) at 25°C, whereas daily egg production is highest (1.80 eggs) at 30°C, when feeding on *T. urticae* along with birch pollen. Females and males require 95.24 and 94.34 degree-days, respectively, to become an adult, and it has the highest intrinsic rate of increase at 30°C, followed by 25°C (Kasap, 2009).

Euseius finlandicus can feed on tetranychid, eriophyid, tyroglyphid, and tarsonemid mites, pollen, fungal spores and hyphae, eggs and larvae of insects, honeydew and plant liquids (Abdallah et al., 2001). It utilizes herbivorous mites, such as *P. ulmi*, *Eotetranychus tiliarium* (Hermann), *Eotetranychus carpini* (Oudemans), *T. urticae*, *A. schlechtendali*, *Aceria olivi* (Zaher and Abou-Awad), *Aceria dioscoridis* (Soliman and Abou-Awad) and *Cisaberoptus kenyae* Keifer (Broufas and Koveos, 2000; Abou-Elella et al., 2014).

The data suggest that pollen can be useful in mass rearing of *E. finlandicus*. Cherry, peach, apricot, walnut, and poppy pollen grains have higher nutritional value for *E. finlandicus* than apple and pear pollen (Broufas and Koveos, 2000). Furthermore, short developmental time and high values of population parameters achieved by *E. finlandicus* feeding on juvenile stages of *Schizotetranychus garmani* Pritchard & Baker indicate great suitability of the prey as food (Puchalska and Kozak, 2016). The eriophyid mites, *A. olivi*, *A. dioscoridis*, and *Cisaberoptus kenyae* Keifer, as prey cause the shortest developmental time and highest oviposition rate for *E. finlandicus* (Abou-Elella et al., 2014).

10.3.1.1.1.18 *Euseius (Amblyseius) victoriensis* (Womersley) *Euseius victoriensis* along with some other phytoseiid mites can suppress tenuipalpid (*Brevipalpus* spp.) and eriophyid (*Colomerus vitis* (Pagenstecher), *Calepitrimerus vitis*) mites in the Riverland district of South Australia (Taylor and James, 1993). *Euseius victoriensis* controls *T. urticae*, *P. oleivora*, and grape leaf blister mite, *C. vitis*. This predator emigrates from deciduous host plants during autumn, and overwinters as nonreproductive individuals (but lacking diapause) on evergreen hosts. Recolonization occurs in spring but may be slow (James, 2000).

10.3.1.1.1.19 *Typhlodromus doreenae* Schieha *Typhlodromus doreenae* can suppress tenuipalpid (*Brevipalpus* spp.) and eriophyid (*Colomerus vitis*, *Calepitrimerus vitis*) mites in the Riverland district of South Australia. They are important components of a sustainable pest management system for grapevine mites, based on the use of predatory mites (Taylor and James, 1993). *Typhlodromus doreenae* is an effective sulfur tolerant predator of bunch mite, *Brevipalpus* spp., and also has been reported as a predator of *Bryobia* mite, *Bryobia rubrioculus* (James and Whitney, 1993). Reproductive females of *T. doreenae* held at 20°C and 10 hours photophase for 1 month failed to oviposit when exposed to summer-like (25°C/15 hours) conditions. Overwintering females are nonreproductive, inactive, aggregative, dorsoventrally flattened, pale colored, and show shelter-seeking behavior that indicates the presence of reproductive diapause. At least some individuals of *T. doreenae* overwinter on the dormant vine (James, 2000). The upper temperature threshold for egg to adult is estimated to be 36°C. No egg to adult survival occurs at 37.5°C (Taylor and James, 1993).

Populations of *T. dareenae* can increase by using pollen of cattail in the presence of mold mites, *T. putrescentiae* and the fungus *Rhizopus stolonifer* (Ehrenb.), while *T. doreenae* dies when supplied with the fungus alone (James and Whitney, 1993).

10.3.1.1.1.20 Typhlodromus pyri Scheuten *Typhlodromus pyri* (Figure 10.5d) is one of the most important biocontrol agents that commonly occur in vineyards and on apple trees in many countries of Europe and North America (Praslička et al., 2011), and it is classified as Type III lifestyle generalist predator. *Typhlodromus pyri* is more active at average spring temperatures (ca. 15°C). Intrinsic rate of increase (*r*) becomes positive from 15°C to 30°C, indicating its population growth potential within this range of temperatures. The net reproductive rate and intrinsic rate of increase are the highest at 25°C (Gadino and Walton, 2012). *Typhlodromus pyri* overwinters as adult females in rough structures, cracks, and crevices of twigs.

This species plays an important role in biological control of several species of Eriophyidae, Tetranychidae, and Tarsonemidae. A single female of *T. pyri* is able to consume about 230 individuals of eriophyoid mites daily and about 18,000 individuals during its whole life (Praslička et al., 2011). *Typhlodromus pyri* has been reported as the most important predator of *P. ulmi*, *Calepitrimerus vitis*, and *T. urticae* (Gadino and Walton, 2012). It can feed on pollen, allowing it to remain in the field even in the absence of prey (McMurtry et al., 2013). The data suggest that rapeseed pollen can be completely unsuitable as alternative food for *T. pyri*. Short developmental time and high values of population parameters have been reported for *T. pyri* was fed with the larvae and protonymphs of *Schizotetranychus schizopus* (Puchalska and Kozak, 2016).

10.3.1.2 Insects

Insects in the orders Coleoptera, Thysanoptera, Heteroptera, Diptera, and Neuroptera feed on mites that are pests in agricultural crops. These predatory insects can be used in two ways in IMM programs: conservation and augmentation. Predatory insects by themselves often are less effective than phytoseiid predators because most of them are generalists (Hoy, 2011).

Several predatory insects are commercially reared and sold for augmentative releases, but they are relatively more expensive than phytoseiids, and research-based guidelines on their release rates and timing may be lacking for specific crops or environments. Cannibalism by these insects makes it difficult and expensive to produce them in large quantities. Highly effective artificial diets are rare and most predatory insects must be reared on live prey, which can be expensive to produce. They are larger than phytoseiids and thus can consume more prey per unit time (higher functional response). Insect predators also have a greater longevity and higher fecundity than phytoseiids and have a potential for a greater numerical response (Hoy, 2011).

10.3.1.2.1 Family Coccinellidae

Coccinellids (ladybirds, ladybugs, lady beetles) are considered beneficial predators of aphids, scale insects, and mealybugs, although a few are plant pests. These obligate predatory beetles to also feed on spider mites. The tribe Stethorini includes the genera *Stethorus* and *Parastethorus* and consists of about 90 described species that are black and known as 'spider mite destroyers'. They are known to be voracious predators with all the motile life stages feeding on all prey stages. They have high host-finding ability, high dispersal potential and long-living adults (Perumalsamy et al., 2010).

Stethorus species are relatively small and remarkably well adapted to live and search for prey in the habitats of plant-feeding mites. Therefore, the addition of a winged predator in biocontrol practices, which could move rapidly to spider mite 'hotspots', would be useful. Some coccinellids, such as *Stethorus punctillum* (Weise), *S. gilvifrons* (Mulsant), and *S. punctum picipes* Casey, have proved to be effective natural enemies of the phytophagous mite species, particularly *Tetranychus piercei* McGregor, *P. citri*, *P. ulmi*, and *T. urticae* (Matter et al., 2011).

Very little is known in relation to rearing *Stethorus* spp. on alternative foods. Several reports suggest that *Stethorus* spp. may feed on nontetranychid mite prey, such as predatory mites, aphids, honeydew, and nectar. *Stethorus punctum* and 17 other coccinellid beetles were reared on a diet containing cane sugar, honey, agar, and royal jelly plus their natural prey in powder form, as well as a modified wheat germ diet plus honey. This diet greatly increases adult survival over 2 weeks compared to sugar water alone. However, none of diets are adequate for reproduction of *S. punctum* (Ebrahimifar et al., 2020).

10.3.1.2.1.1 *Stethorus punctillum* (Weise) *Stethorus punctillum* (Figure 10.5j) is a Palaearctic species, which was first found in North America in the 1940s, and is an important and commercially available predator that suppresses tetranychid mite populations. Commercial mass-rearing methods were developed in the late 1990s (Riddick and Wu, 2011). Examples of pests other than *T. urticae* attacked by this predator are *P. ulmi*, *Oligonychus ununguis* (Jacobi), *Oligonychus ilicis* (McGregor), and *O. afrasiaticus.*

There is difference in adult *S. punctillum* longevity associated with different alternative foods they feed on, in which the longest period recorded is 12.70 days when the predator is fed on a mixture of honey droplets, pollen grains, and royal jelly. The shortest period recorded is 3.75 days when the predator is starved and 3.90 days when it feeds on the aphids (Bakr et al., 2009). The entire life cycle takes 2 weeks at 26°C, and the optimum temperature for rearing *S. punctillum* ranges from 23°C to 25°C. In the case of immature stages, food consumption at the lowest test temperature (15°C) is significantly higher than that at higher temperatures (25°C and 35°C). On the contrary, positive correlation between food consumption and temperature is found in the case of adult predators. A high positive correlation between food consumption and prey density is determined in the case of the 4th instar larvae of the predator, followed by adult predator, while younger instars do not show any reasonable increase with increasing prey density. These results confirm that larval and adult stages of *S. punctillum* exhibit 'Type II' functional response. The 4th instar larvae and adult predators are the most preferable stages in winter and summer crops to control *T. urticae*, respectively (Matter et al., 2011). Second larval instars consume significantly more spider mite larvae in comparison with nymphs. In contrast, third larval instars indicate a strong preference for spider mite eggs. Significantly fewer *T. urticae* larvae are consumed by the fourth larval instars of the predator in comparison to the three other mite stages. Finally, adult predators consume significantly more mite eggs than the other stages offered (Ragkou et al., 2004).

10.3.1.2.1.2 *Stethorus gilvifrons* (Mulsant) *Stethorus gilvifrons* is native to the Mediterranean region and is common in the fields of sugarcane, date palm, and castor bean, and it is a good candidate for biocontrol of numerous spider mites (Imani et al., 2009). Larvae and adults of *S. gilvifrons* feed on different developmental stages of tetranychid mites by sucking their body fluid. However, they prefer the earlier developmental stages, mostly eggs. This mite is a major predator of *O. coffeae* and is often observed feeding on *T. turkestani*, and *E. orientalis* on different host crops, and the closely related Tenuipalpididae.

The intrinsic rate of increase (r) and the net reproductive rate (R_0) of *S. gilvifrons* linearly increase with increasing temperatures from 15°C to 35°C. The highest values of r (0.240 day^{-1}) and R_0 (59.27 females/female) and the lowest mean generation time (T) (17.01 days) and doubling time (DT) (2.88 days) are recorded at 35°C. The optimum temperature for *S. gilvifrons* is 25°C. This predator has high prey-finding ability, high dispersal potential, and the adults are long lived (Taghizadeh et al., 2008a). Although the studies indicated a significant decrease of development time with increasing temperature, the development of *S. gilvifrons* stops at 40°C. In addition, the thermal constant for total immature stage of *S. gilvifrons* is 222.72 degree-day (DD) (Taghizadeh et al., 2008b).

The effects of some diets, including *T. turkestani* eggs, *E. kuehniella* eggs plus date palm pollen, *E. kuehniella* eggs plus maize pollen, and *E. kuehniella* eggs plus bee pollen, have been evaluated on efficacy of *S. gilvifrons*. The results indicated that developmental time was longer and fecundity was lower on all tested factitious diets compared with natural prey, *T. turkestani* eggs. *Stethorus gilvifrons* feeds on *E. kuehniella* eggs plus date palm pollen, which has a higher R_0, r, and lower T than those fed on other factitious diets. Although *T. turkestani* eggs are more nutritious than the factitious diets, the combination of *E. kuehniella* eggs plus date palm pollen holds promise as an alternative food for the mass production of *S. gilvifrons* (Ebrahimifar et al., 2020).

10.3.1.2.1.3 *Stethorus tridens* Gordon *Stethorus tridens* was the most common predator found associated with pests on solanaceous plants during surveys in Brazil. *Stethorus tridens* is a candidate natural enemy of *Tetranychus evansi*, *T. bastosi*, and *T. cinnabarinus* colonies on solanaceous plants. The mean life cycle of *S. tridens* fed on *T. bastosi* infesting physic nut (*Jatropha curcas* L.) was 12.35 days, and the specific fertility came to a maximum of 2.49 eggs; net reproductive rate: 53.05 individuals, and intrinsic rate of increase: 0.23 day^{-1} (Costa et al., 2020). *Stethorus tridens* successfully developed to adulthood between 20°C and 30°C but failed at 10°C, 15°C, and 35°C when fed on *T. evansi*. The optimum and maximum temperatures for development are around 29°C–31°C and 32.9°C, respectively. Oviposition of *S. tridens* is positively correlated with prey consumption and lower threshold prey consumption for *S. tridens* egg laying is 16.3 *T. evansi* mites per day (Britto et al., 2009).

10.3.1.2.1.4 *Stethorus punctum* (LeConte) *Stethorus punctum* is native to North America and was first identified in 1852. This predator species is known generally as the spider mite destroyer, especially *P. ulmi*, and *T. urticae*. It overwinters as an adult under fruit trees. A negative correlation was observed between development time and increasing temperature. No development occurred at 10°C and 40°C. Its mortality percentage was 23.32%, and 89.98% at 25°C, and 35°C, respectively. Among all tested temperatures, only two temperatures (25°C and 30°C) were suitable for *S. punctum* under laboratory condition in terms of mass rearing (Ashraf et al., 2014).

10.3.1.2.1.5 *Stethorus punctum picipes* Casey *Stethorus punctum picipes* is one of the important spider mite predators in conservation biological control (CBC) programs currently being developed for several crops, such as grapes and hops, in the northwestern United States. Various plant-inhabiting mites, including species in the families Stigmaeidae, Eriophyidae, Phytoseiidae, and Tetranychidae, are found in guts of *S. punctum picipes*, indicating that *S. punctum picipes* adults feed on alternative plant-inhabiting mites, in particular, when tetranychids are scarce (Kishimoto et al., 2011).

10.3.1.2.1.6 *Stethorus nigripes* Kapur The Australian predatory ladybird *S. nigripes* is of renewed interest for biological control of spider mites. Adult beetles are small, and move about quite actively, crawling in large numbers along the central rib on the underside of the cornleaf, mainly within the mites' webbing where they feed on mite eggs (preferred), nymphs, and adults. The larvae probe directly through the webbing in search of mites, feed, and they pupate nearby. Therefore, all life history stages of *S. nigripes* are concentrated along the central rib on the underside of the leaf. *Stethorus nigripes* prey up on *Oligonychus pratensis* (Banks), a pest of corn, sorghum, and wheat (Yoder et al., 2003).

10.3.1.2.1.7 *Stethorus aptus* Kapur The ladybird beetle, *S. aptus*, has been reported as a predator on *P. ulmi*, *P. citri*, and *O. coffeae*. The incubation period of the egg ranges from 8 to 10 days and larval developmental period varies from 2 to 4 days. The total development period from egg to adult is completed in 28–30 days. The longevity of the female is more than the male and a mature female lays minimum 4 and maximum 12 eggs per day (Barua et al., 2012). Adults of *S. aptus* consume within a range of 48–56 adults and 82–90 larvae of *O. coffeae*, whereas the 3rd and 4th instar larvae of *S. aptus* consume 20–26 adults and 50–60 larvae of *O. coffeae* per day, respectively (Mehendiran and Khursheed, 2018).

10.3.1.2.1.8 *Stethorus japonicus* Kamiya *Stethorus japonicus* is a native ladybird beetle with a wide distribution in Japan, and it is common in fields of apple, citrus, tea, pear, hydrangea, and kudzu vine. It is considered to be an important predator of numerous spider mites, such as *T. urticae* and *T. kanzawai*. Female and male immature stages of *S. japonicus* consume the same number of *T. urticae* eggs, and adult females showed a Type II functional response to prey density, regardless of the tested temperatures. The lower temperature threshold for development from egg to egg-laying adult is 13°C and the thermal constant is calculated to be 238.7 DD, when this ladybird feeds on a diet of eggs of *T. urticae*, as well as the maximum value of r is 0.241 day^{-1} at 30°C. Reproductive diapause has been induced at photoperiods with light phases shorter than 13 hours at 18°C (Mori et al., 2005).

10.3.1.2.1.9 *Stethorus bifidus* Kapur The endemic *S. bifidus* is found throughout New Zealand and is a specialist predator of mites. *Stethorus bifidus* is a predator of gorse spider mite, *Tetranychus lintearius* (Dufour) (Peterson et al., 2000).

10.3.1.2.1.10 *Stethorus vagans* (Blackburn) *Stethorus vagans* is native to Australia, and has been observed in close association with *T. urticae* in various agricultural and horticultural crops, and ornamental plant systems. This species is capable of detecting and attacking populations of *T. urticae* at very low density, distributed in small and widely scattered patches. There is a strong positive correlation between the rate of development and temperature. Eggs do not develop at 10°C; however, they survive a long period of exposure to this temperature, but when they are subsequently exposed to ≥15°C, more of them hatch. Eggs appeared to develop normally at 35°C; however, larvae die before or immediately after emergence (Khan and Spooner-Hart, 2017).

10.3.1.2.2 Family Thripidae

Thrips (fringed-winged insects) can be predaceous, mycophagous, phytophagous, pollinophagous, or generalist feeders. They contain species that prey on spider mites or other small arthropods, such as thrips, coccids, and whiteflies. Heteropteran predators, with a relatively long developmental time, are limited as predators of rapidly reproducing spider mites, which may develop from egg to adult in less than a week. The family Thripidae is the most speciose family of thrips, with over 290 genera representing just over 2,000 species. Thripidae have two larval and two pupal instars before emerging as adults, and pupation usually occurs at the ground level. The total life cycle is completed in as little as 2 weeks, and adults are relatively long lived. They are considered to be among the more derived of thrips, having evolved many traits key to specializing as cryptophilous phytovores, living in the narrow spaces at the bases of leaves and within flowers.

The genus *Scolothrips* are well-known predators of mites and under the name of 'six-spotted thrips' are used as biocontrol agents. All species of *Scolothrips* appear to be specialized predators of tetranychid mites. Several *Scolothrips* species have been shown to provide effective mite control, especially *Scolothrips indicus* Priesner, *S. takahashi* Priesner, *S. sexmaculatus* (Pergande), and *S. longicornis* Priesner (Pakyari and McNeill, 2020).

10.3.1.2.2.1 *Scolothrips longicornis* Priesner *Scolothrips longicornis* has been reported from India, Iran, North America, and many European countries. Populations of *S. longicornis* are able to develop at a broad range of temperatures and are well adapted to high temperatures in the Mediterranean region. It is considered as an active biocontrol agent of *T. urticae* on bean, cucumber, and eggplant. *Scolothrips longicornis* sometimes has a cannibalism behavior, and it can consume its own first and second instar larvae; however, when *T. urticae* individuals are abundant as an extraguild prey, the predator does not feed anymore on its own larval stages. Therefore, it is concluded that cannibalism is very limited in this species and the lack of the main prey can intensify this behavior (Farazmand et al., 2014).

The consumption rate of this predator in their lifetime is about 300 eggs of *T. urticae* (Pakyari and McNeill, 2020). The net fecundity rate varies from 80.2 eggs/female (at 30°C) to 10.04 eggs/female (at 37°C), and the net fertility rate was the highest at 30°C (69.7 eggs/female) and the lowest at 37°C (5.2 eggs/female). The daily number of eggs laid per female ranges from 0.8 to 3.93 eggs, with a minimum at 15°C and a maximum at 35°C (Pakyari and Enkegaard, 2012). A 12:12 hour (L:D) photoperiod is the most favorable for reproduction and development of *S. longicornis* feeds on *T. urticae*. This photoperiod is ideal to maximize production for mass rearing of this predator (Pakyari and McNeill, 2020).

It is also a biocontrol agent of *Schizotetranychus smirnovi* Wainstein (Tetranychidae), a key pest of almond trees in southwestern Iran. The predator exhibited a Type III functional response when it was offered *S. smirnovi* protonymphs. Switching behavior was observed in S. *longicornis* in which the predator switches from one stage when it becomes rare to another more abundant stage. *Scolothrips longicornis* preys up on all stages of *S. smirnovi* but prefers the larvae and protonymphs (Heidarian et al., 2012). This species is one of the natural enemies associated with colonies of *P. citri*, *E. orientalis*, and *T. cinnabarinus*. Studies show that the functional response of this predator to different densities of eggs and nymphs of *T. urticae* is Type II and per capita searching efficiency decreases significantly with increasing predator density (Farazmand et al., 2013).

10.3.1.2.2.2 Scolothrips takahashii Priesner *Scolothrips takahashii* is a native beneficial thrips in China and some other countries. It is common in fields of cotton, pear, and apple, and is considered to be an important predator and a good candidate for biological control of numerous spider mites. Analysis showed that temperature has significant effects on the predation capacity of adult thrips over the range of 20°C–35°C. Attack rate in females linearly increases with increasing temperature while in males, it is independent of temperature (Ding-Xu et al., 2007). At 25°C, the predation capacity of *S. takahashii* is higher than that of the phytoseiid mite *P. persimilis*, although the intrinsic rate of increase (*r*) of *S. takahashii* is lower than that of *P. persimilis* (Yanagita et al., 2014).

The predatory thrips *S. takahashii* can be used as a biocontrol agent of *T. urticae*. In addition, it has a strong preference for *P. citri* eggs, indicating that *S. takahashii* can suppress *P. citri* by predation on their initial stage (Liu et al., 2020). Because *S. takahashii* can fly and hence has higher mobile ability to immigrate into strawberry fields from the surrounding habitat, it can be employed in biological control programs by enhancing and attracting natural populations. Therefore, it can be a promising natural enemy of spider mites because of its great voracity, high reproduction, and wide distribution in the orchards, but it is not yet commercially available (Yanagita et al., 2014; Liu et al., 2020).

10.3.1.2.2.3 Scolothrips sexmaculatus (Pergande) Six-spotted thrips, *S. sexmaculatus* (Figure 10.5i), is one of the oldest species of thrips known from North America. It feeds on multiple species of spider mites, does not discriminate among life stages of prey, consumes large numbers of prey, has thigmotactic immatures with a high searching capacity within webbing and can be cannibalistic (adults and 2nd instar larvae are cannibalistic) (Haviland et al., 2020).

Adults live 29–67 days depending on temperature with a generation time of 18–51 days. Reproduction is sexual and facultatively arrhenotokous. It has a high reproductive rate and can persist and increase in numbers at temperatures between 23.9°C and 40.6°C. At 20°C, the intrinsic rate of increase is 0.08 day^{-1} because of the long developmental time, while at 30°C, the *r* is 0.26 day^{-1} because of the short developmental time and high age-specific fecundity. The optimal temperature for S. *sexmaculatus* is around 30°C. At 30°C, a population of six-spotted thrips can double its numbers in 27 days compared with 8.7 days at 20°C (Coville and Allen, 1977).

Consumption rate is one of the main parameters to evaluate *S. sexmaculatus* potential as a biocontrol agent of spider mites, which is usually influenced by temperature. The female thrips consume 7–50 eggs, 38 immatures or 16 adults of spider mites per day, and they lay 90 to 166 eggs over their lifespan (Haviland, 2017). In comparing the predation potential of female and male predators, it has been shown that females consume 56.5 eggs, 36.25 immatures, and 19.25 adult spider mites per day at 30°C, while the males consume fewer numbers of the prey. The daily consumption rate of the predator is reduced at 25°C (Salahi et al., 2012).

10.3.1.2.3 Family Anthocoridae

Anthocoridae have soft, elongated oval, and flat bodies that are found in a variety of habitats. The scientific name is a combination of the Greek words *anthos* 'flower' and *koris* 'bug'. Eggs are laid in plant material and hatch in approximately 3–5 days. Nymphs require at least 20 days to progress through five instars. Adults live for approximately 35 days. They can feed on plant material, but they mostly feed on other small soft-bodied arthropods. Anthocorids are often predaceous, both as nymphs and adults. The genera *Anthocoris*, *Orius*, and *Tetraphleps* are predators of small arthropods, although some feed on plants. *Orius* species (often called minute pirate bugs) are generalist predators, feeding on a variety of small arthropods. Adult *Orius* are oval and black with white patches on the wings. Their characteristics make them very efficient predators, because they frequently inhabit the same sites as their prey and they can survive for some time without the presence of prey and they are relatively easy to mass produce. *Orius* species are commercially produced (Hoy, 2011).

10.3.1.2.3.1 *Orius albidipennis* (Reuter) The anthocorid bug, *Orius albidipennis*, is a generalist predator that feeds on *Bemisia tabaci*, *Tetranychus turkestani*, and *Tetranychus atlanticus* McGregor in greenhouse crops. However, generalist predators may exhibit prey preference in the presence of mixed prey species; so, the use of *O. albidipennis* for biological control might be complicated in a situation where both *B. tabaci* and *T. turkestani* are found on the same plant (Banihashemi et al., 2017). It has also been reported that *O. albidipennis* is an effective agent to control *T. urticae* under greenhouse conditions (Arnaouty et al., 2018).

Some characteristics of the species, including its ability to tolerate high temperatures and lack of photoperiod-induced diapauses, make the predator an ideal candidate for mass rearing and augmentative releases against many field and greenhouse pests in subtropical and tropical regions (Rajabpour et al., 2018). The shortest duration of development was achieved when the predator was fed with *Aphis fabae* Scopoli and *Aphis craccivora* Koch. *Aphis fabae* is the most suitable food source because of the highest R_0, r, and finite rate of increase (λ), followed by *A. craccivora*, *Myzus persicae* (Sulzer), and *T. urticae*. These results should contribute to the development of mass-rearing programs for biological control agents (Al-Kherb, 2014). In the current mass-rearing methods for *O. albidipennis* and many other *Orius* bugs, lepidopteran eggs, especially *E. kuehniella* eggs, as well as pollen grains are used as factitious and alternative diets, respectively (Bonte and De Clercq 2010). An artificial diet composed of some natural materials including lamb liver, hen yolk, whey protein, honey, royal jelly, and some specific vitamins have been prepared to rear *O. albidipennis*. The predator's fecundity and survival rate when fed on the mentioned artificial diet is lower than the control (*E. kuehniella* eggs + date palm pollen). Overall, due to lower production costs of the artificial diet, it can be recommended for mass rearing of *O. albidipennis* despite the lower fecundity and survival rate (Rajabpour et al., 2018).

10.3.1.2.3.2 *Orius insidiosus* (Say) The insidious flower bug, *O. insidiosus*, is often released in greenhouses against mites and thrips. *Orius insidiosus* has multiple generations per year. The total developmental time is about 3 weeks, and adults live for approximately 3–4 weeks; nymphs and adults are highly mobile. It is a predator of *T. urticae*, *P. ulmi*, *A. schlechtendali*, and *P. citri* in melons, hops, strawberries, fruit trees, and citrus cropping systems. Furthermore, it feeds on *F. occidentalis* and the eggs of the corn earworm (*Helicoverpa zea* (Boddie)). Under laboratory conditions, *O. insidiosus* can develop from egg to adult in 12.1 days on spidermite prey (Hoy, 2011).

10.3.1.2.3.3 Orius similis Zheng *Orius similis* occurs in southern China where it is an important predator in the agroecosystem. Its adults and nymphs consume a variety of pests, such as aphids, flower thrips, and young larvae and eggs of some lepidopterous insects and spider mites on more than 20 vegetable crops. The most preferred prey species of *O. similis* are flower thrips, aphids, and spider mites. At 28°C, female adults of *O. similis* have the longest oviposition period (21.1 days), the greatest fecundity (40.3 eggs), and the highest potential of population growth ($r = 0.108\,day^{-1}$) when fed on *T. cinnabarinus* (Zhang et al., 2012).

10.3.1.2.3.4 Orius laevigatus (Fieber) *Orius laevigatus* (Figure 10.5h) is a well-known, commercialized, and effective biocontrol agent. It is a predator of *B. tabaci* eggs and *T. urticae*. It has been commercially used for the biological control of the thrips *F. occidentalis* in the Mediterranean regions (Rahman et al., 2020).

Although *Orius* spp. is an effective predator in protected cultivation, it is relatively expensive to mass rear. In commercial mass production systems, *O. laevigatus* is mainly produced on eggs of *E. kuehniella* (Rahman et al., 2020). Bonte and De Clercq (2010) reported that food source and age of the males affect the reproductive potential of their female mates. When both males and females are offered an artificial diet (hen's egg yolk), male age has a significant effect on female oocyte counts. Virgin females that are mated with newly emerged virgin males produce fewer offspring than those mated with 8-day-old virgin males. In contrast, male age does not affect fecundity when the males are fed with *A. kuehniella* eggs. In some cases, this predator has been reared on a mixture of *Artemia* cysts and eggs of *Sitotroga cerealella* (Olivier) (Gelechiidae). Although the *E. kuehniella* eggs are a nutritionally optimal factitious food, their production cost has remained high leading to high market prices. *Orius laevigatus* has been widely used as a successful biocontrol agent in Europe after being reared on eggs of *E. kuehniella* by different commercial insectaries. Moreover, in most commercial insectaries, *O. laevigatus* is still maintained on fresh plants (like green beans) that are used for oviposition and supplementary feeding. The use of plants in production systems for beneficial insects has some major drawbacks. De Puysseleyr et al. (2014) investigated the effect of a plantless rearing system, using wax paper as a walking substrate, water encapsulated in parafilm domes, and an artificial oviposition substrate made of parafilm and moist cotton wool, on the developmental and reproductive fitness of *O. laevigatus*. Females reared without plants have similar predation rates on *F. occidentalis* larvae as their peers maintained on plants.

10.3.1.2.3.5 Orius minutus (Linnaeus) In Korea, Japan, Thailand, North Africa, and the Holarctic Region, *O. minutus* is a dominant predator of aphids, spider mites, and thrips. Although *O. minutus* is widely distributed all over the world, it has not been commercialized. This predator is abundant on various vegetable crops and feeds on eggs, larvae, nymphs, and adults of *T. urticae*, and it can be used as a biocontrol agent against *T. urticae* (Rahman et al., 2020).

10.3.1.2.3.6 Orius strigicollis (Poppius) *Orius strigicollis*, a predaceous flower bug, is a biocontrol agent of *T. urticae* and *T. kanzawai*. The major prey of these flower bugs includes mites, aphids, thrips, whiteflies, and lepidopteran eggs and larvae. *Orius strigicollis* has been commonly used as a biocontrol agent to suppress thrips by virtue of its high search ability and consumption rate. It can feed on eggs of *Cadra cautella* Walker, as an alternative prey (Tuan et al., 2016).

10.3.1.2.4 Family Miridae

The family Miridae is a large and diverse insect family that includes plant bugs, leaf bugs, and grass bugs, with over 10,000 known species. Members of the Miridae family feed on almost all aerial plant tissues and plant products, such as pollen and nectar. Most mirids overwinter as eggs that are laid in soft plant tissue. Nymphs of most species emerge in the spring when there is sufficient new tender foliage on which to feed. Each of five nymphal instars is lasted 5–7 days. Most mirids are studied because they are important agricultural pests (such as *Lygus* bug), and all mirids were once thought to be phytophagous, but we now know that some species are predators. In fact, many phytophagous mirids, perhaps 25%, are facultative predators. In horticultural crops, the genera *Campy loneuropsis*, *Dicyphus*, *Engytatus*, *Macrolophus*, *Nesidiocoris*, and *Tupiocoris* of the subfamily Bryocorinae have been reported to consume a wide range of prey, including aphids, leafminers, lepidopterans, whiteflies, psyllids, spider mites, and thrips. However, to date, only three species from this subfamily, namely, *Dicyphus Hesperus* Knight, *Macrolophus pygmaeus* Rambur, and *Nesidiocoris tenuis* (Reuter) are commercially produced for pest management programs and mainly in tomato crops (Pérez-Hedo et al., 2020).

10.3.1.2.4.1 Macrolophus caliginosus Wagner *Macrolophus caliginosus* is being used in European greenhouses to control multiple pests, including aphids, whiteflies, and mites. This mirid is shown to respond to herbivore-induced plant volatiles produced by plants infested with the two-spotted spider mite or the green peach aphid, *Myzus persicae*. Females usually live for 29 days, producing 0.7 eggs per day, and their development requires 27 days of feeding on two-spotted spider mite as prey; however, life-table data indicate that a spider mite diet is inferior compared to a diet of insect prey (Hoy, 2011).

10.3.1.2.4.2 Macrolophus pygmaeus Rambur *Macrolophus pygmaeus* (Figure 10.5l) is a predatory mirid originating from Europe and the Mediterranean region, which is widely distributed over the Palaearctic region from Finland to Algeria and from the Faroe Islands to Turkmenistan and Tajikistan. *Macrolophus pygmaeus* was first reported in New Zealand from the Auckland botanic gardens in 2007 (Kean et al., 2019). This species is a zoophytophagous bug, i.e., it can feed not only on animals but also on plants; that is why *Macrolophus* bugs can cause damage to cultivated plants. However, their beneficial effect on agrocenoses is much stronger than this minor damage (Pazyuk and Reznik, 2016).

Macrolophus pygmaeus is successfully used for biological control of thrips, aphids, spider mites, and some other pests in greenhouses and open fields, and it can develop successfully on *Myzus persicae* and *Macrosiphum euphorbiae* (Thomas) (Aphididae), which are important pests of eggplant and tomato. It primarily feeds on whiteflies, such as greenhouse whitefly (*Trialeurodes vaporariorum*), a damaging pest of greenhouse tomato crops. This predator has high potential of prey consumption in which adults may consume up to 40 whitefly eggs per day. Predation on whiteflies and other insect pests makes this predator an attractive biocontrol agent for greenhouse tomato growers (Kean et al., 2019). Mass rearing of this predator is usually performed using *E. kuehniella* eggs.

10.3.1.2.4.3 Nesidiocoris tenuis (Reuter) The tomato mirid bug, *N. tenuis* (Figure 10.5k), is a generalist predator, and both adults and nymphs consume a wide range of prey species. It can make a significant contribution to a tomato IPM program by feeding on whiteflies, spider mites, thrips, caterpillars, and leafminers. During the 1990s and early 2000s, most researchers in the Mediterranean region focused on the bug's potential as a biocontrol agent, particularly against tobacco whitefly (*Bemisia tabaci*). However, it is now known that in the absence of insect or mite prey, the predators can cause damage to tomato plants. As a consequence, sometimes this predator and its phytophagous activity have become very controversial.

Nesidiocoris tenuis and *Macrolophus pygmaeus* are commonly used in tomato cropping systems as effective biological agents. The two species are very similar in appearance being of comparable size, shape, and color. Both species feed on the same range of insect and mite prey and can cause similar damage to tomato plants sporadically. However, *N. tenuis* is generally considered to be more robust, more effective as a predator, and more harmful to plants than its relatives. This bug, commonly found on tomatoes, is able to complete development on *B. tabaci*, *F. occidentalis*, tomato leafminer *Tuta absoluta* (Meyrick), and *T. urticae* (Hughes et al., 2009). Nymphal survival and developmental time of this predator has been examined at seven constant temperatures ranging from 14°C to 34°C, and results indicated that egg–adult developmental time decreases sharply with increasing temperature, except at 34°C (17.21 day), for which developmental time is significantly longer than that obtained at 31°C (15.59 day); the lowest and highest percentage of mortality is found at 28°C and 14°C, respectively (Mirhosseini et al., 2018).

A generation is usually completed within a fortnight at 25°C and 60–80 nymphs are produced per female, which gives very rapid population growth. Generation time and reproductive rate decline as the temperature drops, with the lower threshold for development being 12°C–13°C with no indication of ability to diapause. Supercooling points of the adults and nymphs are between 17.6°C and 21.5°C (Hughes et al., 2009). Augmentative release of later instar nymphs of *N. tenuis* against *T. urticae* might assure better mite control. Establishment and colonization of *N. tenuis* can be improved by releasing it onto tomato seedlings prior to transplantation and providing it with alternative food, e.g., eggs of *E. kuehniella*.

10.3.1.2.5 Family Cecidomyiidae

The family Cecidomyiidae (known as gall midges) includes phytophagous, saprophagous, or zoophagous species. These are very small flies and very delicate in appearance. The body and wings have long hairs, which are easily rubbed off. The larvae usually are plant feeders, often producing galls in which they live. A number of species are predators of mites, aphids, scale insects, and other small arthropods (Hoy, 2011).

10.3.1.2.5.1 Feltiella acarisuga (Vallot) *Feltiella acarisuga* (Figure 10.5g) is a spider mite predator with a nearly cosmopolitan distribution except for the Neotropical region. *Feltiella acarisuga* pupae are commercially available from several producers and suppliers of natural enemies. It attacks spider mite, including *Oligonychus* sp., *P. citri*, *T. urticae*, and *T. cinnabarinus* (Hoy, 2011). *Feltiella* larvae prey on all stages of spider mites but generally prefer eggs and young nymphal stages.

Developmental time ranges from 10 days at 27°C to 34 days at 15°C. Developmental time is significantly shorter at 96% RH than at 84% RH at 20°C. There is very poor survival of immatures at 64% RH and none at 36%. Lifespan of adult females decreases with increasing temperature, but temperature has no significant effect on the number of eggs laid. Females deposit 33 eggs during their lifespan of approximately 13 days at 26.7°C and 85% RH. Eggs are deposited on plant tissue infested with tetranychid mites. Upon hatching, the larvae crawl to spider mites and feed on all life stages. The larvae of the predatory gall midge, *F. acarisuga*, diapause as prepupae in tightly woven, brown, silk cells on leaf surfaces. The fully grown larva spins a cocoon on the leaf. As a general rule, this predator requires relative humidities above 50% (Gillespie et al., 2000; Hoy, 2011).

Feltiella acarisuga adults are capable of flying and tracking colonies of *T. urticae*. Larval *F. acarisuga* consume 38, 60, and 87 *T. urticae* eggs as first, second, and third-instar larvae, respectively, under laboratory conditions. The number of spider mites attacked by 3-day-old larvae over 8 hours increases with increasing temperature from 15°C to 27°C. The mean duration of developmental stages of *F. acarisuga* when fed on *T. urticae* eggs is averaged at 13.5 days, and fecundity is attained at 34.9 eggs at 25°C and 60%–75% RH. *Feltiella acarisuga* larvae feed on red spider mites at least five times more than *Phytoseiulus persimilis* (Phytoseiidae) during a day (Refaei and Mohamed, 2013). The larvae of *F. acarisuga* consume an average of 175.4

T. cinnabarinus eggs, and female larvae consume 14% more spider mite eggs (187.6 eggs) than male larvae (165.1 eggs) (Mo and Liu, 2007).

Release of 1,000 individuals of *F. acarisuga* per hectare has been effective in controlling spider mites on tomato, pepper, and cucumber. According to the guidelines of Koppert Biological Systems, the largest producer of entomophages worldwide, the midge *F. acarisuga* should be used in combination with the predatory mite, *P. persimilis*, for more efficient suppression of the spider mite in greenhouses. The two predators are similar in their trophic specialization and consume only spider mites (Razdoburdin and Kozlova, 2019).

10.4 ASSESSMENT OF THE PERFORMANCE AND EFFICACY OF MITE PREDATORS

Before applying a biocontrol agent in integrated management strategies, it is essential to reveal the efficacy of the biocontrol agent. The main methods that are used in the evaluation of predators' effectiveness are life-table parameters, foraging behavior, predation capacity, and multitrophic interactions. The detailed information and methods for evaluation of the mite predators' efficiency and performance have been presented in our previous chapter entitled "Mite **Predators** chapter 11" (Fathipour and Maleknia, 2016).

10.5 THE FUTURE OF BIOLOGICAL CONTROL OF MITE PESTS

Demand for organic and pesticide-free crops is rising worldwide because of increasing public awareness on adverse effects of pesticides, food-borne diseases, and food safety and security. To achieve this goal and to ensure safety in crop production, biocontrol practices should be taken into consideration worldwide. Based on the high demand for safe food, a future high demand for biocontrol agents, especially mite predators, is predicted (Figure 10.6). Therefore, more research and activity on different aspects of biocontrol as an ecofriendly practice is inevitable.

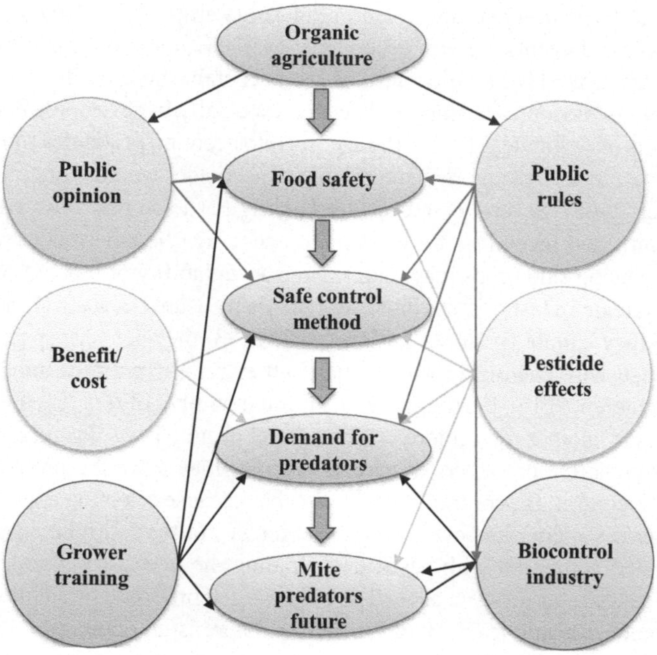

FIGURE 10.6 Factors affecting the future of mite predators.

Commercial biocontrol is an industry where efficient natural enemies are mass reared in biofactories for release in large numbers. This industry includes mass production, shipment, and release of natural enemies. Biocontrol of mite pests is an ecofriendly and economic strategy that decreases pest populations not only in the indoor and small-scale agricultural systems, such as greenhouses, but also in outdoor and large-scale systems, such as orchards, crop fields, and forests. Over recent decades, the focus on crop production has moved from yields to food quality, safety, and sustainability.

In the past decade, biocontrol has been used on a small scale, but demand from consumers and retailers for safe food has changed pest management systems with increased demand and supply of biocontrol agents. However, this high demand for biocontrol agents, especially for use in greenhouses, may cause a shock in market pull and this shock would lead to develop the biocontrol industry by introducing and using new and efficient agents.

Indisputably, increasing public awareness on safe food and organic agriculture in the world will increase the use of biocontrol agents, like mite predators. There are two important challenges in expanding the commercial biocontrol strategies: (1) problems with marketing include benefit–cost ratio and producers not being able to offer production and (2) problems with market demands that are affected by issues like public opinion and consumers' demand. Governments and international organizations have to play an important role to enhance the marketing of biocontrol agents by framing stringent rules to improve the program, thus producing the safe field crops and fruits.

The benefit–cost ratio is an essential tool for appropriate control strategies. This ratio for chemical control will be higher than for biocontrol if we consider the indirect costs of chemical control, like environmental pollution, human health problems, harmful side effects, and risk of pest resistance. Although biocontrol is an ecofriendly strategy that leads to food safety and security, it is still an expensive and time-consuming method initially, but in the long term, it is more effective, self-sustaining, and self-propagating. In high-value crops, such as strawberries, the costs of releasing mite predators can be justified, while in other crops it will only be justifiable if the pest mites become resistant to all of the registered acaricides. Therefore, this may be merely a control option to keep the mite population below the economic injury level. Furthermore, the indirect costs of pesticides can alter the benefit–cost ratio, even in low-value crops (Hoy, 2011).

Commercial biocontrol agents are less acceptable, available, and applicable than chemical pesticides for controlling pests worldwide. Biocontrol is less acceptable because the costs of using synthetic pesticides are lower than biocontrol agents, and people, especially in developing countries, do not consider the indirect costs of pesticides. Taking these costs into account, pesticides should be at least three times more expensive and, as a result, realistic pricing of pesticides would more often lead to choosing biocontrol (Pimentel, 2005). Several countries are levying pesticide taxes, which are partly used to develop IPM programs and incentives for growers to encourage 'low-pesticide' farming. Enactment of strict laws on producing safe crops may increase the acceptability of biocontrol. Biocontrol is less available and the pesticide industry is not interested in this measure because: (1) natural enemies cannot be patented, (2) they cannot be stored for long periods, (3) they act very specifically, (4) they can often not be combined with chemical control, and (5) they require extra training of sales personnel and growers (van Lenteren, 2012). Furthermore, the limited number of retailers to pass bioagents from producers to growers is another reason for poor biocontrol of mite pests. Framing the necessary rules to support and encourage the producers of natural enemies may solve the problems associated with less availability of biocontrol. Biocontrol is less applicable because selecting appropriate agents, using them in crop-producing systems, and evaluating the efficiency of the control measure are quite professional and complicated compared with pesticides. Holding the necessary workshops, technical sessions, and training courses for growers as well as preparing appropriate brochures and agent-specific guidelines may enhance the applicability of biocontrol. There is no doubt that the biological control industry and IPM-based strategies in agricultural production are stronger than the past because of the food safety concerns of the public (Fathipour and Maleknia, 2016).

10.6 CONCLUSIONS

Phytophagous mites cause biotic stress to host plants and adversely affect the marketable production of agricultural crops. Cost-effective, efficient, and safe control measures are needed to limit damage of mite pests and keep their population under economic injury level. Among ecofriendly measures, mite and insect predators have received considerable attention and they play an important role in managing mite population in various agricultural crops. In both natural and agricultural ecosystems, many natural enemies, especially predators, could control herbivore populations. Identification and assessment of the mite predators will help to improve biocontrol programs and select appropriate biocontrol agents to conserve and mass rear them. Although biocontrol guarantees production of safe food, it received less attention because of its lower benefit–cost ratio, and its less availability, acceptability, and applicability compared to other control measures, such as chemical control. The costs of indirect effects of pesticides application including threat of food safety, environmental pollution, human health problems, harmful side effects, and risk of pest resistance are borne by the public. Enforcement of rules will minimize the use of pesticides in agricultural systems and encourage and expand biocontrol strategies in indoor and outdoor cropping systems.

10.7 POINTS TO REMEMBER

i. Phytophagous mites are economically important agricultural pests, which feed on a wide range of host plant species throughout the world.
ii. Extensive use of chemical pesticides resulted in resistance to almost all the available groups of acaricides and elimination of many natural enemies of spider mites; therefore, the development of alternative pest control strategies, such as biological control of mite pests, is necessary.
iii. Biological, technical, and chemical methods must be balanced in order to protect the environment. Biological control provides an environmentally safe, cost-effective, and energy-efficient means of pest control, either alone or as a component of IPM.
iv. Natural enemies of pest mites include predatory mites and insects in which important species belong to the families of Phytoseiidae, Coccinellidae, and Anthocoridae.
v. Predators that are used in biological control programs of pest mites can be divided into two groups: species where very large numbers need to be released per unit area in order to obtain sufficient control (e.g. mites) and species that are released at a low density on very large areas (e.g. coleopterans and heteropterans).
vi. As the second largest group of biological control agents after the hymenopteran, predators of the family Phytoseiidae are commercially produced worldwide, and four species appear on a list of the 25 most used products globally (i.e. *Phytoseiulus persimilis*, *Amblyseius swirskii*, *Neoseiulus cucumeris*, and *N. californicus*).
vii. The predatory insects can be used in two ways: conservation and augmentation. Most predatory insects must be reared on live prey, which can be expensive to produce. Although predatory insects are less effective than phytoseiid predators (because most of them are generalists), they are larger than phytoseiids and therefore can consume more prey per unit time.
viii. Enforcement of rules will minimize the use of pesticides in agricultural systems and encourage and expand biocontrol strategies in IPM programs. There is no doubt that the biological control industry and IPM-based strategies in agricultural production are stronger than the past because of the food safety concerns of the public.

ACKNOWLEDGMENT

The technical support of this research by the Department of Entomology, Tarbiat Modares University, is greatly appreciated.

REFERENCES

Abdallah, A.A., Zhang, Z.Q., Masters, G.J. & Mcneill, S. (2001). *Euseius finlandicus* (Acari: Phytoseiidae) as a potential biocontrol agent against *Tetranychus urticae* (Acari: Tetranychidae): Life history and feeding habits on three different types of food. *Experimental and Applied Acarology*, 25(10–11): 833–847.

Abou-Awad, B.A., El-Sawaf, B.M., Reda, A.S. & Abdel-Khalek, A.A. (2000). Environmental management and biological aspects of the two eriophyoid fig mites *Aceria ficus* (Cotte) and *Rhyncaphytoptus ficifoliae* Keifer in Egypt. *Journal of Pest Science* 73(1): 5–12.

Abou-Awad, B.A., Afia, S.I. & Al-Azzazy, M.M. (2013). Bionomics of the pear bud Mite *Eriophyes pyri* (Pagenstecher) (Acari: Eriophyidae) in Egypt. *Acarines*, 7(1): 31–36.

Abou-Awad, B., Hafez, S.M. & Farahat, B.M. (2014). Bionomics and control of the broad mite *Polyphagotarsonemus latus* (Banks) (Acari: Tarsonemidae). *Archives of Phytopathology and Plant Protection*, 47(5): 631–641.

Abou-Awad, B.A., Abdel-khalek, A.A. & Afia, I. (2018). Life tables, functional and numerical responses of predatory mite *Phytoseius finitimus* (Ribaga) (Acari: Phytoseiidae) to different densities of two eriophyoid mites *Aceria ficus* and *Rhyncaphytoptus ficifoliae*, infesting fig orchards. *Bioscience Research,* 15(4): 3888–3899.

Abou-Elella, G.M., Saber, S.A. & El-Sawi, S.A. (2013). Biological aspects and life tables of the predacious mites, *Typhlodromips swirskii* (Athias-Henriot) and *Euseius scutalis* (Athias-Henriot) feeding on two scale insect species and plant pollen. *Archives of Phytopathology and Plant Protection*, 46(14): 1717–1725.

Abou-Elella, G.M., Hassan, M.F., Nawar, M.S. & Zidana, I.M. (2014). Survival, development and reproduction of *Euseius finlandicus* (Oudemans) (Acari: Phytoseiidae) fed on various kinds of food substances. *Archives of Phytopathology and Plant Protection*, 47(7): 857–868.

Akyazi, R., Sekban, R., Soysal, M., Akyol, D., Colee, J. & Bostan, S.Z. (2020). Ecofriendly control approaches for *Polyphagotarsonemus latus* (Acari: Tarsonemidae) on tea (*Camellia sinensis* L.). *International Journal of Acarology*, 45(1–2): 79–89.

Akyazi, R., Soysal, M. & Altunç, Y.E. (2020). Survival time and size of *Neoseiulus californicus* (Mesostigmata: Phytoseiidae) reared on different prey species and prey life stages. *Journal of Plant Diseases and Protection*, 127: 335–340.

Alatawi, F.J., Mushtaq, H.M.S., Mirza, J.H. & Kamran, M. (2018). Predation efficiency and preference of lab-reared and field-collected populations of predatory mite *Cydnoseius negevi* (Acari: Phytoseiidae) on two mite pest species *Oligonychus afrasiaticus* and *Tetranychus urticae* (Acari: Tetranychidae). *International Journal of Pest Management,* 56(4): 363–369.

Al-Azzazy, M.M. & Alhewairini, S.S. (2020). Effect of temperature and humidity on development, reproduction, and predation rate of *Amblyseius swirskii* (Phytoseiidae) fed on *Phyllocoptruta oleivora* (Eriophyidae) and *Eutetranychus orientalis* (Tetranychidae). *International Journal of Acarology*, 46(5): 304–312.

Al-Azzazy, M.M., Al-Rehiayani, S.M. & Abdel-Baky, N.F. (2018). Life tables of the predatory mite *Neoseiulus cucumeris* (Acari: Phytoseiidae) on two pest mites as prey, *Aculops lycopersici* and *Tetranychus urticae*. *Archives of Phytopathology and Plant Protection*, 51(11–12): 637–648.

Alipour, Z., Fathipour, Y., Farazmand, A. & Khanamani, M. (2019). Resistant rose cultivar affects life table parameters of two-spotted spider mite and its predators *Phytoseiulus persimilis* and *Amblyseius swirskii* (Phytoseiidae). *Systematic & Applied Acarology*, 24(9): 1620–1630.

Al-Kherb, W.A. (2014). Biological characteristics of *Orius albidipennis* (Hemiptera: Anthocoridae) reared on insect and mite preys. *International Journal of Agricultural Research*, 9(2): 110–118.

Ansari-Shiri, H., Fathipour, Y., Hajiqanbar, H., Riahi, E. & Riddick, E.W. (2022). Quality control of the predatory mite *Amblyseius swirskii* during long-term rearing on almond *Prunusamygdalis* pollen. *Arthropod-Plant Interactions*, 16(6): 645–655.

Arnaouty, S.A.E., Kortam, M.N., Afifi, A.I. & Heikal, I.H. (2018). *Orius albidipennis* (Rueter) as an effective biocontrol agent against *Tetranychus urticae* Koch on pepper crops in greenhouse in Egypt. *Egyptian Journal of Biological Pest Control*, 28: 42.

Ashraf, A., Ahmad, S. & Ali, H. (2014). Effect of temperature on immature stages of small black ladybird beetle *Stethorus punctum*, Leconte (Coleoptera: Coccinillidae) and percent mortality. *Entomologia Generalis*, 35(2): 129–136.

Bakr, R.F.A., Genidy, N.A.M., Gesraha, M.A., Farag, N.A. & Elbehery, H.H.A. (2009). Biological study of the coccinellid predator, *Stethoruspunctillum* under different constant temperatures. *Egyptian Academic Journal of Biological Sciences*, 2(1): 1–7.

Banihashemi, A., Seraj, A.A., Yarahmadi, F. & Rajabpour, A. (2017). Effect of host plants on predation, prey preference and switching behaviour of *Oriusalbidipennis* on *Bemisia tabaci* and *Tetranychus turkestani*. *International Journal of Tropical Insect Science*, 37, 176–182.

Barua, A., Babu, A. & Rajkhowa, R.C. (2012). Study on life cycle parameters of *Stethorus aptus* (Kapur) (Coleoptera: Coccinellidae): A new predator of tea red spider mite *Oligonychus coffeae* (Nietner) (Tetranychidae). *International Journal of Scientific Research*, 2(12): 566–567.

Bazgir, F., Shakarami, J. & Jafari, S. (2020). Prey stage preference, mutual interference and switching of *Amblyseius swirskii* (Acari: Phytoseiidae) on *Eotetranychus frosti* (Acari: Tetranychidae) and *Cenopalpus irani* (Acari: Tenuipalpidae). *International Journal of Pest Management*, 66(3): 262–270.

Beard, J.J. (1999). Taxonomy and biological control: *Neoseiulus cucumeris* (Acari: Phytoseiidae), a case study. *Australian Journal of Entomology,* 38(2): 51–59.

Bonte, M. & De Clercq, P. (2010). Influence of male age and diet on reproductive potential of *Orius laevigatus* (Hemiptera: Anthocoridae). *Annals of the Entomological Society of America*, 103(4): 597–602.

Britto, E.P.J., Gondim Jr, M.G.C., Torres, J.B., Fiaboe, K.K.M., Moraes, G.J. & Knapp, J.L. (2009). Predation and reproductive output of the ladybird beetle *Stethorus tridens* preying on tomato red spider mite *Tetranychus evansi*. *BioControl*, 54(3): 363–368.

Britto, E.P.J., Gago, E. & de Moraes, J.G. (2012). How promising is *Lasioseius floridensis* as a control agent of *Polyphagotarsonemus latus*? *Experimental and Applied Acarology*, 56: 221–223.

Broufas, G.D. & Koveos, D.S. (2000). Effect of different pollens on development, survivorship and reproduction of *Euseius finlandicus* (Acari: Phytoseiidae). *Environmental Entomology*, 29(4): 743–749.

Calvo, F.J., Knapp, M., van Houten, Y.M., Hoogerbrugge, H. & Belda, J.E. (2015). *Amblyseius swirskii*: What made this predatory mite such a successful biocontrol agent? *Experimental and Applied Acarology*, 65(4): 419–433.

Carrillo, D., Peña, J.E., Hoy, M.A. & Frank, J.H. (2010). Development and reproduction of *Amblyseius largoensism* (Acari: Phytoseiidae) feeding on pollen, *Raoiella indica* (Acari: Tenuipalpidae), and other microarthropodsinhabiting coconuts in Florida, USA. *Experimental and Applied Acarology*, 52(2): 119–129.

Carrillo, D., de Coss, M.E., Hoy, M.A. & Peña, J.E. (2012). Variability in response of four populations of *Amblyseius largoensis* (Acari: Phytoseiidae) to *Raoiella indica* (Acari: Tenuipalpidae) and *Tetranychus gloveri* (Acari: Tetranychidae) eggs and larvae. *Biological Control*, 60(1): 39–45.

Chora, A., Kropczynska-Linkiewicz, D., Sas, D. & Escudero-Colomar, L.A. (2016). Distribution of *Amblydromalus limonicus* in northeastern Spain and diversity of phytoseiid mites (Acari: Phytoseiidae) in tomato and other vegetable crops after its introduction. *Experimental and Applied Acarology*, 69(4): 465–478.

Costa, J.F., Matos, C.H.C., de Oliveira, C.R.F. & dos Santos, G.A. (2020). Biology and life table of *Stethorus tridens* fed *Tetranychus bastosi* on physic nut. *Bulletin of Insectology*, 73(1): 111–116.

Coville, P.L. & Allen, W.W. (1977). Life table and feeding habits of *Scolothrips sexmaculatus* (Thysanoptera: Thripidae). *Annals of the Entomological Society of America*, 70(1): 11–16.

Dalir, S., Hajiqanbar, H., Fathipour, Y. & Khanamani, M. (2021a). A comprehensive picture of foraging strategies of *Neoseiulus cucumeris* and *Amblyseius swirskii* on western flower thrips. *Pest Management* Science, 77(12): 5418–5429.

Dalir, S., Hajiqanbar, H.R., Fathipour, Y. & Khanamani, M. (2021b). Age-dependent functional and numerical responses of *Neoseiulus cucumeris* (Acari: Phytoseiidae) on two-spotted spider mite (Acari: Tetranychidae). *Journal of Economic Entomology,* 114(1): 50–61.

De Puysseleyr, V., Höfte, M. & De Clercq, P. (2014). Continuous rearing of the predatory anthocorid *Orius laevigatus* without plant materials. *Journal of Applied Entomology*, 138(1–2): 45–51.

Delisle, J.F., Brodeur, J. & Shipp, L. (2015). Evaluation of various types of supplemental food for two species of predatory mites, *Amblyseiusswirskii* and *Neoseiuluscucumeris* (Acari: Phytoseiidae). *Experimental and Applied Acarology*, 65(4): 483–494.

Ding-xu, L., Juan, T. & Zuo-rui, S. (2007). Functional response of the predator *Scolothrips takahashii* to hawthorn spider mite, *Tetranychus viennensis*: Effect of age and temperature. *BioControl*, 52(1): 41–61.

Duarte, M.V.A., Venzon, M., Bittencourt, M.C.S., Rodriguez-Cruz, F.A., Pallini, A. & Janssen, A. (2015). Alternative food promotes broad mite control on chilli pepper plants. *Biocontrol*, 60(6): 817–825.

Ebrahimifar, J., Shishehbor, P., Rasekh, A. & Riddick, E.W. (2020). Effect of factitious diets on development and reproduction of the ladybird beetle *Stethorus gilvifrons*, a predator of tetranychid mites. *BioControl*. doi: 10.1007/s10526-020-10033-y.

Eini, N., Jafari, S., Fathipour, Y., & Zalucki, M.P. (2022). How pollen grains of 23 plant species affect performance of predatory mite *Neoseiulus californicus*. *BioControl*, 67(2): 173–187.

Fadaei, E., Hakimitabar, M., Seiedy, M. & Sarraf Moaieri, H. (2018). Effects of different diets on biological parameters of the predatory mite *Amblyseius swirskii* (Acari: Phytoseiidae). *International Journal of Acarology*, 44(7): 341–346.

Farazmand, A., Fathipour, Y. & Kamali, K. (2013). Functional response and mutual interference of *Scolothrips longicornis* (Thysanoptera: Thripidae) on two spotted spider mite; integrated control of plant-feeding mites. *Bulletin*, 93: 39–44.

Farazmand, A., Fathipour, Y. & Kamali, K. (2014). Cannibalism in *Scolothrips longicornis* (Thysanoptera: Thripidae), *Neoseiulus californicus* and *Typhlodromus bagdasarjani* (Acari: Phytoseiidae). *Systematic & Applied Acarology*, 19(4): 471–480.

Fathipour, Y. & Maleknia, B. (2016). Mite predators. In: *Ecofriendly Pest Management for Food Security* (Ed. Omkar). Elsevier: San Diego, CA, pp. 329–366.

Fathipour, Y., Karimi, M., Farazmand, A. & Talebi, A.A. (2017). Age-specific functional response and predation rate of *Amblyseius swirskii* (Phytoseiidae) on two-spotted spider mite. *Systematic and Applied Acarology*, 22(2): 159–169.

Fathipour, Y., Karimi, M., Farazmand, A. & Talebi, A.A. (2018). Age-specific functional response and predation capacity of *Phytoseiulus persimilis* (Phytoseiidae) on the two-spotted spider mite. *Acarologia*, 58(1): 31–40.

Fathipour, Y., Maleknia, B., Bagheri, A., Soufbaf, M. & Reddy G.V.P. (2020). Functional and numerical responses, mutual interference, and resource switching of *Amblyseius swirskii* on two-spotted spider mite. *Biological Control*, 146(7): 104266.

Fathipour, Y., Maleknia, B., Bagheri, A., Soufbaf, M. & Zalucki, M.P. (2019). Spider mite host plant resistance traits improve the predatory performance of *Phytoseiulus persimilis* on cucumber, despite negative life history impacts. *Biological Control*, 138: 104064.

Fathipour, Y., Maleknia, B., Bagheri, A., Soufbaf, M., Zalucki, M.P. (2021). Functional and numerical responses of *Neoseiulus barkeri* (Acari: Phytoseiidae) on two-spotted spider mite: the effect of patch condition and additional food source. *Systematic and Applied Acarology*, 26(3): 543–556.

Furtado, I.P., de Moraes, G.J., Kreiter, S., Tixier, M.S. & Knapp, M. (2007). Potential of a Brazilian population of the predatory mite *Phytoseiulus longipes* as a biological control agent of *Tetranychus evansi* (Acari: Phytoseiidae, Tetranychidae). *Biological Control*, 42(2): 139–147.

Gadino, A.N. & Walton, V.M. (2012). Temperature-related development and population parameters for *Typhlodromus pyri* (Acari: Phytoseiidae) found in Oregon vineyards. *Experimental and Applied Acarology*, 58(1): 1–10.

Gerson, U., Smiley, R.L. & Ochoa, R. (2003). *Mites (Acari) for Pest Control*. Blackewell: Oxford, p. 560.

Ghazy, N.A., Suzuki, T., Shah, M., Amano, H. & Ohyama, K. (2012). Using high relative humidity and low air temperature as a long-term storage strategy for the predatory mite *Neoseiulus californicus* (Gamasida: Phytoseiidae). *Biological Control*, 60(3): 241–246.

Gillespie, D.R., Opit, G. & Roitberg, B. (2000). Effects of temperature and relative humidity on development, reproduction, and predation in *Feltiella acarisuga* (Vallot) (Diptera: Cecidomyiidae). *Biological Control*, 17(2): 132–138.

Goleva, I. & Zebitz, C.P.W. (2013). Suitability of different pollen as alternative food for the predatory mite *Amblyseius swirskii* (Acari, Phytoseiidae). *Experimental and Applied Acarology*, 61(3): 259–283.

Gómez-Moya, C.A., Gondim Jr, M.G.C., Moraes, G.J. & de Morais, E.G.F. (2018). Effect of relative humidity on the biology of the predatory mite *Amblyseius largoensis* (Acari: Phytoseiidae). *International Journal of Acarology*, 44(8): 400–411.

Gravandian, M., Fathipour, Y., Hajiqanbar, H., Riahi, E. & Riddick E.W. (2022). Long-term effects of cattail *Typhalatifolia* pollen on development, reproduction, and predation capacity of *Neoseiulus cucumeris, a predator of Tetranychus urticae. BioControl,* 67(2): 149–160.

Hadadi, A., Fathipour, Y., Hajiqanbar, H. & Riahi, E. (2022). Long-term effects of cattail pollenon development, population growth potential, and predation capacityof *Amblyseius swirskii* (Acari: Phytoseiidae). *Biocontrol Science and Technology*, 32(12): 1403–1416.

Haviland, D. (2017). Evaluation of sixspotted thrips, *Scolothrips sexmaculatus*, for biological control of spider mites in California almonds. In: *Proceedings of the 5th International Symposium on Biological Control of Arthropods* (Eds. P.G. Mason, D.R. Gillespie & C. Vincent), Langkawi, September 11–15. pp. 290–291.

Haviland, D.R., Rill, S.M. & Gordon, C.A. (2020). Evaluation of sticky traps for monitoring *Scolothrips sexmaculatus* (Thysanoptera: Thripidae) and *Stethorus punctum* (Coleoptera: Coccinellidae) as predators of spider mites in California almonds. *Journal of Economic Entomology*. doi: 10.1093/jee/toaa283.

Heidarian, M., Fathipour, Y. & Kamali, K. (2012). Functional response, switching, and prey-stage preference of *Scolothrips longicornis* (Thysanoptera: Thripidae) on *Schizotetranychus smirnovi* (Acari: Tetranychidae). *Journal of Asia-Pacific Entomology*, 15(1): 89–93.

Hoy, M.A. (2011). *Agricultural Acarology: Introduction to Integrated Mite Management*. CRC Press: Boca Raton, FL, p. 410.

Hoy, M.A., Castro, D. & Cahn, D. (1982). Two methods for large-scale production of pesticide–resistant strains of the spider mite predator *Metaseiulus occidentalis* (Nesbitt) (Acarina, Phytoseiidae). *Journal of Applied Entomology*, 94(1–5): 1–9.

Hughes, E., Bale, J.S. & Sterk, G. (2009). Thermal biology and establishment potential in temperate climates of the predatory mirid *Nesidiocoris tenuis* Gwennan. *BioControl*, 54(6): 785–795.

Imani, Z., Shishehbor, P. & Sohrabi, F. (2009). The effect of *Tetranychus turkestani* and *Eutetranychus orientalis* (Acari: Tetranychidae) on the development and reproduction of *Stethorus gilvifrons* (Coleoptera: Coccinellidae). *Journal of Asia-Pacific Entomology*, 12(4): 213–216.

Jafari, S., Fathipour, Y. & Faraji F. (2012). The influence of temperature on the functional response and prey consumption of *Neoseiulus barkeri* (Phytoseiidae) on two-spotted spider mite. *Journal of Entomological Society of Iran*, 31(2): 39–52.

Jafari, S., Fathipour, Y., Faraji, F. & Bagheri, M. (2010). Demographic response to constant temperatures in *Neoseiulus barkeri* (Phytoseiidae) fed on *Tetranychus urticae* (Tetranychidae). *Systematic and Applied Acarology*, 15: 83–99.

James, D.G. (2000). Reproductive diapause in *Typhlodromus doreenae* Schicha (Acari: Phytoseiidae). *International Journal of Acarology*, 26(1): 101–103.

James, D. & Whitney, J. (1993). Cumbungi pollen as a laboratory diet for *Amblyseius victoriensis* (womersley) and *Typhlodromus doreenae* schicha (acari: phytoseiidae). *Australian Journal of Entomology*, 32(1): 5–6.

Kasap, I. (2009). Influence of temperature on life table parameters of the predaceous mite *Euseius finlandicus* (Oudemans) (Acari: Phytoseiidae). *Turkish Journal of Agriculture and Forestry*, 33(1): 29–36.

Kean, J.M., Mansfield, S., Hardwick, S. & Barton, D.M. (2019). A risk-based detection survey for the predatory mirid *Macrolophus pygmaeus* in New Zealand. *Bulletin of Entomological Research*, 110(3): 370–378.

Khan, I. & Spooner-Hart, R. (2017). Temperature-dependent development of immature stages of predatory ladybird beetle *Stethorus vagans* (coleoptera: coccinellidae) at constant and fluctuating temperatures. *Acta Zoologica Academiae Scientiarum Hungaricae*, 63(1): 83–96.

Khanamani, M., Fathipour, Y., Talebi, A.A. & Mehrabadi, M. (2017a). How pollen supplementary diet affect life table and predation capacity of *Neoseiulus californicus* on two-spotted spider mite. *Systematic and Applied Acarology*, 22(1): 135–147.

Khanamani, M., Fathipour, Y., Talebi, A.A. & Mehrabadi, M. (2017b). Evaluation of different artificial diets for rearing the predatory mite *Neoseiulus californicus* (Acari: Phytoseiidae): Diet-dependent life table studies. *Acarologia*, 57(2): 407–419.

Kishimoto, H., Maeda, T., Wright, L.C. & James, D.G. (2011). Identification of prey consumed by *Stethorus punctum* picipes (Casey) (Cleoptera: Coccinellidae) in tree fruit and vines in Washington State, USA. *International Journal of Acarology*, 37(1): 216–220.

Lee, M.H. & Zhang, Z.Q. (2016). Habitat structure and its influence on populations of *Amblydromalus limonicus* (Acari: Phytoseiidae). *Systematic & Applied Acarology*, 21(10): 1361–1378.

Li, Y.J., Liu, Q.Y., Chang, J., Jia, Y.H. & Meng, R.X. (2019). Effects of temperature os. *Acarologia*, 59(4): 475–483.

Liu, J.F. & Zhang, Z.Q. (2017). Development, survival and reproduction of a New Zealand strain of *Amblydromalus limonicus* (Acari: Phytoseiidae) on *Typhaorientalis* pollen, *Ephestia kuehniella* eggs, and an artificial diet. *International Journal of Acarology*, 43(2): 153–159.

Liu, M.X., Chu, W.Q., Xu, C., Zheng, Q.M., Song, W.B., Li, Y.Y. & Liu, H. (2020). Extraguild prey availability reduced cannibalism and reciprocal intraguild predation of *Neoseiulus barkeri* (Acari: Phytoseiidae) and *Scolothrips takahashii* (Thysanoptera: Thripidae). *Systematic & Applied Acarology*, 25(5): 775–786.

Majolo, F. & Ferla, N.J. (2014). Life history of *Phytoseiulus macropilis* (Acari: Phytoseiidae) feeding on *Mononychellus planki* (Acari: Tetranychidae) on common bean leaves (Phaseoulus vulgaris L.). *International Journal of Acarology*, 40(4): 332–336.

Maleknia, B., Fathipour, Y. Soufbaf M. (2016). Intraguild predation among three phytoseiid species, *Neoseiulus barkeri, Phytoseiulus persimilis* and *Amblyseius swirskii*. *Systematic and Applied Acarology*, 21(4): 417–426.

Marafeli, P.P., Reis, P.R., Silveira, E.C., Souza-Pimentel, G.C. & Toledo, M.A. (2014). Life history of *Neoseiulus californicus* (McGregor, 1954) (Acari: Phytoseiidae) fed with castor bean (*Ricinus communis* L.) pollen in laboratory conditions. *Brazilian Journal of Biology*, 74(3): 691–697.

Matter, M.M., El-Shershaby, M.M.A., Farag, N.A. & Gesraha, M.A. (2011). Impact of temperature and prey density on the predacious capacity and behaviour of *Stethorus punctillum* Weise. *Archives of Phytopathology and Plant Protection*, 44(2): 127–134.

McGregor, R., Crisp, K. & Castiglia, C. (2020). Feeding lifestyles of the Phytoseiidae revisited: Searching for a factitious rearing host for *Neoseiulus fallacis* (Acari: Phytoseiidae). *BioControl*, 65(5): 593–599.

McMurtry, J.A. & Croft, B.A. (1997). Life-styles of phytoseiid mites and their roles in biological control. *Annual Review of Entomology*, 42: 291–321.

McMurtry, J.A., de Moraes, G.J. & Sourassou, N.F. (2013). Revision of the lifestyles of phytoseiid mites (Acari: Phytoseiidae) and implications for biological control strategies. *Systematic and Applied Acarology*, 18(4): 297–320.

Mehendiran, G. & Khursheed, S. (2018). A new host and distribution record for the black coccinellid, *Stethorus aptus* Kapur (Coccinellidae: Coleoptera). *Egyptian Journal of Biological Pest Control*, 28(53): 1–3.

Mendes, J.A., Barros, M.E.N., Lima, D.B. & Melo, J.W.S. (2017). Cannibalism in *Amblyseius largoensis* (Muma) (Acari: Phytoseiidae), an important natural enemy of coconut mite pests. *International Journal of Acarology*, 43(5): 387–392.

Mendes, J.A., Lima, D.B., Sousa Neto, E.P., Gondim Jr, M.G.C. & Melo, J.W.S. (2018). Functional response of *Amblyseius largoensis* to *Raoiella indica* eggs is mediated by previous feeding experience. *Systematic & Applied Acarology*, 23(10): 1907–1914.

Mirhosseini, M.A., Fathipour, Y., Soufbaf, M. & Reddy, G.V.P. (2018). Thermal requirements and development response to constant temperatures by *Nesidiocoris tenuis* (Hemiptera: Miridae), and implications for biological control. *Environmental Entomology*, 47(2): 467–476.

Mo, T.L. & Liu, T.X. (2007). Biology, life table and predation of *Feltiella acarisuga* (Diptera: Cecidomyiidae) feeding on *Tetranychus cinnabarinus* eggs (Acari: Tetranychidae). *Biological Control*, 39(3): 418–426.

Momen, F. & Abdel-khalek, A.A. (2008). Influence of diet on biology and life-table parameters of the predacious mite *Euseius scutalis* (A.H.) (Acari: Phytoseiidae). *Archives of Phytopathology and Plant Protection*, 41(6): 418–430.

Mori, K., Nozawa, M., Arai, K. & Gotoh, T. (2005). Life-history traits of the acarophagous lady beetle, *Stethorus japonicus* at three constant temperatures. *BioControl*, 50(1): 35–51.

Mortazavi, N., Fathipour, Y. & Talebi, A.A. (2019). The efficiency of *Amblyseius swirskii* in control of *Tetranychus urticae* and *Trialeurodes vaporariorum* is affected by various factors. *Bulletin of Entomological Research*, 109(3): 365–375.

Mutisya, D.L., El-Banhawy, E.M., Kariuki, C.W., Khamala, C.P., Fiaboe, K.K.M. & Kungu, M.M. (2012). Effect of the cassava green mite, *Mononychellus progresivus*, on the development and reproduction of the introduced predatory mite, *Phytoseiulus longipes* (Acari: Tetranychidae; Phytoseiidae), at different temperatures. *Systematic & Applied Acarology*, 17(4): 378–383.

Nemati, A. & Riahi, E. (2019). Does feeding on pollen grains affect the performance of *Amblyseius swirskii* (Acari: Phytoseiidae) during subsequent generations? *Bulletin of Entomological Research*, 110(4): 449–456.

Nguyen, D.T., Vangansbeke, D., Lu, X. & De Clercq, P. (2013). Development and reproduction of the predatory mite *Amblyseius swirskii* on artificial diets. *BioControl*, 58(3): 369–377.

Nguyen, D.T., Vangansbeke, D. & De Clercq, P. (2015). Performance of four species of phytoseiid mites on artificial and natural diets. *Biological Control*, 80: 56–62.

Onzo, A., Houedokoho, A.F. & Hanna, R. (2012). Potential of the predatory mite, *Amblyseius swirskii* to suppress the broad mite, *Polyphagotarsonemus latus* on the gboma eggplant, *Solanum macrocarpon*. *Journal of Insect Science*, 12(7): 1–11.

Pakyari, H. & Enkegaard, A. (2012). Effect of different temperatures on reproductive performance of *Scolothrips longicornis* reared on *Tetranychus urticae* eggs. *International Journal of Acarology*, 38(7): 571–575.

Pakyari, H. & McNeill, M.R. (2020). Effects of photoperiod on development and demographic parameters of the predatory thrips *Scolothrips longicornis* fed on *Tetranychus urticae*. *Bulletin of Entomological Research*, 110(5): 620–629.

Pappas, M.L., Xanthis, C., Samaras, K., Koveos, D.S. & Broufas, G.D. (2013). Potential of the predatory mite *Phytoseius finitimus* (Acari: Phytoseiidae) to feed and reproduce on greenhouse pests. *Experimental and Applied Acarology*, 61(4): 387–401.

Patenaude, S., Tellier, S. & Fournier, V. (2020). Cyclamen mite (Acari: Tarsonemidae) monitoring in eastern Canada strawberry (Rosaceae) fields and its potential control by the predatory mite *Neoseiulus cucumeris* (Acari: Phytoseiidae). *The Canadian Entomologist*, 152(2): 1–12.

Pazyuk, I.M. & Reznik, S.Y. (2016). Influence of photoperiod on development and maturation of *Macrolophus pygmaeus* (Hemiptera, Miridae). *Entomological Review*, 96(3): 274–279.

Pérez-Hedo, M., Riahi, C. & Urbaneja, A. (2020). Use of zoophytophagous mirid bugs in horticultural crops: Current challenges and future perspectives. *Pest Management Science,* 77(1): 33–42.

Perumalsamy, K., Selvasundaram, R., Roobakkumar, A., Rahman, V.J. & Muraleedharan, N. (2010). Life table and predatory efficiency of *Stethorus gilvifrons* (Coleoptera: Coccinellidae), an important predator of the *red* spider mite, *Oligonychus coffeae* (Acari: Tetranychidae), infesting tea. *Experimental and Applied Acarology*, 50(2): 141–150.

Peterson, P.G., McGregor, P.G. & Springett, B.P. (2000). Density dependent prey-feeding time of *Stethorus bifidus* (Coleoptera: Coccinellidae) on *Tetranychus lintearius* (Acari: Tetranychidae). *New Zealand Journal of Zoology*, 27(1): 41–44.

Pimentel, D. (2005). Environmental and economic costs of the application of pesticides primarily in the United States. *Environment, Development and Sustainability*, 7: 229–252.

Pirayeshfar, F., Safavi, S.A., Sarraf Moayeri, H.R. & Messelink, G.J. (2020). The potential of highly nutritious frozen stages of *Tyrophagus putrescentiae* as a supplemental food source for the predatory mite *Amblyseius swirskii. Biocontrol Science and Technology*, 30(5): 403–417.

Pochubay, E., Tourtois, J., Himmelein, J. & Grieshop, M. (2015). Slow-release sachets of *Neoseiulus cucumeris* predatory mites reduce intraguild predation by Dalotia coriaria in greenhouse biological control systems. *Insects*, 6(2): 489–507.

Praslička, J., Schlarmannová, J., Matejovičová, B. & Tancík, J. (2011). The predatory mite *Typhlodromus pyri* (Acari: Phytoseiidae) as a biocontrol agent of *Eriophyes pyri* (Acari: Eriophyidae) on pear. *Biologia*, 66(1): 146–148.

Pratt, P.D., Rosetta, R. & Croft, B.A. (2002). Plant-related factors influence the effectiveness of *Neoseiulus fallacis* (Acari: Phytoseiidae), a biological control agent of spider mites on landscape ornamental plants. *Journal of Economic Entomology*, 95(6): 1135–1141.

Puchalska, E.K. & Kozak, M. (2016). *Typhlodromus pyri* and *Euseius finlandicus* (Acari: Phytoseiidae) as potential biocontrol agents against spider mites (Acari: Tetranychidae) inhabiting willows: Laboratory studies on predator development and reproduction on four diets. *Experimental and Applied Acarology*, 68(1): 39–53.

Ragkou, V.S., Athanassiou, C.G., Kavallieratos, N.G. & Tomanovid, Z. (2004). Daily consumption and predation rate of different *Stethorus punctillum* instars feeding on *Tetranychus urticae*. *Phytoparasitica*, 32(2): 154–159.

Rahimi, A., Moradi, F., Sadeghi, A., Fathipour, Y. & Maroufpoor, M. (2022). Impact of constant temperatures on population characteristics of *Amblyseuis swirskii* (Acari: Phytoseiidae). *Systematic and Applied Acarology*, 27(9): 1775–1786.

Rahman, Md.A., Sarker, S., Ham, E., Lee, J.S. & Lim, U.T. (2020). Development and Fecundity of *Orius minutus* (Hemiptera: Anthocoridae) and *O. laevigatus* Reared on *Tetranychus urticae* (Acari: Tetranychidae). *Journal of Economic Entomology*, 113(4): 1735–1740.

Rahmani, H. & Afshari, N. (2020). Intraguild predation and cannibalism *between* two phytoseiid Mites, *Neoseiulus californicus* and *Amblyseius andersoni. Journal of Agricultural Science and Technology*, 22(2): 477–487.

Rajabpour, A., Seraj, A.A. & Mehrnia, A. (2018). Developing an artificial diet for rearing *Orius albidipennis* Reuter (Het., Anthocoridae). *Journal of Plant Protection Research*, 58(3): 276–281.

Ranabhat, N.B., Goleva, I. & Zebitz, C.P.W. (2014). Life tables of *Neoseiuluscucumeris* exclusively fed with seven different pollens. *BioControl*, 59: 195–203.

Razdoburdin, V.A. & Kozlova, E.G. (2019). Interactions of acariphagous arthropods in the system "host plant spider mite *Tetranychus urticae* Koch (Acarina, Tetranychidae) predatory mite *Phytoseiulus persimilis* (Parasitiformes, Phytoseiidae) and predatory midge *Feltiella luboviae* Fedotova et Kozlova (Diptera, Cecidomyiidae)" on cucumber cultivars. *Entomological Review*, 99(9): 1231–1238.

Refaei, G.S. & Mohamed, A.A. (2013). Biological characters of *Feltiella acarisuga* (vallot) (Diptera: Cecidomyiidae) when fed on eggs of *Tetranychus urticae* koch [Acari: Tetranychidae]. *Egyptian Journal of Agricultural Research*, 91(1): 119–124.

Renkema, J.M., LeFors, J.A. & Johnson, D.T. (2017). First report of broad mite (Acari: Tarsonemidae) on commercial strawberry in Florida. *Florida Entomologist*, 100(4): 804–806.

Rezaie, M. (2017). Effect of three pollen grains on life table parameters of *Neoseiulus californicus* (Acari: Phytoseiidae). *International Journal of Engineering and Applied Sciences*, 4(9): 21–24.

Riahi, E., Fathipour, Y., Talebi, A.A. & Mehrabadi, M. (2016). Pollen quality and predator viability: Life table of *Typhlodromus bagdasarjani* on *seven* different plant pollens and two-spotted spidermite. *Systematic and Applied Acarology*, 21(10): 1399–1412.

Riahi, E., Fathipour, Y., Talebi, A.A. & Mehrabadi, M. (2017a). Linking life table and consumption rate of *Amblyseiusswirskii* (Acari: Phytoseiidae) in presence and absence of different pollens. *Annual Entomological Society of America*, 110(2): 244–253.

Riahi, E., Fathipour, Y., Talebi, A.A. & Mehrabadi, M. (2017b). Natural diets vs factitious prey: Comparative effects on development, fecundity and life table of *Amblyseius swirskii* (Acari: Phytoseiidae). *Systematic and Applied Acarology*, 22(5): 711–723.

Riahi, E., Fathipour, Y., Talebi, A.A. & Mehrabadi, M. (2017c). Attempt to develop cost-effective rearing of *Amblyseius swirskii* (Acari: Phytoseiidae): Assessment of different artificial diets. *Journal of Economic Entomology*, 110(4): 1525–1532.

Riahi, E., Fathipour, Y., Talebi, A.A. & Mehrabadi, M. (2018). Factitious prey and artificial diets: Do they all have the potential to facilitate rearing of *Typhlodromus bagdasarjani* (Acari: Phytoseiidae)? *International Journal of Acarology*, 44(2–3): 121–128.

Riddick, E.W. & Wu, Z. (2011). Lima bean–lady beetle interactions: Hooked trichomes affect survival of *Stethorus punctillum* larvae. *BioControl*, 56(1): 55–63.

Salahi, M., Basheer, A.A. & Asslan, L. (2012). Effect of temperature on daily consumption rate of the predatory thrips, *Scolothrips sexmaculatus* Pergande (Thysanoptera: Thripidae). *Egyptian Journal of Biological Pest Control*, 22(1): 51–54.

Samaras, K., Pappas, M.L., Fytas, E. & Broufas, G.D. (2015). Pollen suitability for the development and reproduction of *Amblydromalus limonicus* (Acari: Phytoseiidae). *BioControl*, 60(6): 773–782.

Samaras, K., Pappas, M.L., Fytas, E. & Broufas, G.D. (2019). Pollen provisioning enhances the performance of *Amblydromalus limonicus* on an unsuitable prey. *Frontiers in Ecology and Evolution*, 7: 122.

Sarwar, M. (2016). Comparative life history characteristics of the mite predator *Neoseiuluscucumeris* (Oudemans) (Acari: Phytoseiidae) on mite and pollen diets. *International Journal of Pest Management*, 62(2): 140–148.

Schmidt-Jeffris, R.A. & Beers, Z.H. (2015). Comparative biology and pesticide susceptibility of *Amblydromella caudiglans* and *Galendromus occidentalis* as spider mite predators in apple orchards. *Experimental and Applied Acarology*, 67(1): 35–47.

Schmidt-Jeffris, R.A. & Beers, E.H. (2017). Effects of foliar apple trichomes on *Galendromus occidentalis* (Nesbitt) (Acari: Phytoseiidae): Preferences, fecundity and prey consumption. *Agricultural and Forest Entomology*, 19(1): 93–100.

Soltaniyan, A., Kheradmand, K., Fathipour, Y. & Shirdel, D. (2020). Supplementation of natural prey with pollen grains exerts an influence on the life table parameters of *Neoseiulus californicus*. *Bulletin of Entomological Research*, 110(4): 535–541.

Song, Z.W., Nguyen, D.T., Li, D.S. & De Clercq, P. (2019). Continuous rearing of the predatory mite *Neoseiulus californicus* on an artificial diet. *BioControl*, 64(2): 125–137.

Taghizadeh, R., Fathipour, Y. & Kamali, K. (2008a). Influence of temperature on life-table parameters of *Stethorus gilvifrons* (Mulsant) (Coleoptera: Coccinellidae) fed on *Tetranychus urticae* Koch. *Journal of Applied Entomology*, 132(8): 638–645.

Taghizadeh, R., Fathipour, Y. & Kamali, K. (2008b). Temperature-dependent development of Acarophagous ladybird, *Stethorus gilvifrons* (Mulsant) (Coleoptera: Coccinellidae). *Journal of Asia-Pacific Entomology*, 11(3): 145–148.

Taylor, A. & James, D.G. (1993). Effect of temperature on development and survival of *Typhlodromus doreenae* Schicha (Acari: Phytoseiidae). *International Journal of Acarology*, 19(2): 185–188.

Tuan, S.J., Lin, Y.H., Peng, S.C. & Lai, W.H. (2016). Predatory efficacy of *Orius strigicollis* (Hemiptera: Anthocoridae) against *Tetranychus urticae* (Acarina: Tetranychidae) on strawberry. *Journal of Asia-Pacific Entomology*, 19(1): 109–114.

Vacante, V. (2010). *Citrus Mites: Identification, Bionomy and Control*. CABI: Wallingford, 392 p.

van Lenteren, J.C. (2012). The state of commercial augmentative biological control: Plenty of natural enemies, but a frustrating lack of uptake. *BioControl*, 57(1): 1–20.

Vangansbeke, D., Nguyen, D.T., Audenaert, J., Verhoeven, R., Gobin, B., Tirry, L. & De Clercq, P. (2014). Performance of the predatory mite *Amblydromaluslimonicus* on factitious foods. *BioControl*, 59(1): 67–77.

Yanagita, H., Morita, S., Kunimaru, K. & Takemoto, H. (2014). Capability of *Scolothrips takahashii* (Thysanoptera: Thripidae) as a control agent of *Tetranychus urticae* (Acari: Tetranychidae) for protecting strawberry plug plants in summer. *Applied Entomology and Zoology*, 49(3): 437–441.

Yanar, D. & Marjorie, A.H. (2015). The fitness of *Metaseiulus occidentalis* (Acari: Phytoseiidae) adult females after different intervals of storage at low temperatures. *Florida Entomologist*, 98(2): 541–546.

Yang, J., Lv, J., Liu, J., Xu, X. & Wang, E. (2019). Prey preference, reproductive performance, and life table of *Amblyseius tsugawai* (Acari: Phytoseiidae) feeding on *Tetranychus urticae* and *Bemisia tabaci*. *Systematic and Applied Acarology*, 24(3): 404–413.

Yazdanpanah, S., Fathipour, Y., Riahi, E. & Zalucki, M.P. (2021a). Mass production of *Neoseiulus cucumeris* (Acari: Phytoseiidae): an assessment of 50 generations reared on almond pollen. *Journal of Economic Entomology*, 114(6): 2255–2263.

Yazdanpanah, S., Fathipour, Y. & Riahi, E. (2021b). Pollen grains are suitable alternative food for rearing the commercially used predatory mite *Neoseiuluscucumeris* (Acari: Phytoseiidae). *Systematic and Applied Acarology*, 26(5): 1009–1020.

Yazdanpanah, S., Fathipour, Y., Riahi, E. & Zalucki, M.P. (2022a). Cost-effective and efficient factitious prey for mass production of *Neoseiulus cucumeris* (Acari: Phytoseiidae): assessing its quality compared with natural prey. *Egyptian Journal of Biological Pest Control*, 32: 16.

Yazdanpanah, S., Fathipour, Y., Riahi, E. & Zalucki, M.P. (2023). Effects of diet switching from almond pollen to natural prey on predation capacity of *Neoseiulus cucumeris*. *Journal of Applied Entomology*, 147(1): 72–84.

Yazdanpanah, S., Fathipour, Y., Riahi, E. & Zalucki, M.P. (2022b). Generation-dependent functional and numerical responses of *Neoseiulus cucumeris* (Acari: Phytoseiidae) long-term reared on almond pollen. *Biocontrol Science and Technology*, 32(4): 484–496.

Yazdanpanah, S., Fathipour, Y., Riahi, E. & Zalucki, M.P. (2022c). Modeling temperature-dependent development rate of *Neoseiuluscucumeris* (Acari: Phytoseiidae) fed on two alternative diets. *Environmental Entomology*, 51(1): 145–152.

Yazdanpanah, S., Fathipour, Y., Riahi, E. & Zalucki, M.P. (2022d). Pollen alone or a mixture of pollen types? Assessing their suitability for mass rearing of *Neoseiulus cucumeris* over 20 generations. *Journal of Insect Science*, 22(4): 6.

Yoder, J.A., Pollock, D.A. & Benoit, J.B. (2003). Moisture requirements of the ladybird beetle *Stethorus nigripes* in relation to habitat preference and biological control. *Entomologia Experimentalis et Applicata*, 109(1): 83–87.

Zhang, Y., Zhang, Z.Q., Lin, J. & Ji, J. (2000). Potential of *Amblyseius cucumeris* (Acari: Phytoseiidae) as a biocontrol agent against *Schizotetranychus nanjingensis* (Acari: Tetranychidae) in Fujian, China. *Systematic and Applied Acarology*, 4: 109–124.

Zhang, S.C., Zhu, F., Zheng, X.L., Lei, C.L. & Zhou, X.M. (2012). Survival and developmental characteristics of the predatory bug *Orius similis* (Hemiptera: Anthocoridae) fed on *Tetranychus cinnabarinus* (Acari: Tetranychidae) at three constant temperatures. *European Journal of Entomology*, 109(4): 503–508.

11 Aquatic Insects

K. A. Subramanian
Zoological Survey of India, Southern Regional Centre

CONTENTS

LEARNING OBJECTIVES

1. To provide an overview of current status and diversity of aquatic insect predators.
2. To learn their potential and conservation in the management of aquatic insect vectors.

11.1 INTRODUCTION

Inland freshwater ecosystems occupy less than 1% of the earth's surface but disproportionately harbour high diversity of animal species. About 10% of all animal species recorded are from the inland freshwaters and among this, 60% are aquatic insects (Balian et al., 2008). Aquatic insects are represented from 12 insect orders and they spend one or more of their life stages in freshwater. It is proposed that the diversity of freshwater insects may well be over 200,000 species constituting 80% of freshwater animal biodiversity (Dijkstra et al., 2014). Aquatic insects are key components of freshwater ecosystems with multiple ecosystem linkages and play many important roles. Different taxa of aquatic insects are primary consumers, detritivores, herbivores and predators. Most of the aquatic insects upend their life stage as egg, larva or nymph in freshwaters. Many taxa, such as Ephemeroptera, Odonata, Trichoptera, Diptera, etc., have immature aquatic stages and terrestrial adult stages. Aquatic representatives of Coleoptera and Heteroptera have their immature and adult stages in water.

Several species of aquatic Diptera are vectors of human and veterinary diseases and are a major public health concern. Worldwide, aquatic insect vectors of human disease transmit malaria, lymphatic filariasis, trypanosomiasis, onchocerciasis and arboviruses, such as dengue virus and Zika virus (Durden & Mullen, 2018). Biological control of vectors utilizing microorganisms, pathogens, parasites and predators without damage to the ecosystem is globally advocated (Thomas, 2018a). Since vaccines or drugs are not available for most of the vector-borne diseases, vector control is still

DOI: 10.1201/9781003370864-11

the primary form of prevention, especially in the developing countries (Benelli et al., 2016). Due to ecological concerns and development of insecticide resistance, biocontrol strategies are advocated for sustainable vector control programmes. In this context, predatory aquatic insects form a primary defence against many vectors and have high potential in vector control programmes. In this chapter, the importance of aquatic insect predators as potential control agents of vectors is discussed.

11.2 DIVERSITY OF AQUATIC INSECTS

Global freshwater animal diversity is estimated to be 126,000 species and majority of them are insects (60.4%), vertebrates (14.5%), crustaceans (10%), Arachnids (5%), molluscs (5%), Rotifera (1.6%), Annelida (1.4%) Nematoda (1.4%), Platyhelminthes (Turbellaria: 1%) and minor groups, such as Collembola (Balian et al., 2008). Unlike terrestrial ecosystems, where Coleoptera dominates, the freshwater ecosystems are dominated by Diptera both in terms of species richness and abundance. In India, over 1,500 species are recorded from the freshwater habitats in comparison with the aquatic beetles, represented by only 776 species. In terms of species diversity, Diptera is followed by Trichoptera. The global and Indian diversity of aquatic insects is summarized in Table 11.1. In general, high diversity and endemism is prevalent in taxa inhabiting lotic habitats than the ones in lentic habitats (Dijkstra et al., 2014). Orders, such as Ephemeroptera (mayflies), Odonata (dragonflies and damselflies), Plecoptera (stoneflies), Trichoptera (caddisflies) and Megaloptera (alderflies), are restricted to freshwater habitats and their larval stages have the ability to absorb dissolved oxygen. The vector species of aquatic and semiaquatic insects are restricted to different families of Diptera.

11.3 AQUATIC INSECT HABITATS AND ADAPTATIONS

Aquatic insects utilize a spectrum of habitats for completing their life cycle and have important implications in the bionomics of vectors and their predators. The morphology, behaviour and life history strategies of aquatic insects are linked with their habitats. Distribution of aquatic insects across habitats and geographic region is scale dependent (Koroiva & Pepinelli, 2019). Long-term geological processes, such as the continental drift, formation of rivers and mountains, bioclimate and global climate change, have been some of the factors influencing the large-scale distribution of aquatic insects. At local scale, dispersal, availability of suitable habitats, substrate heterogeneity, land use, aquatic vegetation and abiotic environmental variables determine the structure of aquatic insect communities (Koroiva & Pepinelli, 2019).

TABLE 11.1
Species Diversity across Aquatic Insect Orders

Sl. No.	Taxa	Global Species Diversity	Indian Species Diversity
1	Collembola[a]	8,162	339[b]
2	Ephemeroptera	3,046	172
3	Odonata	6,302	494
4	Plecoptera	3,625	146
5	Neuroptera	118	5
6	Megaloptera	348	30
7	Heteroptera	4,940	325
8	Coleoptera	13,649	776
9	Diptera	43,271	1,588
10	Trichoptera	16,267	1,301
11	Lepidoptera	740	80
12	Hymenoptera	150	10

[a,b] Semiaquatic species and estimate of diversity not available.

In lotic systems, such as streams and rivers, there is a predictable change in aquatic insect communities from headwaters to downstream. This change in community structure is attributed to differences in energy sources and nutrients, which was conceptualized as the River Continuum Concept (Vannote et al., 1980). Recently, spatial structure of river networks are considered to be one of the determining factors for biodiversity (Siqueira et al., 2014).

Aquatic insects have evolved many morphological adaptations for living in lotic and lentic habitats, especially for breathing, food acquisition and locomotion. The dorsoventral flattening of body found in many mayflies, stonefly nymphs, naucorid bugs, Psephenidae, etc. are adaptations for surviving fast currents, which provide least resistance to water current and more surface area for contact with the substrate. Insects, such as Blephariceridae (Diptera) and some aquatic beetles, have developed suction cups or use silk, hooks, grapples or cases made of pebbles and debris (Simuliidae, Trichoptera) to withstand heavy water current.

In lotic systems, due diverse habitat structure, physicochemical properties and seasonal dynamism aquatic insects have developed diverse morphological, physiological and life history traits. For example, in Gerridae (Heteroptera), the legs are modified as long and slender appendages to use surface tension of water to stride through the water surface. In whirligig beetles (Coleoptera: Gyrinidae), each compound eye is divided into a ventral and dorsal unit to aid simultaneous vision above and below water. The respiratory adaptations of aquatic insects are as diverse as their habitat. Some of the important mechanisms include (1) simple diffusion over thin integument (Ephemeroptera, Odonata), (2) temporary use of an air bubble (Heteroptera, Coleoptera), (3) using plastron (Naucoridae) or blood gill (Diptera), (4) oxygen storage in haemoglobin molecules (Diptera) and (5) breathing via tubes or siphons (Diptera). Detailed respiratory mechanisms across insect orders are provided in Table 11.2.

The life history traits, such as oviposition, nymphal or larval duration, emergence and dispersal, are fine-tuned with the habitat dynamics and bioclimatic conditions. Many vector species have three important biological traits, *viz.*, (1) association with humans, (2) egg resistance to desiccation and (3) ability to breed in small water collections. Selection of insect biocontrol agents needs to consider these factors for success vector control programmes.

11.4 AQUATIC INSECT FUNCTIONAL GROUPS

Based on their morphology, the mode of existence of aquatic insects in wetland habitats are diverse. The common habits of aquatic insects are skaters, divers, swimmers, clingers, sprawlers and burrowers. They utilize the food source in diverse ways and can be broadly classified into different functional groups based on the feeding mechanism. These different modes of feeding

TABLE 11.2
Functional Feeding Groups of Aquatic Insects

Sl. No.	Functional Feeding Group (FFG)	Feeding Mechanism and Food	Orders
1	Shredders	Chewers of living and dead plant material	Ephemeroptera, Plectoptera, Coleoptera, Trichoptera, Lepidoptera, Diptera
2	Collectors	Decomposing organic matter and detritivore filter feeders	Ephemeroptera, Hemiptera, Coleoptera, Trichoptera, Diptera
3	Scrapers	Substrate attached algae	Ephemeroptera, Hemiptera, Coleoptera, Trichoptera, Lepidoptera, Diptera
4	Macrophyte piercers	Living plants	Neuroptera, Trichoptera
5	Predators	Living animals	Ephemeroptera, Plectoptera, Hemiptera, Coleoptera, Trichoptera, Diptera
6	Parasites	Living animals	Diptera and Hymenoptera

TABLE 11.3
Habitats, Functional Feeding Groups, Respiratory Mechanisms and Prevalence of Vector Species in Aquatic Insect Orders

Sl. No.	Taxa	Habitat	Functional Feeding Group (FFG)	Respiratory Mechanism	Vectors
1	Ephemeroptera	Lotic and lentic	Collectors, gatherers, filter feeders, grazers	Tracheal gills	NIL
2	Odonata	Lotic and lentic	Predators at larval and adult stages	Tracheal gills	NIL
3	Plecoptera	Lotic	Predators at larval stage	Tracheal gills, cutaneous	NIL
4	Neuroptera		Piercers at larval stage	Tracheal gills	NIL
5	Megaloptera	Lotic	Predators at larval stage	Tracheal gills	NIL
6	Heteroptera	Lotic and lentic	Predators at nymphal and adult stages	Atmospheric oxygen, temporary air storage, permanent air storage	NIL
7	Coleoptera	Lotic and lentic	Predators at larval and adult stages	Tracheal gills, atmospheric oxygen, plant breathers, temporary air storage, permanent air storage, spiracular gills, haemoglobin	NIL
8	Diptera	Lotic and lentic	All FFG	Tracheal gills, atmospheric oxygen, plant breathers, spiracular gills, cutaneous, haemoglobin.	Many vector species
9	Trichoptera	Lotic and lentic	All FFG	Tracheal gills, cutaneous	NIL
10	Lepidoptera	Lotic and lentic	Shredders and scrapers	Tracheal gills, permanent air storage	NIL
11	Hymenoptera	Lotic and lentic	Parasites		NIL

mechanisms among aquatic insects are provided in Table 11.2. Understanding of habits, habitat distribution and functional feeding group of vectors and their potential insect predators is very important to develop locally suitable vector control measures. A summary of habitat distribution, functional feeding groups and prevalence of vectors across aquatic insect orders is provided in Table 11.3.

11.5 AQUATIC INSECTS AS VECTORS

Aquatic insect vectors are restricted to the order Diptera. Different families of Diptera with diverse habitats and host preferences transmit vector-borne diseases in humans and domesticated animals. Aquatic dipterans, such as mosquitoes, blackflies, biting midges, tabanids, etc., transmit protozoal, viral and bacterial diseases. A summary of the important vector-borne diseases transmitted by aquatic insect vectors is provided in Table 11.4. The larvae of these vectors are aquatic or semiaquatic and adults are terrestrial, which depend upon vertebrate host for their blood meal. Understanding the ecology, habitat and host preference of vector is crucial in selecting appropriate predators for their control. A brief summary of ecology and biology of aquatic insect vectors is provided below.

TABLE 11.4
Aquatic Insect Vectors, Human and Veterinary Diseases Transmitted and Predators

Sl. No.	Aquatic Dipteran Vectors	Important Diseases Transmitted	Predators
1	Mosquitoes (Culicidae)	**Viruses:** Yellow fever, dengue, Japanese encephalitis, Chikungunya fever, Zika encephalitis **Protozoans:** Malaria **Filarial Nematodes:** Wuchererian filariasis, Bancroftian filariasis, dog heartworm	**Larvae**: Odonata nymphs, adult and nymphs of aquatic heteropteran and Coleoptera. **Adult**: Adult Odonata
2	Black flies (Simuliidae)	**Filarial Nematodes:** Human onchocerciasis (river blindness), bovine onchocerciasis	**Larvae**: Odonata nymphs, adult and nymphs of aquatic heteropteran and Coleoptera **Adult**: Adult Odonata
3	Biting midges (Ceratopogonidae)	**Viruses:** Bluetongue disease, epizootic haemorrhagic disease, **Filarial Nematodes:** Equine onchocerciasis, mansonellosis	**Larvae**: Odonata nymphs, adult and nymphs of aquatic heteropteran and Coleoptera **Adult**: Adult Odonata
4	Horse flies and deer flies (Tabanidae)	**Viruses:** Equine infectious anaemia, hog cholera **Bacteria:** Tularemia **Protozoans:** Surra (livestock trypanosomiasis) **Filarial Nematodes:** Loiasis, Elaeophorosis	**Larvae**: Odonata nymphs, adult and nymphs of aquatic heteropteran and Coleoptera **Adult**: Adult Odonata

Source: Modified after Durden and Mullen (2018).

11.5.1 Culicidae (Mosquitoes)

Mosquitoes are ubiquitous and found throughout the world except in Antarctica. They use a wide range of wetland habitats with the highest species diversity in the tropical forests. About 3,500 species are known globally (Durden & Mullen, 2018). Of these, 393 species are recorded from India; of these, 31 species are known to be vectors (Bhattacharyya et al., 2014). The mosquitoes use a wide range of aquatic habitats for completing their life cycle. The aquatic habitat is species-specific. These include tidal pools, salt marshes, rain pools, floodstreams, swamps, lakes, tree holes, leaf axils, fruit husks, discarded tyres, etc. The eggs are laid in water and larval, and pupal stages are spent there. Eggs of some species have the ability to withstand desiccation. The adults feed on nectar and females require blood meal from vertebrate hosts for egg development. The female mosquitoes are host-specific and their activity patterns, life cycle and biting frequencies are fine-tuned with host ecology and behaviour and play an important role in disease transmission.

11.5.2 Simuliidae (Blackflies)

Blackflies are small serious vectors of humans, livestock and wild animals. They have wide geographic distribution except in Antarctica. Globally, over 2,200 species are reported; of these, 69 species in one genera are known from India (Currie & Adler, 2008; Mitra et al., 2017). They are vectors of human onchocerciasis and mansonellosis, bovine onchocerciasis and avian leucocytozoonosis (Durden & Mullen, 2018). Immature stages of blackflies are aquatic and characterized by labral fans for filter feeding and silk glands, which are used for anchoring in the substrate. They are adapted to live in flowing water and species are habitat- specific. Larval and pupal duration are species-specific and usually completed in a week. Mature males and females feed on nectar. The females require blood meal to mature eggs and feed on mammals and birds.

11.5.3 Ceratopogonidae (Biting Midges)

Biting midges are minute blood sucking flies represented globally by 123 genera and 6,267 species (Durden & Mullen, 2018; Wagner et al., 2008). In India, 12 genera are recorded; of this 73 species of *Culicoides* with 13 species are known vectors (Chandra et al., 2017; Mukhopadhyay et al., 2016). They are vectors of viral, protozoan and nematodal diseases of humans and livestock. They are found in a wide range of aquatic and semiaquatic habitats, such as marshes, ponds, rivers, streams, tree holes, tidal marshes and mangroves. The females oviposit eggs in moist habitats and larvae develop there. Depending on the species, the larvae feed on small aquatic and semiaquatic invertebrates, bacteria, fungus, detritus, algae, diatoms and other organic matter. Adult males and females feed on nectar; however, female adults also feed on vertebrate blood, such as mammals, reptiles and birds.

11.5.4 Tabanidae (Horseflies and Deerflies)

Tabanids are large strong flying dipterans with brightly coloured patterned eyes. They are widely distributed except in Antarctica and high diversity is found in the tropics. Globally, 4,455 species are known in 144 genera (Durden & Mullen, 2018). Indian fauna comprises 247 species under 14 genera and 3 subfamilies. They transmit many viral, bacterial and protozoan diseases among livestock. They are the vectors of Trypanosomiasis (Surra) and are reported to be transmitted by 20 species of tabanids in India (Maity et al., 2016). Females lay a mass of eggs in vegetation overhanging wetlands or moist soil. The larvae hatch and develop in moist soil or water. They are carnivorous and feed on a variety of small invertebrates. When fully developed, they move to dry ground to pupate. Tabanids are known to feed on nectar and extra floral nectaries. Only females bite and they require blood meal from vertebrate host for egg development. Their behaviour of taking multiple blood meal is an important factor in disease transmission.

11.6 AQUATIC INSECTS IN VECTOR CONTROL

Many species of aquatic insects are predators of immature or adult stages of the vectors (Table 11.3). Successful predators in vector control programmes ideally should maintain: (1) close interaction with prey population, (2) capacity to sustain population, (3) be climatically compatible, and (4) low potential for negative impacts (Shaalan & Canyon, 2009). Studies have confirmed that the natural enemies cause significant reduction in vector population and are an indispensable part of integrated vector control programmes (Legner, 1994). Several species of predatory arthropods are considered important in the control of mosquitoes (James, 1967; Service, 1977; Collins & Washino, 1985; McDonald & Buchanan, 1981). Among them, only a few are deployed to control mosquitoes. The larvae and adults of odonates were considered to have high potential for mosquito control (Mathavan, 1976; Sebastian et al., 1980, 1990; Urabe et al., 1986; Sebastian et al., 1990; Singh et al., 2003; Bond et al., 2006; Chatterjee et al., 2007; Mandal et al., 2008; Sathe and Bhusnar, 2010; Venkatesh and Tyagi, 2013). Odonata families, such as Coenagrionidae, Libellulidae, Gomphidae, Aeshnidae and Macromedia, breed in marshes and other lentic wetlands. Among them, many species such as *Bradinopyga geminata*, *Brachythemis contaminata*, *Orthetrum sabina*, *Trithemis aurora* and *Ceriagrion coromandelianum* breed in garden ponds and other small water collections around human habitation and have high potential in anthropophilic vectors. However, the larvae and adults of odonates are difficult to rear and handle, which have limited their use.

Aquatic Heteroptera, such as the Notonectidae, Belostomatidae, Corixidae, Notonectidae and Naucoridae, prey upon mosquito larvae. Some species of these families can survive in moderately polluted wetlands, paddy fields, aquaculture ponds and garden ponds. This makes them an ideal candidate for the control of larval stages of vectors, especially the mosquitoes (Chesson, 1984; Blaustein et al., 1995; Blaustein, 1998; Eitam et al., 2002; Eitam and Blaustein, 2004). Adults and larvae of predatory aquatic Coleoptera of genera, such as *Dytiscus*, *Laccophilus*, *Agabus* and *Rhantus*, are found to be important predators of mosquito larvae (Baldwin et al., 1955;, Borland, 1971; Bay, 1974; Lundkvist et al., 2003).

Tree-hole breeding predatory mosquito genus *Toxorhynchites*, whose larvae are predators of other mosquitoes, has high potential for controlling container breeding vectors, such as *Aedes albopictus* and *Ae. Aegypti*. Globally, 88 species of Toxorhynchites are recorded; of these, nine species are reported from India (Krishnamoorthy et al., 2013). They are suitable for the control of medically important vector species due to their nectarivorous nature and lack of blood feeding behaviour. They have been successfully used in the control of many container breeding mosquitoes (Donald et al., 2020). Under laboratory conditions in multiple prey choice experimental setups, aquatic hemipteran and odonate predators, such as *Diplonychus annulatus*, *D. rusticus* (Belastomatidae), *Anisops bouvieri* (Notonectidae), *Ceriagrion coromandelianum* (Coenagrionidae) and *Brachydiplax chalybea* (Libellulidae), were found to be more efficient in controlling the mosquito larvae (Saha et al., 2014).

11.7 CONSERVATION OF AQUATIC INSECT PREDATORS

Chemical insecticides are important control agents for disease vectors. In recent years, the spread of insecticide resistance, damage to ecosystems and nontarget organisms, climate change and the emergence of new diseases have created the need for exploring sustainable alternative strategies for vector control. Globally, inland freshwater ecosystems are the most threatened. Recently, it is reported that, on average, the abundance of 944 monitored freshwater species has globally declined by 84% since 1970 (WWF, 2020). Global and regional studies show that freshwater biodiversity is threatened due to habitat destruction, flow modification, agrochemicals, agriculture runoff, over exploitation, invasive species and large infrastructure projects (Allen et al., 2010; Dudgeon, 2006; Dudgeon et al., 2006; WWF, 2008). Recently, global status of freshwater biodiversity and its status, threats and conservation have been extensively reviewed (Dudgeon, 2020). These studies emphasize that the freshwater biodiversity including aquatic insect predators, such as several species of dragonflies, are threatened globally with extinction due to reasons mentioned above.

Conservation of freshwater biodiversity is very crucial in maintaining healthy populations of natural predators. As highlighted earlier, several species of natural predators coexist with vectors. Freshwater biodiversity conservation needs to be integrated into vector control programmes for sustaining natural predators. In this context, providing artificial habitats, such as dragonfly ponds, were found to increase local species diversity (Simaika et al., 2016) and thereby ensure possible enhancement of their predatory services including control of vector species. Tropical countries like India have a good diversity of predatory aquatic insects inhabiting in all freshwater habitats and has a huge potential for vector control programmes. Hence, it is important to develop short-, mid- and long-term strategies for freshwater biodiversity conservation including promoting natural predator populations at local and regional levels to manage vector populations along with other integrated vector management tools.

11.8 CONCLUSIONS

Aquatic insect vectors of human and livestock diseases are a serious global concern with direct socioeconomic impacts. Ongoing depletion of freshwater biodiversity, emergence of new vector-borne diseases, climate change and concern of impact of chemical vector control measures have emphasized the importance of natural aquatic insect predators in vector control. Local and regional freshwater biodiversity conservation measures coupled with the creation of artificial habitats for natural predators can supplement integrated vector control programmes.

11.9 POINTS TO REMEMBER

1. More than 60% of inland freshwater biodiversity is contributed by insects.
2. Unlike terrestrial ecosystems where coleopterans dominate, freshwater ecosystems are dominated by dipterans.
3. All the aquatic insect vectors are represented from the order Diptera and inhabit diverse inland freshwater ecosystems.
4. Aquatic insect vectors transmit many human, livestock and wildlife diseases.

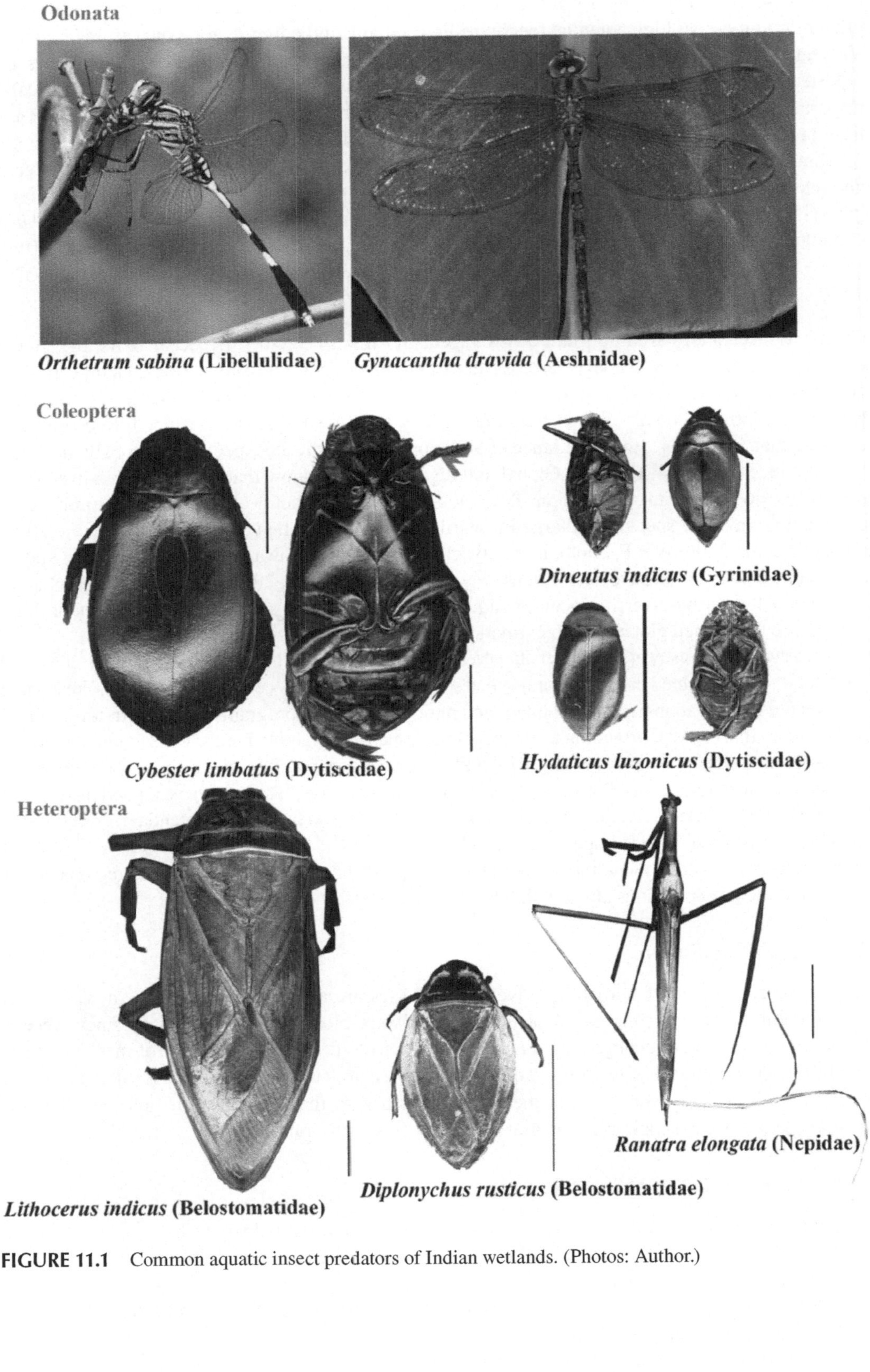

FIGURE 11.1 Common aquatic insect predators of Indian wetlands. (Photos: Author.)

5. Aquatic insect predators, such as Odonata, Coleoptera and Hemiptera, are found to be effective in controlling the vector population.
6. Global depletion of freshwater biodiversity has significantly affected the populations of natural aquatic insect predators.
7. Freshwater biodiversity conservation is crucial for sustaining the population of predators for effective vector management.
8. Current integrated vector management tools need to incorporate principles of freshwater biodiversity conservation.

ACKNOWLEDGEMENTS

The author sincerely thanks Prof. Omkar, Department of Zoology, University of Lucknow, for encouraging and providing him an opportunity to prepare this manuscript. He is also grateful to the Director, Zoological Survey of India, for providing facilities for research.

REFERENCES

Allen, D. J., Molur, S., and Daniel, B. A. (Compilers). (2010). *The Status and Distribution of Freshwater Biodiversity in the Eastern Himalaya*. Zoo Outreach Organisation, Cambridge, UK and Gland, Switzerland: IUCN, and Coimbatore, India, 100 pp.

Almond, R. E. A., Grooten M., and Petersen, T. (Eds). (2020). *Living Planet Report 2020 - Bending the Curve of Biodiversity Loss*. WWF, Gland, Switzerland, 153 pp.

Baldwin, W. F., James, H. G., and Welch, H. E. (1955). A study of predators of mosquito larvae and pupae with a radio-active tracer. *Canadian Entomologist*, *87*: 350–356. [*Review of Applied Entomology B*, *45*: 95].

Balian, E. V., Segers, H., Lévèque, C., and Martens, K. (2008). The Freshwater Animal Diversity Assessment: An overview of the results. *Hydrobiologia*, *595*(1): 627–637.

Bay, E. C. (1974). Predatory-prey relationships among aquatic insects. *Annual Review of Entomology*, *19*: 441–453.

Benelli, G., Jeffries, C. L., and Walker, T. (2016). Biological control of mosquito vectors: Past, present, and future. *Insects*, *7*(4), 1–18.

Bhattacharyya, D. R., Rajavel, A. R., Natarajan, R., Mohapatra, P. K., Jambulingam, P., and Mahanta, J. (2014). Faunal richness and the checklist of Indian mosquitoes (Diptera : Culicidae). *Check List*, *10*(6), 1342–1358.

Blaustein, L. (1998). Influence of the predatory backswimmer, *Notonecta maculata,* on invertebrate community structure. *Ecological Entomology*, *23*: 246–252.

Blaustein, L., Kotler, B. P., and Ward, D. (1995). Direct and indirect effects of a predatory backswimmer *(Notonecta maculata)* on community structure of desert temporary pools. *Ecological Entomology*, *20*: 311–318.

Bond, J. C., Novelo-Gutierrez, R., Ulloa, A, Rojas, J. C., Quiroz-Martinez, H., and Williams, T. (2006). Diversity, abundance and disturbance response of odonata associated with breeding sites of *Anopheles seudopunctipennis* (Diptera: Culicidae) in southern Mexico. *Community and Ecosystem Ecology*, *06*: 1561–1568.

Borland, S. M. (1971). Biology and life history of *Laccophilus terminalis* Sharp (Coleoptera: Dytiscidae), an aquatic predator of mosquito larvae. M. S. Thesis, University of California, Riverside, 24 p.

Chandra, K., Gopi, K. C., Rao, D. V., Valarmathi, K., and Alfred, J. R. B. (2017). *Current Status of Freshwater Biodiversity of India: An Over View Faunal Diversity of India*. Zoological Survey of India, Kolkata, 624 pp.

Chatterjee, S. N., Ghosh, A., and Chandra, G. (2007). Eco-friendly control of mosquito larvae by *Brachytron pratense* larva. *Journal of Environmental Health*, *69*: 44–49.

Chesson, J. (1984). Effect of notonectids (Hemiptera: Notonectidae) on mosquitoes (Diptera: Culicidae): predation or selective oviposition. *Environmental Entomology*, *13*: 531–538.

Collins, F. H. and Washino, R. K. (1985). Insect predators. In: H. C. Chapman (Ed.), *Biological Control of Mosquitoes. Amer. Mosq. Contr. Assoc. Bull. 6*: 25–42.

Currie, D. C. and Adler, P. H. (2008). Global diversity of black flies (Diptera: Simuliidae) in freshwater. *Hydrobiologia*, *595*(1): 469–475.

Dijkstra, K. D. B., Monaghan, M. T., and Pauls, S. U. (2014). Freshwater biodiversity and aquatic insect diversification. *Annual Review of Entomology*, *59*: 143–163.

Donald, C. L., Siriyasatien, P., and Kohl, A. (2020). Toxorhynchites species: A review of current knowledge. *Insects*, *11*(11): 1–17.

Dudgeon, D. (2006). The impacts of human disturbance on stream benthic invertebrates and their drift in North Sulawesi, Indonesia. *Freshwater Biology*, *51*(9): 1710–1729.

Dudgeon, D. (2020). *Freshwater Biodiversity Status, Threats and Conservation* (1st ed.). Cambridge University Press, 533 pp.

Dudgeon, D., Arthington, A. H., Gessner, M. O., Kawabata, Z. I., Knowler, D. J., Lévêque, C., Naiman, R. J., Prieur-Richard, A. H., Soto, D., Stiassny, M. L. J., and Sullivan, C. A. (2006). Freshwater biodiversity: Importance, threats, status and conservation challenges. *Biological Reviews of the Cambridge Philosophical Society*, *81*(2): 163–182.

Durden, L. A. and Mullen, G. R. (2018). Introduction. In: *Medical and Veterinary Entomology*. Academic Press, London, 769 pp.

Eitam, A. and Blaustein, L. (2004). Oviposition habitat selection by mosquitoes in response to predator *(Notonecta maculata)* density. *Physiological Entomology*, *29*: 188–191.

Eitam, A., Blaustein, L., and Mangel, M. (2002). Effects of *Anisops sardea* (Hemiptera:Notonectidae) on oviposition habitat selection of mosquitoes and other dipterans and on community structure in artificial pools. *Hydrobiologia*, *485*: 183–189.

James, H. G. (1967). Seasonal activity of mosquito predators in woodland pools in Ontario. *Mosquito News*, *27*: 453–457.

Koroiva, R. and Pepinelli, M. (2019). Distribution and habitats of aquatic insects. In R. G. K. Del-Claro (Ed.), *Aquatic Insects*, 11–33. Springer Nature, Switzerland.

Krishnamoorthy, R., Munirathninam, A., Dhananjeyan, K. J., Hiriyan, J., Mariappan, T., Samuel, P. P., and Venkatesh, A. (2013). Description of a new species, *Toxorhynchites (Toxorhynchites) tyagii* (Diptera: Culicidae), from Nilgiri hills, Western Ghats, southern India. *Zootaxa*, *3701*(4): 447–459.

Legner, E. F. (1994). Biological control of diptera of medical and veterinary importance. *Journal of Vector Ecology*, *20*(1): 59–120.

Lundkvist, E., Landin, J. M., and Svensson, C. (2003). Diving beetles (Dytiscidae) as predators of mosquito larvae (Culicidae) in field experiments and in laboratory tests of prey preference. *Bulletin of Entomological Research*, *93*: 219–226.

Maity, A., Naskar, A., Sengupta, J., Hazra, S., Parui, P., and Homechaudhuri, S. B. (2016). An annotated checklist of Horseflies (Diptera: Tabanidae) from India with remarks on Surra disease vectors. *The Journal of Zoology Studies*, *3*(4): 50–81.

Mandal, S. K, Ghosh, A., Bhattacharjee, I., and Chandra, G. (2008). Biocontrol efficiency of odonate nymphs against larvae of the mosquito, *Culex quinquefasciatus* Say, 1823. *Acta Tropica*, *106*: 109–114.

Mathavan, S. (1976). Satiation time and predatory behaviour of the dragonfly larvae *Mesogomphus lineatus*. *Hydrobiologica*, *50*: 55–64.

McDonald, G. and Buchanan, G. A. (1981). The mosquito and predatory insect fauna inhabiting fresh-water ponds, with particular reference to *Culex annulirostris* Skuse (Diptera: Culicidae). *Australian Journal of Ecology*, *6*: 21–27.

Mitra, B., Roy, S., Roy, S., Halder, S., Chattopadhyay, U., and Bhaumik, S. (2017). A current status on diversity and distribution of black flies (Insecta: Diptera: Simuliidae) in India. *Ambient Science*, *4*(1). DOI:10.21276/ambi.2017.04.1.rv01

Organisation, Cambridge, UK and Gland, Switzerland: IUCN, and Coimbatore, India, 128 pp.

Mukhopadhyay, E., Mazumdar, A., Joardar, S. N., Saha, G. K., and Banerjee, D. (2016). An annotated checklist of Culicoides Latreille, 1809 (Insecta: Ceratopogonidae: Diptera) with incorporation of a vector species list from India. *Journal of Vector Ecology*, *41*(2): 279–284.

Saha, N., Aditya, G., and Saha, G. K. (2014). Prey preferences of aquatic insects: Potential implications for the regulation of wetland mosquitoes. *Medical and Veterinary Entomology*, *28*(1): 1–9.

Sathe, T. V. and Bhusnar, A. R. (2010). Biodiversity of mosquitovorous dragonflies (Order: Odonata) from Kolhapur district including Western Ghats. *Biological Forum-An International Journal*, *2*: 38–41.

Sebastian, A., Sein, M. M., Thu, M. M., and Corbet, P. S. (1990). Suppression of *Aedes aegypti* (Diptera: Culicidae) using augmentative release of dragonfly larvae (Odonata: Libellulidae) with community participation in Yangon, Myanmar. *Bulletin of Entotnoioqy Research*, *80*: 223–232.

Sebastian, A., Thu, M. M., Kyaw, M., and Sein, M. M. (1980). The use of dragonfly larvae in the control of *Aedes aegypti*. *Southeast Asian Journal of Tropical Medicine and Public Health*, *11*: 104–107.

Service, M. W. (1977). Mortalities of the immature stages of species B of the *Anopheles gambiae* complex in Kenya: Comparison between rice fields and temporary pools, identification of predators and effects of insecticidal spraying. *Journal of Medical Entomology*, *13*: 535–545.

Shaalan, E. A. S. and Canyon, D. V. (2009). Aquatic insect predators and mosquito control. *Tropical Biomedicine*, *26*(3): 223–261.

Simaika, J. P., Samways, M. J., and Frenzel, P. P. (2016). Artificial ponds increase local dragonfly diversity in a global biodiversity hotspot. *Biodiversity and Conservation*, *25*(10): 1921–1935.

Singh, R. K., Dhiman, R. C., and Singh, S. P. (2003). Laboratory studies on the predatory potential of dragonfly nymphs on mosquito larvae. *Journal of Communicable Diseases*, *35*: 96–10.

Siqueira, T., Durães, L. D., and Roque, F. D. O. (2014). Ecological modelling applied to entomology. In: C. P. Ferreira, W. A. C. Godoy (Eds.), *Entomology in Focus*, 109–126. Springer International Publishing, Switzerland.

Thomas, M. B. (2018a). Biological control of human disease vectors: a perspective on challenges and opportunities. *BioControl*, *63*(1): 61–69.

Urabe, K., Ikemoto, T. and Aida, C. (1986). Studies on *Sympetrum frequens* (Odonata: Libellulidae) nymphs as natural enemies of the mosquito larvae, *Anopheles sinensis* in rice fields. III. Estimation of the prey consumption rate in the rice fields. *Japanese Journal of Applied Entomology and Zoology*, *30*: 129–135.

Vannote, R. L., Minshall, G. W., Cummins, K. W., Sedell, J. R., and Cushing, C. E. (1980). The River Continuum Concept. *Canadian Journal of Fisheries and Aquatic Sciences*, *37*(1): 130–137.

Venkatesh, A. and Tyagi, B. K. (2013). Predatory potential of *Bradinopyga geminata* and *Ceriagrion coromandelianum* on dengue vector *Aedes aegypti* under controlled conditions (Anisoptera; Libellulidae; Zygoptera: Coenagrionidae; Diptera: Culicidae). *Odonatologica*, *42*: 139–149.

Wagner, R., Barták, M., Borkent, A., Courtney, G., Goddeeris, B., Haenni, J. P., Knutson, L., Pont, A., Rotheray, G. E., Rozkošný, R., Sinclair, B., Woodley, N., Zatwarnicki, T., and Zwick, P. (2008). Global diversity of dipteran families (Insecta Diptera) in freshwater (excluding Simulidae, Culicidae, Chironomidae, Tipulidae and Tabanidae). *Hydrobiologia*, *595*(1): 489–519.

WWF. (2008). *Rivers at Risk*, 3–48. https://d2ouvy59p0dg6k.cloudfront.net/downloads/riversatriskfullreport.pdf

12 Weed Bio-Control Agents

Uzma Afaq
Integral University

Omkar
University of Lucknow

CONTENTS

LEARNING OBJECTIVES

1. To have a good understanding of weeds and factors that make them the most costly group of agricultural pests.
2. To understand the categories of weed management tactics and the use of ecofriendly methods to maintain the quality of the environment.
3. To learn the biocontrol of weeds using natural enemies.
4. To learn the beneficial characteristics of insects in the management of weeds.
5. To learn how insects have been used as biocontrol agents of weeds with examples of successful management of major weeds.
6. To learn the antagonistic potential of insect bioagents against weeds with examples.

12.1 INTRODUCTION

There is no universally accepted definition for a weed; the European Weed Research Society defined a weed as "any plant or vegetation, excluding fungi, interfering with the objectives or requirements of people." Weeds have always been with us since the advent of settled agriculture some 10,000 years ago. The battle between humans and weeds is endless; even today, there are acres of valuable land subjugated by weeds. They exude harmful phytotoxins, which reduce the growth of many crop plants. Worldwide, weeds are the most costly group of agricultural pests, as weeds cause more yield loss and add to farmer's production cost than insect pests, pathogens, nematodes, birds, rodents, etc. The Weed Science Society of America has identified 17 important early publications on weeds (1895–1965), 12 of which dealt with their destruction, control or eradication. Chemical, mechanical, cultural and biocontrol methods have been the core for the management of weeds, particularly in agricultural fields. However, biological control of weeds using natural enemies and native organisms is an important approach in the weed management techniques. Insects have been most frequently used as biocontrol agents of weeds and this is likely to continue. The practice has

DOI: 10.1201/9781003370864-12

been expanding from use primarily on rangelands and aquatic systems into other environments. Sometimes non-native plants which have become weeds in new region can be controlled by introducing the host-specific biocontrol agents from the weed's native range, this is known as classical biocontrol and it is analogous to the classical biocontrol of insect pests. The first documented case of Commonwealth Biological Control (CBC) of a weed species dates back 182 years, when mistakenly cochineal insect *Dactylopius ceylonicus* (Green), (Wild cochineal insect) was brought from Brazil to India for dye production in 1795. It was deliberately moved from northern to southern India in 1836 and from India to Sri Lanka in 1865 for the biocontrol of *Opuntia monacantha* (Willd.) Haw. (Drooping prickly pear) (Goeden, 1988). The first global record of CBC releases for weeds was published under the title 'Biological control of weeds: a world catalog of agents and their target weeds' (henceforth 'the catalog') in the early 1980s (Julien, 1982). The catalog comprises data on bioagents introduced for targeted weed destruction along with the date of introduction and name of the country. It also provides information on the success or failure of the introduced biocontrol program. The catalog was subsequently updated and expanded with its new editions. Another form of biological control, known as augmentation, involves increase in the natural population of biocontrol agents of weed and thus entails mass rearing of the insects for its inundative release. But, it is unlikely to be cost-effective for most insect–plant systems.

12.2 WEEDS: FOOD RESOURCE FOR INSECTS

Weeds are an important food resource for numerous phytophagous insects. The host relationship between insects and plants is highly variable, ranging from very specialized to generalized feeding behaviors. Insect feeding on weeds can have both positive and negative effects on crop productivity; however, positive outcome is more as compared to the negative outcome. In a positive sense, insect feeding on weeds reduces the competition of crop plants and makes water, soil nutrients and sunlight more available to the crop plants. This reduced competition ultimately results in immense amplification of the crop yield. Numerous phytophagous insects feed completely on weeds. For example, the sesiid moth, *Carmenta haematica* (Ureta), attacks only snakeweeds, *Gutierrezia* and *Grindelia* spp. (Cordo et al., 1995).

12.3 RATIFICATION OF INSECTS FOR WEED CONTROL

They are best-known and proven biocontrol agents to solve the worst weed problems. Insect which is selected for weed biocontrol must be capable, either directly or indirectly, of determining the abundance of its plant host and thus of determining its own abundance as interaction. The bioagents of weeds have to be target-specific destructive organisms, free of predators or parasites, and must possess the ability to readily adapt to climatic conditions. It must be competent to reproduce sufficiently with the increase of its host species (Table 12.1).

12.4 ANTAGONISTIC POTENTIAL OF INSECT BIOAGENTS AGAINST WEEDS

Insects, which are capable of causing damage to different parts of the weed plants, are more likely to control the population of weeds, as they may kill the weed plants completely and can contribute to an overall reduction in their growth, vigor and reproductive potential. For this purpose, if one species alone cannot kill the weed adequately, other species can be used with it and the synergistic action of both the species might possibly enhance the level of control by inflicting more damage to the weed plants; for instance, management of water hyacinth by *Neochetina eichhorniae* (mottled water hyacinth weevil) and *Orthogalumna terebrantis* Wallwork (water hyacinth mite) (Delt'osse, 1978). Defoliators and flower feeder species are less drastic than the insect species that spoils the root or shoot system of the plant. These killers can kill the plant within a short period of time and

TABLE 12.1
Common Weeds and Their Biological Control Agents

Weed	Common Name	Biological Control Agents	Common Name
Parthenium hysterophorus	Gajar Ghas, Carrot Grass	*Zygogramma bicolorata* Pallister	Mexican beetle
Lantana camara	Lantana	*Crocidosema lantana, Octotoma scabripennis*	Lantana flower-cluster moth or Lantana tortricid moth, *Lantana leaf beetle, Leaf-mining chrysomelids*
Opuntia spp.	Prickly pear	*Cactoblastis cactorum, Dactylopius* spp.	The cactus moth, cochineals
Alternanthera philoxeroides	Alligator weed	*Agasicles hygrophila, Disonycha argentinensis* Jacob	Alligator weed *Flea beetle*
Cyperus rotundus	*Coco-grass, Java grass, Nut grass*	*Bactra verutana*	Javelin moth
Lythrum salicaria	Purple loosestrife	*Hylobius transversovittatus, Galerucella* spp.	Loosestrife root weevil, Loosestrife beetle
Eichhornia crassipes (Mart.)	*Water hyacinth*	*Neochetina eichhorniae* Warner	Water hyacinth weevil
Centaurea maculosa Lam.	Spotted knapweed	*Agapeta zoegana Linnaeus*	Sulfur knapweed root moth
Centaurea diffusa	Diffuse knapweed	*Bangasternus fausti* (Reitter)	Broad-nosed seed head weevil
Cytisus scoparius	The common broom or *Scotch broom*	*Bruchidius villosus* (Fabricius)	Broom seed beetle
Jacobaea vulgaris	The tansy ragwort	*Botanophila seneciella* (Meade)	Ragwort seed fly
Linaria genistifolia ssp. *Dalmatica*	Balkan toadflax, broadleaf toadflax, Dalmatian *toadflax*	*Calophasia Iunula* (Hufnagel)	Toadflax brocade moth
Cirsium arvense	Canada thistle	*Cassida rubiginosa Muller*	Thistle tortoise beetle
Cryptostegia grandiflora R. Br.	Rubber vine	*Euclasta whalleyi* Popescu-Gorj and Constantinescu	Rubber vine moth
Pistia stratiotes L.	Water lettuce, Nile cabbage, Chok, Jawg	*Neohydronomus affinis* Hustache	Water lettuce weevil
Ambrosia artemisiifolia L.	Common ragweed, Ragweed, Annual Ragweed	*Epiblema strenuana* (Walker)	Ragweed borer
Carduus acanthoides L.	Plumeless thistle, Spiny plumeless thistle, Bristly thistle	*Rhinocyllus conicus* (Frölich)	Thistle-head weevil
Carduus pycnocephalus L.	Slender winged thistle, Italian thistle	*Cheilosia grossa* (Fallen)	Syrphid fly
Echium plantagineum L.	Paterson's curse, Salvation Jane	*Dialectica scalariella* (Zeller)	Echium leaf miner

(*Continued*)

TABLE 12.1 (*Continued*)
Common Weeds and Their Biological Control Agents

Weed	Common Name	Biological Control Agents	Common Name
Acanthocereus tetragonus (L.) Hummelinck	Sword pear	*Hypogeococcus festerianus* (Lizer y Trelles)	Cactus mealybug
Halogeton glomeratus (M. Bieb.) C. A. Mey.	Halogeton	*Coleophora parthenica* Meyrick	Russian thistle stem miner moth
Tamarix spp.	Saltcedar, tamarisk	*Diorhabda sublineata* (Lucas)	Subtropical tamarisk beetle

show more drastic effect in their control. Weeds, which propagate by seeds and fruits, are more vulnerable to the seed feeding biocontrol agents. The foremost hurdle of the phytophagous insects is their low population. These insects are themselves subject to parasitism, predation and pathogenic diseases, and as a consequence various species are found at low populations, predominantly at the most susceptible stages in the life cycle of the weed plants. For example, seeds of individual *Striga angustifolia* (Witch weed) plant can be destroyed completely by the larvae of the noctuid, *Eulocastra argentisparsa*, but their natural populations are always very low. A microsporidian disease is known to infect them. *Eulocastra* is incapable of establishing its population early enough to significantly reduce the total seed output of *Striga* plants in an infested region.

12.5 MAJOR WEEDS AND THEIR CONTROL BY INSECTS

I. *Lantana camara*: *Lantana camara* Linnaeus (Lantana) (Figure 12.1) is a fast-growing perennial woody shrub native to Central and South America. It is now a major toxic invasive weed in many countries, invading both natural and agricultural ecosystems. It is regarded as one of the most serious weeds of various crops, like coffee, oil palm, coconuts and cotton. Lantana is also toxic to sheep and cattle owing to the presence of triterpenoids in the leaves and seeds of the weed plants (Kumar et al., 2016). This was the first weed ever targeted for classical biological control at the turn of the 20th century, and after that 36 insect species have been released in 33 countries to control the weed population. Koebele was sent to Mexico and Central America for the collection of Lantana insects so that they can be introduced in Hawaii. Among all the insects introduced, only eight became established: Lantana seed fly, *Agromyza lantanae* Froggatt; Lace bug, *Teleonemia scrupulosa* (Stal); *Lantana* flower-cluster moth, *Epinotia lantana* (Busck); Lantana Plume moth,

(a) (b)

FIGURE 12.1 (a) *Lantana camara* Linnaeus and (b) *Cremastobombycia lantanella* Busck.

Platyptilia pusillidactyla (Walker); Lantana leaf miner, *Cremastobombycia lantanella* Busck (Figure 12.1); Stem-galling fly, *Butreta xanthochaeta* Aldrich; and two butterflies, *Thecla echion* Linnaeus (Red-spotted hairstreak) and *Thecla bazochi* Godart (Lycaenid *butterfly*). G. Wilson Fernandes and Jean C. Santos (2008) revealed the three species of longhorn beetles associated with *L. camara* in November 2004: *Trachyderes succintus duponti* Aurivillius, 1912; and *Dorcacerus barbatus* (Olivier, 1790). *Dorcacerus barbatus* represented more than 95% of all individuals found. To their knowledge, this is the first record of these three cerambycid species on *L. camara*.

II. *Eichhornia crassipes*: Water hyacinth, *Eichhornia crassipes* (Mart.) (Figure 12.2) Solms-Laub., is a free-floating, annual or perennial toxic aquatic weed native to South America. It is one of the world's most prevalent invasive aquatic weeds known to cause major ecological and socioeconomic changes. It commonly forms dense interlocking mats and alters the clarity of water and declines the production of phytoplankton, dissolved oxygen, nitrogen, phosphorous, heavy metals and concentrations of other contaminants. Once introduced in 1884 for its beautiful flowers, water hyacinth has invaded most of the Southern United States and many tropical and subtropical regions around the world. It entered India in the early nineteenth century into the State of West Bengal, and was treated as an ornamental plant owing to the beautiful bluish-violet flowers but now, it is known as "Bengal terror," 'blue devil' and noxious weed. It not only affects the economic uses of surface waters, including fishing, irrigation and navigation, but also has unpleasant effects on aquatic life. For the ecofriendly management of water hyacinth through biocontrol agent, *Neochetina eichhorniae* Warner (the water hyacinth weevil) (Figure 12.2), a natural enemy of *E. crassipes* in Argentina and South America was taken into consideration. The adult female measured about 5.00 mm and adult male about 4.9 mm in length (Gore et al., 2020). Both are cryptic, dull-brown in color and become mottled when wet. The total life cycle comprises about 92 days, with four stages, i.e., eggs, larva, pupa and adult. The larval stage comprises five instars with duration of about 46 days. The pupal stage comprises 32 days found in the root zone under water. There are usually three generations in a year. They feed on the leaves of *E. crassipes*, creating characteristic rectangular scars by eliminating the epidermal layers of the laminae and petioles. The females are also known to squash through the epidermis and dig out holes into which she lays eggs singly. Females can lay up to five eggs per day and can live for more than 50 days. The larval instars tunnel internally through the petioles and into the crown of the plants, causing them to rot, die and submerge.

III. *Parthenium hysterophorus*: The most popular among the various weeds is *Parthenium hysterophorus* Linnaeus (Gajar ghas) (Figure 12.3), which has become a *Weed of International Significance*. This weed is rated as one of the ten worst weeds in India because of its invasiveness, easy spread and some other economic and environmental impacts (Bhan et al., 1999). It is widely dispersed in wastelands, pastures, degraded soil, along water canal, railway track, barren lands, foothills and agricultural fields almost throughout the country infesting more

(a) (b)

FIGURE 12.2 (a) *Eichhornia crassipes* (Mart.) and (b) *Neochetina eichhorniae* Warner.

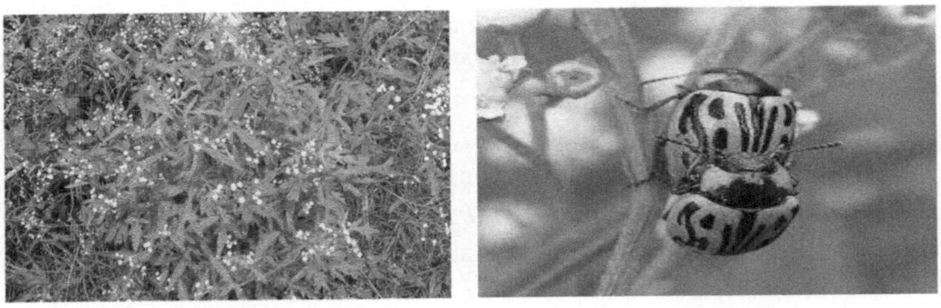

(a) (b)

FIGURE 12.3 (a) *Parthenium hysterophorus* Linnaeus and (b) *Zygogramma bicolorata* Pallister.

than 5 million hectares. Therefore, it is mainly recognized as a weed of 'no man's land'. A program of biocontrol of Parthenium weed was initiated in Australia in 1977 and since then, eight species of insects and one fungal pathogen have been reported and utilized (McFadyen, 1992; Dhileepan et al., 1996; Dhileepan and McFadyen, 1997). Among these, the leaf-feeding beetle, *Zygogramma bicolorata* Pallister (Coleoptera: Chrysomelidae) (Figure 12.3), is the most prominent bioagent. The adults of *Z. bicolorata* measure 5–6 mm in length, are off-white in color and the rest of the body and appendages are dark brown, except the pronotum, which has off-white patches at the anterior edges and the elytra, with undulating dark brown longitudinal markings (Jayanth and Bali, 1993; Siddhapara et al., 2012). The posterior margin in the case of the female is entire, while it is slightly serrated at the tip of the male. Females lay eggs either singly or in clusters on the leaves, flower heads, stem surfaces and terminal and axillary buds. The newly hatched instars feed voraciously on new terminal and axillary buds. The grubs molt thrice to become fully grown. The fully grown larvae on the undersurface of leaves cease feeding a few hours to one day before they bury themselves into the soil for pupation. The larval and pupal periods were observed to last for 11–13 and 10–12 days, respectively, with the total developmental period lasting for 27–29 days. The whole life cycle takes 6–8 weeks and there may be up to four generations per year, depending on rainfall and food availability. Adult beetles can live up to 2 years and usually spend around 6 months diapausing in the soil (about 3 cm deep) during autumn and winter. Chemical analysis of *P. hysterophorus* has revealed that it is a toxic weed; and *Z. bicolorata* has immense potential to detoxify and metabolize the toxins of this weed. Defoliation and voracious feeding by *Z. bicolorata* causes severe damage to the Parthenium plant. It damages the meristems of the plant, which results in shorter height of the primary stem and changes in the branching pattern of the plant, flower production and root and shoot biomass decline with an increase in the duration of defoliation.

IV. *Opuntia* spp.: Prickly Pears belong to the genus *Opuntia*, several species of which are considered weeds or potential weeds in most states of Australia. The genus *Opuntia* appears to be native to Mexico and Guatemala, where it is widely used as fodder, forage, fruit and a green vegetable. This species is mostly found in subtropical, semi-arid and warmer temperate environments. It is a weed of roadsides, railways, pastures, grasslands, open woodlands, rangelands, disturbed sites and waste areas. They are shrubs or small trees with succulent, cylindrical or flattened stems armed with sharp spines. Wild cochineal insects, *Dactylopius indicus* Green (=*D. ceylonicus* n.n.) and Cactus moths, *Cactoblastis cactorum* (Berg) (Figure 12.4), were introduced for the biological control of this weed. Insects introduced prior to *C. cactorum*, including Leaf-footed bug, *Chelinidea tabulata* (Burmeister); Tomentose cochineal scale, *Dactylopius tomentosus* Lamarck; and desert spider mite, *Tetranychus desertorum* Banks (= *T. apuntiae* Banks), made speedy progress from 1925 to 1927. From 1930 onward, *C. cactorum* completely overshadowed all other species of insects used for the biocontrol of prickly pears. According to David & Muthukrishnan (1953), *Opuntia vulgaris* Miller (Figure 12.4) and *O. elatior* Miller (=*O. nigricans* Haworth) were introduced

(a) (b)

FIGURE 12.4 (a) *Opuntia vulgaris* Miller and (b) *Cactoblastis cactorum* (Berg).

in India, in the year 1787, for the commercial culture of the cochineal insect, valued for its dye, while *Opuntia dillenii* (Ker Gawler) was probably introduced much earlier. *O. vulgaris* had long been under effective control by the cochineal insects, *Dactylopius indicus* Green, which apparently arrived along with the introduced *Opuntias.*

V. *Alternanthera philoxeroides* (Mart.): Commonly known as the Alligator weed (Figure 12.5), it is a South American immigrant that has invaded tropical and mild-temperate regions around the world. It is a noxious weed and a prohibited plant in Arizona, California, Florida and South Carolina (USDA, NRCS, 1999). Alligator weed is a perennial aquatic or terrestrial herb with long, fibrous roots. Stems are pink and hollow and can attain lengths of 1 m with opposite narrow elliptical leaves. Usually, it grows in thick dense mats along the shoreline of lakes and streams, but it can also grow terrestrially during dry periods. The alligator weed flea beetle, *Agasicles hygrophila* Selman (Figure 12.5), was the first insect ever studied for biological control of this aquatic weed. The beetle kills the alligator weed by destroying its stored food and interfering with photosynthesis by removing leaf tissues. Both adults and larvae feed on the leaves of alligator weed, often defoliating the stems.

VI. *Striga hermonthica* Benth.: Commonly known as purple witch weed or giant witch weed (Figure 12.6), it is a herbaceous hemiparasitic annual plant. It can attain 30–100 cm in height, the most vigorous form is found in Sudan and Ethiopia. It is destructive to major crops, such as sorghum (*Sorghum bicolor*) and rice (*Oryza sativa).* These hemiparasites attach to and penetrate the roots of the crop plants and then feed on the nutrients of its host, thereby stunting the growth of the host plant. The weevils, *Smicronyx* spp., are known to reduce the seed production from gall-forming, but there has been no successful development of a biological control program from these weevils. *S. albovariegatus* and the noctuid moth, *Eulocastra argentisparsa*, were introduced to Ethiopia from India but the attempts apparently failed. Meanwhile, conclusions from a mathematical modeling project have suggested that *Smicronyx* spp. would in any case be unlikely to have a significant impact on giant witch weed population dynamics. Few other potentially useful organisms for its

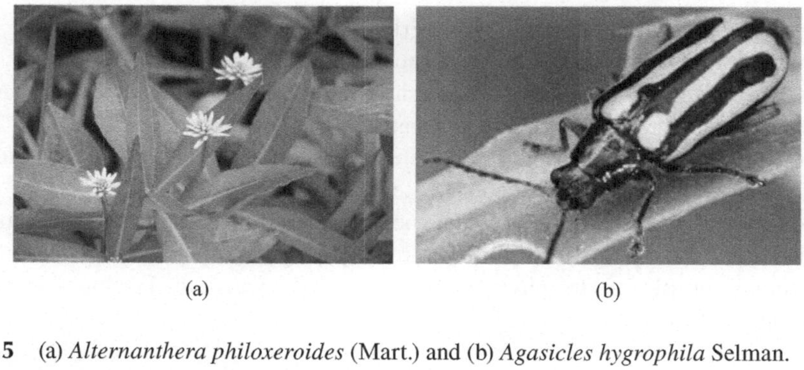

(a) (b)

FIGURE 12.5 (a) *Alternanthera philoxeroides* (Mart.) and (b) *Agasicles hygrophila* Selman.

(a) (b)

FIGURE 12.6 (a) *Striga hermonthica* Benth. and (b) *Precis tugela* Trimen, 1879.

management include the following: the African leaf *butterfly* or eared commodore, *Precis tugela* Trimen, 1879 (Figure 12.6) is considered as a useful agent for the control of this weed. The larvae of the butterfly feed on leaves, buds and capsules of several *Striga* species (ICIPE Biennial Scientific Report, 2004–2005).

12.6 CONCLUSIONS

Numerous studies have verified the effectiveness of insects in weed management, both in terms of biocontrol agent impact on the target weed as well as economic returns from the decline in weed populations. Use of insect biocontrol agents against weeds is one of the best-known alternatives for chemical control. It involves the designed use of insects to control the economic losses caused by weeds that occur worldwide. Management of weed by these biocontrol agents, whether natural enemies or native species, involves a complex of interacting morphological, phenological, biochemical and other factors in both host-plant and host-control agents. The host relationship between insects and plants is the most essential requirement for an insect to be used as a weed control agent, and it is highly variable, ranging from very specialized to generalized feeding behaviors. Egg laying behavior and nutritional suitability of the plant for growth and reproduction of the insect are vital criteria in determining insect–plant association. Presence of more than one species of insect on target weed at the same time but exhibiting complementary and not competitive feeding and oviposition behavior can have a synergistic effect in weed control. The bioagents of weeds have to be target-specific destructive organisms, free of parasites, predators and pathogenic organisms, as well as must possess the ability to readily adapt to climatic conditions. Additional research should focus on how to improve the establishment rates of biocontrol agents and identify the insect that can result in heavy impacts on the target weeds.

12.7 POINTS TO REMEMBER

i. Weeds are the most destructive group of agricultural pests.
ii. Weeds are an important food resource for numerous phytophagous insects.
iii. Insects feeding on weeds can have both positive and negative effects on crop productivity; however, positive outcome is more as compared to the negative outcome.
iv. Biological control of weeds using natural enemies and native phytophagous organisms is an important approach in the weed control techniques.
v. The bioagents of weeds have to be target-specific destructive organisms, free of predators or parasites and must possess the ability to readily adapt to climatic conditions.
vi. The foremost hurdle of the phytophagous insects in weed management is their low population.

REFERENCES

Bhan V.M., Kauraw L.P. and Chile A. (1999). Biological suppression of weeds with pathogens. In: *Biological Suppression of Plant Diseases, Phytoparasitic Nematodes and Weed* (eds. Singh S.P. & Hussaini S.S.). Project Directorate of Biological Control: Bangalore, India, pp. 192–210.

Cordo P., Gurfinkel V.S., Bevan, L. and Kerr G.K. (1995). Proprioceptive consequences of tendon vibration during movement. *J. Neurophysiol.*, 74: 1675–1688.

David A.L. and Muthukrishnan T.S. (1953). Summaries of *Opuntia* control by insects. The prickly pear cochineal: Observation on its natural hosts and enemies in South Africs. *Indian J. Ent.,* 15(3): 219–224.

Delt'osse E.S. (1978). Effect on water hyacinth of *Neocmtinaeich Jwrniae* (Col.: Curculionidae) combined with *Orthogalumna terebrantis* (Acari: Galumnidae). *Entomophaga*, 23: 379–387.

Dhileepan K. and McFadyen R.E. (1997). Biological control of Parthenium in Australia: Progress and prospects. In: *First International Conference on Parthenium Management* (eds. Mahadeveppa M. and Patil V.C.). University of Agricultural Sciences: Dharwad, India, pp. 40–44.

Dhileepan K., Madigan B., Vitelli M., McFadyen R., Webster K. and Trevino M. (1996). A new initiative in the biological control of Parthenium. In: *Proceedings of the Eleventh Australian Weed Conference* (ed. Shepherd R.C.H.). Weed Society of Victoria: Australia, pp. 309–312.

Fernandes G.W. and Santos J.C. (2009). Feeding and mating behavior of *Dorcacerus barbatus* (Olivier, 1790) (Coleoptera: Cerambycidae) on *Lantana camara* L. (Verbenaceae). *Lundiana*, 9(2): 81–84.

Goeden R.D. (1988). A capsule of biological control of weeds. *Biocontrol. News Inf.*, 9(2): 55–61.

Gore P., Ganguli J. and Gauraha R. (2020). Studies on the biology of the water hyacinth weevil, *Neochetina eichhorniae* hustache (Coleoptera: curculionidae) at Raipur, Chhattisgarh. *J. Entomol. Zool. Stud.,* 8(2): 915–919.

ICIPE Biennial Scientific Report (2004–2005). Habitat management strategies for control of stemborers and Striga weed in cereal-based farming systems. pp. 40–58.

Jayanth K.P. and Bali G. (1993). Biological studies on *Zygogramma bicolorata* Pallister (Coleoptera: Chrysomelidae), a biocontrol agent of *Parthenium hysterophorus* L. (Asteraceae). *Biol. Control.*, 7(2): 93–98.

Julien M.H. (1982). *Biological Control of Weeds: A World Catalogue of Agents and Their Target Weeds*, 1st edn. CABI International: Wallingford.

Kumar R., Katiyar R., Kumar S., Kumar T. and Singh V. (2016). Lantana camara: An allien weed, its implication on animal health and strategies to control. *J. Exp. Biol. Agric. Sci.,* 4(3S): 321–337.

McFadyen, R.E.C. (1992). Biological control against Parthenium weed in Australia. *Crop. Prot.*, 11: 400–407.

Siddhapara M.R., Patel M.B. and Patel H.V. (2012). Biology of *Zygogramma bicolorata* Pallister (Coleoptera: Chrysomelidae) and their feeding potential on Parthenium and sunflower. *Madras Agric. J.*, 99: 841–844.

USDA, NRCS. (1999). United States Department of Agriculture, Natural Resources Conservation Service. Plants Database http://plants.usda.gov.

Index

Note: **Bold** page numbers refer to tables; *italic* page numbers refer to figures.